Methods in Enzymology

Volume 284

LIPASES

Part A

Biotechnology

METHODS IN ENZYMOLOGY

Methods in Enzymology

Volume 284

Lipases

Part A
Biotechnology

EDITED BY

Byron Rubin

LIPOMED
SAN DIEGO, CALIFORNIA

Edward A. Dennis

DEPARTMENT OF CHEMISTRY AND BIOCHEMISTRY
UNIVERSITY OF CALIFORNIA, SAN DIEGO
LA JOLLA, CALIFORNIA

ACADEMIC PRESS
San Diego London Boston New York Sydney Tokyo Toronto

This book is printed on acid-free paper. ♾

Academic Press
15 East 26th Street, 15th Floor, New York, New York 10010, USA
http://www.apnet.com

Academic Press Limited
24-28 Oval Road, London NW1 7DX, UK
http://www.hbuk.co.uk/ap/

International Standard Book Number: 0-12-182185-4

PRINTED IN THE UNITED STATES OF AMERICA
97 98 99 00 01 02 MM 9 8 7 6 5 4 3 2 1

Table of Contents

Section I. Primary and Tertiary Structure

Section II. Isolation, Cloning, Expression, and Engineering

Contributors to Volume 284

Article numbers are in parentheses following the names of contributors.
Affiliations listed are current.

PATRICK ADLERCREUTZ (12), *Department of Biotechnology, Center for Chemistry and Chemical Engineering, Lund University, S-221 00 Lund, Sweden*

LILIA ALBERGHINA (14), *Dipartimento Fisiologia e Biochimica Generali, Sezione Biochimica Comparata, Universitá degli Studi di Milano, 20133 Milano, Italy*

EUNICE ASHIZAWA (18), *Genencor International, Palo Alto, California 94304*

LARS BLÄCKBERG (10), *Department of Physiological Chemistry, University of Umeå, Umeå, Sweden*

KIM BORCH (19), *Enzyme Design, Novo Nordisk A/S, DK-2880 Bagsvaerd, Denmark*

MATTHEW BOSTON (18), *Genencor International, Palo Alto, California 94304*

RICHARD BOTT (18), *Genencor International, Palo Alto, California 94304*

YVES BOURNE (5), *AFMB-CNRS, 13402 Marseille cédex 20, France*

CHRISTIAN CAMBILLAU (5), *AFMB-CNRS, 13402 Marseille cédex 20, France*

STEPHANE CANAAN (15), *Laboratoire de Lipolyse Enzymatique, UPR 9025 de l'IFR-1 du CNRS, 13402 Marseille cédex 20, France*

IB GROTH CLAUSEN (19), *Enzyme Design, Novo Nordisk A/S, DK-2880 Bagsvaerd, Denmark*

JUAN ANTONIO CONTRERAS (16), *Department of Cell and Molecular Biology, Section for Molecular Signalling, Lund University, S-221 00 Lund, Sweden*

MIROSLAW CYGLER (1, 4), *Biotechnology Research Institute, National Research Council of Canada, Montréal, Québec H4P 2R2, Canada*

STEVE DANKO (18), *Genencor International, Palo Alto, California 94304*

RICHARD C. DAVIS (9, 13), *Lipid Research Laboratory, West Los Angeles VA Medical Center, Los Angeles, California 90073, and Department of Medicine, University of California, Los Angeles, Los Angeles, California 90024*

FINN DRABLØS (2, 3, 7), *SINTEF Unimed, N-7034 Trondheim, Norway*

RUI-DONG DUAN (10), *Department of Cell Biology, University Hospital, S-221 85 Lund, Sweden*

LILIANE DUPUIS (15), *Laboratoire de Lipolyse Enzymatique, UPR-9025 de l'IFR-1 du CNRS, 13402 Marseille cédex 20, France*

MARIE PIERRE EGLOFF (5), *AFMB-CNRS, 13402 Marseille cédex 20, France*

MAARTEN R. EGMOND (6), *Unilever Research Laboratory, Vlaardingen, The Netherlands*

ASSOU EL BATTARI (20), *Inserm U 260, 13385 Marseille cédex 5, France*

THOMAS GITLESEN (12), *Department of Biotechnology, Center for Chemistry and Chemical Engineering, Lund University, S-221 00 Lund, Sweden*

LUTZ HAALCK (21), *Institute of Chemical and Biochemical Sensor Research, D-48149 Münster, Germany*

JOHN S. HILL (9, 13), *Lipid Research Laboratory, West Los Angeles VA Medical Center, Los Angeles, California 90073, and Department of Medicine, University of California, Los Angeles, Los Angeles, California 90024*

CECILIA HOLM (16), *Department of Cell and Molecular Biology, Section for Molecular Signalling, Lund University, S-221 00 Lund, Sweden*

ALISHA JARNAGIN (18), *Genencor International, Palo Alto, California 94304*

DOMINIQUE LOMBARDO (20), *Inserm U 260, 13385 Marseille cédex 5, France*

MARINA LOTTI (14), *Dipartimento Fisiologia e Biochimica Generali, Sezione Biochimica Comparata, Universitá degli Studi di Milano, 20133 Milano, Italy*

MARK E. LOWE (8, 17), *Departments of Pediatrics and of Molecular Biology and Pharmacology, Washington University School of Medicine, St. Louis, Missouri 63110*

PAULO MARTEL (7), *Instituto de Tecnologia Quimica e Tecnologica, P-2781 Oeiras, Portugal*

CHRISLAINE MARTINEZ (5), *AFMB-CNRS, 13402 Marseille cédex 20, France*

ERIC MAS (20), *Inserm U 260, 13385 Marseille cédex 5, France*

IGNATIOUS NCUBE (12), *Department of Biochemistry, University of Zimbabwe, Harare, Zimbabwe*

FRITZ PALTAUF (21), *Department of Biochemistry and Food Chemistry, Technical University Graz, A-8010 Graz, Austria*

SHAMKANT ANANT PATKAR (19), *Enzyme Design, Novo Nordisk A/S, DK-2880 Bagsvaerd, Denmark*

EVAMARIA I. PETERSEN (3, 7), *SINTEF Unimed, N-7034 Trondheim, Norway*

MARIA TERESA NEVES PETERSEN (3, 7), *SINTEF Unimed, N-7034 Trondheim, Norway*

STEFFEN B. PETERSEN (2, 3, 7), *SINTEF Unimed, N-7034 Trondheim, Norway*

JÜRGEN PLEISS (21), *Institute of Technical Biochemistry, University of Stuttgart, D-70569 Stuttgart, Germany*

A. J. POULOSE (18), *Genencor International, Palo Alto, California 94304*

JOHN S. READ (12), *Department of Biochemistry, University of Zimbabwe, Harare, Zimbabwe*

CAROL REQUADT (18), *Genencor International, Palo Alto, California 94304*

MIREILLE RIVIÈRE (15), *Laboratoire de Lipolyse Enzymatique, UPR 9025 de l'IFR-1 du CNRS, 13402 Marseille cédex 20, France*

M. LUISA RÚA (11), *Departemento de Bioquimica, Universidad de Vigo, 32003 Ourense, Spain*

MARIE-ODILE SADOULET (20), *Inserm U 260, 13385 Marseille cédex 5, France*

ROLF D. SCHMID (11, 21), *Institute of Technical Biochemistry, University of Stuttgart, D-70569 Stuttgart, Germany*

CLAUDIA SCHMIDT-DANNERT (11), *Institute of Technical Biochemistry, University of Stuttgart, D-70569 Stuttgart, Germany*

MICHAEL C. SCHOTZ (9, 13, 16), *Lipid Research Laboratory, West Los Angeles VA Medical Center, Los Angeles, California 90073, and Department of Medicine, University of California, Los Angeles, Los Angeles, California 90024*

JOSEPH D. SCHRAG (1, 4), *Biotechnology Research Institute, National Research Council of Canada, Montréal, Québec H4P 2R2, Canada*

FRITZ SPENER (21), *Institute of Chemical and Biochemical Sensor Research, D-48149 Münster, Germany*

PETER STADLER (21), *Department of Biochemistry and Food Chemistry, Technical University Graz, A-8010 Graz, Austria*

BERIT STERNBY (10), *Department of Medicine, University Hospital, S-221 85 Lund, Sweden*

ALLAN SVENDSEN (19), *Enzyme Design, Novo Nordisk A/S, DK-2880 Bagsvaerd, Denmark*

MARIANNE THELLERSEN (19), *Enzyme Design, Novo Nordisk A/S, DK-2880 Bagsvaerd, Denmark*

CARLA J. VAN BEMMEL (6), *Unilever Research Laboratory, Vlaardingen, The Netherlands*

HERMAN VAN TILBEURGH (5), *CBM-CNRS, Montpellier, France*

ROBERT VERGER (16), *Laboratoire de Lipolyse Enzymatique, UPR 9025 de l'IFRC1 du CNRS, 13402 Marseille cédex 20, France*

CATHERINE WICKER-PLANQUART (15), *Laboratoire de Lipolyse Enzymatique, UPR*

9025 de l'IFR-1 du CNRS, 13402 Marseille cédex 20, France

HOWARD WONG (9, 13), *Lipid Research Laboratory, West Los Angeles VA Medical Center, Los Angeles, California 90073, and Department of Medicine, University of California, Los Angeles, Los Angeles, California 90024*

SHAN WU (18), *Genencor International, Palo Alto, California 94304*

DAWN YANG (9, 13), *Lipid Research Laboratory, West Los Angeles VA Medical Center, Los Angeles, California 90073, and Department of Medicine, University of California, Los Angeles, Los Angeles, California 90024*

Preface

The pace of lipase research has been accelerating. The powerful tools of molecular biology have been brought to bear, more new lipase amino acid sequences and three-dimensional structures are appearing, and new approaches for handling their complicated interfacial kinetics are being reported. In addition, more ways are being discovered and used to control lipase activity and for harnessing their catalytic prowess to pull greater efficiency into older chemical processes. Indeed, studies of heterogeneous lipase catalysis, long passed over by many academic researchers in favor of more experimentally tractable homogeneous, single-phase enzyme systems, are moving closer to the level of depth previously reserved for proteases, their hydrolytic cousins.

To the usual problems of abundance and purity that enzymologists and structural biologists generally face, lipases present the additional difficulty associated with multiphase systems. Unlike proteases, the substrates that are hydrolyzed by lipases are most efficiently presented to the enzyme in a separate, lipid phase. The presence of a suitable second phase appears to bring about an increase in lipase activity and, in some cases, effect a change in their three-dimensional structures. Part of the expanding interest in lipases derives from the increasing applications for these enzymes and from the success of new techniques for studying them.

Previous volumes of *Methods in Enzymology* have dealt specifically with phospholipids, their degradation (Volume 197, Phospholipases), and their biosynthesis (Volume 209, Phospholipid Biosynthesis). The recent explosion of interest in lipases led us to develop Volumes 284 and 286. The first, Biotechnology (Volume 284), includes sequencing, cloning, and structural studies of lipases and, the second, Enzyme Characterization and Utilization (Volume 286), includes the purification of novel lipases, kinetics and assay issues, aspects of lipid metabolism, and the use of lipases in organic synthesis.

Research in the lipase field has been dominated by European scientists and stimulated by the European Community Bridge Program. In addition, there has been a great deal of research emphasis on this field in industrial laboratories. Thus, the authorship of this volume is truly international and includes a diverse mixture of academic and industrial scientists.

Expert secretarial assistance from Mary Kincaid, Ophelia Chiu, and Vina Wong helped enormously with the development of this volume. Editorial assistance from Shirley Light is greatly appreciated.

Byron Rubin
Edward A. Dennis

METHODS IN ENZYMOLOGY

VOLUME I. Preparation and Assay of Enzymes
Edited by SIDNEY P. COLOWICK AND NATHAN O. KAPLAN

VOLUME II. Preparation and Assay of Enzymes
Edited by SIDNEY P. COLOWICK AND NATHAN O. KAPLAN

VOLUME III. Preparation and Assay of Substrates
Edited by SIDNEY P. COLOWICK AND NATHAN O. KAPLAN

VOLUME IV. Special Techniques for the Enzymologist
Edited by SIDNEY P. COLOWICK AND NATHAN O. KAPLAN

VOLUME V. Preparation and Assay of Enzymes
Edited by SIDNEY P. COLOWICK AND NATHAN O. KAPLAN

VOLUME VI. Preparation and Assay of Enzymes (*Continued*)
Preparation and Assay of Substrates
Special Techniques
Edited by SIDNEY P. COLOWICK AND NATHAN O. KAPLAN

VOLUME VII. Cumulative Subject Index
Edited by SIDNEY P. COLOWICK AND NATHAN O. KAPLAN

VOLUME VIII. Complex Carbohydrates
Edited by ELIZABETH F. NEUFELD AND VICTOR GINSBURG

VOLUME IX. Carbohydrate Metabolism
Edited by WILLIS A. WOOD

VOLUME X. Oxidation and Phosphorylation
Edited by RONALD W. ESTABROOK AND MAYNARD E. PULLMAN

VOLUME XI. Enzyme Structure
Edited by C. H. W. HIRS

VOLUME XII. Nucleic Acids (Parts A and B)
Edited by LAWRENCE GROSSMAN AND KIVIE MOLDAVE

VOLUME XIII. Citric Acid Cycle
Edited by J. M. LOWENSTEIN

VOLUME XIV. Lipids
Edited by J. M. LOWENSTEIN

VOLUME XV. Steroids and Terpenoids
Edited by RAYMOND B. CLAYTON

VOLUME XVI. Fast Reactions
Edited by KENNETH KUSTIN

VOLUME LV. Biomembranes (Part F: Bioenergetics)
Edited by SIDNEY FLEISCHER AND LESTER PACKER

VOLUME LVI. Biomembranes (Part G: Bioenergetics)
Edited by SIDNEY FLEISCHER AND LESTER PACKER

VOLUME LVII. Bioluminescence and Chemiluminescence
Edited by MARLENE A. DELUCA

VOLUME LVIII. Cell Culture
Edited by WILLIAM B. JAKOBY AND IRA PASTAN

VOLUME LIX. Nucleic Acids and Protein Synthesis (Part G)
Edited by KIVIE MOLDAVE AND LAWRENCE GROSSMAN

VOLUME LX. Nucleic Acids and Protein Synthesis (Part H)
Edited by KIVIE MOLDAVE AND LAWRENCE GROSSMAN

VOLUME 61. Enzyme Structure (Part H)
Edited by C. H. W. HIRS AND SERGE N. TIMASHEFF

VOLUME 62. Vitamins and Coenzymes (Part D)
Edited by DONALD B. MCCORMICK AND LEMUEL D. WRIGHT

VOLUME 63. Enzyme Kinetics and Mechanism (Part A: Initial Rate and Inhibitor Methods)
Edited by DANIEL L. PURICH

VOLUME 64. Enzyme Kinetics and Mechanism (Part B: Isotopic Probes and Complex Enzyme Systems)
Edited by DANIEL L. PURICH

VOLUME 65. Nucleic Acids (Part I)
Edited by LAWRENCE GROSSMAN AND KIVIE MOLDAVE

VOLUME 66. Vitamins and Coenzymes (Part E)
Edited by DONALD B. MCCORMICK AND LEMUEL D. WRIGHT

VOLUME 67. Vitamins and Coenzymes (Part F)
Edited by DONALD B. MCCORMICK AND LEMUEL D. WRIGHT

VOLUME 68. Recombinant DNA
Edited by RAY WU

VOLUME 69. Photosynthesis and Nitrogen Fixation (Part C)
Edited by ANTHONY SAN PIETRO

VOLUME 70. Immunochemical Techniques (Part A)
Edited by HELEN VAN VUNAKIS AND JOHN J. LANGONE

VOLUME 71. Lipids (Part C)
Edited by JOHN M. LOWENSTEIN

VOLUME 72. Lipids (Part D)
Edited by JOHN M. LOWENSTEIN

VOLUME 73. Immunochemical Techniques (Part B)
Edited by JOHN J. LANGONE AND HELEN VAN VUNAKIS

VOLUME 74. Immunochemical Techniques (Part C)
Edited by JOHN J. LANGONE AND HELEN VAN VUNAKIS

VOLUME 75. Cumulative Subject Index Volumes XXXI, XXXII, XXXIV–LX
Edited by EDWARD A. DENNIS AND MARTHA G. DENNIS

VOLUME 76. Hemoglobins
Edited by ERALDO ANTONINI, LUIGI ROSSI-BERNARDI, AND EMILIA CHIANCONE

VOLUME 77. Detoxication and Drug Metabolism
Edited by WILLIAM B. JAKOBY

VOLUME 78. Interferons (Part A)
Edited by SIDNEY PESTKA

VOLUME 79. Interferons (Part B)
Edited by SIDNEY PESTKA

VOLUME 80. Proteolytic Enzymes (Part C)
Edited by LASZLO LORAND

VOLUME 81. Biomembranes (Part H: Visual Pigments and Purple Membranes, I)
Edited by LESTER PACKER

VOLUME 82. Structural and Contractile Proteins (Part A: Extracellular Matrix)
Edited by LEON W. CUNNINGHAM AND DIXIE W. FREDERIKSEN

VOLUME 83. Complex Carbohydrates (Part D)
Edited by VICTOR GINSBURG

VOLUME 84. Immunochemical Techniques (Part D: Selected Immunoassays)
Edited by JOHN J. LANGONE AND HELEN VAN VUNAKIS

VOLUME 85. Structural and Contractile Proteins (Part B: The Contractile Apparatus and the Cytoskeleton)
Edited by DIXIE W. FREDERIKSEN AND LEON W. CUNNINGHAM

VOLUME 86. Prostaglandins and Arachidonate Metabolites
Edited by WILLIAM E. M. LANDS AND WILLIAM L. SMITH

VOLUME 87. Enzyme Kinetics and Mechanism (Part C: Intermediates, Stereochemistry, and Rate Studies)
Edited by DANIEL L. PURICH

VOLUME 88. Biomembranes (Part I: Visual Pigments and Purple Membranes, II)
Edited by LESTER PACKER

VOLUME 89. Carbohydrate Metabolism (Part D)
Edited by WILLIS A. WOOD

VOLUME 90. Carbohydrate Metabolism (Part E)
Edited by WILLIS A. WOOD

VOLUME 91. Enzyme Structure (Part I)
Edited by C. H. W. HIRS AND SERGE N. TIMASHEFF

VOLUME 92. Immunochemical Techniques (Part E: Monoclonal Antibodies and General Immunoassay Methods)
Edited by JOHN J. LANGONE AND HELEN VAN VUNAKIS

VOLUME 93. Immunochemical Techniques (Part F: Conventional Antibodies, Fc Receptors, and Cytotoxicity)
Edited by JOHN J. LANGONE AND HELEN VAN VUNAKIS

VOLUME 94. Polyamines
Edited by HERBERT TABOR AND CELIA WHITE TABOR

VOLUME 95. Cumulative Subject Index Volumes 61–74, 76–80
Edited by EDWARD A. DENNIS AND MARTHA G. DENNIS

VOLUME 96. Biomembranes [Part J: Membrane Biogenesis: Assembly and Targeting (General Methods; Eukaryotes)]
Edited by SIDNEY FLEISCHER AND BECCA FLEISCHER

VOLUME 97. Biomembranes [Part K: Membrane Biogenesis: Assembly and Targeting (Prokaryotes, Mitochondria, and Chloroplasts)]
Edited by SIDNEY FLEISCHER AND BECCA FLEISCHER

VOLUME 98. Biomembranes (Part L: Membrane Biogenesis: Processing and Recycling)
Edited by SIDNEY FLEISCHER AND BECCA FLEISCHER

VOLUME 99. Hormone Action (Part F: Protein Kinases)
Edited by JACKIE D. CORBIN AND JOEL G. HARDMAN

VOLUME 100. Recombinant DNA (Part B)
Edited by RAY WU, LAWRENCE GROSSMAN, AND KIVIE MOLDAVE

VOLUME 101. Recombinant DNA (Part C)
Edited by RAY WU, LAWRENCE GROSSMAN, AND KIVIE MOLDAVE

VOLUME 102. Hormone Action (Part G: Calmodulin and Calcium-Binding Proteins)
Edited by ANTHONY R. MEANS AND BERT W. O'MALLEY

VOLUME 103. Hormone Action (Part H: Neuroendocrine Peptides)
Edited by P. MICHAEL CONN

VOLUME 104. Enzyme Purification and Related Techniques (Part C)
Edited by WILLIAM B. JAKOBY

VOLUME 105. Oxygen Radicals in Biological Systems
Edited by LESTER PACKER

VOLUME 106. Posttranslational Modifications (Part A)
Edited by FINN WOLD AND KIVIE MOLDAVE

VOLUME 107. Posttranslational Modifications (Part B)
Edited by FINN WOLD AND KIVIE MOLDAVE

VOLUME 108. Immunochemical Techniques (Part G: Separation and Characterization of Lymphoid Cells)
Edited by GIOVANNI DI SABATO, JOHN J. LANGONE, AND HELEN VAN VUNAKIS

VOLUME 109. Hormone Action (Part I: Peptide Hormones)
Edited by LUTZ BIRNBAUMER AND BERT W. O'MALLEY

VOLUME 110. Steroids and Isoprenoids (Part A)
Edited by JOHN H. LAW AND HANS C. RILLING

VOLUME 111. Steroids and Isoprenoids (Part B)
Edited by JOHN H. LAW AND HANS C. RILLING

VOLUME 112. Drug and Enzyme Targeting (Part A)
Edited by KENNETH J. WIDDER AND RALPH GREEN

VOLUME 113. Glutamate, Glutamine, Glutathione, and Related Compounds
Edited by ALTON MEISTER

VOLUME 114. Diffraction Methods for Biological Macromolecules (Part A)
Edited by HAROLD W. WYCKOFF, C. H. W. HIRS, AND SERGE N. TIMASHEFF

VOLUME 115. Diffraction Methods for Biological Macromolecules (Part B)
Edited by HAROLD W. WYCKOFF, C. H. W. HIRS, AND SERGE N. TIMASHEFF

VOLUME 116. Immunochemical Techniques (Part H: Effectors and Mediators of Lymphoid Cell Functions)
Edited by GIOVANNI DI SABATO, JOHN J. LANGONE, AND HELEN VAN VUNAKIS

VOLUME 117. Enzyme Structure (Part J)
Edited by C. H. W. HIRS AND SERGE N. TIMASHEFF

VOLUME 118. Plant Molecular Biology
Edited by ARTHUR WEISSBACH AND HERBERT WEISSBACH

VOLUME 119. Interferons (Part C)
Edited by SIDNEY PESTKA

VOLUME 120. Cumulative Subject Index Volumes 81–94, 96–101

VOLUME 121. Immunochemical Techniques (Part I: Hybridoma Technology and Monoclonal Antibodies)
Edited by JOHN J. LANGONE AND HELEN VAN VUNAKIS

VOLUME 122. Vitamins and Coenzymes (Part G)
Edited by FRANK CHYTIL AND DONALD B. MCCORMICK

VOLUME 123. Vitamins and Coenzymes (Part H)
Edited by FRANK CHYTIL AND DONALD B. MCCORMICK

VOLUME 124. Hormone Action (Part J: Neuroendocrine Peptides)
Edited by P. MICHAEL CONN

VOLUME 125. Biomembranes (Part M: Transport in Bacteria, Mitochondria, and Chloroplasts: General Approaches and Transport Systems)
Edited by SIDNEY FLEISCHER AND BECCA FLEISCHER

VOLUME 143. Sulfur and Sulfur Amino Acids
Edited by WILLIAM B. JAKOBY AND OWEN GRIFFITH

VOLUME 144. Structural and Contractile Proteins (Part D: Extracellular Matrix)
Edited by LEON W. CUNNINGHAM

VOLUME 145. Structural and Contractile Proteins (Part E: Extracellular Matrix)
Edited by LEON W. CUNNINGHAM

VOLUME 146. Peptide Growth Factors (Part A)
Edited by DAVID BARNES AND DAVID A. SIRBASKU

VOLUME 147. Peptide Growth Factors (Part B)
Edited by DAVID BARNES AND DAVID A. SIRBASKU

VOLUME 148. Plant Cell Membranes
Edited by LESTER PACKER AND ROLAND DOUCE

VOLUME 149. Drug and Enzyme Targeting (Part B)
Edited by RALPH GREEN AND KENNETH J. WIDDER

VOLUME 150. Immunochemical Techniques (Part K: *In Vitro* Models of B and T Cell Functions and Lymphoid Cell Receptors)
Edited by GIOVANNI DI SABATO

VOLUME 151. Molecular Genetics of Mammalian Cells
Edited by MICHAEL M. GOTTESMAN

VOLUME 152. Guide to Molecular Cloning Techniques
Edited by SHELBY L. BERGER AND ALAN R. KIMMEL

VOLUME 153. Recombinant DNA (Part D)
Edited by RAY WU AND LAWRENCE GROSSMAN

VOLUME 154. Recombinant DNA (Part E)
Edited by RAY WU AND LAWRENCE GROSSMAN

VOLUME 155. Recombinant DNA (Part F)
Edited by RAY WU

VOLUME 156. Biomembranes (Part P: ATP-Driven Pumps and Related Transport: The Na,K-Pump)
Edited by SIDNEY FLEISCHER AND BECCA FLEISCHER

VOLUME 157. Biomembranes (Part Q: ATP-Driven Pumps and Related Transport: Calcium, Proton, and Potassium Pumps)
Edited by SIDNEY FLEISCHER AND BECCA FLEISCHER

VOLUME 158. Metalloproteins (Part A)
Edited by JAMES F. RIORDAN AND BERT L. VALLEE

VOLUME 159. Initiation and Termination of Cyclic Nucleotide Action
Edited by JACKIE D. CORBIN AND ROGER A. JOHNSON

VOLUME 160. Biomass (Part A: Cellulose and Hemicellulose)
Edited by WILLIS A. WOOD AND SCOTT T. KELLOGG

VOLUME 161. Biomass (Part B: Lignin, Pectin, and Chitin)
Edited by WILLIS A. WOOD AND SCOTT T. KELLOGG

VOLUME 162. Immunochemical Techniques (Part L: Chemotaxis and Inflammation)
Edited by GIOVANNI DI SABATO

VOLUME 163. Immunochemical Techniques (Part M: Chemotaxis and Inflammation)
Edited by GIOVANNI DI SABATO

VOLUME 164. Ribosomes
Edited by HARRY F. NOLLER, JR., AND KIVIE MOLDAVE

VOLUME 165. Microbial Toxins: Tools for Enzymology
Edited by SIDNEY HARSHMAN

VOLUME 166. Branched-Chain Amino Acids
Edited by ROBERT HARRIS AND JOHN R. SOKATCH

VOLUME 167. Cyanobacteria
Edited by LESTER PACKER AND ALEXANDER N. GLAZER

VOLUME 168. Hormone Action (Part K: Neuroendocrine Peptides)
Edited by P. MICHAEL CONN

VOLUME 169. Platelets: Receptors, Adhesion, Secretion (Part A)
Edited by JACEK HAWIGER

VOLUME 170. Nucleosomes
Edited by PAUL M. WASSARMAN AND ROGER D. KORNBERG

VOLUME 171. Biomembranes (Part R: Transport Theory: Cells and Model Membranes)
Edited by SIDNEY FLEISCHER AND BECCA FLEISCHER

VOLUME 172. Biomembranes (Part S: Transport: Membrane Isolation and Characterization)
Edited by SIDNEY FLEISCHER AND BECCA FLEISCHER

VOLUME 173. Biomembranes [Part T: Cellular and Subcellular Transport: Eukaryotic (Nonepithelial) Cells]
Edited by SIDNEY FLEISCHER AND BECCA FLEISCHER

VOLUME 174. Biomembranes [Part U: Cellular and Subcellular Transport: Eukaryotic (Nonepithelial) Cells]
Edited by SIDNEY FLEISCHER AND BECCA FLEISCHER

VOLUME 175. Cumulative Subject Index Volumes 135–139, 141–167

VOLUME 176. Nuclear Magnetic Resonance (Part A: Spectral Techniques and Dynamics)
Edited by NORMAN J. OPPENHEIMER AND THOMAS L. JAMES

VOLUME 177. Nuclear Magnetic Resonance (Part B: Structure and Mechanism)
Edited by NORMAN J. OPPENHEIMER AND THOMAS L. JAMES

VOLUME 266. Computer Methods for Macromolecular Sequence Analysis
Edited by RUSSELL F. DOOLITTLE

VOLUME 267. Combinatorial Chemistry
Edited by JOHN N. ABELSON

VOLUME 268. Nitric Oxide (Part A: Sources and Detection of NO; NO Synthase)
Edited by LESTER PACKER

VOLUME 269. Nitric Oxide (Part B: Physiological and Pathological Processes)
Edited by LESTER PACKER

VOLUME 270. High Resolution Separation and Analysis of Biological Macromolecules (Part A: Fundamentals)
Edited by BARRY L. KARGER AND WILLIAM S. HANCOCK

VOLUME 271. High Resolution Separation and Analysis of Biological Macromolecules (Part B: Applications)
Edited by BARRY L. KARGER AND WILLIAM S. HANCOCK

VOLUME 272. Cytochrome P450 (Part B)
Edited by ERIC F. JOHNSON AND MICHAEL R. WATERMAN

VOLUME 273. RNA Polymerase and Associated Factors (Part A)
Edited by SANKAR ADHYA

VOLUME 274. RNA Polymerase and Associated Factors (Part B)
Edited by SANKAR ADHYA

VOLUME 275. Viral Polymerases and Related Proteins
Edited by LAWRENCE C. KUO, DAVID B. OLSEN, AND STEVEN S. CARROLL

VOLUME 276. Macromolecular Crystallography (Part A)
Edited by CHARLES W. CARTER, JR., AND ROBERT M. SWEET

VOLUME 277. Macromolecular Crystallography (Part B)
Edited by CHARLES W. CARTER, JR., AND ROBERT M. SWEET

VOLUME 278. Fluorescence Spectroscopy
Edited by LUDWIG BRAND AND MICHAEL L. JOHNSON

VOLUME 279. Vitamins and Coenzymes, Part I
Edited by DONALD B. MCCORMICK, JOHN W. SUTTIE, AND CONRAD WAGNER

VOLUME 280. Vitamins and Coenzymes, Part J
Edited by DONALD B. MCCORMICK, JOHN W. SUTTIE, AND CONRAD WAGNER

VOLUME 281. Vitamins and Coenzymes, Part K
Edited by DONALD B. MCCORMICK, JOHN W. SUTTIE, AND CONRAD WAGNER

VOLUME 282. Vitamins and Coenzymes, Part L (in preparation)
Edited by DONALD B. MCCORMICK, JOHN W. SUTTIE, AND CONRAD WAGNER

VOLUME 283. Cell Cycle Control
Edited by WILLIAM G. DUNPHY

Section I

Primary and Tertiary Structure

[1] Structure as Basis for Understanding Interfacial Properties of Lipases

By MIROSLAW CYGLER and JOSEPH D. SCHRAG*

Introduction

Lipases are water-soluble enzymes that catalyze the hydrolysis of ester bonds in triacylglycerols. In 1958, Sarda and Desnuelle[1] investigated the lipolytic activity of pancreatic lipase and carefully documented the "interfacial activation" phenomenon—the enhancement of activity by the presence of micelles. They made a suggestion that pancreatic lipase works most efficiently on the two-dimensional surface of the micelle. In this and in a subsequent article,[2] Desnuelle and co-workers postulated that the lipase becomes activated when adsorbed at the water–lipid interface and that this activation process is associated with a conformational change in the enzyme. This interfacial activation phenomenon was subsequently observed for many other lipases of various origins and became a characteristic feature used to distinguish "true" lipases from esterases (e.g., see Ref. 3; reviewed in Ref. 4). The more recent discovery of lipolytic enzymes that are not interfacially activatable despite clear sequence homology to other lipases[5–7] showed that the classification based solely on the interfacial activatability is too simplistic. Ransac *et al.*[4] discussed this question thoroughly and proposed to define lipase as "a carboxylesterase able to catalyze hydrolysis of long chain acylglycerols (p. 270)." Nevertheless, the activity of the majority of the known lipases increases in the presence of a water–lipid interface, and it is the molecular nature of this activation that has fascinated research-

[1] L. Sarda and P. Desnuelle, *Biochim. Biophys. Acta* **30,** 513 (1958).

[2] P. Desnuelle, L. Sarda, and G. Ailhard, *Biochim. Biphys. Acta* **37,** 570 (1960).

[3] B. Borgström and H. L. Brockman, "Lipases." Elsevier, Amsterdam, 1984.

[4] S. Ransac, F. Carrière, E. Rogalska, R. Verger, F. Marguet, G. Buono, E. P. Melo, J. M. S. Cabral, M. P. Egloff, H. van Tilbeurgh, and C. Cambillau, "Molecular Dynamics of Biomembranes" (J. A. F. Op den Kamp, ed.), Vol. H96, pp. 265–304. NATO ASI series. Springer-Verlag, Berlin, 1996.

[5] C. Martinez, P. De Geus, M. Lauwereys, G. Matthyssens, and C. Cambillau, *Nature* (*London*) **356,** 615 (1992).

[6] E. Lesuisse, K. Schanck, and C. Colson, *Eur. J. Biochem.* **216,** 155 (1993).

[7] A. Hjorth, F. Carrière, C. Cudrey, H. Wöldike, E. Boel, D. M. Lawson, F. Ferrato, C. Cambillau, G. G. Dodson, L. Thim, and R. Verger, *Biochemistry* **32,** 4702 (1993).

ers. Various biophysical and biochemical methods have been developed to study the activation phenomena and many of them, and their results, are described in other chapters of this volume. Although these investigations provided much information on the macroscopic properties of lipases, their stereoselectivity, the influence of the environment (solvent, interface properties, and parameters) on the reaction outcome, etc., the first direct evidence that a conformational rearrangement does, indeed, occur at the water–lipid interface came from crystallography. The three-dimensional structures of the first three, seemingly different, lipases were determined in short succession: the 27-kDa *Rhizomucor miehei* lipase,[8] the 46-kDa human pancreatic lipase,[9] and the 60-kDa *Geotrichum candidum* lipase.[10] Two surprising findings emanated from these studies. The first finding, recognized immediately in the original papers, was that the active site, composed of a Ser-His-Asp/Glu catalytic triad reminiscent of the serine proteases,[11] is not exposed on the protein surface but rather is covered by surface loops and is not accessible to the substrate. Yet there was no doubt about the enzymatic competence of the enzymes used for crystallization. The second surprising finding was the commonality of the basic fold of these lipases within their catalytic domains.[12,13] This fold, which has also been found in many esterases and other hydrolytic enzymes, has been termed the *α/β-hydrolase fold* (see [4] in this volume[13a]).

That three different lipases crystallized from aqueous solutions in enzymatically incompetent conformations suggested that the observed conformations are not an artifact of the crystallization conditions, but instead represent the biologically relevant inactive forms that are prevalent in solution. On the basis of these crystal structures it was obvious that at least some of the surface loops would have to change their conformations or positions significantly for the substrate to be able to access the active site.[8–10]

This prediction was quickly confirmed experimentally. First, the struc-

[8] L. Brady, A. M. Brzozowski, Z. S. Derewenda, E. Dodson, G. Dodson, S. Tolley, J. P. Turkenburg, L. Christiansen, B. Huge-Jensen, L. Norskov, L. Thim, and U. Menge, *Nature (London)* **343,** 767 (1990).

[9] F. K. Winkler, A. D'Arcy, and W. Hunziker, *Nature (London)* **343,** 771 (1990).

[10] J. D. Schrag, Y. G. Li, S. Wu, and M. Cygler, *Nature (London)* **351,** 761 (1991).

[11] D. M. Blow, *Nature (London)* **343,** 694 (1990).

[12] D. L. Ollis, E. Cheah, M. Cygler, B. Dijkstra, F. Frolow, S. M. Franken, M. Harel, S. J. Remington, I. Silman, J. Schrag, J. L. Sussman, K. H. G. Verschueren, and A. Goldman, *Protein Eng.* **5,** 197 (1992).

[13] M. Cygler, J. D. Schrag, and F. Ergan, *Biotechnol. Genet. Eng. Rev.* **10,** 143 (1992).

[13a] J. D. Schrag and M. Cygler, *Methods Enzymol.* **284,** [4], 1997 (this volume).

tures of the *Rhizomucor miehei* lipase complexed with two different inhibitors were reported[14,15]) and showed a large rearrangement of a single loop (lid or flap) relative to the uncomplexed structure. This movement opened access to the active site and exposed the nucleophilic serine. A similar effect was caused by a phospholipid molecule bound in the active site of the pancreatic lipase–colipase complex.[16] In this case, the conformational change was larger and involved two loops. Another direct confirmation of the conformational rearrangement came from the determination of the structure of *Candida rugosa* lipase, a homolog of *G. candidum* lipase, crystallized under two different conditions and displaying two different conformations for a large surface loop: one that covers the active site and another in which the loop is moved to the side, exposing the active site.[17,18] This was further corroborated by the structures of *C. rugosa* lipase complexed with phosphonyl and sulfonyl inhibitors.[19] Another example of a conformational rearrangement of the lid has been observed in *Pseudomonas* lipases.[20]

To date, the 3-dimensional structures of more than 10 lipases determined by X-ray crystallography have been described in the literature and work on more is near completion. Apart from those mentioned above they are *Pseudomonas glumae* lipase,[21] *Chromobacterium viscosum* lipase (nearly identical to *P. glumae* lipase,[22] *Candida antarctica* lipase B,[23] *Humicola lanuginosa* lipase,[24] *Fusarium solani* cutinase,[5] *Rhizopus delemar* lipase,[24]

[14] A. M. Brzozowski, U. Derewenda, Z. S. Derewenda, G. G. Dodson, D. M. Lawson, J. P. Turkenburg, F. Bjorkling, B. Huge-Jensen, S. A. Patkar, and L. Thim, *Nature* (*London*) **351,** 491 (1991).

[15] U. Derewenda, A. M. Brzozowski, D. M. Lawson, and Z. S. Derewenda, *Biochemistry* **31,** 1532 (1992).

[16] H. van Tilbeurgh, M. P. Egloff, C. Martinez, N. Rugani, R. Verger, and C. Cambillau, *Nature* (*London*) **362,** 814 (1993).

[17] P. Grochulski, Y. Li, J. D. Schrag, F. Bouthillier, P. Smith, D. Harrison, B. Rubin, and M. Cygler, *J. Biol. Chem.* **268,** 12843 (1993).

[18] P. Grochulski, Y. Li, J. D. Schrag, and M. Cygler, *Protein Sci.* **3,** 82 (1994).

[19] P. Grochulski, F. Bouthillier, R. J. Kazlauskas, A. N. Serreqi, J. D. Schrag, E. Ziomek, and M. Cygler, *Biochemistry* **33,** 3494 (1994).

[20] J. D. Schrag, Y. Li, M. Cygler, D. Lang, T. Burgdorf, H.-J. Hecht, D. Schomburg, T. Rydel, J. Oliver, L. Strickland, M. Dunaway, S. Larson, and A. McPherson, *Structure* **5,** 187 (1997).

[21] M. E. Noble, A. Cleasby, L. N. Johnson, M. R. Egmond, and L. G. Frenken, *FEBS Lett.* **331,** 123 (1993).

[22] D. Lang, B. Hoffmann, L. Haalck, H.-J. Hecht, F. Spener, R. D. Schmid, and D. Schomburg, *J. Mol. Biol.* **259,** 704 (1996).

[23] J. Uppenberg, M. T. Hansen, S. Patkar, and T. A. Jones, *Structure* **2,** 293 (1994).

[24] U. Derewenda, L. Swenson, R. Green, Y. Wei, G. G. Dodson, S. Yamaguchi, M. J. Haas, and Z. S. Derewenda, *Nat. Struct. Biol.* **1,** 36 (1994).

a second isoform of *C. rugosa* lipase (referred to as a cholesterol esterase[25]), *Penicillium camembertii* lipase,[24] *Pseudomonas cepacia* lipase,[20] *Penicillium* sp. UZLM 4 lipase (J. D. Schrag and M. Cygler, unpublished, 1997), and a *Pseudomonas mendaccino* lipase.[26]

Here we briefly describe common features of the lipase fold, discuss the characteristics of the "closed" vs "open" conformation, and explore the differences between lipases. Last, we describe substrate-binding sites and discuss possible interactions of lipases with the lipid layer.

Lipase Fold

Lipases with known three-dimensional structures span a wide range of molecular weights, from ~19 kDa (cutinase) to ~60 kDa (*G. candidum* lipase). All of them, with the exception of pancreatic lipases, contain only one domain. These catalytic domains belong to the α,β doubly wound protein fold[27] and are formed from a parallel β sheet and a number of helices that flank the sheet on both sides. The minimal fragment of this fold common to all lipases is a subset of the α/β-hydrolase fold as described by Ollis *et al.*[12] and contains a five-stranded β sheet and two α helices (B and C in Fig. 1). In addition, helix A exists in all lipases except for those of the *R. miehei* family, and helix D usually exists in the form of a short, distorted turn that is positioned in space between that of helix D and E of the α/β-hydrolase fold. The β strands correspond to strands $\beta3$ to $\beta7$ of the α/β-hydrolase fold (see [4] in this volume[13a]). The positions of the catalytic residues in this nomenclature are as follows: serine after strand $\beta5$, histidine after strand $\beta8$, and acid after strand $\beta7$. Pancreatic lipase is an exception to this arrangement, with the acid after strand $\beta6$.[9,28] Although the order of triad residues, their location, and the lipase fold have been conserved in evolution, the topological location of the lid-forming loops differs between lipases (see p. 14). In some lipases the lid is located N terminal to the nucleophile, whereas in others it is C terminal to the nucleophile. In addition, whereas in some lipases only one loop undergoes rearrangement, in others one or even two additional loops move together

[25] D. Ghosh, Z. Wawrzak, V. Z. Pletnev, N. Li, R. Kaiser, W. Pangborn, H. Jörnvall, M. Erman, and W. L. Duax, *Structure* **3,** 279 (1995).

[26] R. Sarma, J. Dauberman, A. J. Poulose, J. Van Beilen, S. Power, B. Shew, G. Gray, S. Norton, and R. Bott, *in* "Lipases: Structure, Mechanism and Genetic Engineering." (L. Alberghina, R. D. Schmid, and R. Verger, eds.) Vol. 16, pp. 71–76. GBF Monographs, CEC-GBF International Workshop, Braunschweig, September 13–15. VCH Verlagsgesellschaft mbH, Weinheim, Federal Republic of Germany, 1991.

[27] J. S. Richardson, *Adv. Protein Chem.* **34,** 167 (1981).

[28] J. D. Schrag, F. K. Winkler, and M. Cygler, *J. Biol. Chem.* **267,** 4300 (1992).

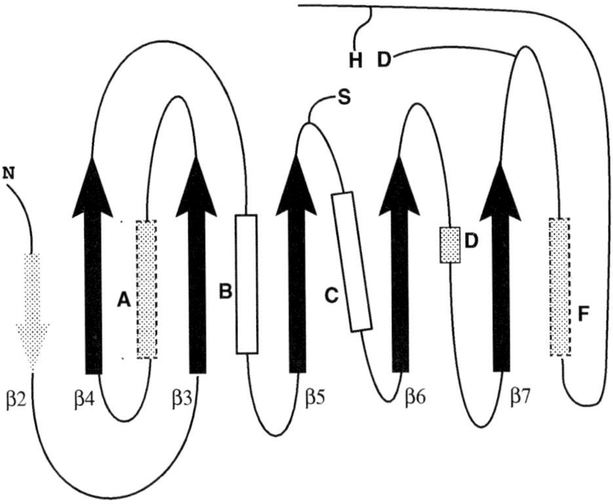

FIG. 1. The common lipase fold. Arrows indicate β strands and rectangles indicate α helices. β Strands are numbered according to the nomenclature of the α/β-hydrolase fold (Ollis *et al.*[12]). Secondary structural elements shown in black or white (strands $\beta3$–$\beta7$ and helices B and C) occur in all lipases; those shown in gray (strand $\beta2$ and helices A, D, and F) occur in most. Helices A and F, shown with dashed border lines, are on the concave side of the β sheet; the other helices are on the convex side. Helix D is often composed of only one (distorted) turn.

with the lid loop over significant distances to create the active conformation. Cutinase is an example of a lipase without a lid, and in this structure all of the loops near the active site are relatively short.[5]

Two Classes of Conformers: Closed and Open

A comparison of three-dimensional structures of related molecules obtained from X-ray diffraction experiments can provide a wealth of information on the flexibility and/or mobility of various parts of the proteins, going beyond what can be learned from a static view of a single molecule. To that effect, the crystallization process may be used to investigate the structural features of a protein fold that are sensitive to external factors and differentiate them from those features that are independent of the protein environment.

One of the important features of an enzyme is the shape of the substrate-binding site and its accessibility to the substrate. Structures of the majority of enzymes show that their active sites are on the surface of the molecule, accessible to the solvent. This is not always the case for lipases. The various lipase structures determined to date can be divided into two categories:

those with the active site accessible from the solvent (open form) and those with an inaccessible active site (closed form). For some lipases both forms have been experimentally observed (see Table I), which immediately raises the question of the relevance of these forms for the enzymatic activity of lipases. The fact that the open and closed conformations are observed not only within one homologous family of lipases but across families suggests that these two states are important for the function of these enzymes.

Within homologous lipases all the open forms are quantitatively similar, as are the closed forms. When the structures of the native lipases (uncomplexed) are analyzed, we find that there is some correlation between the crystallization conditions and the conformational state of the lipase. This is summarized in Table I. The polyethylene glycol conditions in general promote the crystallization of the closed form, whereas a substantial content of alcohol (e.g., 2-methyl-2,4-pentanediol) or detergent in the crystallization solution is conducive to the crystallization of lipases in an open state. The polyethylene glycol is believed to affect protein solubility during crystallization by partial exclusion of water from around the protein,[29] but the local environment of the protein remains water based. The addition of alcohol or detergent, however, changes significantly the microscopic properties of the solution and influences protein environment. Another factor influencing the equilibrium between the closed and open forms is the presence of substrate-like (inhibitor) molecules. When they are bound in the active site region, either covalently or noncovalently, only the open conformation is observed.

Closed (Inactive) Conformation

The closed conformations are presumed to predominate in aqueous media as indicated by the lower lipolytic activity of most lipases in the absence of an interface. At the same time, the persistence of a low level of activity shows that other conformations allowing access of the substrate to the active site are also at least transiently present in solution. The closed conformations are characterized by an unoccupied active site, which is shielded from the solvent by one or more loops forming the lid. The contacting surfaces of the lid and the remaining part of the protein are complementary, with the exception of an area near the nucleophile, where a cavity under the lid is frequently found. This cavity usually contains some water molecules. Overall, the external surfaces of the closed forms of lipases are reasonably hydrophilic, typical for water-soluble proteins. A closer examination of the distribution of polar atoms (nitrogen and oxygen) on

[29] T. Arakawa and S. Timasheff, *Biochemistry* **24,** 6756 (1985).

the molecular surface shows that in some lipases, but not all, the surface has a more hydrophobic character on the lid-containing face. This is shown in Fig. 2 for *C. rugosa* and *H. lanuginosa* lipases.

Open Conformation

The features of the lipase surface in the open conformation are quite different from those of the closed conformation. The movement of the lid not only opens access to the active site, but at the same time exposes a large hydrophobic surface (Fig. 3a). In *R. miehei* lipase the activation rearrangement is associated with ~700 Å^2 of hydrophobic area being exposed.[15] At the same time, some of the hydrophilic surface previously exposed becomes buried. The decrease in the hydrophilic surface is ~450 Å^2. An even larger area, more than 1000 Å^2, of newly exposed hydrophobic surface has been observed for the *C. rugosa* lipase.[18] Similarly, a significant increase in hydrophobic surface around the active site was reported for pancreatic lipase.[16] In all open conformations of lipases, the nucleophile is at the bottom of a depression created by the lid movement. In the observed conformation of *C. antarctica* lipase B, the path leading to the active site is rather narrow (Fig. 3b).[23] The more hydrophobic character of the face containing the active site can also be noticed in cutinase, an enzyme that does not have a lid (Fig. 3c) and whose structure represents the active state of this enzyme. In particular, loop 175–190 on one side of the nucleophile, Ser-120, contains many hydrophobic residues.

Oxyanion Hole

An important component of the catalytic machinery of hydrolases, first postulated for serine proteases,[30,31] is the oxyanion hole, a constellation of properly located hydrogen bond donors (usually main-chain NH groups) that helps to stabilize the intermediate state arising during the reaction, in which the carbonyl oxygen bears a partial negative charge. In serine proteases, this oxyanion hole exists in a competent form in the substrate-free enzyme. This is not the case with many lipases. In most of them at least one of the residues that contribute to the oxyanion hole is a part of one of the moving loops and attains its proper position and orientation only in the open conformation of the lipase. Among the presently known structures, the *C. rugosa* lipase is somewhat of an exception. Its oxyanion hole is already preformed in the closed conformation and does not change on lid

[30] D. M. Blow, J. J. Birktoft, and B. S. Hartley, *Nature (London)* **221,** 337 (1969).
[31] J. D. Robertus, J. Kraut, R. A. Alden, and J. J. Birktoft, *Biochemistry* **11,** 4293 (1972).

TABLE I

CRYSTALLIZATION CONDITIONS VERSUS CONFORMATIONAL STATE OF LIPASE[a]

Lipase	Crystallization conditions	Conformational state
Candida antarctica B	1. 20% (w/v) PEG 4000, 20% (v/v) 2-propanol, 0.6% (w/v) β-OG, 100 m*M* sodium acetate (pH 3.6)	Partially open
	2. 1.0–1.3 *M* $(NH_4)_2SO_4$, 10% (v/v) dioxane, 0.5% (w/v) β-OG, 100 m*M* sodium citrate (pH 4.0)	Partially open
Geotrichum candidum	15–18% (w/v) PEG 8000 (pH 6–8.8)	Closed
Candida rugosa		
Form I	35–40% (v/v) MPD, 30 m*M* sodium acetate (pH 5.5), 50 m*M* $CaCl_2$	Open
Form II	10% (w/v) PEG 8000, 0.15 *M* cacodylate (pH 6.5)	Closed
Cholesterol esterase–lipase	25% (w/v) PEG 3350, $(Na,K)PO_4$ buffer (pH 7.3)	Open
Cholesterol esterase–lipase–linoleate	25% (w/v) PEG 3350, $(Na,K)PO_4$ buffer (pH 7.3), 0.04 m*M* cholesterol linoleate	Open
Human pancreatic		
Lipase	8% (w/v) PEG 8000, 1% (w/v) β-OG, 0.15 *M* LiCl, 50 m*M* sodium cacodylate (pH 5.5)	Closed
Lipase–colipase	2% (w/v) PEG 8000, 0.2–0.5 *M* NaCl, 0.1 *M* MES (pH 6.0), trace of β-OG	Closed
Lipase–colipase–phospholipid	As above, but in the presence of mixed micelles of 1,2-dididecanoyl-*sn*-3-glycerophosphorylcholine and sodium taurodeoxycholate	Open
Lipase–colipase–inhibitor	2% (w/v) PEG 8000, 0.1 *M* MES (pH 6), 0.4 *M* NaCl, trace of β-OG	Open

Rhizomucor miehei	2–2.8 *M* (Na,K)PO_4 (pH 8)	Closed
Rhizomucor miehei–inhibitor	8–12% (v/v) PEG 600, 20 m*M* Tris (pH 8)	Open
Humicola lanuginosa	4% (w/v) PEG 6000, 0.1 *M* sodium acetate (pH 4.0), 1 m*M* K_2PdCl_4, 3 m*M* spermine	Closed, mobile
Humicola lanuginosa–inhibitor	50–70% saturated KH_2PO_4, 20 m*M* Tris (pH 8)	Open
Rhizopus delemar	40% ammonium sulfate, 2% (v/v) MPD, 0.1 *M* sodium acetate (pH 6.0), 0.05 m*M* *N*,*N*-dimethyl octadecylamine-*N*-oxide	One molecule closed, other partially open
Penicillium camemberti	10% (w/v) PEG 2000, 0.1 *M* sodium acetate (pH 4.5), 2 m*M* DTT	Closed
Penicillium sp. UZLM 4	25% (w/v) PEG 1450, 0.1 *M* HEPES (pH 7.5)	Disordered
Pseudomonas cepacia	1. 25–35% (v/v) *n*-propanol, 50 m*M* Tris-HCl (pH 8.5)	Open
	2. 60% (v/v) MPD, 0.2 *M* sodium citrate, 50 m*M* HEPES (pH 7.5), 0.36% (w/v) β-OG	Undetermined
Pseudomonas glumae	27–29% (w/v) PEG 8000, 10% (v/v) acetone, trace amounts β-DG, 0.1 *M* Tris (pH 9)	Closed
Chromobacterium viscosum	10–14% (w/v) PEG 4000, 10–14% (v/v) MPD, 0.1 *M* citrate phosphate (pH 6.4), 0.25% (v/v) *n*-octyl-β-D-glycopyranoside	Closed

[a] Abbreviations: PEG, polyethylene glycol; MPD, 2-methyl-2,4-pentanediol; β-DG, β-dodecylglucoside; DTT, dithiothreitol.

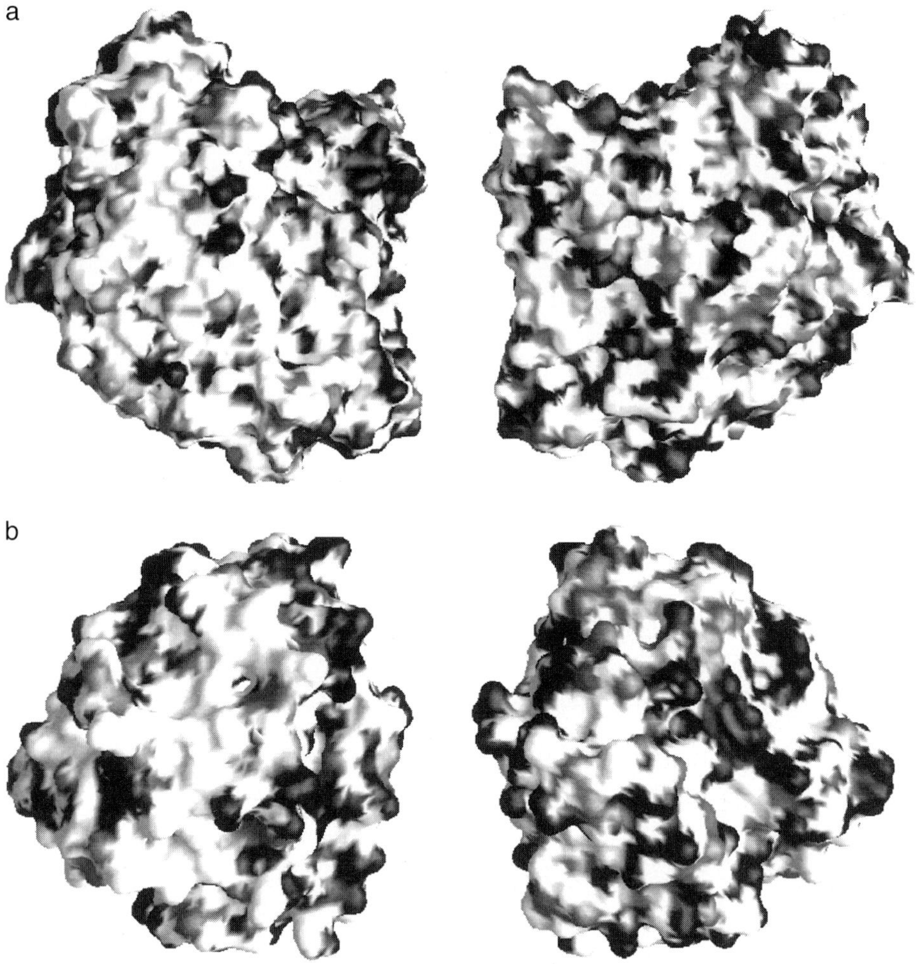

FIG. 2. The distribution of polar (nitrogen and oxyen) and nonpolar (carbon and sulfur) atoms on the surfaces of closed conformations of lipases. Molecular surface corresponding to polar atoms is shown in dark shades. Two sides of the molecule are shown: on the left is the side containing the lid (near the center); on the right is the opposite side (rotation by 180° around vertical axis). Notice much more light-colored surface on the side containing the lid (left). (a) Closed form of *C. rugosa* lipase; (b) closed form of *Humicula lanuginosa* lipase.

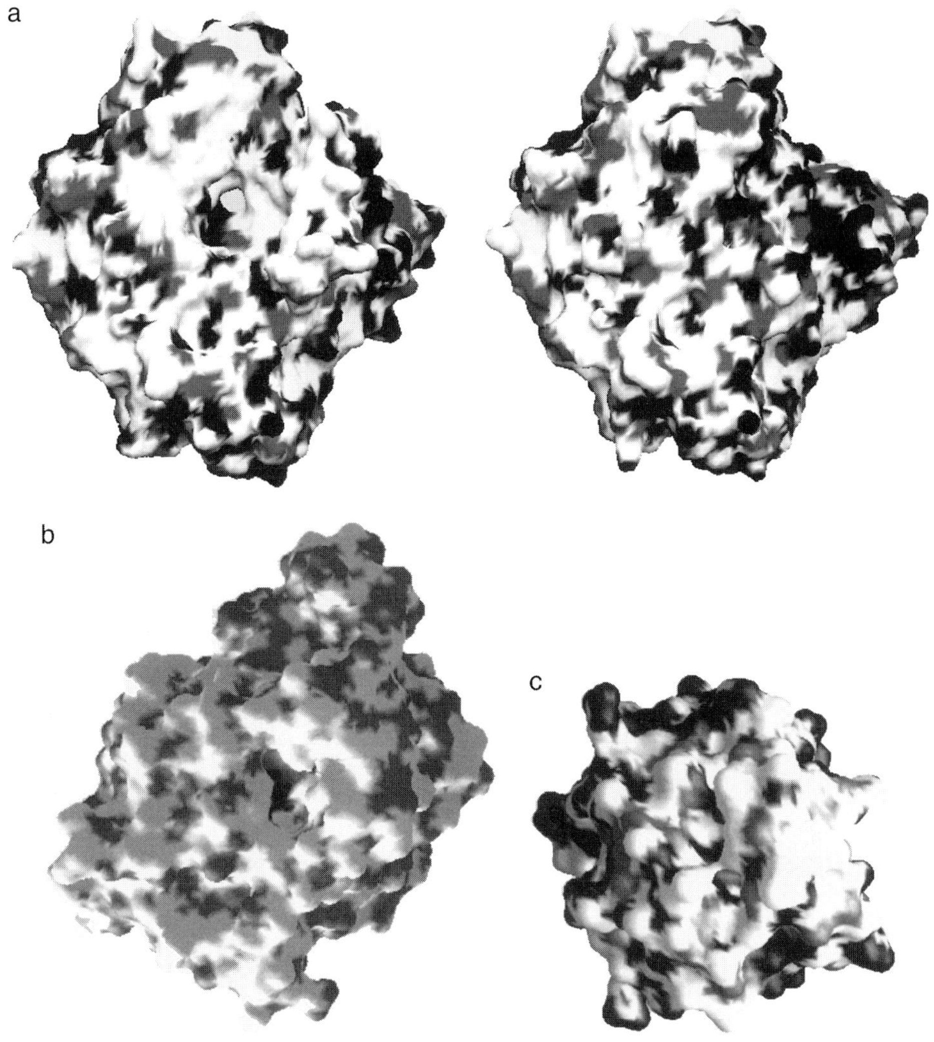

FIG. 3. The distribution of polar (nitrogen and oxygen) and nonpolar (carbon and sulfur) atoms on the active site-containing surfaces of open and/or active conformations of lipases. Molecular surface corresponding to polar atoms is shown in dark shades. (a) *Candida rugosa* lipase in open (left) and closed (right) form. There is more lightly colored (hydrophobic) surface in the open form and the entrance to the tunnel is visible near the center. (b) *Candida antarctica* lipase B: The active site is located at the bottom of a deep depression and is accessible from the solvent even in the native form. (c) *Fusarium solani* cutinase: Active site in the center.

opening. However, a closely related lipase from *G. candidum* is predicted to undergo a more complex rearrangement, likely resulting in a remodeling of the oxyanion hole in the process.[10,32] Obviously, the lipolytic enzymes that do not undergo a conformational change, e.g., *F. solani* cutinase, have a properly assembled oxyanion hole in the absence of substrate.[5]

Lid and Its Topological Location in Various Lipases

Although the topological location of the members of the catalytic triad is highly conserved among lipases with known three-dimensional structures (see above), the topological location of the lid varies among the lipases and its length and complexity increase in general with the size of the molecule. Within a family of lipases sharing amino acid sequence homology the lid is formed by the corresponding structural elements. However, the location of the lid varies from one family to another. The native structure of cutinase,[5] the smallet of the lipolytic enzymes with known three-dimensional structure, and that of its complexes with inhibitors[5] suggest that it has no lid at all. Indeed, no enhancement of enzymatic activity by the presence of the lipid–water interface was measured.[5] The main biological function of this enzyme is the hydrolysis of cutin and, hence, this enzyme may well be considered an esterase with a broader activity that also includes lipids. The activity and enantioselectivity of this enzyme toward lipid substrates has been mapped by using a wide variety of synthetic triglyceride analogs.[33]

The existence of an active site access-controlling lid in *C. antarctica* lipase B is somewhat uncertain. The structural data for this 33-kDa enzyme without and with an inhibitor or a detergent molecule show little difference in the overall structure of the enzyme and, in particular, in the features near the active site. The entrance to the active site is bordered by two helices: a short one encompassing residues 142–146 and a long C-terminal helix (residues 268–287). Both helices show higher mobility than the rest of the structure as expressed by their high temperature factors and by the fact that the short helix has in some cases been found to be disordered.[23,34] This helix, located after strand $\beta 6$ and on the C-terminal side of nucleophile Ser-105, was suggested as the most likely candidate for the lid. Although in the reported structure of the free enzyme the active site is accessible

[32] J. D. Schrag and M. Cygler, *J. Mol. Biol.* **230,** 575 (1993).
[33] M. L. Mannesse, R. C. Cox, B. C. Koops, H. M. Verheij, G. H. de Haas, M. R. Egmond, H. T. van der Hijden, and J. de Vlieg, *Biochemistry* **34,** 6400 (1995).
[34] J. Uppenberg, N. Ohrner, M. Norin, K. Hult, G. J. Kleywegt, S. Patkar, V. Waagen, T. Anthonsen, and T. A. Jones, *Biochemistry* **34,** 16838 (1995).

from the solvent (Fig. 3b), it was suggested on the basis of the observed limited size of the binding site that some conformational adjustment around the active site may have to occur on binding of a triglyceride substrate, which is significantly bulkier than the inhibitors investigated.[23] These authors concluded that this lipase may not exist in a fully closed conformation. Kinetic measurements of activity of *C. antarctica* lipase B showed a lack of interfacial activation,[35] which would agree with the observed lack of a major conformational rearrangement in the enzyme. The entrance to the active site has a significantly nonpolar character, with ~450 Å^2 of its surface being hydrophobic (Fig. 3b).

The *R. miehei, H. lanuginosa, R. delemar,* and *P. camembertii* lipases have highly homologous sequences and their structures are similar.[24] We have characterized another member of this structural class, a lipase from *Penicillium* sp. UZLM 4,[36] and determined its three-dimensional structure (J. D. Schrag and M. Cygler, unpublished data, 1997). These lipases have molecular masses close to 30 kDa. Their lids are formed by a single loop, which is approximately 15 residues long. In those structures in which the lid was well ordered in the crystal, its middle segment was folded into an α helix. Residues Ser-82 to Pro-96 constitute the lid in *R. miehei* lipase (homolog of *H. lanuginosa*), situated on the N-terminal side of the active site serine (Fig. 4). Interestingly, the location of the lid corresponds topologically to strand β4 (and the following helix), which is missing in these lipases as compared to other α/β-hydrolase fold enzymes. This coincidence adds some support to the hypothesis that in these lipases strand β4 was lost from the β sheet of the α/β-hydrolase fold during evolution (see [4] in this volume[13a]).

Two distinct positions of the lid, the closed and open conformations, have been observed for *R. miehei* lipase.[8,14] The closed form was found for the native enzyme, whereas an open form was observed for the complexes of this lipase with small inhibitors. During the reorientation, the α helix of the lid is displaced by ~10 Å and the movement can be considered approximately as a rigid body motion with two pivot points, Ser-83 and Val-95, near the ends of the loop. Accompanying this rearrangement is a small shift in the position of Ser-82, and in particular its NH moiety, to create the proper geometry for the oxyanion hole. At the same time, the side chain of this serine rotates to a different rotamer orientation and may also be involved in the hydrogen bonding to the oxyanion.[15] These rearrangements opened access to the active site serine and exposed a number of hydrophobic residues, creating a significantly larger hydrophobic

[35] M. Martinelle, M. Holmquist, and K. Hult, *Biochim. Biophys. Acta* **1258,** 272 (1995).
[36] K. Gulomova, E. Ziomek, J. D. Schrag, K. Davranov, and M. Cygler, *Lipids* **31,** 379 (1996).

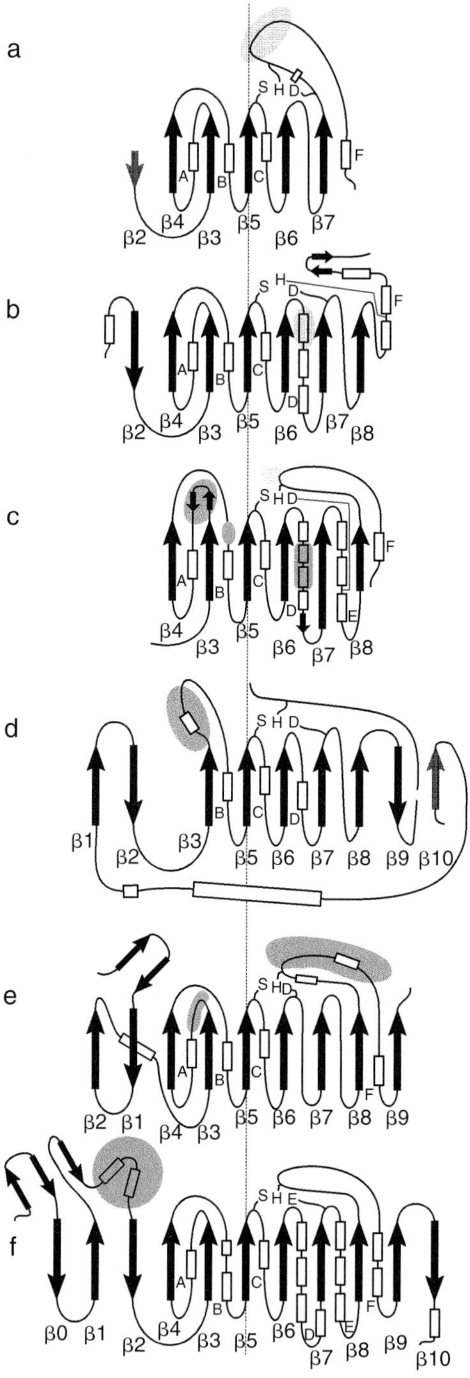
a
S H D
A B C F
β2 β4 β3 β5 β6 β7
b
H S D F
A B C D
β2 β4 β3 β5 β6 β7 β8
c
S H D
A B C F D E
β4 β3 β5 β6 β7 β8
d
S H D
B C D
β1 β2 β3 β5 β6 β7 β8 β9 β10
e
S H D
A B C F
β2 β1 β4 β3 β5 β6 β7 β8 β9
f
S H E
A B C D E F
β0 β1 β2 β4 β3 β5 β6 β7 β8 β9 β10

area for substrate binding and interaction with the interface. The previously exposed, hydrophilic side of the lid folds back onto the enzyme surface and forms many new favorable contacts.

The structures of other enzymes in this group have been reported only in the absence of inhibitor. Nevertheless, important information on the flexibility of the lid has been obtained. In the case of *H. lanuginosa* and *Penicillium* spp. lipases the lid was poorly ordered, without significant contacts in the central part with the rest of the protein (see Ref. 37; and J. D. Schrag and M. Cygler, unpublished data, 1997). *Rhizopus delemar* lipase provided even more interesting information. These crystals contained two independent molecules in the asymmetric unit, allowing a comparison of the conformations of their lids. Whereas the conformations of the two molecules were identical in most regions, those of the lids were significantly different. In one of the molecules the lid was in a closed conformation, similar to the corresponding conformation of *R. miehei* lipase, whereas the lid of the second molecule displayed a conformation intermediate between that of the open and closed conformations of *R. miehei* lipase.[37] Modification in the lid of *H. lanuginosa*[38,39] and *R. delemar* lipases[40] by site-directed mutagenesis provided direct evidence of the importance of a number of residues within the lid in terms of substrate binding and specificity.

Lipases from *Pseudomonas* species can be subdivided into at least two homologous families.[41] Structural data are presently available for only one of these families, that represented by *P. glumae.* These enzymes are likely to undergo a more complex rearrangement than those of the previously discussed lipases. This statement is based on a comparison between two

[37] U. Derewenda, L. Swenson, Y. Wei, R. Green, P. M. Kobos, R. Joerger, M. J. Haas, and Z. S. Derewenda, *J. Lipid Res.* **35,** 524 (1994).

[38] M. Holmquist, M. Martinelle, P. Berglund, I. G. Clausen, S. Patkar, A. Svendsen, and K. Hult, *J. Protein Chem.* **12,** 749 (1993).

[39] M. Holmquist, I. G. Clausen, S. Patkar, A. Svendsen, and K. Hult, *J. Protein Chem.* **14,** 217 (1995).

[40] R. D. Joerger and M. J. Haas, *Lipids* **29,** 377 (1994).

[41] J. Gilbert, *Enzyme Microb. Technol.* **15,** 634 (1993).

FIG. 4. Topological location of the lid in various lipases. The lid is marked by gray areas. The folding diagrams are aligned and the vertical dotted line marks the $\beta 5$ strand of the nucleophilic elbow. (a) Cutinase: There is no lid; shaded area shows the loop extending to one side of the active site. (b) *Candida antarctica* lipase B: Likely no lid; shaded area shows the location of a helix-containing loop that could play the role of a lid. (c) *Pseudomonas* lipase: The lid contains two helices adjacent to helix D; positions of two other loops participating in the rearrangement are also marked (the two short strands are formed only in the open form). (d) *Humicola lanuginosa* lipase: Helix-containing lid. (e) Pancreatic lipase. (f) *Candida rugosa* lipase.

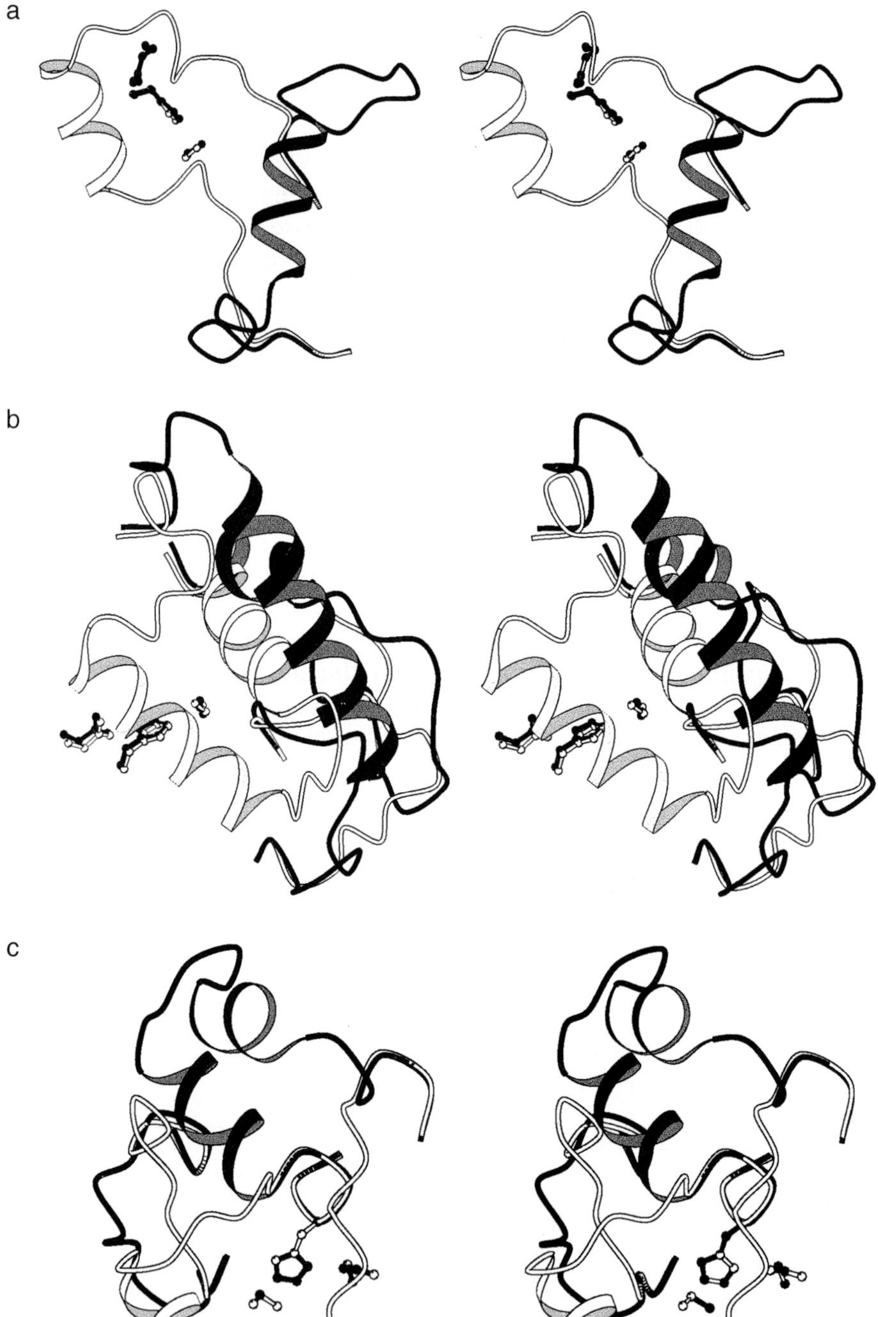
a
b
c

closely related lipases, *P. glumae*[21,22] and *P. cepacia*[20,42] The two lipases share 84% amino acid identity and consequently their three-dimensional structures are similar. If one excludes three surface loops from the comparison, the differences between these two lipases are of the same order as those between the four independent copies of *P. glumae* lipase in the crystal. Whereas all four molecules of *P. glumae* lipase (and *C. viscosum*) adopted closed conformations (inaccessible active site Ser-87), *P. cepacia* lipase crystallized in an open conformation. The three loops that differ in conformation encompass residues 17–27, 49–53, and 128–166 (*P. cepacia* numbering) (Fig. 4). The sequence differences between the two lipases within these loops are few, and most of them are conservative replacements. The largest loop, which forms the lid covering the active site in *P. glumae,* also displays the largest difference in position and in its secondary structure. Not only does its center move by more than 15 Å, but the helix in the middle of this loop also becomes elongated in the open state. The two shorter loops differ in position by more than 5 Å at their extremities in the two lipases. The topological location of the lid is after strand $\beta 6$ (C terminal to nucleophile), whereas the other two loops are between $\beta 3$ and $\beta 4$ (N terminal to nucleophile).

Do these observed conformational differences in two related *Pseudomonas* lipases represent changes that occur on activation? A high level of sequence identity and a limited number of (mostly) conservative mutations in the lid loop strongly suggest that under similar conditions the two lipases should have similar structures. The level of identity noted here is, for example, higher than in the homologs of *R. miehei* lipase, which show structural similarity that includes the lid. The changes that seem to be required to open access to the active site in *Pseudomonas* lipases involve a concerted movement of three loops. The movement of only the lid in *P. glumae* lipase to the conformation observed in the open form of *P. cepacia* lipase is prohibited owing to steric collisions with the other two loops in their original positions. Importantly, during this concerted movement Leu-17 becomes reoriented to form a proper geometry for the oxyanion hole. The positions of the rearranged loops are shown in Fig. 5a.

[42] M. Cygler, EC BRIDGE Lipase T-Project, International Workshop, Bendor Island, France, September 14–17, 1994. [Abstracts]

FIG. 5. Superposition of the lid and accompanying movable loops in the closed and open forms. The catalytic triad residues are shown in full to help in orientation. The open conformation is shown in black and the closed conformation is shown in white. (a) *Candida rugosa* lipase. (b) In *P. glumae* (closed) and *P. cepacia* (open) lipases. (c) Pancreatic lipase–colipase system: The colipase is located at the top and contacts the loop in an open conformation.

Pancreatic lipase has been the subject of extensive protein crystallography studies. Its three-dimensional structure has been determined in the absence[9] and presence of a colipase[43] and as ternary complexes with colipase and noncovalently bound phospholipids[16] or phosphonyl inhibitors.[44] The presence of colipase on its own does not affect in any significant way the lipase structure.[43] However, when the inhibitor is added to this binary complex the lipase undergoes a conformational rearrangement that exposes the active site, allows the inhibitor to bind, and leads to the formation of a covalent bond with the active site serine. Here, the rearrangement involves movement of two loops (Fig. 4). The larger one, encompassing residues 237 to 260, forms the lid covering the active site in the closed form. It is located after strand $\beta 8$, on the C-terminal side of the nucleophile and shortly before the active site histidine. Its center changes position by more than 30 Å. The second loop, composed of residues 75–84, occurs after strand $\beta 3$ on the N-terminal side of the nucleophile. The movement of this loop is smaller than that of the lid; its center travels approximately 13 Å. The lid in the open conformation makes numerous contacts with the colipase (Fig. 5b) and uncovers a large hydrophobic area. The proper geometry of the oxyanion hole is formed during this reorganization by the movement of Phe-77 near the edge of the shorter of the moving loops. In contrast to *R. miehei,* but similar to the *Pseudomonas* and *C. rugosa* lipases, the movement of the lid involves a change of its secondary structure and cannot be described as a rigid body motion, although two hinge points are clearly identifiable. Significant sequence homology between pancreatic, hepatic, and lipoprotein lipases[45] suggests that similar rearrangements will also occur in these other enzymes. The involvement of the lid in catalytic events and in binding to the interface of the lipases has been shown by site-directed mutagenesis.[46–48] The properties of a lipase isolated from the pancreas of guinea pig differed somewhat from those of other pancreatic lipases and showed no interfacial activation.[7] Comparison of its sequence with those of other pancreatic lipases indicated clearly that this enzyme lacks the lid and, consequently, no substantial rearrangement at the interface is likely to take place.

[43] H. van Tilbeurgh, L. Sarda, R. Verger, and C. Cambillau, *Nature (London)* **359,** 159 (1992).

[44] M. P. Egloff, F. Marguet, G. Buono, R. Verger, C. Cambillau, and H. van Tilbeurgh, *Biochemistry* **34,** 2751 (1995).

[45] M. C. Komaromy and M. C. Schotz, *Proc. Natl. Acad. Sci. U.S.A.* **84,** 1526 (1987).

[46] K. A. Dugi, H. L. Dichek, G. D. Talley, H. B. Brewer, Jr., and S. Santamarina-Fojo, *J. Biol. Chem.* **267,** 25086 (1992).

[47] M. L. Jennens and M. E. Lowe, *J. Biol. Chem.* **269,** 25470 (1994).

[48] K. A. Dugi, H. L. Dichek, and S. Santamarina-Fojo, *J. Biol. Chem.* **270,** 25396 (1995).

The last group of lipases with three-dimensional structures known at present includes large enzymes (approximately 60–65 kDa), e.g., lipases from the yeast *C. rugosa* and the fungus *G. candidum*. These lipases belong to a large family of lipases–esterases also encompassing cholinesterases, carboxylesterases, cholesterol-esterases, and other esterases.[13] Owing to their large size, these lipases contain more strands in the central β sheet and more α-helices than in the canonical α/β-hydrolase fold. Although *G. candidum* lipase has been crystallized in three different crystal forms, albeit under similar crystallization conditions, in all cases only the closed conformation was observed. *Candida rugosa* lipase, however, produced crystals under varying conditions in the presence and absence of the inhibitors and two crystal forms were obtained. The conformation of the lipase differed between the crystal forms even when no inhibitors were present.[17,18] The closed conformation has features similar to those of the closed conformations of other lipases, namely, the active site is inaccessible from the solvent. In the open conformation, the active site is accessible and lies at the bottom of a large depression created by the lid movement (Fig. 5c). The rearrangement involves only one loop encompassing residues 65 to 93,[18] the center of which moves by more than 25 Å. This loop contains two α helices and its topological location is between strands $\beta 1$ and $\beta 2$, N terminal to the nucleophile. The opening of the lid involves a complex movement during which the helices partially unwind on one side and extend on the other, and a proline (Pro-92) at the C-terminal end of this loop undergoes a *cis*-to-*trans* isomerization. The lid in the open conformation forms one of the walls of the substrate-binding site. A new set of interactions is formed between the formerly exposed surface of the lid and the adjacent lipase surface. The same open conformation has been found in the presence of covalently bound inhibitors, strongly suggesting that this structure represents the active, substrate-binding conformation. The existence of this conformation even in the absence of inhibitors indicates that it corresponds to a low-energy state. The multitude of contacts between the lid and the rest of the structure in the closed and open conformations and the additional involvement of a carbohydrate in the stabilization of the open conformation suggest that these two states are the low-energy end points of the activation pathway, which proceeds through transition states of higher energy.[18,25] The *G. candidum* lipase is likely to undergo a similar rearrangement of the loop equivalent to the lid in *C. rugosa* lipase. However, a detailed comparison of the two lipases led us to suggest that, in this case, there is a concerted movement of two additional loops, in particular the loop containing residues that provide the main-chain NH groups for the oxyanion hole.[17] Some evidence in support of this model comes from a comparison of *G. candidum* lipase in the three crystal forms. The largest differences between the structures are in the loop residing under the lid and contributing to the oxyanion hole formation (J. D. Schrag and M. Cygler, unpublished data, 1997). The *cis*-to-*trans* isomerization of proline is a feature likely to be unique to *C. rugosa* lipase, as no corresponding proline exists in *G. candidum* lipase.

Lid Mobility

In a number of lipase structure determinations a portion of the lid was found to be disordered and was not located in the three-dimensional structure. To date this was reported exclusively for the lipases with structures similar to that of *R. miehei,* in which the lid consists of one medium-size loop with a helix in the middle (the case of *C. antarctica* lipase B is unclear because the presence of a lid is in question). The lid in *H. lanuginosa* was partially disordered; only the position of the main chain, not those of the side chains, was visible.[37] Similarly, in a *Penicillium* species lipase the lid is disordered (J. D. Schrag and M. Cygler, unpublished data, 1997). Even more interestingly, *R. delemar* lipase displayed two different conformations of the lid in the same crystal,[24] further indicating the conformational flexibility of the lid. Had the high mobility of the lid been observed in only one of these lipases one could try to explain this by the specifics of that system, e.g., the lack of crystal contacts, specific amino acid sequence, and crystallization conditions. However, the observation of disorder or multiple conformations of the lid in a number of these lipases crystallized under different conditions and in different crystal lattices indicates that the observed flexibility of the lid is likely inherent to this family of homologous lipases. Furthermore, the lids of these lipases likely also undergo a dynamic conformational rearrangement in solution with more conformational states than just one open and one closed.[37]

In other lipases only two well-defined states have been observed that most likely represent the end points on the transition pathway between the active and inactive states. A high mobility of the lid in lipases where a concerted movement of a few loops is required for opening seems rather unlikely. An interesting case is pancreatic lipase, in which the open conformation has been observed only in the presence of the colipase. In the absence of bile salts this lipase does not require the colipase for activity and a similar lid opening must also occur in the absence of the colipase. Because many of the observed interactions of the open conformation of the lid occur with the colipase, it is possible that without this cofactor the lid might display higher mobility. A disorder of the helix ($\alpha 5$) at the entrance to the active site was also observed in *C. antarctica* lipase B under some crystallization conditions.[23] This mobility was observed not only when the lipase is free but also when an inhibitor molecule is present in the active site and reflects the high mobility of this segment in solution. The presently known structures cover, then, a wide spectrum from highly mobile lids with many intermediate states of comparable energies to lids for which there seem to be well-defined end-point conformations of significantly lower energy than in the transition states.

Substrate-Binding Sites

Comparison of the structures of various lipase–inhibitor complexes permits delineation of some of the common features of their substrate-binding sites. Figure 6 shows the superposition of the catalytic triad and inhibitors from three different complexes. The orientations of the carbonyl oxygen and of the fatty acyl chain corresponding to the acid product of the triglyceride hydrolysis are similar in all lipases. More difficult is a comparison of the alcohol-binding sites, as most of the inhibitors have either a small substituent there or are completely devoid of this portion. Nevertheless, comparison of various lipase–inhibitor complexes indicates that the scissile fatty acyl chain maintains a similar orientation and position relative to the nucleophile elbow supersecondary structural element and serine–histidine diad in all lipases. This chain is held firmly near the hydrolyzed bond.[16,19,44] Although there are no direct experimental data regarding the position of the intact triglyceride molecule in the binding site of any of the lipases (see recently published data on cutinase-triglyceride analogue complex)[48a], there is some information suggesting that the second fatty acyl chain emerges from the structure of the complexes where either a phospholipid[16] or a second molecule of the inhibitor[19,23,44] was bound nearby. On the basis of these data it was suggested that, at least in some cases, the triglyceride binds to a lipase in a tuning fork conformation.[19,44] This predicted positioning of a triglyceride molecule in the lipase-binding site poses questions as to the orientation of the lipase at the interface, its penetration into the lipid layer, and the sequestration of the lipid molecule to the active site.

Candida rugosa and, most likely, *G. candidum* lipases display a more complex substrate-binding mode than other lipases. This is due to the rather unusual shape of their binding sites. Whereas in other lipases the scissile acyl chain lies on the surface of the protein, in *C. rugosa* lipase the polypeptide chain folds over this site, which now forms a deep tunnel penetrating toward the center of the molecule.[19,25] The mouth of this tunnel is near the catalytic Ser-209. Its walls are lined with hydrophobic residues and no solvent molecules were found in the native structure. It seems certain that the triglyceride must adopt the tuning fork conformation to be accommodated by this lipase. How the triglyceride acyl chain enters this tunnel is not clear. Even less clear is how the acid product leaves the tunnel, and

[48a] S. Loughi, M. Mannesse, H. M. Verheij, G. H. de Haas, M. Egmond, E. Knoops-Mouthuy, and C. Cambillan, *Prot. Sci.* **6,** 275 (1997).

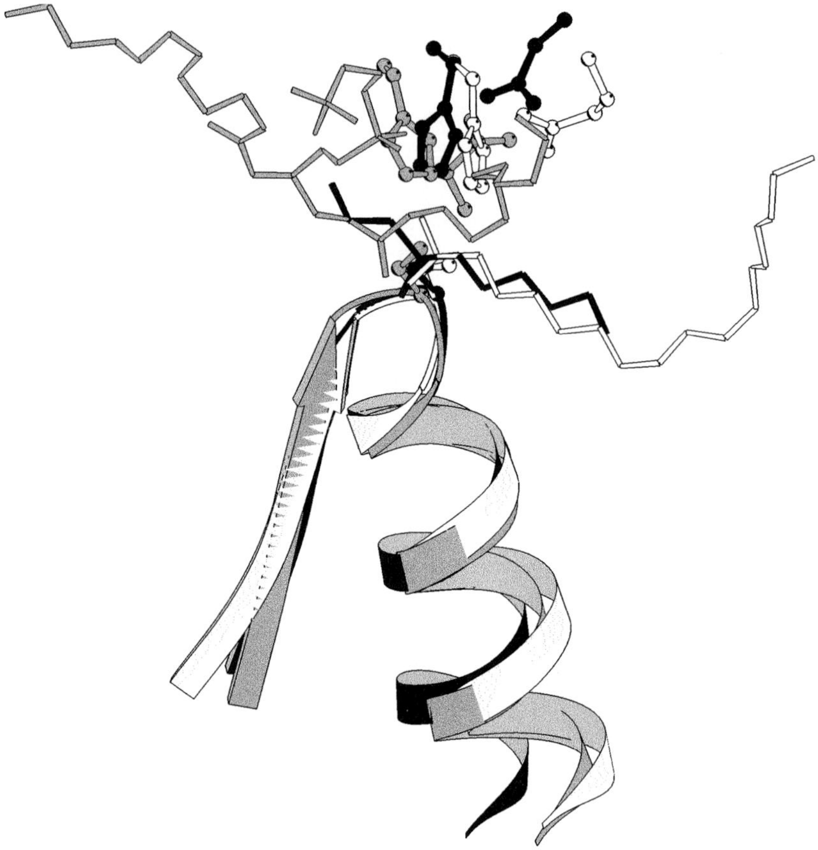

FIG. 6. Superposition of the nucleophilic elbow, catalytic triad, and a bound inhibitor of three lipases: *C. rugosa* (white), pancreatic (gray), and *R. miehei* (black), showing a similar disposition of the scissile fatty acyl chain (extending to the right) relative to the catalytic machinery in these lipases.

what constitutes the driving force for this step. The size of the tunnel indicates that it can accommodate only one acyl chain and that the positions of the other two fatty acyl chains of the substrate are on the surface of this lipase. A possible location of the second chain was revealed by crystallographic studies performed in the presence of a 10-fold excess of the inhibitor hexadecylsulfonyl chloride. In this case two molecules of the inhibitor were bound to the lipase, the first in the tunnel and the second following the groove between two aromatic side chains, Phe-344 and Phe-448. The possible position and orientation of a triglyceride substrate in the binding site of *C. rugosa* lipase was investigated by a Monte Carlo search for low-energy conformations combined with energy minimization (H. Zuegg and M. Cygler, unpublished data, 1997). These calculations showed that apart from the scissile acyl chain in the tunnel there is sufficient space for the remaining two fatty acyl chains in the binding site in the open form. These two fatty acyl chains are bound on the lipase surface. They follow, in general, the hydrophobic patches but show significant flexibility beyond the first few atoms with no specific contacts to the protein. One of these areas corresponds to the site identified by crystallography.

Orientation of Lipase on Membrane

The surfaces of the catalytically competent, open conformations of the various lipases show a distinctive characteristic—the area surrounding the catalytic site has a hydrophobic character. This is the side of the lipase surface that is directed toward the lipid layer. No direct structural data exist to show the interactions of the enzyme with the lipids at the interface. Biophysical data, e.g., from experiments on monolayers, indicated that a lipase adsorbs to the interface and likely penetrates partially into the lipid layer (for a review, see Ref. 4). An extensive, flat hydrophobic surface of the human pancreatic lipase–colipase complex seems well suited to be a surface interacting with lipids.[43] Similarly, a side view of the open form of *C. rugosa* lipase is suggestive of a partial penetration of the lid into the lipid layer (Fig. 7). Crystallization experiments in the presence of detergents showed that in *C. antarctica* lipase B, pancreatic, and *C. rugosa* lipases, the detergent molecules are sequestered from the crystallization solution and

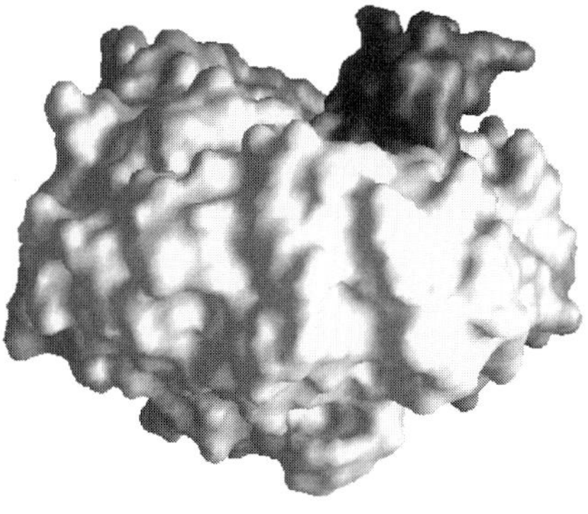

FIG. 7. Side view of the open form of *C. rugosa* lipase, suggesting a possible penetration of the lid into the lipid layer. The surface of the lid is presented in gray, whereas the rest of the lipase surface is shown in lighter tones. The entrance to the active site is in the center of the top surface, to the left of the lid.

occupy the hydrophobic substrate-binding cleft. Similar behavior is likely common to all lipases and is most probably representative of the sequestration of the lipid substrate molecules from the interface once the lipase attains its proper position and orientation at the lipid–water interface. The increased activity in the presence of a lipid–water interface is not only the result of shifting the equilibrium from a closed to an open state, but also of the adsorption of a lipase molecule on the micelle surface. The lower activities of lipases in organic solvents are explained both by poor adsorption and equilibrium favoring the closed state.[49]

Stereospecificity

Lipases display selectivity for their substrates at three different levels: fatty acyl chain type, position of the scissile chain on the glycerol backbone, and stereoisomer of the ester substrate. All three types of selectivities have been widely studied for various lipases, with the general conclusion being that lipases do not display (strictly speaking) well-defined specificities but rather show selectivity for certain substrates. Extensive characterization by Rogalska and co-workers of the stereopreference of many lipases, using the monolayer method, showed that the enantiomeric excess varies between lipases and that it depends on the surface tension (see Refs. 50 and 51; reviewed in Ref. 4). The enantiopreferences of many lipases for a wide variety of esters have also been widely studied, resulting in a number of semiempirical rules for prediction of the outcome of hydrolytic reactions (e.g., see Refs. 52 and 53). These rules, especially for the alcohol leaving groups, could be applied to all lipases irrespective of their origin. The molecular basis for the latter rule became better understood when the structures of two enantiomers of menthyl hexylphosphonate covalently complexed with *C. rugosa* lipase were determined.[54] The structural results suggested that the discrimination between the enantiomers occurs owing to the shape of the binding site near the nucleophile, which is to a large extent formed from the loops that assemble the catalytic triad and the oxyanion hole. The binding of the less well-suited enantiomer in this site disturbed the position of the catalytic histidine, thus making the catalysis

[49] A. Louwrier, G. J. Drtina, and A. M. Klibanov, *Biotech. Bioeng.* **50,** 1 (1996).
[50] E. Rogalska, C. Cudrey, F. Ferrato, and R. Verger, *Chirality* **5,** 24 (1993).
[51] E. Rogalska, S. Nury, I. Douchet, and R. Verger, *Chirality* **7,** 505 (1995).
[52] D. H. G. Crout and M. Christen, *in* "Modern Synthetic Methods" (R. Scheffold, ed.), Vol. 5, pp. 1–114. Springer-Verlag, Berlin, 1989.
[53] R. J. Kazlauskas, *Trends Biotechnol.* **12,** 464 (1994).
[54] M. Cygler, P. Grochulski, R. J. Kazlauskas, J. D. Schrag, F. Bouthillier, B. Rubin, A. N. Serreqi, and A. K. Gupta, *J. Am. Chem. Soc.* **116,** 3180 (1994).

less efficient. The generalization to other lipases was based on the similarity of their fold and on the assembly and position of the catalytic residues.

A number of other attempts to explain the observed enantioselectivities of various lipases have been published, in which the authors used the known three-dimensional structures of the enzymes and modeled the substrates into their binding sites. Most of these calculations agreed with the observed enantiomeric excess (e.g., see Refs. 34 and 55–57). However, many more structural data are needed to gain an understanding of the specificity of lipases. This is especially important for reactions catalyzed in organic solvents, in which the stereoselectivity may be different than in water.

Conclusion

The determination of the three-dimensional structures of many lipases provided a view of these enzymes at a molecular level. The large number of lipase structures available for analysis furnished the means to search for common patterns that would shed light on the mode of action of these enzymes. The observation of a common fold provided evidence of an evolutionary relationship among many lipases, which diverged in their sequences beyond a level recognizable by current sequence comparison algorithms. The observation of multiple conformers for most of the structurally determined lipases, which could be divided into two groups (with an accessible active site or with the active site covered by a lid), gave a strong indication that these two states are essential for the interfacial activation mechanism of lipases, and that this mechanism is likely to be common to all lipases that undergo such activation. The common fold, nearly identical disposition of the catalytic triad, and similarity in the binding mode of transition state analogs of various lipases also result in the similar pattern of enantiopreference for many ester substrates.

Although much has been learned about the molecular details of lipase action, much is yet to be determined. Especially important is the characterization of the interaction of the lipases at the lipid–water interface, mapping of the binding site of triglycerides, gaining an understanding of the determinants of substrate selectivity, and in particular engineering of lipases for specific enantioselective syntheses.

Acknowledgments

The authors thank Mr. René Coulombe for help in preparation of the figures. This chapter adapted by permission of the National Research Council of Canada.

[55] M. Norin, F. Haeffner, A. Achour, T. Norin, and K. Hult, *Protein Sci.* **3,** 1493 (1994).
[56] M. Holmquist, F. Haeffner, T. Norin, and K. Hult, *Protein Sci.* **5,** 83 (1996).
[57] J. Zuegg, H. Hönig, J. D. Schrag, and M. Cygler, *J. Mol. Catalysis,* **3,** (Section B), 83 (1997).

[2] Identification of Conserved Residues in Family of Esterase and Lipase Sequences

By FINN DRABLØS and STEFFEN B. PETERSEN

Introduction

The core of our knowledge about enzymes is based on experiments in which data pertaining to specificity, activity, or stability have been collected. With the introduction of methods for structure determination we now have access to a powerful tool for explaining some of these experimental data in terms of structural properties. Genetic engineering and site-directed mutagenesis have added further to the versatility of the tool kit. It is now possible to probe the importance of single residues in and around the active site through protein engineering, and thus test functional or structural hypotheses that stem from careful studies of the three-dimensional (3D) structural information obtained from X-ray or nuclear magnetic resonance (NMR) data.

However, the most exciting laboratory for studying the effect of mutations is found in nature itself. Since the first system for transfer of information from one generation of an organism to the next came into existence, some 3.7 billion years ago, the information has constantly been subject to modification, and depending on the success of the modification, the organism may have been fit to pass the modified information on to new generations. By this process all organisms constantly adapt by natural selection to an often changing environment, leading to retention of beneficial changes. Thus, by analyzing this information we may learn which structural positions are critical for function or folding, how proteins can be adapted to a new environment, or how the specificity of a protein can be changed.

This chapter shows how we can do a general sequence analysis, using lipase and esterase data as an example. A flow chart showing the basic steps of the analysis is shown in Fig. 1. This analysis does not demonstrate all of the techniques that are available; in particular, we do not discuss phylogenetic methods. However, we do show some general techniques that we have found useful for generating and analyzing large alignments. We show how to apply these techniques, and what type of information they may give us. However, which complementary techniques to use, which data to include in the analysis, and how to actually do it depend very much on what type of information one is looking for. Therefore, this chapter should be read mainly as a description of an interesting approach to protein se-

0076-6879/97 $25.00

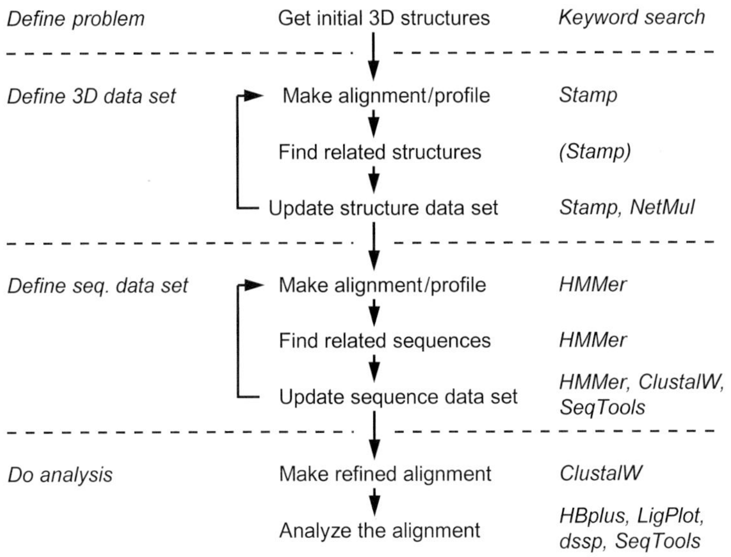

FIG. 1. A flow chart showing the general steps of sequence analysis used in this chapter. The general purpose of each section is shown to the left, the flow chart itself is shown in the middle, and the tools used for this specific analysis are listed to the right. These tools represent only a selection of several possible alternatives; some of these alternatives are described in the text. SeqTools is a collection of tools for sequence analysis developed in our laboratory.

quence analysis. We believe that some of the properties we identify are of general interest, but they certainly do not answer all relevant questions about conserved properties in lipases and esterases.

Defining Three-Dimensional Data Set

The current analysis is based on the fact that 3D structures of relevant proteins are available. This is not a necessary condition for doing sequence analysis, but it is definitely a useful one. In particular, if we want to align sequences with a low degree of sequence similarity, it may be difficult to obtain a good alignment on the basis of sequence data alone. If we know the 3D structure of a representative subset of the sequences, then an alignment of the 3D structures will provide us with an excellent starting point for setting up a good sequence alignment.

In addition, 3D structures are of course necessary if we want to correlate the conservation patterns that we see in the alignment with structural features.

Identification of Relevant Three-Dimensional Structures

The first step is to identify relevant 3D protein structures that are available. The initial approach is to browse the Brookhaven database[1,2] [for example, at the PDB (Protein Data Base) web site] for structures, using key words that represent the structure of interest. This will normally give a representative list. However, in some cases important entries may be missed by this approach, because the annotation does not include any of the key words selected. One alternative (or supplementary) approach may be to use a representative protein sequence and search a sequence library generated for PBD protein structures (e.g., the NRL_3D library[3]), using a standard search program like Blast[4] or Fasta.[5,6] Even this approach may miss some distant homologs if the sequence similarity is low. Therefore, it may be useful also to look for similarity at the structural level, independent of sequence similarity. This can be found, e.g., in fssp[7] files or in other classifications based on structural similarities.[8] However, for this type of sequence analysis, it is normally best to be careful and include only homologous proteins. There are several cases in which proteins without any apparent sequence similarity have been shown to share similar folds. In these cases it may be difficult to differentiate between truly homologous proteins (i.e., proteins that have evolved by divergence from a common ancestral protein) and generally similar but unrelated proteins, whose similarity is caused by convergent evolution. Inclusion of nonhomologous sequences generally makes it difficult to obtain a good alignment, and we have tried to avoid inclusion of such sequences in this study.

For this analysis we did a keyword search of PDB entries at the Brookhaven web server with "lipase" and "esterase" as keywords. This returned a total of 51 entries. Nine entries were then removed because they represented nonrelevant proteins (Fab fragments, alkaline phosphatase, procolipase, phospholipase). The remaining 42 entries were retrieved and their

[1] F. C. Bernstein, T. F. Koetzle, G. J. B. Williams, E. F. Meyer, M. D. Brice, J. R. Rodgers, O. Kennard, T. Shimanouchi, and M. Tasumi, *J. Mol. Biol.* **112,** 535 (1977).

[2] E. Abola, F. C. Bernstein, S. H. Bryant, T. F. Koetzle, and J. Weng, *in* "Crystallographic Databases—Information Content, Software Systems, Scientific Applications" (F. H. Allen, G. Bergerhoff, and R. Sievers, eds.), p. 107. Data Commission of the International Union of Crystallography, Bonn/Cambridge/Chester, 1987.

[3] N. Pattabiraman, K. Namboodiri, A. Lowrey, and B. P. Gaber, *Protein Seq. Data Anal.* **3,** 387 (1990).

[4] S. F. Altschul, W. Gish, W. Miller, E. W. Myers, and D. J. Lipman, *J. Mol. Biol.* **215,** 403 (1990).

[5] W. R. Pearson and D. J. Lipman, *Proc. Natl. Acad. Sci. U.S.A.* **85,** 2444 (1988).

[6] W. R. Pearson, *Methods Enzymol.* **183,** 63 (1990).

[7] L. Holm, C. Ouzounis, C. Sander, G. Tuparev, and G. Vriend, *Protein Sci.* **1,** 1691 (1992).

[8] S. J. Wodak, *Nat. Struct. Biol.* **3,** 575 (1996).

corresponding SwissProt[9] entries were identified by extracting the protein sequences from the PDB entries and searching the SwissProt database with Blast, keeping the highest scoring match for each sequence. (The 42 entries are listed in Table I.[10–34]) A subset of 13 PDB entries was then selected,

[9] A. Bairoch and B. Boeckmann, *Nucleic Acids Res.* **20,** 2019 (1992).

[10] J. L. Sussman, M. Harel, F. Frolow, C. Oefner, A. Goldman, L. Toker, and I. Silman, *Science* **253,** 872 (1991).

[11] M. Harel, I. Schalk, L. Ehret Sabatier, F. Bouet, M. Goeldner, C. Hirth, P. H. Axelsen, I. Silman, and J. L. Sussman, *Proc. Natl. Acad. Sci. U.S.A.* **90,** 9031 (1993).

[12] D. Ghosh, Z. Wawrzak, V. Z. Pletnev, N. Li, R. Kaiser, W. Pangborn, H. Jornvall, M. Erman, and W. L. Duax, *Structure* **3,** 279 (1995).

[13] P. Grochulski, Y. Li, J. D. Schrag, F. Bouthillier, P. Smith, D. Harrison, B. Rubin, and M. Cygler, *J. Biol. Chem.* **268,** 12843 (1993).

[14] C. Martinez, P. De Geus, M. Lauwereys, G. Matthyssens, and C. Cambillau, *Nature (London)* **356,** 615 (1992).

[15] A. Nicolas, M. Egmond, C. T. Verrips, J. de Vlieg, S. Longhi, C. Cambillau, and C. Martinez, *Biochemistry* **35,** 398 (1996).

[16] M. Harel, G. J. Kleywegt, R. B. Ravelli, I. Silman, and J. L. Sussman, *Structure* **3,** 1355 (1995).

[17] Y. Bourne, C. Martinez, B. Kerfelec, D. Lombardo, C. Chapus, and C. Cambillau, *J. Mol. Biol.* **238,** 709 (1994).

[18] J. Uppenberg, N. Ohrner, M. Norin, K. Hult, G. J. Kleywegt, S. Patkar, V. Waagen, T. Anthonsen, and T. A. Jones, *Biochemistry* **34,** 16838 (1995).

[19] H. van Tilbeurgh, M. P. Egloff, C. Martinez, N. Rugani, R. Verger, and C. Cambillau, *Nature (London)* **362,** 814 (1993).

[20] M. P. Egloff, F. Marguet, G. Buono, R. Verger, C. Cambillau, and H. van Tilbeurgh, *Biochemistry* **34,** 2751 (1995).

[21] M. Cygler, P. Grochulski, R. Kazlauskas, J. Schrag, F. Bouthillier, B. Rubin, A. Serreqi, and A. Gupta, *J. Am. Chem. Soc.* **116,** 3180 (1994).

[22] P. Grochulski, F. Bouthillier, R. J. Kazlauskas, A. N. Serreqi, J. D. Schrag, E. Ziomek, and M. Cygler, *Biochemistry* **33,** 3494 (1994).

[23] Y. Bourne, P. Taylor, and P. Marchot, *Cell* **83,** 503 (1995).

[24] M. E. Noble, A. Cleasby, L. N. Johnson, M. R. Egmond, and L. G. Frenken, *FEBS Lett.* **331,** 123 (1993).

[25] J. Uppenberg, M. T. Hansen, S. Patkar, and T. A. Jones, *Structure* **2,** 293 (1994).

[26] L. Brady, A. M. Brzozowski, Z. S. Derewenda, E. Dodson, G. Dodson, S. Tolley, J. P. Turkenburg, L. Christiansen, B. Huge-Jensen, L. Norskov, L. Thim, and V. Menge, *Nature (London)* **343,** 767 (1990).

[27] J. D. Schrag and M. Cygler, *J. Mol. Biol.* **230,** 575 (1993).

[28] U. Derewenda, L. Swenson, R. Green, Y. Wei, G. G. Dodson, S. Yamaguchi, M. J. Haas, and Z. S. Derewenda, *Nat. Struct. Biol.* **1,** 36 (1994).

[29] U. Derewenda, L. Swenson, Y. Wei, R. Green, P. M. Kobos, R. Joerger, M. J. Haas, and Z. S. Derewenda, *J. Lipid Res.* **35,** 524 (1994).

[30] P. Grochulski, Y. Li, J. D. Schrag, and M. Cygler, *Protein Sci.* **3,** 82 (1994).

[31] C. Martinez, A. Nicolas, H. van Tilbeurgh, M. P. Egloff, C. Cudrey, R. Verger, and C. Cambillau, *Biochemistry* **33,** 83 (1994).

[32] Z. S. Derewenda, U. Derewenda, and G. G. Dodson, *J. Mol. Biol.* **227,** 818 (1992).

[33] U. Derewenda, A. M. Brzozowski, D. M. Lawson, and Z. S. Derewenda, *Biochemistry* **31,** 1532 (1992).

TABLE I

LIPASE AND ESTERASE STRUCTURES RETRIEVED FROM PROTEIN DATA BASE AND SWISSPROT

Compound	Source	Het groups	Code	Resolution (Å)	Ref.	SwissProt code
Acetylcholinesterase	*Torpedo californica*	Acetylcholine	1ACE	2.8	Sussman *et al.*[10]	ACES_TORCA
Acetylcholinesterase	*Torpedo californica*	Tacrine	1ACJ	2.8	Harel *et al.*[11]	ACES_TORCA
Acetylcholinesterase	*Torpedo californica*	Edrophonium	1ACK	2.8	Harel *et al.*[11]	ACES_TORCA
Acetylcholinesterase	*Torpedo californica*	Decamethonium	1ACL	2.8	Harel *et al.*[11]	ACES_TORCA
Acetylcholinesterase	*Torpedo californica*	*m*-(*N*,*N*,*N*-Trimethylammonio)-trifluoroacetophenone; SO_4^{2-}	1AMN	2.8	Harel *et al.*	ACES_TORCA
Cholesterol esterase	*Candida cylindracea*	*N*-Acetyl-D-glucosamine; cholesteryl linoleate; PO_4^{3-}	1CLE	2.00	Ghosh *et al.*[12]	LIP3_CANRU
Lipase	*Candida rugosa*	*N*-Acetyl-D-glucosamine	1CRL	2.06	Grochulski *et al.*[13]	LIP1_CANRU
Cutinase	*Fusarium solani pisi*[a]		1CUS	1.25	Martinez *et al.*[14]	CUTI_FUSSO
Esterase	*Streptomyces scabies*		1ESC	2.1	Derewenda and Wei	ESTA_STRSC
Esterase	*Streptomyces scabies*	Methyl phosphinic acid	1ESD	2.3	Derewenda and Wei	ESTA_STRSC
Esterase	*Streptomyces scabies*	Diethyl phosphonate	1ESE	2.4	Derewenda and Wei	ESTA_STRSC
Cutinase N84A	*Fusarium solani pisi*[a]		1FFA	1.69	Nicolas *et al.*[15]	CUTI_FUSSO
Cutinase N84D	*Fusarium solani pisi*[a]		1FFB	1.75	Nicolas *et al.*[15]	CUTI_FUSSO
Cutinase N84L	*Fusarium solani pisi*[a]		1FFC	1.75	Nicolas *et al.*[15]	CUTI_FUSSO
Cutinase N84W	*Fusarium solani pisi*[a]		1FFD	1.69	Nicolas *et al.*[15]	CUTI_FUSSO
Cutinase S42A	*Fusarium solani pisi*[a]		1FFE	1.69	Nicolas *et al.*[15]	CUTI_FUSSO
Acetylcholinesterase with fasciculin-II	*Electrophorus electricus; Dendroaspis angusticeps*	Zn^{2+}; *N*-acetyl-D-glucosamine	1FSS	3.0	Harel *et al.*[16]	ACES_TORCA; TXF7_DENAN

Lipase	*Equus caballus* pancreas	Ca^{2+}	1HPL	2.3	Bourne *et al.*[17]	LIPP_HORSE
Lipase	*Candida antarctica*	*N*-Acetyl-D-glucosamine; *N*-hexylphosphonate ethyl ester	1LBS	2.6	Uppenberg *et al.*[18]	LIPB_CANAR
Lipase	*Candida antarctica*	*N*-Acetyl-D-glucosamine; methylpenta(oxyethyl) heptadecanate	1LBT	2.5	Uppenberg *et al.*[18]	LIPB_CANAR
Lipase with colipase	*Homo sapiens* pancreas; *Sus scrofa* pancreas	β-Nonylglucoside; Ca^{2+}; diundecyl phosphatidylcholine	1LPA	3.04	van Tilbeurgh *et al.*[19]	COL2_PIG; LIPP_HUMAN
Lipase with colipase	*Homo sapiens* pancreas; *Sus scrofa* pancreas	Methoxyundecylphosphinic acid; Ca^{2+}; β-octylglucoside	1LPB	2.46	Egloff *et al.*[20]	COL2_PIG; LIPP_HUMAN
Lipase	*Candida rugosa*	(1*R*)-Menthyl hexyl phosphonate; Ca^{2+}; *N*-acetyl-D-glucosamine	1LPM	2.2	Cygler *et al.*[21]	LIP1_CANRU
Lipase	*Candida rugosa*	Ca^{2+}; dodecane sulfonate; *N*-acetyl-D-glucosamine	1LPN	2.2	Grochulski *et al.*[22]	LIP1_CANRU
Lipase	*Candida rugosa*	1-Hexadecanosulfonic acid; *N*-acetyl-D-glucosamine; Ca^{2+}	1LPO	2.2	Grochulski *et al.*[22]	LIP1_CANRU
Lipase	*Candida rugosa*	1-Hexadecanosulfonic acid; *N*-acetyl-D-glucosamine; Ca^{2+}	1LPP	2.05	Grochulski *et al.*[22]	LIP1_CANRU
Lipase	*Candida rugosa*	(1*S*)-Menthyl hexyl phosphonate; *N*-acetyl-D-glucosamine; Ca^{2+}	1LPS	2.2	Cygler *et al.*[21]	LIP1_CANRU

(continued)

TABLE I (*continued*)

Compound	Source	Het groups	Code	Resolution (Å)	Ref.	SwissProt code
Acetylcholinesterase with fasciculin-II	*Mus musculus; Dendroaspis angusticeps*	*N*-Acetyl-D-glucosamine	1MAH	3.2	Bourne *et al.*[23]	ACES_MOUSE; TXF7_DENAN
Lipase	*Pseudomonas glumae*	Ca^{2+}	1TAH	3.0	Noble *et al.*[24]	LIP_PSEGL
Lipase	*Candida antarctica* b	*N*-Acetyl-D-glucosamine	1TCA	1.55	Uppenberg *et al.*[25]	LIPB_CANAR
Lipase	*Candida antarctica* b	*N*-Acetyl-D-glucosamine; β-octylglucoside	1TCB	2.1	Uppenberg *et al.*[25]	LIPB_CANAR
Lipase	*Candida antarctica* b	*N*-Acetyl-D-glucosamine; β-octylglucoside	1TCC	2.5	Uppenberg *et al.*[25]	LIPB–CANAR
Lipase	*Rhizomucor miehei*		1TGL[b]	1.9	Brady *et al.*[26]	LIP–RHIMI
Lipase	*Geotrichum candidum*	*N*-Acetyl-D-glucosamine	1THG	1.8	Schrag and Cygler[27]	LIP1–GEOCN
Lipase	*Penicillium camembertii*		1TIA[b]	2.0	Derewenda *et al.*[28]	MDLA_PENCA
Lipase	*Humicola lanuginosa*		1TIB	1.84	Derewenda *et al.*[29]	
Lipase	*Rhizopus delemar*		1TIC[b]	2.6	Derewenda *et al.*[29]	LIP–RHIDL
Lipase	*Candida rugosa*	*N*-Acetyl-D-glucosamine	1TRH	2.1	Grochulski *et al.*[30]	LIP1_CANRU
Cutinase	*Fusarium solani pisi*[a]	Diethyl *p*-nitrophenyl phosphate	2CUT	1.9	Martinez *et al.*[31]	CUTI_FUSSO
Lipase	*Rhizomucor miehei*		3TGL	1.9	Derewenda *et al.*[32]	LIP_RHIMI
Lipase	*Rhizomucor miehei*	Diethyl phosphate	4TGL	2.6	Derewenda *et al.*[33]	LIP_RHIMI
Lipase	*Rhizomucor miehei*	*N*-Hexylphosphonate ethyl ester	5TGL[b]	3.0	Brzozowski *et al.*[34]	LIP_RHIMI

[a] Recombinant form expressed in *Escherichia coli.*
[b] C_α coordinates only.

representing the structure with highest resolution of each unique protein sequence in the data set. We did not do an iterative retrieval of structures, which is indicated as a possibility in the flow chart (Fig. 1). The keyword search returned the entries that were needed for this analysis, and to our knowledge no tools are available yet for searching libraries of 3D structures with consensus data. One alternative is to search the library with the individual structures that have been identified so far, e.g., with the *scan* procedure of Stamp.[35] This is what we have indicated as an option in the flow chart. Browsing one of the existing classifications of protein structures will probably give essentially the same information.

Initial Alignment of Three-Dimensional Structures

For well-conserved sequences it is easy to set up a structural alignment by hand, using a standard modeling package with options for superposition of macromolecules based on RMS (root-mean-square) deviations. This can be achieved by identifying some well-conserved motifs, e.g., by an initial sequence alignment, and superimposing those motifs. However, for less well-conserved sequences it is normally easier and less biased to use an automatic approach.

In this analysis we have used the Stamp[35] program, although there are other alternatives as well.[36,37] Stamp will structurally superimpose 3D data sets, and then use this superposition to identify structurally conserved regions (SCRs) in the proteins. The global superposition is generated by first doing a pairwise structural alignment of all structures in the data set, computing a score value for each alignment, and then using this pairwise score matrix as a similarity matrix for clustering. The clustering is used to ddefine the order in which structures are added to the alignment, so that the most similar structures are aligned first.

The algorithm used by Stamp requires the protein structures under comparison to be approximately superimposed initially, which often works well for similar proteins. However, an initial attempt to align all structures directly by using the *roughfit* procedure of Stamp failed, with low score values. We therefore used cutinase (1cus) as a "probe" structure, as cutinase may be regarded as the smallest representative of the typical lipase/esterase

---

[34] A. M. Brzozowski, U. Derewenda, Z. S. Derewenda, G. G. Dodson, D. M. Lawson, J. P. Turkenburg, F. Bjorkling, B. Huge Jensen, S. A. Patkar, and L. Thim, *Nature* (*London*) **351,** 491 (1991).

[35] R. B. Russell and G. J. Barton, *Proteins* **14,** 309 (1992).

[36] L. Holm and C. Sander, *Trends Biochem. Sci.* **20,** 478 (1995).

[37] T. Madej, J.-F. Gibrat, and S. H. Bryant, *Proteins* **23,** 356 (1995).

(α/β-hydrolase) fold. We used the *scan* procedure of Stamp to align each structure in the library against the cutinase structure, and then used this information as a starting point for the final Stamp alignment. This did improve the result slightly, although the score of the final alignment was still too low (Sc 1.05) compared to the recommended value of 2.0. After the final step 46 sequence positions were identified as structurally conserved. These positions are found in helices A2 and A3 and strands B1, B3, B4, and B5, and in the loops connecting A2 to B3 and B5 to A5 in the cutinase α/β fold (using the annotation of the PDB file); because these regions include the active site serine environment, which is known to be characteristic of lipases and esterases, as well as the active site histidine, this is a strong indication that all structures share a common basic fold. This is a somewhat surprising result, as it is known that the *Streptomyces scabies* esterase (1esc) has a 3D fold that is different from the typical α/β fold.[38] Therefore, this analysis shows that although this structure is different it still contains the essential features of the lipase/esterase core. However, for more general sequence analysis the data set is structurally too disperse, and it is probably better to look at a subset of these data.

Identification of Structural Subsets

The information necessary for identification of subsets is already in the Stamp data. As described above, the Stamp alignment is made by first aligning all possible pairs of structures to make a similarity matrix, followed by global superposition of structures by cluster analysis of this matrix. This effectively generates a hierarchical tree of the data set as shown in Fig. 2a. From this representation we see that the structures 1thg, 1cle, 1crl, 1mah, and 1ace seem to form the largest and most compact subset. An alternative representation can be achieved by transforming the Stamp score values from the pairwise structural alignment into a distance matrix, and use this matrix as input to a principal coordinates analysis. This will decompose the matrix into an eigenvector representation that can be used to make a 2D representation of the "distance" between individual structures. The decomposition of this data set by using the NetMul[39] program is shown in Fig. 2b. Again we see that the largest and most compact cluster is formed by the structures listed above, and this subset of the data set was therefore selected for the final alignment.

[38] Y. Wei, J. L. Schottel, U. Derewenda, L. Swenson, S. Patkar, and Z. S. Derewenda, *Nat. Struct. Biol.* **2,** 218 (1995).

[39] J. Thioulouse and F. Chevenet, *Comput. Stat. Data Anal.* **21,** 369 (1996).

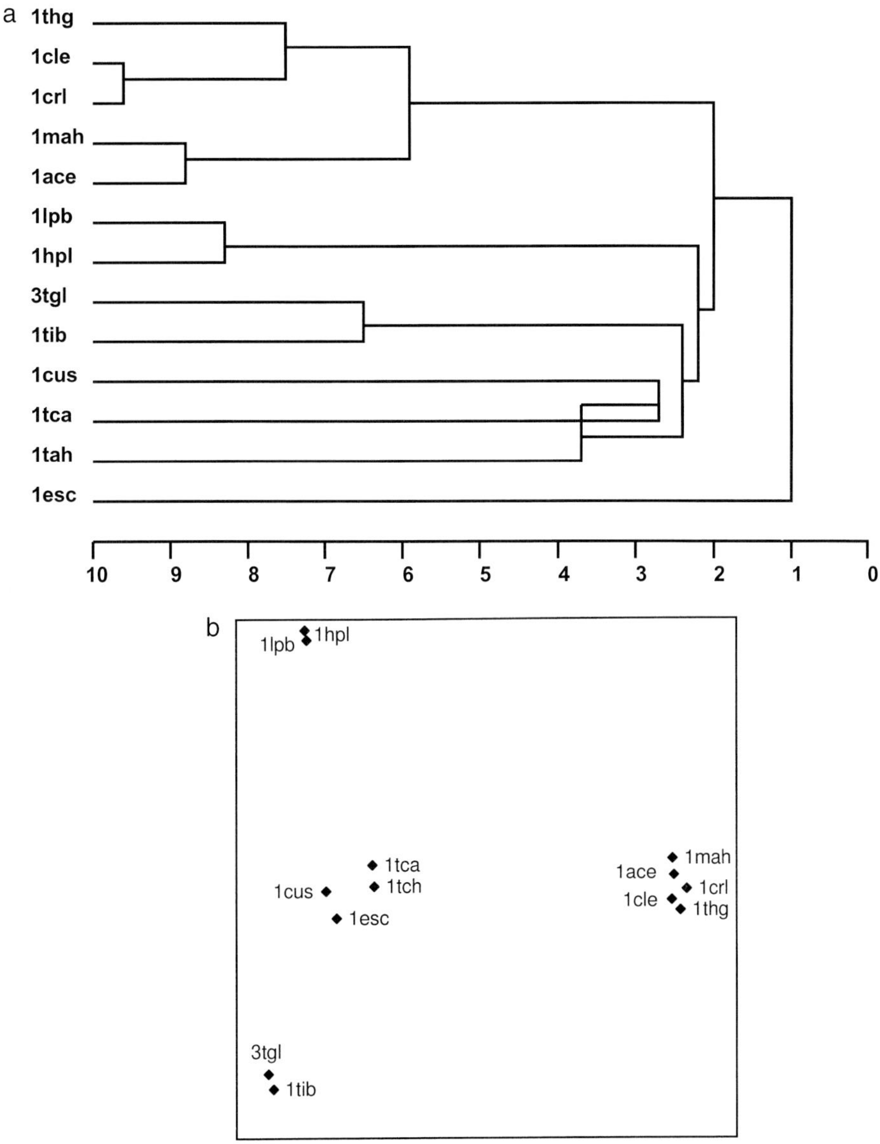

FIG. 2. Structural similarity of lipase structures: The pairwise score values for structural alignment of lipase structures by Stamp are presented as a hierarchical cluster analysis (a) and a principal coordinates plot (b). The 2D plot was generated using the NetMul[39] online multivariate data analysis system.

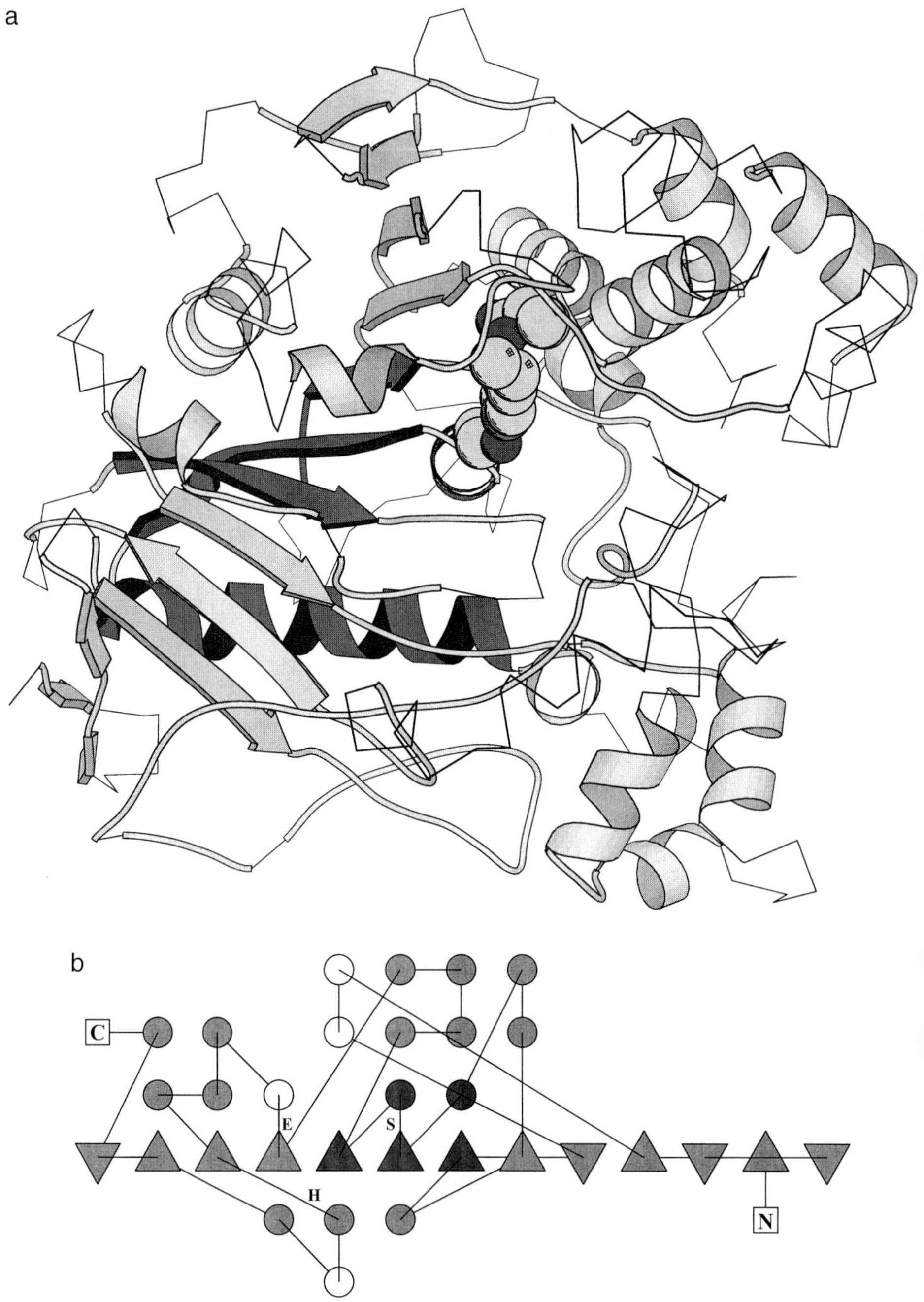
a
b
C
E
S
H
N

Final Alignment of Three-Dimensional Structures

The subset consisting of 1thg, 1cle, 1crl, 1mah, and 1ace was structurally aligned by Stamp using the *roughfit* procedure. This generated a well-defined multiple alignment with a high score value (Sc 7.59) and a low RMS value (RMS 1.69), with 381 sequence positions identified as structurally conserved. This shows that the degree of structural conservation is high. However, if we look at the degree of sequence conservation identified in the pairwise Stamp alignment, this is around 25% sequence identity for several of the sequence pairs, showing that at least parts of the sequence alignment represent a nontrivial problem.

The conserved regions are shown in Fig. 3,[40,41] using 1cle as an example. We see that the structurally conserved regions (SCRs) are found in most parts of the structure, in particular close to the active site and in the central β-sheet structure. It is interesting to note that a number of regions lacking well-defined secondary structure are also identified as SCRs. This shows that regions with a secondary structure that often is classified as "random coil" do not necessarily have a random conformation. The high degree of structural conservation seems to indicate that such regions do have specific and essential functions in lipase/esterase structure or activity. However, a detailed analysis of this is outside the scope of this chapter.

Defining Sequence Data Set

We now want to identify other protein sequences that show homology to this data set. The main purpose of this is to introduce as much variation as possible into the alignment (without reducing the general quality of the alignment) to make it easier to pinpoint the residues that are most crucial

[40] P. J. Kraulis, *J. Appl. Crystallogr.* **24,** 946 (1991).
[41] T. P. Flores, D. S. Moss, and J. M. Thornton, *Protein Eng.* **7,** 31 (1994).

FIG. 3. Structurally conserved regions in esterases and lipases. (a) A simplified ribbon-type representation of cholesterol esterase (1cle) showing regions identified as structurally conserved by Stamp, using all different lipase and esterase structures (dark gray) and using a subset of similar structures (light gray). The backbone for regions not identified as conserved is shown as a thin line. The residues of the catalytic triad are shown as CPK models, with serine located directly above the well-conserved helix in the center of this structure. [Drawn using Molscript[40] with input from the Stamp tool kit.] In (b) a Tops[41] version of the same cholesterol esterase is shown, with the positions of the active site residues indicated. Here the "lid" region found in many lipase structures can be seen directly over the active site. We see that almost all secondary structures are conserved; most of the variation is found in the loop regions.

for lipase/esterase structure and activity. One possible approach to this is to search a protein sequence database with the sequence of each individual structure, and then merge the output from each search into a nonredundant database. However, it may be both an easier and more sensitive approach to make a search profile based on the generated structural alignment, and use this profile for searching the database in a single pass. By this approach we include information about position-specific sequence variability into the search process.

The Profilesearch program by Gribskov *et al.*[42] is probably the best known implementation of a profile search. For this analysis we have used an alternative approach based on a hidden Markov model[43,44] (HMM), as implemented in the HMMer tool kit.[45,46] This approach is sensitive and easy to use, although it may be slow for complex alignments.

Searching Sequence Data Base

An HMM was built from the final structural alignment by using the *hmmb* program with "maximum discrimination" training. This model was then used for searching the SwissProt database using the *hmmsw* program. This program will do a Smith–Waterman[47] search, i.e., find local alignments with respect to both model and sequence. From the HMMer score values for the output (Fig. 4, first model) we find that there are close to 60 entries that are significant hits. To make an improved "next-generation" model all entries with a score value >100, sequence length >400, not annotated as a fragment, and not represented in the structural alignment were retrieved (in total, 23 entries).

Adding New Sequences to Initial Alignment

A new HMM was built from the structural alignment and the sequences retrieved during the first library search, by forcing the *hmmt* program to leave the structural alignment undisturbed. We tried first to do the HMM training by using the initial HMM (from the structural alignment) as a "hint" (which will allow it to change during the model-building process), based on the assumption that the Stamp alignment outside SCRs will be suboptimal, at least compared to standard multiple alignment, and that

[42] M. Gribskov, A. D. McLachlan, and D. Eisenberg, *Proc. Natl. Acad. Sci. U.S.A.* **84,** 4355 (1987).

[43] A. Krogh, M. Brown, I. S. Mian, K. Sjolander, and D. Haussler, *J. Mol. Biol.* **235,** 1501 (1994).

[44] S. R. Eddy, *Curr. Opin. Struct. Biol.* **6,** 361 (1996).

[45] S. R. Eddy, G. Mitchison, and R. Durbin, *J. Comput. Biol.* **2,** 9 (1995).

[46] S. R. Eddy, *Intelligent Systems for Molecular Biology* (*ISMB*) **3,** 114 (1995).

[47] T. Smith and M. Waterman, *J. Mol. Biol.* **147,** 195 (1981).

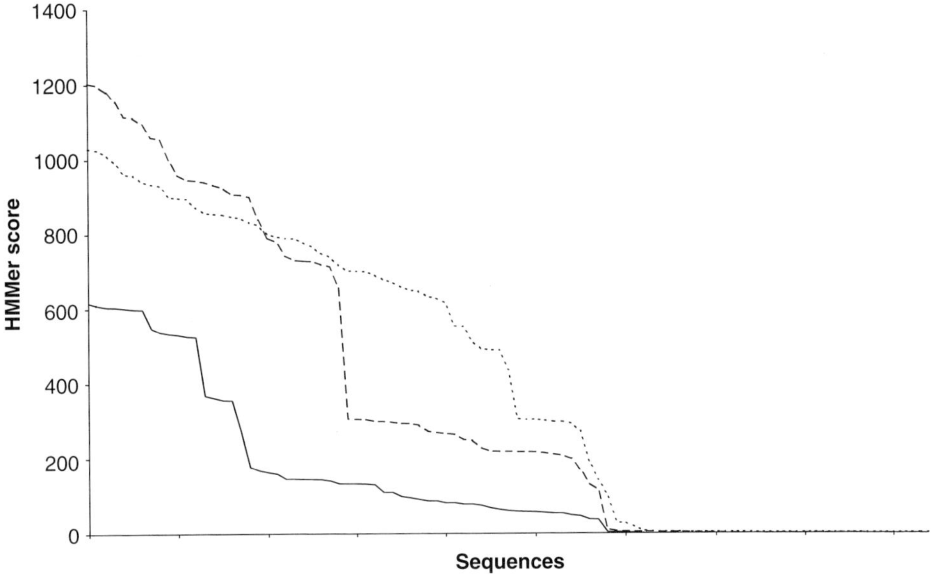

FIG. 4. Distribution of HMMer score values collected by searching the SwissProt database with three different models. (—) The initial model, based on the 3D structural alignment; (– – –) the second model, in which closely related sequences have been included; (· · ·) the final model, in which all significantly similar sequences have been included, with some minor exceptions. A full description of these data can be found in text.

allowing HMMer to modify the alignment in these regions would make them better suited for searching. However, it turned out that HMMer was not able to retain a reasonable alignment inside several SCRs, probably because of low sequence similarity, and forcing HMMer to retain the structural alignment seemed to generate better models.

Iteration of Search Process

The library search of the SwissProt database was then repeated using the model from the extended data set. This search generated score values (Fig. 4, second model) with a clear distinction between the sequences included in the model or sequences with high homology to these, sequences related to sequences in the model, and unrelated sequences. Again a library of sequences was generated from the output, this time all entries with score values >16 (which is close to the limit for significant hits), sequence lengh >100, not annotated as a fragment, and not represented in the structural alignment. A total of 46 entries was retrieved. Some initial alignments showed that four sequences lacked the GxSxG motif known to be typical

TABLE II
SELECTED LIPASE AND ESTERASE PROTEIN SEQUENCES[a]

Score	SwissProt code	Description	Source	PDB
1029.97	ACES_MOUSE	Acetylcholinesterase	*Mus musculus*	1MAH
1025.18	ACES_RAT	Acetylcholinesterase	*Rattus norvegicus*	
1014.95	ACES_HUMAN	Acetylcholinesterase	*Homo sapiens*	
990.90	ACES_BOVIN	Acetylcholinesterase	*Bos taurus*	
960.78	CHLE_HUMAN	Cholinesterase	*Homo sapiens*	
959.68	CHLE_MOUSE	Cholinesterase	*Mus musculus*	
941.46	CHLE_RABIT	Cholinesterase	*Oryctolagus cuniculus*	
934.79	ACES_TORCA	Acetylcholinesterase	*Torpedo californica*	1ACE
930.47	ACES_TORMA	Acetylcholinesterase	*Torpedo marmorata*	
899.80	EST6_DROMA	Esterase 6	*Drosophila mauritiana*	
897.15	EST6_DROSI	Esterase 6	*Drosophila simulans*	
896.05	EST6_DROME	Esterase 6	*Drosophila melanogaster*	
872.55	ESTB_DROPS	Esterase 5b	*Drosophila pseudoobscura*	
857.65	EST1_HUMAN	Liver carboxylesterase	*Homo sapiens*	
854.67	ACES_CHICK	Acetylcholinesterase	*Gallus gallus*	
853.32	ESTP_RAT	Pi 6.1 esterase	*Rattus norvegicus*	
847.56	ESTC_DROPS	Esterase 5c	*Drosophila pseudoobscura*	
843.19	EST1_RAT	Liver 60-kDa esterase E1	*Rattus norvegicus*	
831.81	EST_MOUSE	Carboxylesterase	*Mus musculus*	
824.71	ESTA_DROPS	Esterase 5a	*Drosophila pseudoobscura*	
800.77	LIP3_CANRU	Lipase 3	*Candida rugosa*	1CLE
794.13	ESTP_DROME	Esterase P	*Drosophila melanogaster*	
789.87	LIP4_CANRU	Lipase 4	*Candida rugosa*	
788.62	LIP1_CANRU	Lipase 1	*Candida rugosa*	1CRL
777.32	LIP2_CANRU	Lipase 2	*Candida rugosa*	
766.28	EST1_RABIT	Liver 60-kDa carboxylesterase 1	*Oryctolagus cuniculus*	
749.83	LIP5_CANRU	Lipase 5	*Candida rugosa*	

of most lipases[48,49] (GLT_DROME, NRT_DROME, THYG_BOVIN, THYG_HUMAN). These sequences were removed from the library, and a final HMM was generated, as described above. The SwissProt databank was searched with this final model (Fig. 4, third model; Table II). We see that the model now seems to describe all entries included in the model, as they generally have a high score value (although the score value for the most closely related sequences has dropped). A small number of sequences that are not included in the model do have a significant score; these are

[48] D. L. Ollis, E. Cheah, M. Cygler, B. Dijkstra, F. Frolow, S. M. Franken, M. Harel, S. J. Remington, I. Silman, J. Schrag, J. L. Sussman, K. H. G. Verschueren, and A. Goldman, *Protein Eng.* **5,** 197 (1992).

[49] H. W. Anthonsen, A. Baptista, F. Drabløs, P. Martel, S. B. Petersen, M. Sebastião, and L. Vaz, *in* "Biotechnology Annual Review" (M. R. El-Gewely, ed.), p. 315. Elsevier Science, Amsterdam, 1995.

TABLE II (*continued*)

Score	SwissProt code	Description	Source	PDB
739.53	ACE1_CAEEL	Acetylcholinesterase 1	*Caenorhabditis elegans*	
717.68	SASB_ANAPL	Fatty acyl-CoA hydrolase	*Anas platyrhynchos*	
702.53	ESTF_MYZPE	Esterase FE4	*Myzus persicae*	
701.36	ESTE_MYZPE	Esterase E4	*Myzus persicae*	
700.36	EST2_RABIT	Liver 60-kDa carboxylesterase 2	*Oryctolagus cuniculus*	
691.98	BAL_HUMAN	Bile salt-activated lipase	*Homo sapiens*	
679.25	BAL_RAT	Bile salt-activated lipase	*Rattus norvegicus*	
671.26	ACES_DROME	Acetylcholinesterase	*Drosophila melanogaster*	
659.17	LIP1-GEOCN	Lipase 1	*Geotrichum candidum*	1THG
650.08	LIP2_GEOCN	Lipase 2	*Geotrichum candidum*	
647.60	EST1_CAEBR	Gut esterase	*Caenorhabditis briggsae*	
633.36	ESTS_DROVI	Esterase S	*Drosophila virilis*	
628.03	EST1_CAEEL	Gut esterase	*Caenorhabditis elegans*	
553.66	D2_DICDI	cAMP-regulated D2 protein	*Dictyostelium discoideum*	
551.74	CRYS_DICDI	Crystal protein	*Dictyostelium discoideum*	
511.82	PNBA_BACSU	*p*-Nitrobenzyl esterase	*Bacillus subtilis*	
491.92	EST2_CAEEL	Esterase CM06B1	*Caenorhabditis elegans*	
489.16	EST1_CULPI	Esterase B1 precursor	*Culex pipiens*	
488.29	ESTJ_HELVI	Juvenile hormone esterase	*Heliothis virescens*	
435.70	PCD_ARTOX	Phenmedipham hydrolase	*Arthrobacter oxidans*	
303.33	THYG_BOVIN	Thyroglobulin	*Bos taurus*	
271.59	THYG_HUMAN	Thyroglobulin	*Homo sapiens*	
190.02	GLT_DROME	Glutactin	*Drosophila melanogaster*	
140.68	NRT_DROME	Neurotactin	*Drosophila melanogaster*	

[a] Protein sequences retrieved by the final hidden Markov model, sorted according to HMMer score. Only sequences with score values >16 and not annotated as fragments are shown. The last four sequences lack the GxSxG motif and are not included in the final alignment.

sequences showing a significant similarity to lipases and esterases, but they lack the GxSxG motif. They are probably of lipase or esterase origin, but have acquired new or modified functionality. It was decided to leave these sequences out of the alignment to avoid potential noise problems, which can be introduced by the inclusion of nontypical sequences.

The final model also recognizes a small number of new sequences not included in previous models, in particular the palmitoyl-CoA hydrolase (PMCH_RAT). However, these sequences were only fragments and not relevant for inclusion in the final alignment.

Making Multiple Sequence Alignment

The HMM that we have generated can be converted directly into an alignment by using the HMMer tool kit. This was done for the final model, using the *hmma* program. A number of conserved positions could be identified. However, in particular for the first part of the alignment, large sections did not seem to be consistent with the structural alignment. Therefore, an alternative approach was used to make the final multiple alignment. The initial structural alignment was used as a profile in ClustalW,[50,51] and then all sequences in the final library were aligned on the basis of information in this profile. Visual inspection showed that this generated a reasonable multiple alignment for most of the sequence space, and this alignment was used without further refinement.

This alignment includes both closely related sequences and more distant homologs. To generate an even more robust alignment with respect to structural information another alignment based on the library for the second HMM was also generated. This includes only sequences that are similar to the sequences for which we have 3D structural information.

Identification of Conserved Residues

After having defined the multiple alignments we want to identify "important" positions in the sequences on the basis of the information found in these alignments. Several different approaches are available for doing this, such as the Kabat score,[52] conservation index,[53] or entropy measures like the one used in hssp.[54] For this analysis we have chosen to use a simple criterion, focusing on conserved positions. A position is classified as conserved if all residues found at that position in the alignment belong to the same class. We then need to define reasonable classes of residues. The simplest approach may be to treat each residue type as an individual class, i.e., look only for total conservation. However, this is normally too restrictive to be useful; for example, it is known that both aspartate and glutamate are used as the acidic residue of the typical lipase/esterase catalytic triad.

There are several ways that we can group residues, based on codon usage, physical data, mutation data, etc. This is reflected, e.g., in the number of different mutation matrices that have been developed. Useful discussions can be found in Johnson and Overington[55] and in Vogt *et al.*[56] It is important

[50] D. G. Higgins and P. M. Sharp, *Gene* **73,** 237 (1988).

[51] D. G. Higgins, A. J. Bleasby, and R. Fuchs, *Comput. Appl. Biosci.* **8,** 189 (1992).

[52] E. Kabat, T. Wu, H. Bilofsky, M. Reid-Miller, and H. Perry, "Sequences of Proteins of Immunological Interest." National Institute of Health, Bethesda, Maryland, 1983.

[53] W. A. Hide, L. Chan, and W. H. Li, *J. Lipid Res.* **33,** 167 (1992).

[54] C. Sander and R. Schneider, *Proteins* **9,** 56 (1991).

[55] M. S. Johnson and J. P. Overington, *J. Mol. Biol.* **233,** 716 (1993).

[56] G. Vogt, T. Etzold, and P. Argos, *J. Mol. Biol.* **249,** 816 (1995).

to realize that it may be relevant to assign individual residues to several classes, depending on how these are defined.

For the current analysis we have used the classification ((IVLM), (GAST), (FYW), (NEDQ), (KR), (P), (C), (H)), which means that, e.g., isoleucine, valine, leucine, and methionine are accepted as equivalent residues. This classification is close to what we obtain if we do a principal coordinates analysis on the Blosum 62 score matrix,[57] it is somewhat related to the ClustalW classification scheme, and it is inspired by the work of Johnson and Overington.[55] The classification also reflects similarities in standard volume of residues, in particular for the nonpolar residues. This is consistent with the study of Gerstein *et al.*[58] showing that there are significant constraints on how much the volume of residues at buried positions can vary during evolution, reflecting the fact that most proteins have a tightly packed core.

However, it is important to remember that this classification is only one of several possible choices, and the choice of classification scheme (or scoring scheme in general) may be influenced by the type of information being searched for.

To give some room for sequencing errors and other, more atypical deviations, at most one residue at each position of the small alignment and two residues for the large alignment can belong to a different class than the majority of the residues without breaking the conservation at that position.

Definition of Conserved Positions

Both alignments were analyzed by the approach described above, and the result for the large alignment is shown in Table III,[59] together with possible explanations (see below). A surprisingly high number of positions seem to be conserved (66 in total), given the high number of quite dissimilar sequences included in the alignment. There also seems to be a difference in the distribution of conserved positions, with more conserved positions in the first part of the alignment. To study this observation in more detail we tried to identify domains in the relevant structures.

It is not evident how to split these structures into domains. We have used the Protein Domain Assignment server[60,61] to analyze 1crl, 1ace, and 1thg. This classifies 1thg as consisting of two domains, splitting it just before the active site serine; 1crl is also classified as a two-domain fold, with a

[57] S. Henikoff and J. G. Henikoff, *Proc. Natl. Acad. Sci. U.S.A.* **89,** 10915 (1992).

[58] M. Gerstein, E. L. Sonnhammer, and C. Chothia, *J. Mol. Biol.* **236,** 1067 (1994).

[59] R. J. Kazlauskas, *Trends Biotechnol.* **12,** 464 (1994).

[60] S. A. Islam, J. Luo, and M. J. Sternberg, *Protein Eng.* **8,** 513 (1995).

[61] M. J. Sternberg, H. Hegyi, S. A. Islam, J. Luo, and R. B. Russell, *Intelligent Systems for Molecular Biology (ISMB)* **3,** 376 (1995).

TABLE III

CONSERVED POSITIONS FROM MULTIPLE ALIGNMENT

1cle[a]	1thg	1ace	1mah	Residues	Function[b]
12	13	15	17	IIIVLLLVVLLVVVVVIVVVVVVIIIIVVVVVVVLLLLLIIIIIILLV	np cluster
14	15	17	19	GGGGGGGGGGGRGGGGGGGGGGGGGGGGLTGGGGGGGGGGGGGGGGG	
23	24	30	32	FFFFFFFFFYFFFFFFFFFFFFFFFFFFFFFFFYYYYYYYFFFYYFW	np cluster
25	26	32	34	GGGGGGGGGGGGGGGGGGGGGGGGGGGGGGGAAGGSSSASSGGGSSG	
26	27	33	35	IIIIIIIIIIIIIIIIIIVVIIVIIIILIIIIIIIIIILLIIVIIVI	np cluster
27	28	34	36	PPPPPPPPPPPPPPPPPPPPPPPPPPPPPPPPPPPPPPPPPPPPPPP	np cluster, E → C
28	29	35	37	FFFFFFFFFYFFFFYYYFFFFFFFFFFFFFFFYYYYYYYYYYYYYYY	np cluster
29	30	36	38	AAAAAAAAAAAAAAAAAAAAAAAAAAAAAAAAAAAAAAAAAAAAAAA	
32	33	39	41	PPPPPPPPPPPPPTKPPPPPPPPPPPPPPPPPPPPPPPPPPPPPPPP	np cluster
41	42	48	50	PPPPPPPPPPPPPPPPPPPPPPPPPPPPPPPPPPPPPPPPPPLPPAP	
60	61	67	69	CCCCCCCCCCCCCCCCCCCCCCCCCCCCCCCCCCCCCCCCCCCCCAC	Cys bridge
95	103	92	94	EEEEEEEEEEEEEEEEEEEEEEEEEEEEEEEEAEEEEEEEEEEEEEE	H bond (37)
96	104	93	95	DDDDDDDDDDDDDDDDDDDDDDDDDDDDDDDDDGDDDDDDDDADDDD	H bond (94 back)
97	105	94	96	CCCCCCCCCCCCCCCCCCCCCCCCCCCCCCCCSCCCCCCCCCCCCCC	Cys bridge
98	106	95	97	LLLLLLLLLLLLLLLLLLLLLLLLLLLLLLLLLLLLLLLLLLLLLLL	np cluster
100	108	97	99	LIILLLLLILLLLLLLLLLLLLLLIIILLLLLLILVVVVVLLAVVLV	np cluster
102	110	99	101	VVVIVVVVVVVIIIVVVIIIIVIVVVVVVVVIVVIVIVVIVVIIILV	np cluster
116	124	111	114	VVVVVVVVIVVVVVVVVVVVVVVVVVVVVVVVVVVVVVVVVVIVVVV	np cluster
117	125	112	115	MMMMLMLLLLLMMMLMMMMMMMMMMMMFMILLMMVVVVVVIILVVLM	np cluster
120	128	115	118	IIIIVIIIIIIIIIIIIIIIIIIIIIIIIVIIIIIIILIMIIILIII	
122	130	117	120	GGGGGGGGGGGGGGGGGGGGGGGGGGGGGGGGGGGGGGGGGGGGGGG	E → C, ligand
123	131	118	121	GGGGGGGGGGGGGGGGGGGGGGGGGGGGGGGGGGGGGGGGGGGGGGG	E → T, ligand

124	132	119	122	AGGGGGGGGGGAAGGGGGGGGGAGGGGAAGGGGAAAAAAGGGAGGA	Ligand, oxyanion
128	136	123	126	GGGGGGGGGGGGGGGGGGGGGGGGGGGGGGGGGCGGGGGGGGGGGG	T → C
150	158	142	145	VIIVVIVVIVVVIIIIIVVVVVVVIIIVIIIIVVIIIIIIVVILIVI	
152	160	144	147	VVVVVVVVAVVVVVVVVVVVVVVVVVVVVVVAVVVVVVVVVVVIVVIV	np cluster
156	164	148	151	YYYYYYYYYYYYYYYYYYYYYYYYYYYYYYYYYYYYYYYYYYYYTYY	np cluster, H bond (121 back)
157	165	149	152	RRRRRRRRRRRRRRRRRRRRRRRRRRRRRRRRRRRRRRRRRRRRERR	H bond (181)
159	167	151	153	GASGGSGGGGGGGGGGGGGGGGGAAAGGGGGGGGGGGGGGGNGEGG	C → H
162	170	154	157	GGGGGGGGGGGGGGGGGGGGGGGGGGGGGGGGGGGGGGGGGGGGGG	H → C
163	171	155	158	FFFFFFFFFFFFFFFFFFFLFFLFFFFYFFFFFFFFFFFFFFFFFF	np cluster
176	184	167	170	NNNNNNNNNNNNNNNNNN-NNNNNNNNNNNNNNNNNNNNNNNNNNN	
178	186	169	172	GGGGGGGGGGGGGGGGGGDAGGAGGGGGGGGGGGGGGGGGGGGGWG	C → H
181	189	172	175	DDDDDDDDDDDDDDDDDDFDDDDDDDDDDDDDDDDDDDDDDDDDDD	H bond (157)
182	190	173	176	QQQQQQQQQQQQQQQQQQLQQQQQQQQQQQQQQQQQQQQQQQQQQQ	np cluster, H bond (214)
185	193	176	179	GGGAAAAAAAAAAAAAAAVAAAAGAGGAAAAAAAAAAAAAAAALAAAA	
186	194	177	180	LMMLLMLLILLLIILLLLNLLLLLMLMLLLLLILLLLLLLLLLLILL	np cluster
188	196	179	182	WWWWWWSWWWWWWWWWWWWWWWWWWWWWWWWWFFWWWWWWWWWWWWWW	np cluster
189	197	180	183	VVVVVVVVLVVVVVVVVVVVVVVVVVVVIVVVVVVVIIIIIIIIVIIIV	np cluster
192	200	183	186	NNNNNNNNNNNNNNNNNNNNNNNNNNNNNNNNNNNHNNNNNNNNNNNN	H bond (188 back)
196	204	187	190	FFFFFFFFFFFFFFFFFFFFFFFFFFFFFFFFFFFFFFFFFFFFFF	np cluster
197	205	188	191	GGGGGGGGGGGGGGGGGGGGGGGGGGGRGGGGGGGGGGGGGGGGGG	T → E
198	206	189	192	GGGGGGGGGGGGGGGGGGGGGGGGGGGGGGGGGGGGGGGGGGGGGG	T → E
199	207	190	193	DDDDDDDNDDDDDNNNNDNNNNDDDDDDNDDDDEEEEEEDDDMDDD	H bond (201 back)
203	211	194	197	VVVVVIVIMVVVIIVIVVVVVVVVVVVVVIIIVVIVVIIIVVIIIIV	np cluster
205	213	196	199	IIIILLLLLLLLILILIIIIIIIIIIIIIIILVLLLVVVIIIILLLV	np cluster
207	215	198	201	GGGGGGGGGGGGGGGGGGGGGGGGGGGGGGGGGGGGGGGGGGGGGG	E → T
209	217	200	203	SSSSSSSSSSSSSSSSSSSSSSSSSSSSSSSSSSSSSSSSSSSSSS	Active site, H bond (449)

(continued)

TABLE III (*continued*)

1cle[a]	1thg	1ace	1mah	Residues	Function[b]
210	218	201	204	AAAAAAAAAAAAAAAAAAAAAASAAAAAAAAAAAAAAAAAAAAATGA	np cluster, oxyanion
211	219	202	205	GGGGGGGGGGGGGGGGGGGGGGGGGGGGGGGGGGGGGGGGGGGGGGG	T → H
212	220	203	206	ASSGAAAASAARGAAAAGGGGGGASSSGAAAAAGGGGGGGAAAGGAG	
214	222	205	208	SSSSSSSSSSSSSSSSSSSSSSSSSSSSSSSSSSSSSSSSSSASSSS	H bond (182)
237	245	222	225	AGGAAAAAGAAAAAAAAAAAAAAAAAAGAAAAARFAAAAAGAAAAAAA	
241	249	226	229	SSSSSSSSSSSSSSSSSSSSSSSSSSSSSSSSSSSSSSSGSSSSSSS	H bond (341)
242	250	227	230	GGGGGGGGGGGGGGGGGGGGLGGGGGGGTSAAGGGGGGGGGGGGGPG	
286	297	274	281	LLLLLLLVILLLLLLIIILLLLMLLLLLLMIIIIILLLLLIIIIILIFTL	np cluster
336	349	322	329	GGGGGGGGGGGGGGGGGGGGGGGGGGGGGGGGGGGSSSTTTSSGTSGG	
341	354	327	334	EEEEEEEEEEEEDDEEEEEEEEEEEEEEEEEEEEEEDDDDDDEEEEDNEE	Active site, H bond (241), H bond (449)
411	424	396	403	SGGGGGGGGGGGTTGGGAASAAGTGGGATTSSSGTTTTTTSSSTTGT	
412	425	397	404	DDDDDDDDDDDDDDDDDDDDDDDDDDDDDDDDDDDDDDDDDDDDNDD	H bond (381), H bond (518)
432	445	419	426	WYYYYYYYHYYYYYFFFYYYYFYWYYYYYYWWFYYYYYYYYYYYYRW	np cluster
434	447	421	428	YFFYYYYYYYYYYYYYYYYYYYYYYFFFYYYYYYYYYYFFFYYYFYYY	np cluster
449	463	440	447	HHHHHHHHHHHHHHHHHHHHHHHHHHHHHHHHHHHHHH-HHHHHHMHH	Active site, H bond (209), H bond (341)
452	466	443	450	EDDEEEEEEEEEEDDEEEEEEEEEEDDDDEEEEEEEEEDDDDDEEEEDDDEE	H bond (240)
474	488	476	483	FFFFFFFFFFFFFFFFFFFFFFFFFFFFFFFFFFFFFFFFFFFFFF	np cluster
489	503	492	500	WWWWWWWWWWWWWWWWWWWWWWWWWWWWWWWWWWGWFFFFFLWWWFFWW	np cluster

[a] The numbering for 1crl is identical to 1cle.

[b] np cluster, Side chain participates in a nonpolar cluster; E → C, position is close to the indicated transition between different secondary structures (E, extended; H, helix; T, turn; C, coil); Cys bridge, residue is involved in cysteine bridge formation; H bond, residue is involved in hydrogen bond formation. Partner is indicated behind; (94 back), partner is backbone of residue 94; ligand, residue is involved in ligand binding; active site, residue is part of active site; oxyanion, residue is involved in oxyanion stabilization in the active site.[59]

split just before the active site glutamate; and 1ace is classified as a more complex three-domain fold, with splitting just before both serine and glutamate of the active site. From the structure as it is drawn in Fig. 3 a splitting somewhere between the active site residues serine and glutamate seems reasonable, defining it as a two-domain fold where active site residues are found in the crevice between the two domains. This is similar to what is found, e.g., in the trypsin family.[62]

If we use the domains for 1crl as a basis, there are 17% conserved positions (56 of 332) in the first domain, and only 5% (10 of 202) in the second domain. This can be compared to the structurally conserved positions identified by Stamp, i.e., 69% conserved positions (228 of 332) in the first domain, and 59% (119 of 202) in the second domain. This shows that although the second domain seems to be less well conserved, the 3D fold of the protein is retained. This has some implications for the sequence analysis. If we do this sequence alignment with ClustalW without taking the 3D structural information into account, most conserved positions in the first domain are still identified in the new alignment, although there are some differences in the location of gaps and in the details of the alignment. However, in the second domain almost none of the residues are identified as conserved in the new alignment, meaning that, e.g., the active site histidine is no longer identified as conserved. This shows the importance of including structural information in the alignment, in particular when the sequence similarity is low. If structural information is not available it may be useful to compare several alignment methods, as we have previously shown with this data set that different alignment methods may identify different positions as conserved, although the total number of positions identified by each method is similar (our unpublished data, 1994).

It is also informative to see which residue types are conserved, as shown in Fig. 5. Here the distribution of totally conserved residues and residues conserved according to the classes defined above is compared to the total distribution of residues in the data set. We see that for the absolute conservation cysteine, proline, and in particular glycine are overrepresented. These residues often play a crucial role in defining the structure of proteins, indicating that the conservation at this level is mainly associated with structure. The same seems to be true for the conservation according to classes of residues. Here residues such as isoleucine, leucine, and valine show a high degree of conservation, indicating that for the hydrophobic interactions involved in forming the core of the protein the general residue type is important, including the relative volume of the residue.

[62] C. Branden and J. Tooze, "Introduction to Protein Structure." Garland Publishing, New York, 1991.

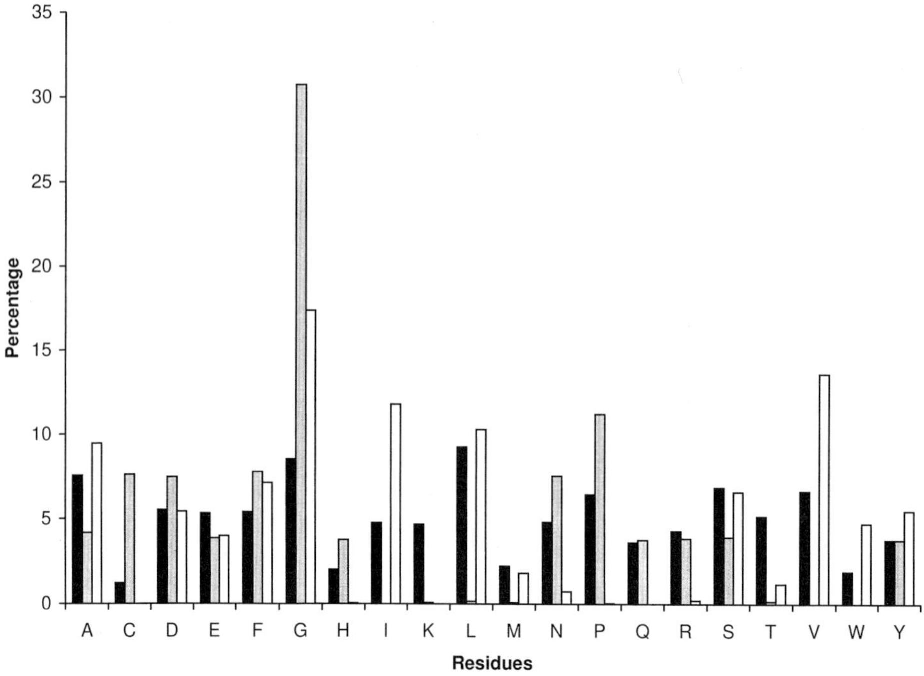

FIG. 5. Relative distribution of residue types in the data set. Black bars: The whole data set used for multiple alignment. Gray bars: Positions identified as totally conserved. White bars: Positions with only conservative mutations, as defined in text.

Explaining Conservation Patterns

Although this type of analysis can give some information about the general driving forces for conservation, a more detailed analysis is needed to identify why specific residues are conserved. In the following sections we use specific tools to look at individual types of interactions and properties that we believe may be of importance for a detailed understanding of conservation patterns. We give some examples of how to use this information. A detailed study of all conserved interactions is outside the scope of this chapter. However, Cygler *et al.*[63] have done a comprehensive analysis of several interactions identified by a similar study, and the reader should consult that article for more information.

For this analysis we use *Candida cylindracea* cholesterol esterase (1cle) as a sample structure. This structure is selected because it has reasonable

[63] M. Cygler, J. D. Schrag, J. L. Sussman, M. Harel, I. Silman, M. K. Gentry, and B. P. Doctor, *Protein Sci.* **2,** 366 (1993).

resolution (2.0 Å), it has a good score in the HMMer search (showing that the sequence is reasonably "typical"), and it has an interesting ligand in the crystal structure (cholesteryl linoleate). For the discussion of specific residues, the numbers given refer to 1cle; corresponding numbers for the other structures can be found in Table III.

Active Site Residues. It is known that the active site for lipases and esterases is a catalytic triad that generally consists of Ser-His-(Asp/Glu).[63] The serine is in most cases well conserved in an GxSxG motif located in a turn between a β strand and an α helix. This motif is characteristic of lipases and esterases.[48,64] Also, the other two residues of the triad are well conserved, although it is known that at least the acidic residue can move to a different sequence position (maintaining a similar relative orientation with respect to histidine) without loss of activity.[49,65] In the current alignment all three positions are easily identified as Ser-209, Glu-341, and His-449. The sequence alignment for the regions close to the active site residues is shown in Fig. 6.[66] We see that the region around the active site is well conserved for serine and well conserved for aspartate/glutamate, although there is some variation that seems to be characteristic of subclasses of sequences, in particular for some of the esterases. For the region around histidine we see a number of insertions close to the active site for some of the esterases; otherwise this region is well conserved, too. The position of the insertions may not be correct; we see that several sequences have a glycine at position −3, counted from the histidine, and it is possible that this indicates a better alignment. However, owing to a lack of stronger evidence we have kept the alignment as it was generated by ClustalW.

The active site serine is interesting also from an evolutionary point of view. Serine can be coded by two different codon types, TC(A/C/G/T) and AG(C/T). In addition, *Candida rugosa* uses the CTG codon for serine.[67] It has been suggested that codon usage for the active site serine can be used for studying the evolutionary origin of different lipases and esterases,[68] as one would expect that a change of codon usage would need a simultaneous change of two nucleotides to happen without loss of activity. Initial studies seemed to indicate a possible difference between lipases and esterases with respect to codon usage for the active site serine.[68] More recent

[64] S. Brenner, *Nature (London)* **334,** 528 (1988).

[65] J. D. Schrag, F. K. Winkler, and M. Cygler, *J. Biol. Chem.* **267,** 4300 (1992).

[66] G. J. Barton, *Protein Eng.* **6,** 37 (1993).

[67] Y. Kawaguchi, H. Honda, J. Taniguchi Morimura, and S. Iwasaki, *Nature (London)* **341,** 164 (1989).

[68] S. B. Petersen and F. Drabløs, *in* "Lipases—Their Structure, Biochemistry and Application" (P. Woolley and S. B. Petersen, eds.), p. 23. Cambridge University Press, Cambridge, 1994.

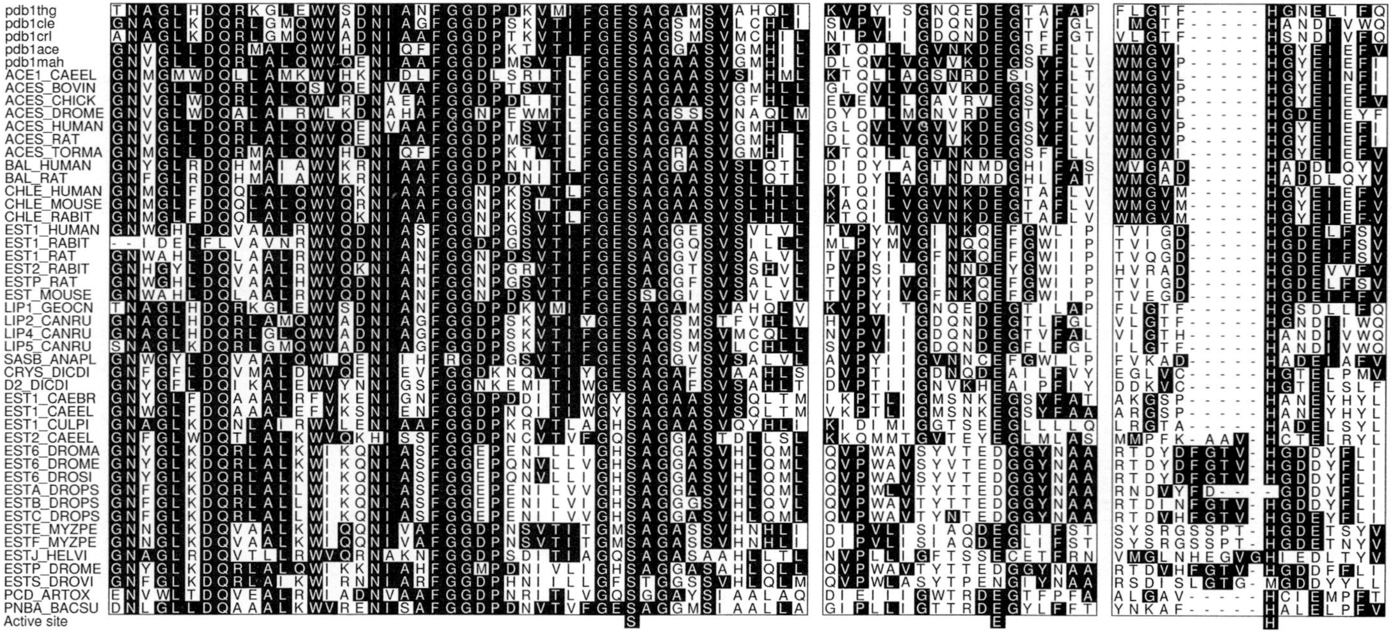

FIG. 6. Sequence alignment for the regions close to active site residues. The boxes represent the structurally conserved regions as identified with Stamp. The active site residues are identified below the alignment. The alignment was generated with Alscript[66] from ClustalW data.

studies did not seem to support this theory,[49] but the amount of structural information included was limited, and no clear conclusion could be drawn.

From the high degree of structural similarity it seems reasonable to assume that the 3D structures used for this alignment do have a common evolutionary origin. It is unlikely that such a high degree of similarity could develop through convergent evolution of two intrinsically unrelated amino acid sequences. If we look at the codon usage for the active site serine of these sequences (as found in the EMBL Nucleotide Sequence Database) we see that all three codon types are represented (1thg—TCC, 1cle—CTG, 1crl—CTG, 1ace—AGT, 1mah—AGT). The same is also true if we look at other sequences included in the alignment. This shows that there is codon variability for the active site serine also within a set of sequences with common evolutionary origin, and this codon is therefore not a unique probe for identification of potentially different origin of sequences with a low degree of sequence similarity.

Ligand Binding. The residues involved in ligand binding in 1cle are shown in Fig. 7,[69] including two of the active site residues. As expected for a nonpolar ligand almost all interactions are hydrophobic, typically involving residues such as isoleucine, leucine, and phenylalanine. If we compare this to Table III we see that only Gly-122, Gly-123, and Gly-124 are identified as conserved. An important reason for having glycines in these positions seems to be that a side chain would make it difficult to achieve tight ligand binding. If we look only at the small alignment of closely related sequences, Phe-133, Ala-450, and Ile-453 are identified as conserved in addition to the glycines, showing the importance of the immediate environment of the active site. However, most of the residues involved in binding of this ligand are not conserved.

This result is as one would expect, for two reasons. First, the enzymes in this data set have different substrate specificity, and this is achieved by variation of the residues involved in ligand binding. Second, a more technical reason is that most residues involved in ligand binding are found in loop regions. To make a good sequence alignment for loop regions is always difficult, both because of structural variability and because these regions often include gaps, and most programs for multiple alignment seem to have problems in regions with gaps. Therefore, we should generally not expect to see a high degree of conservation for residues involved in ligand binding, and even when there is conservation, it may be difficult to identify.

Geometric Constraints. As seen in Fig. 5, there is a relatively high fraction of cysteine, proline, and glycine residues found at the conserved positions. The cysteine residues are involved in formation of disulfide bridges. In this

[69] A. C. Wallace, R. A. Laskowski, and J. M. Thornton, *Protein Eng.* **8,** 127 (1995).

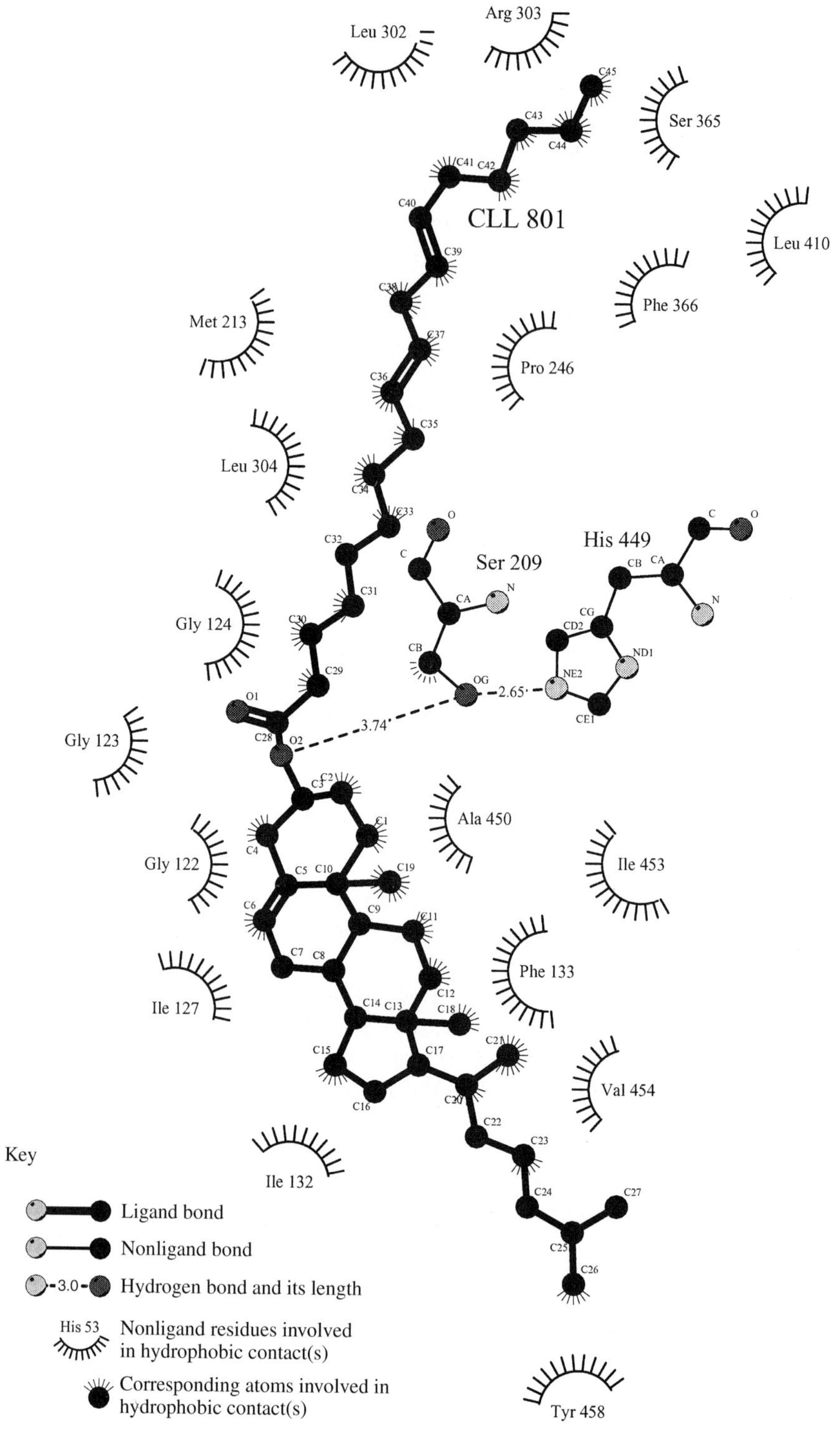

CLL 801
Leu 302
Arg 303
Ser 365
Leu 410
Phe 366
Met 213
Pro 246
Leu 304
Ser 209
His 449
Gly 124
Gly 123
Ala 450
Ile 453
Gly 122
Phe 133
Ile 127
Val 454
Ile 132
Tyr 458
3.74
2.65
Key
Ligand bond
Nonligand bond
3.0
Hydrogen bond and its length
His 53
Nonligand residues involved in hydrophobic contact(s)
Corresponding atoms involved in hydrophobic contact(s)

molecule there are two such bridges. However, only the first seems to be well conserved, involving Cys-60 and Cys-97.

The proline and glycine residues are mainly involved in defining the global fold of the protein, as they are generally found in transitions between different elements of secondary structure. According to the dssp[70] secondary structure assignment Gly-25 and Pro-27 are at the end of a β strand; Pro-32 and Pro-41 are found just before and after a structurally conserved turn region; and Gly-122, Gly-123, Gly-124, and Gly-128 are found in a turn between a β strand and an α helix. Gly-159 and Gly-162 are found at the start and the end of an α helix; Gly-178 is found at the beginning of another α helix; Gly-197 and Gly-198 are found in a turn between an α helix and a β strand; Gly-207 and Gly-211 are found in a tight turn between a β strand and an α helix, shaping the environment of the active site serine; Gly-242 is in a turn between a β strand and an α helix; and Gly-336 is close to the end of the β strand leading up to the active site Glu-341. Thus these residues are generally found in turn regions between the larger building block elements of secondary structure. In addition to this, some glycine residues may be necessary because of space constraints in ligand binding, as described above.

Hydrogen Bonding and Salt Bridges. For the identification of hydrogen bonds we have used the HBplus[71] program, and focused on hydrogen bonds involving side chains of conserved residues. The most obviously important hydrogen-bonding network involves the active site residues, with His-449 bonded to Ser-209 and Glu-341. The Glu-341 is also hydrogen bonded to Ser-241, which is almost completely conserved. Other interactions also involve two conserved side chains: Arg-157 is bonded to Asp-181 and Gln-182 is bonded to Ser-214. Some interactions involve conserved side chains hydrogen bonded to the backbone of the protein; this is the case for Asp-96, which is bonded to the backbone of Ser-94, and also for Tyr-156 (bonded to the backbone of Phe-121), Asn-192 (bonded to the backbone of Trp-188), and Asp-199 (bonded to the backbone of Ser-201). Finally, some conserved side chains are involved in interactions with side chains that show some degree of variation, such as Glu-95 bonded to Arg-37, which is almost completely conserved in all sequences except the bile salt-activated

[70] W. Kabsch and C. Sander, *Biopolymers* **22,** 2577 (1983).

[71] I. K. McDonald and J. M. Thornton, *J. Mol. Biol.* **238,** 777 (1994).

FIG. 7. Binding of cholesteryl linoleate to cholesterol esterase. The binding of cholesteryl linoleate in 1cle as identified and plotted by LigPlot.[69] The LigPlot parameters for cholesteryl linoleate were added locally.

lipases. Asp-412 is bonded to Tyr-381, which seems to be conserved in several sequences. However, this residue is found close to a loop region that is difficult to align. Asp-452 is bonded to Gln-240, which seems to be conserved only in the small alignment. Thus these residues are involved either in interactions that are less crucial to the function of the protein, or the interactions may be specific to only a subset of the proteins, or changes in the general fold of the protein modify the function of selected residues.

Figure 8 shows several of these interactions as they are found in 1cle. We see that these interactions are involved in keeping the correct relative orientation of several structural elements, in particular the α helices and the central β sheet structure close to the active site, and the long loop structure close to the end of the central β sheet. It is interesting to note that most of these residues are found in loop regions or close to the end of secondary structure elements.

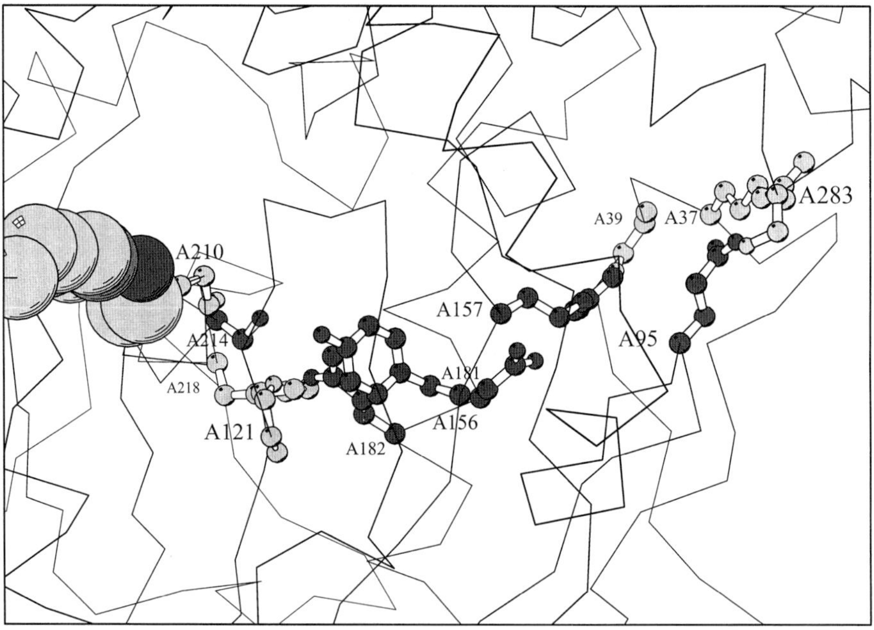

FIG. 8. Conserved hydrogen-bonding network in cholesterol esterase: A central hydrogen-bonding network in 1cle, identified with HBplus. The orientation of 1cle is identical to that in Fig. 3, and active site residues are shown in CPK representation. Conserved residues are shown in black, variable residues are shown in gray. However, most residues identified as variable are in fact conserved in a significant subset of sequences.

Nonpolar Interactions. It has been shown that nonpolar interactions contribute significantly to the stability of the protein core.[72] Therefore, the identification of important nonpolar interactions will be a useful contribution to understanding conservation patterns in proteins. As indicated in the discussion about how residue types are conserved, we would expect these conservations to be within classes of nonpolar residues, rather than complete conservations. We see that the nonpolar conservations identified in Table III generally fit this pattern; a large number of them involve isoleucine, valine, and leucine residues.

However, to find out which residues are in fact involved in significant nonpolar interactions we need a better approach. In our laboratory we have developed a simple tool kit (ResClus, *pdb_np_cont* and *pdb_np_clus*) for clustering of nonpolar residues by pairwise contact areas. These tools are part of our local tool kit for sequence analysis (SeqTools).

In the first step we compute for each nonpolar atom the area covered by some other nonpolar atom from a different residue. For these computations we generally use the definitions from the 3D–1D profile approach of Lüthy *et al.*[73] with respect to atom classification and radius. We then summarize these values on a residue level, which gives us residue–residue nonpolar contact areas.

In the second step we use these contact areas as a "similarity" matrix. We do a hierarchical clustering on this matrix, by first joining the residues with the highest contact area. This gives a tree structure showing the general clustering of nonpolar interactions. If we do a complete clustering on all residues we will in most cases obtain one large cluster involving all nonpolar interactions. However, if we specify cutoff levels so that contact areas below a specific level are ignored, and clusters below a certain size (after applying the area cutoff value) are removed, then we will see several clusters with large pairwise contact areas, representing tight clusters of pairwise interactions.

If we do this type of analysis for 1cle, with an area cutoff value at approximately 50% of the largest contact area and ignoring clusters with fewer than five residues, we obtain nine clusters of varying size. Almost all of the conserved nonpolar residues are found in the two largest clusters. If we look at the largest cluster (Fig. 9) we see that most conserved residues are found in a subregion of this cluster. The locations of the corresponding residues in 1cle are shown in Fig. 10. We see that the residues involved in this cluster are found close to the hydrogen-bonding network shown in Fig. 8, and the secondary structure elements involved are essentially the same.

[72] A. R. Fersht and L. Serrano, *Curr. Opin. Struct. Biol.* **3,** 75 (1993).

[73] R. Lüthy, A. D. McLachlan, and D. Eisenberg, *Proteins* **10,** 229 (1991).

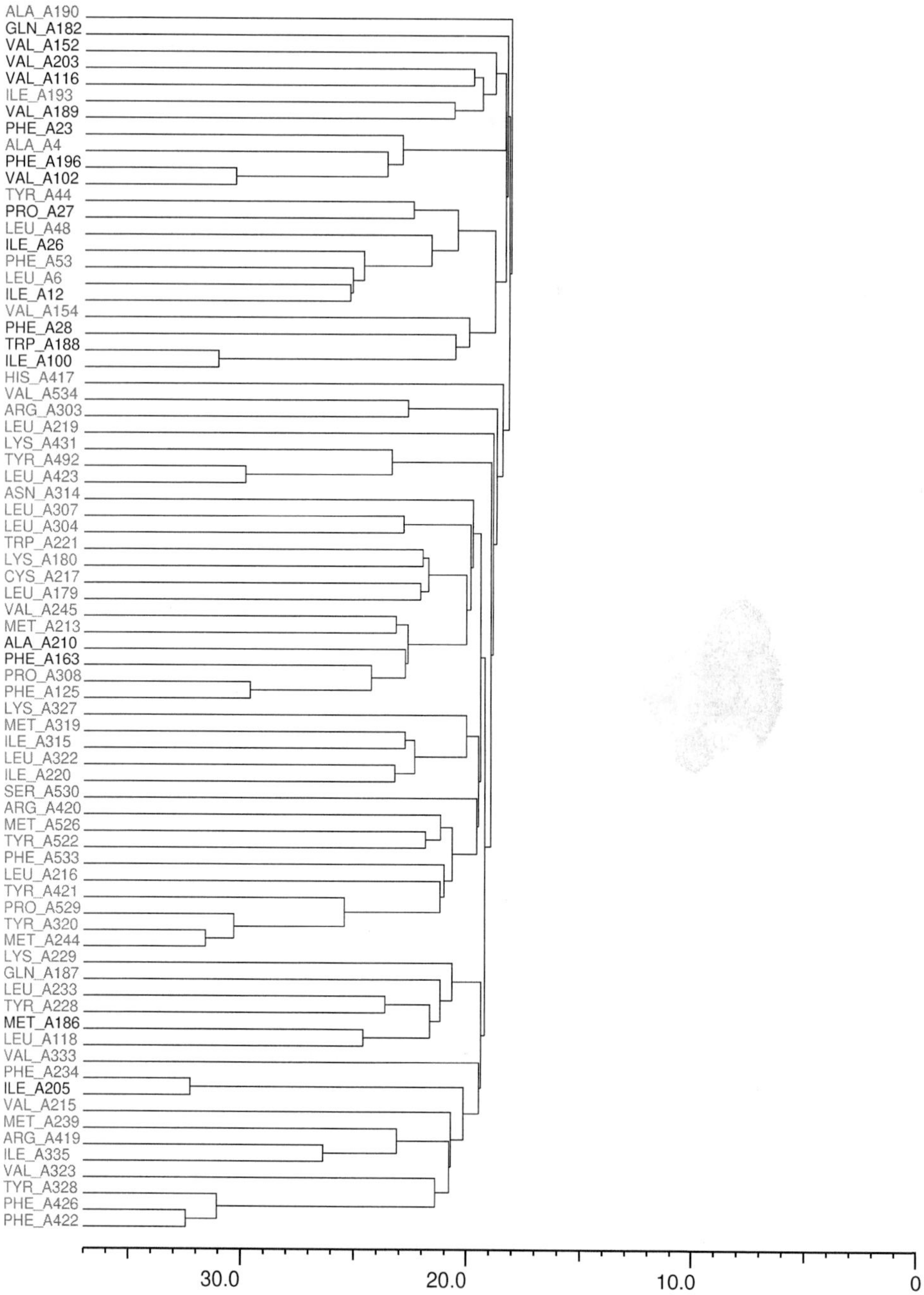

FIG. 9. Nonpolar contacts in cholesterol esterase: Hierarchical clustering of nonpolar contacts in 1cle, as described in text. Only the largest cluster is shown. Residues corresponding to conserved positions are shown in black, other residues are shown in gray.

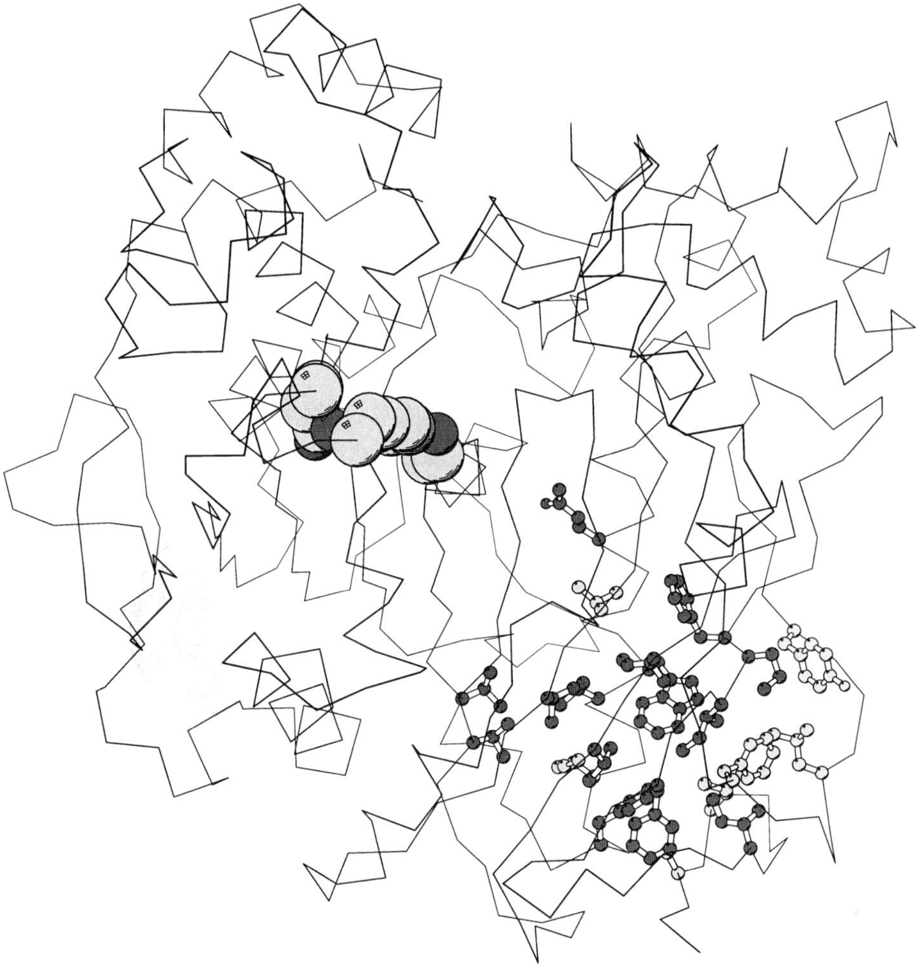

FIG. 10. Conserved nonpolar cluster in cholesterol esterase: Residues belonging to the well-conserved subcluster seen in the upper part of Fig. 9. Conserved residues are shown in black; variable residues are shown in gray. We see that the conserved residues form a tight cluster, with the variable residues on the "surface" of this cluster.

However, compared to the hydrogen-bonding network most of the residues involved in nonpolar interactions are found in the core of the protein, involving the central parts of the secondary structure elements.

However, although conserved residues are important in nonpolar interactions, most of the interactions show a high degree of variability with respect to accepted residues. This may reflect both differences in local structure as well as a more general flexibility with respect to side-chain packing.

Conclusion

This chapter has demonstrated a general approach to sequence analysis. The analysis has identified a number of conserved positions in a selected subset of lipase and esterase sequences, and it has also been demonstrated that most identified conservations can be explained by simple observations with respect to properties such as hydrogen bonding and nonpolar interactions.

However, it should be remembered that some explanations may be more complex than what we have hinted at here; some conservations have been identified, but not explained, and some sites identified as variable should really have been identified as conserved if the analysis had been done at an even more detailed level. This is mainly because of two reasons. First, this analysis does not take dynamic properties into account. Such properties may include dynamic movement of domains and individual residues necessary for facilitating the catalytic reactions, but it may also include the dynamics of protein folding. Individual residues may be of key importance at specific steps along the folding pathway to achieve the correct fold.[74] However, that role may be less obvious as soon as the final fold has been reached. Information related to dynamic properties is generally not accessible from static X-ray structures, and although some information can be found by NMR experiments, these types of requirements are generally difficult and time consuming to identify.

Second, some elements of variability may be more restricted than can be seen directly from the alignment, because they are involved in substrate specificity, adapted to specific solvent or temperature conditions, or involved in correlated complementary mutations. This is consistent with mutation studies of β-lactamase, showing that the accepted variability at individual positions in the gene of a single organism seems to be less than the variability identified by multiple alignment studies of the corresponding gene family.[75,76] Some information about such restrictions may be gained by subdividing the alignment according to, e.g., substrate specificity, or by looking at correlation in the mutations for structural neighbors. However, this type of analysis is outside the scope of this chapter.

[74] E. Shakhnovich, V. Abkevich, and O. Ptitsyn, *Nature* (*London*) **379,** 96 (1996).
[75] W. Huang, J. Petrosino, M. Hirsch, and P. S. Shenkin, *J. Mol. Biol.* **258,** 688 (1996).
[76] P. S. Shenkin, personal communication (1996).

Finally, this type of analysis still has an element of trial-and-error in it, and the research area itself is in rapid development. It is not possible to give a complete and detailed procedure for how it should be done; rather, one must try different approaches and use the one that seems to give the most useful information. Therefore this chapter tries to focus on some useful approaches, and the type of information that can be found by using these tools, possibly at the cost of more detailed experimental procedures. It is hoped that this chapter can serve as a source of inspiration for future projects in sequence analysis.

Acknowledgments

The work described in this chapter has been supported by the European Commission (BIO2-CT94-3016) and the Norwegian Research Council (29345/213). F. D. thanks Tom Flores for valuable assistance with the Tops program.

[3] Identification of Important Motifs in Protein Sequences: Program MULTIM and Its Applications to Lipase-Related Sequences

By STEFFEN B. PETERSEN, FINN DRABLØS, MARIA TERESA NEVES PETERSEN, and EVAMARIA I. PETERSEN

Introduction

It is generally agreed that the first early signs of life on Earth emerged about 3.7 billion years ago. What caused this to happen is unclear, and we may never understand all the details of this wonderful process. Today we are presented with a wealth of information in the shape of the ever-growing amount of biological sequences, either in the form of DNA, RNA, or protein sequences. This information represents collections of families of protein sequences, some of them obviously related through evolution. As an example, today we have more than 250 trypsin-like sequences that we believe all share the same protein fold, but differ markedly in their sequence composition—some of them have even lost one or more of their active site residues, i.e., they probably have adapted to a totally new function. Such examples are well known from the other protein families, e.g., lysozyme and lactalbumin share both sequence and 3D fold, but have widely different functions today. As a consequence of the availability of massive amounts of sequence information, many tools for analyzing such sequences have

0076-6879/97 $25.00

been developed, either aiming at identifying (weak) homology between pairs of sequences or aiming at analyzing a group of related sequences in a so-called multiple alignment approach. As a matter of fact, there are so many methods available that it can be difficult to choose between them. Part of the problem is often the intrinsic complexity of the applied method: Which penalty should one adapt for the introduction of gaps in the aligned sequences? Which score matrix should be used? Why? In addition, several other parameters may need to be specified before one can use the search or alignment program.

In this chapter, we present a novel methodology called MULTIM (multiple motifs); it is based on a combination of graphic visualization of the results and a simple, yet versatile, method for the identification of conserved motifs in the protein sequences being compared. Its true value resides in its ability to present the results in an easy-to-grasp graphic picture that, in addition, is well suited for documentation and/or presentation purposes. We illustrate the qualities of MULTIM on the lipase and esterase family of sequences, as well as other selected sequences. Last, we discuss the present limitations of MULTIM.

MULTIM Program

MULTIM (version 3.0) was developed to identify common motifs in protein sequences.[1] The program is distributed free of charge by the authors. It is available both UNIX and DOS versions. The algorithm was tailored for an exhaustive search while keeping the number of comparisons as low as possible. Perhaps more important is that the method is insensitive to insertions and deletions. The latter is a serious drawback with most other sequence alignment methods. These methods apply algorithms that assign a score for all alignments considered and, when evaluating insertions and deletions, specific penalties (often negative scores) are adapted. These penalties increase as the size of the insertion or deletion increases. Consequently, large gaps are treated poorly by these methods, but well by MULTIM, provided that MULTIM can identify conserved motifs on both sides of the gap.

MULTIM (version 3.0) consists of three programs.

MOTIF: Identifies motifs that are conserved in all sequences

FILTER: Weights the motifs based on probability

CONNECT: Makes the graphic output

In Fig. 1 a flow chart is shown that illustrates the different programs, input and output data, and the sequence in which they must be executed.

[1] S. B. Petersen and F. Drablos, "MULTIM," version 3.0. 1992 (contact either one of the authors for a copy of the program).

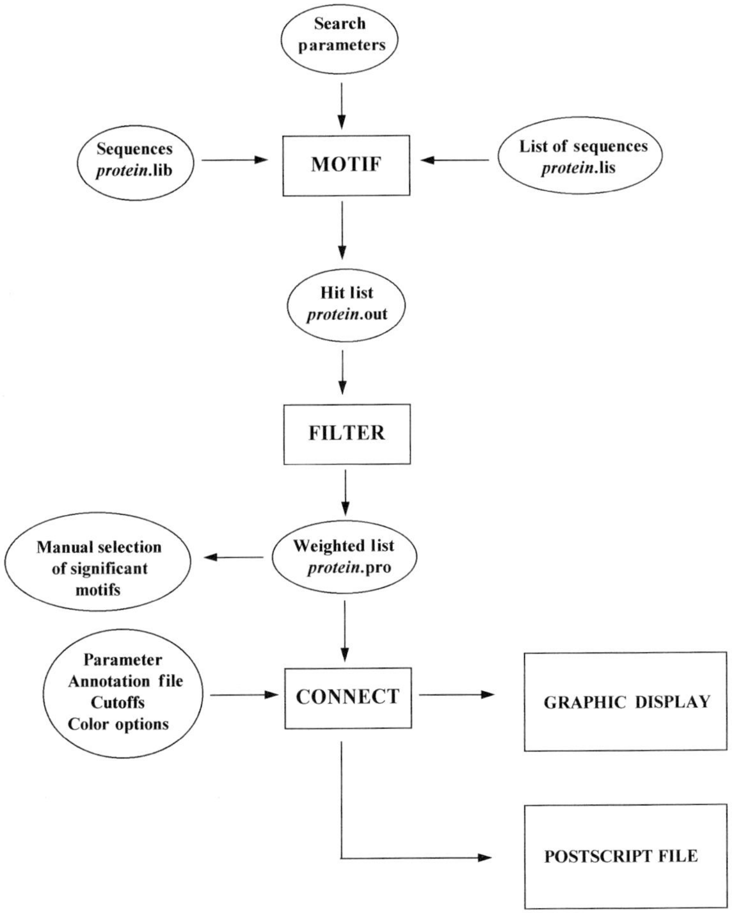

FIG. 1. Flow chart representing the steps of MULTIM necessary to display or print the motifs found by the program. Besides the two input files (*protein*.lib, amino acid sequences and *protein*.lis, list of sequences), search parameters (window size and number of conserved residues) must be specified.

For a successful run of the programs, three files (⟨mot_path.dat⟩, ⟨*.lib⟩, and ⟨*.lis⟩) are necessary.

⟨mot_path.dat⟩ The ⟨mot_path.dat⟩ file contains the complete path(s) to the library files ⟨*.lib⟩, one per line. Because they are searched in sequential order, it is preferable to have the most used library file on the top of this list.

⟨*.lib⟩ The file contains the library of protein sequences. The first line of each protein entry starts with a "⟩" and is followed by a descriptor that identifies the protein. After the descriptor, the sequence starts on a new line and continues with 80 characters per line until the end, which is denoted by an asterisk (*). The following is a typical entry in a library file.

```
>CUTI_ALTBR  cutinase

MMNLNLLLSKPCQASTTRNELETGSSDACPRTIFIFARGSTEAGNMGALVGPFTANALESAYGASNVWVQGVGGP

YTAGLVENALPAGTSQAAIREAQRLFNLAASKCPNTPITAGGYSQGAAVMSNAIPGLSAAVQDQIKGVVLFGYTK

NLQNGGRIPNFPTSKTTIYCETGDLVCNGTLIITPAHLLYSDEAAVQAPTFLRAQIDSA*

>CUTI COLGL  cutinase

MKFLSVLSLAITLAAAAPVEVETGVALETRQSSTRNELETGSSSACPKVIYIFARASTEPGNMGISAGPIVADAL

ERIYGANNVWVQGVGGPYLADLASNFLPDGTSSAAINEARRLFTLANTKCPNAAIVSGGYSQGTAVMAGSISGLS

TTIKNQIKGVVLFGYTKNLQNLGRIPNFETSKTEVYCDIADAVCYGTLFILPAHFLYQTDAAVAAPRFLQARIG*
```

⟨*.lis⟩ In the file ⟨*.lis⟩ the sequence names that should be analyzed are listed, one per line. The actual sequence should be located in the ⟨*.lib⟩ file, which is found using the path given in ⟨moth_path.dat⟩. The following is a typical entry in a list file.

```
CUTI_ALTBR
CUTI_COLGL
```

MOTIF

The MULTIM program defines a motif as a short string of amino acids, out of which all or some are conserved, e.g.,

```
MTSVGK
xTxxGK
```

In the preceding example, the window length is six residues and the motif consists of at least three residues, all located in the six-residue window. Note that MULTIM automatically generates all possible motifs conforming to the choices made for the number of conserved residues and window size. For details see Table I.

The user starts the Motif program by giving the following command line:

```
motif ⟨n⟩ ⟨motif_length⟩ ⟨*.lis⟩ (⟨-options⟩)
```

where ⟨n⟩ is the number of residues that as a minimum should be conserved within the motif length (⟨motif_length⟩). These motifs should be retained

TABLE I
NUMBER OF MOTIFS IDENTIFIED BY MOTIF IN A WINDOW SIZE OF Y WITH n RESIDUES CONSERVED[a]

Number of conserved residues	Window size	Number of motifs considered per window position
2	3	3
2	4	6
3	4	4
2	5	10
3	5	10
4	5	5
2	6	15
3	6	20
4	6	15
3	8	56
4	8	70
6	12	924

[a] $Y!/[n!(Y - n)!]$, where Y is the window size and n is the number of conserved residues.

only if at least one such motif is found in all sequences whose names are given in the file ⟨*.lis⟩. MOTIF will report the location and local sequence for all such hits. The default mode of comparison is based on residue identity. The output is found in the output file ⟨*.out⟩. The following is a typical fragment in an output file.

```
CODE   CUTI_FUSSO  [  27 -  32 ]  EARQLG   1.0000000000
CODE   CUTI_MAGGR  [  30 -  35 ]  EARQLN   1.0000000000
CODE   CUTI_ALTBR  [  96 - 101 ]  EAQRLF   1.0000000000
CODE   CUTI_COLGL  [ 113 - 118 ]  EARRLF   1.0000000000
CODE   CUTI_COLCA  [ 117 - 122 ]  EAKRLF   1.0000000000
CODE   CUTI_ASCRA  [ 112 - 117 ]  EAVRLF   1.0000000000
CODE   CUTI_ASPOR  [ 103 - 108 ]  EAQGLF   1.0000000000
CONS                              EA--L-   1.0000000000
```

The following options may be added to the command line:

-s Use similarity for the motif search instead of identity, which is the default. The sequences in the library file will be searched for concurrent motifs based on similarity using a default similarity matrix.

a ⟨matrix_file⟩ Read an alternative similarity matrix from a file. The default similarity matrix used for this program is the Pam250 matrix from the FASTA package (version 1.7)[2,3] with a cutoff of 2.5.

-c ⟨cut_off⟩ Use an alternative cutoff value when comparing residues. This determines when different residues are treated as equal.

FILTER

If we, for example, are searching for motifs of length 4 in a sequence consisting of 10 alanines, 1 glycine, and 1 proline, the probability that the motif AAAA will match the sequence at some position is very high, whereas the probability that the motif AAGP will match is much lower. If we make the assumption that structurally or functionally relevant motifs in proteins are generally nontrivial, we may use a test for uniqueness as a tool for filtering our frequent hits from the MULTIM alignment. In other words, we make the assumption that AAAA is less important as a motif than AAGP, and therefore we reduce the significance of matches against AAAA, compared to matches against AAGP.

This strategy has been implemented in the filter program. For a given general motif (which may include "wild-card" positions, such as GxSxG), we compute the probability of finding the motif in a randomized version of the same sequence. That is, we compute the probability p that this motif is found by chance in a given sequence. The probabilities from individual sequences in the alignment are multiplied together to give the probability for the total alignment. The value of $(1 - p)$ can be used as a measure of the "uniqueness" or "significance" of a given motif, and this is added to the MOTIF output by FILTER.

To run FILTER a ⟨*.out⟩ file is used:

```
filter ⟨*.out⟩
```

The output of this program is written to a ⟨*.pro⟩ file, where the file name is the same as the one used in ⟨*.out⟩, except for the extension. The significance values listed by the program can give an indication on the distribution of more or less significant motifs. The following is a typical fragment in a filtered output file.

[2] W. R. Pearson and D. J. Lipman, *Proc. Natl. Acad. Sci. U.S.A.* **85,** 2444 (1988).

[3] W. R. Pearson, *in* "Molecular Evolution: Computer Analysis of Protein and Nucleic Acid Sequences" (R. F. Doolittle, ed.) pp. 63–98. Academic Press, San Diego, California, 1990.

```
CODE   CUTI_FUSSO  [  27 -  32 ]  EARQLG   0.8827855665
CODE   CUTI_MAGGR  [  30 -  35 ]  EARQLN   0.8685090250
CODE   CUTI_ALTBR  [  96 - 101 ]  EAQRLF   0.9000709435
CODE   CUTI_COLGL  [ 113 - 118 ]  EARRLF   0.8775581315
CODE   CUTI_COLCA  [ 117 - 122 ]  EAKRLF   0.8840850909
CODE   CUTI_ASCRA  [ 112 - 117 ]  EAVRLF   0.8769608437
CODE   CUTI_ASPOR  [ 103 - 108 ]  EAQGLF   0.9284829967
CONS                              EA--L-   0.0029923538
```

CONNECT

The CONNECT program generates the graphic output in the form of a PostScript file. As input it reads either ⟨*.out⟩ or ⟨*.pro⟩ files. One invokes the program in the following way:

1. connect ⟨*.pro⟩ (⟨-options⟩) > ⟨*.ps⟩
2. connect ⟨*.pro⟩ (⟨-options⟩) > lpt1:

In option 1 the graphic output is sent to a PostScript file ⟨*.ps⟩. In option 2 the graphic output is redirected to the printer, here denoted with lpt1:. The printer must be capable of accepting standard PostScript files. In the printout of the PostScript file each sequence of the ⟨*.lis⟩ file is represented by a horizontal bar and common motifs in different sequences are represented by thin lines in a pairwise manner. The first sequence in the ⟨*.lis⟩ file is aligned against all other sequences in the file. The sequence bar displays a vertical line every 50 amino acids.

CONNECT has the following options.

-M Do not scale the sequence length to fit on the page. By default the longest sequence is scaled to fill the page and all other sequences are scaled proportionally.

-c Make color output. The color scale adopted gives strong colors for the most significant motifs. The color output can also be printed in gray tones on most black-and-white PostScript printers.

-i ⟨medium⟩ ⟨low⟩ Set limits for medium and low significance of motifs. Default values are 0.5 and 0.25. Motifs with high significance are drawn in black, medium significance in green, and low significance in yellow. If a black-and-white output is used, analogous gray tones will result.

-a ⟨*.ann⟩ Read annotation data from the connection file: ⟨*.ann⟩. The annotation file can be used to highlight important residues, such as active site residues. It may also be used to draw attention to a sequence fragment by coloring uniformly the corresponding range in the horizontal sequence bar.

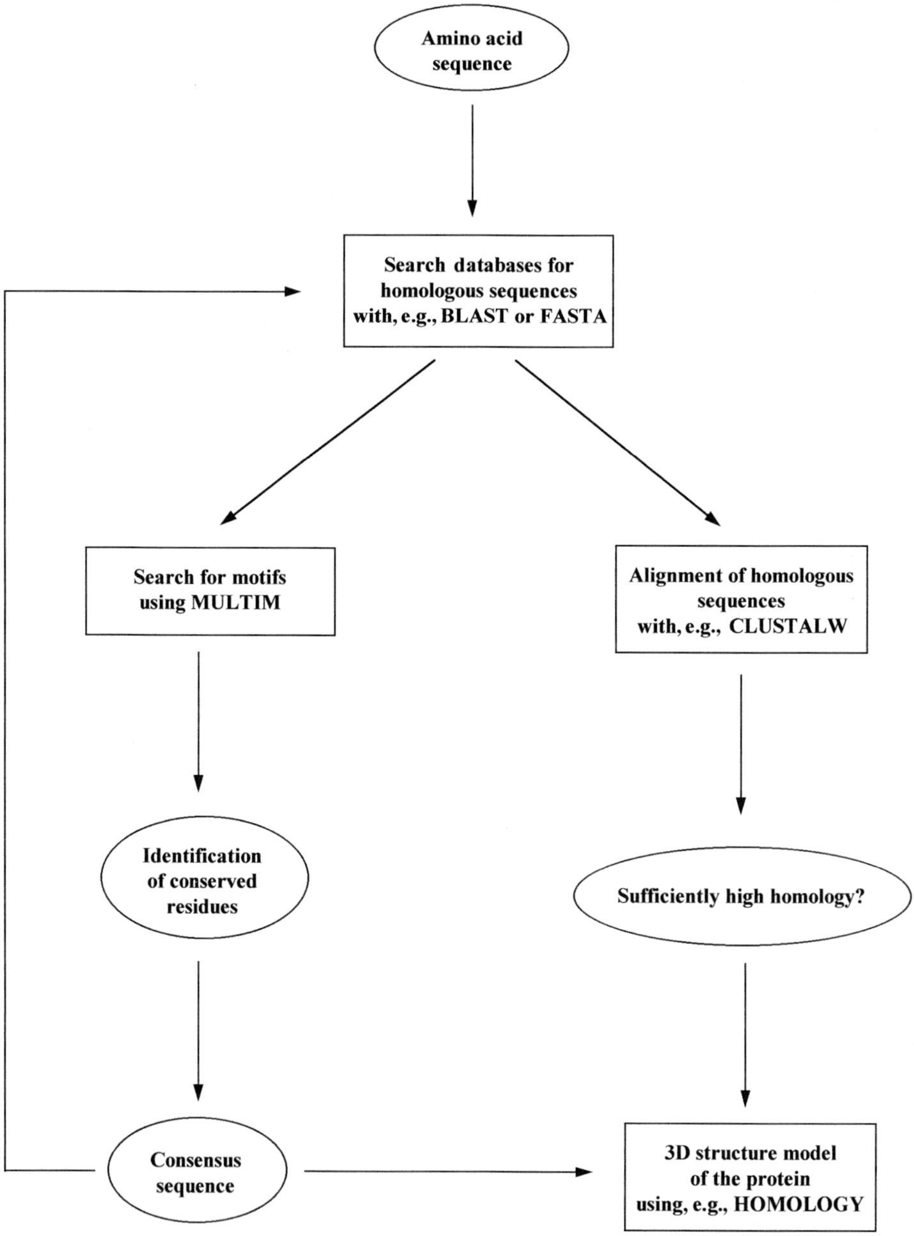

FIG. 2. Flow chart showing a possible route for obtaining all relevant sequences homologous to a query sequence. First, a sequence database is searched with the original sequence of interest using, e.g., BLAST[4] or FASTA.[2,3] A number of hits may result, and these are aligned

Applications of MULTIM

Amino acid sequences contain a large amount of information that may help us to understand protein structure–function relationships. The first step in obtaining this information is to search available databases for homologous sequences (Fig. 2). To accomplish this different programs, such as BLAST[4] or FASTA,[2,3] can be used. These programs align the query sequence against sequences in the libraries and identify significant homologies. The identified homologous sequences can then be aligned with each other in a so-called multiple alignment. One program that can be used for this task is CLUSTALW,[5] which aligns multiple sequences with high sensitivity. The alignment is performed according to the branching order in a guide tree, which has been calculated by the program. This program can help to find patterns that characterize a protein family and detect homology between new sequences and existing families of sequences. If the homology found involves a sequence for which the corresponding three-dimensional structure is known, a three-dimensional structure can be modeled. To search for homology, modeling programs such as HOMOLOGY[6] can be used. From a three-dimensional model of a protein important information about structure–function relationships can be obtained.

However, if the sequences differ with respect to the presence of large insertions or deletions, a commonly used program like CLUSTALW may fail to find the correct alignment (see Fig. 10). In contrast, MULTIM will still give good results provided that identical or similar motifs can be found in all sequences. The conserved motifs that MULTIM identifies can highlight important structural and functional elements of a protein, if the three-

[4] S. F. Altschul, W. Gish, W. Miller, E. W. Myers, and D. J. Lipman, *J. Mol. Biol.* **215,** 403 (1990).
[5] J. D. Thompson, D. G. Higgins, and T. J. Gibson, *Nucleic Acids Res.* **22,** 4673 (1994).
[6] Biosym/MSI, "Homology," version 95.0. Biosym/MSI, San Diego, CA, 1995.

---

with MULTIM[1] and possibly also with CLUSTALW[5] and by manual inspection of the *.pro file a consensus sequence is established. The sequence database can now be searched again, using the consensus sequence instead of the original sequence. It is likely that new hits will emerge, because less overall homology is required (all residues that are not conserved appear in the consensus sequence as an "X" and in sequence comparisons this character is most often treated as neutral, thus not influencing the score). If the three-dimensional structure is known for one of the homologous sequences, the three-dimensional structure can be modeled for others using, e.g., the program HOMOLOGY.[6] The conserved positions identified in the consensus sequence as well as with CLUSTALW can be mapped onto the protein and thus give three-dimensional structural information about where conserved residues are located in the protein. Such information is crucial for understanding structure–function relationships.

dimensional structure for one of the aligned sequences is known. We illustrate here the use and versatility of MULTIM in analyzing several examples, most of them related to lipases and esterases.

In Fig. 3A is shown a MULTIM analysis of the highly homologous family of lipoprotein lipases, ranging from guinea pig to human. This type of protein is likely to have evolved recently, concurrent with the appearance of the circulatory system in animals, and an amazingly high degree of overall homology is seen. Nevertheless, the graphic output highlights sequence ranges where mutability appears to be much higher. Those regions are depleted of the connecting lines that MULTIM uses to indicate the presence of conserved motifs in the sequence family. As is seen in Fig. 3A, all sequences are for convenience compared with the lipoprotein lipase sequence of human origin. This sequence name must appear as #1 in the ⟨*.lis⟩ file. Each sequence is represented with a horizontal bar, which has tic marks for every 50 amino acids. At about position 130–150 MULTIM detects no motifs. This does not mean that single amino acids cannot be conserved in this range; rather, it means that no motif with six of nine residues conserved could be identified as being common for all sequences.

In Fig. 3B, MULTIM has been used to analyze a set of protein sequences including lipoprotein, hepatic, as well as pancreatic lipases. Members of this set of sequences share much less homology than do members of the lipoprotein lipase family. Nevertheless, the appearance of parallel, strongly colored connecting lines clearly indicates that all sequences share the same protein fold. The central region from residue ~ 150 to ~ 280 in LIPP_HUMAN displays the majority of the connecting lines. This region also contains the active site residues Ser-169, Asp-193, and His-280.[7] The fact that all active site residues are found in the region where most conserved motifs are located confirms the importance of this range of residues. This observation is further strengthened when one is correlating the MULTIM patterns with the known human disease-related mutations in human lipoprotein lipase (Table II, and Fig. 3B). In Fig. 3B, the top sequence bar corresponds to lipoprotein lipase. Here, the exact locations for the known disease-related mutations are highlighted with an asterisk (*), and the ranges for which a MULTIM motif will include the relevant position are indicated as gray boxes going upstream from the exact positions. Many, but not all, of these boxes coincide with the conserved motifs identified by MULTIM. Thus, it is reasonable to assume that the molecular basis for the deleterious effect of these mutations is a perturbation of either functionally or structurally important parts of the protein.

In Fig. 4, MULTIM has been used to analyze nine acetylcholinesterase

[7] F. K. Winkler, A. D'Arcy, and W. Hunziker, *Nature* (*London*) **343,** 771 (1990).

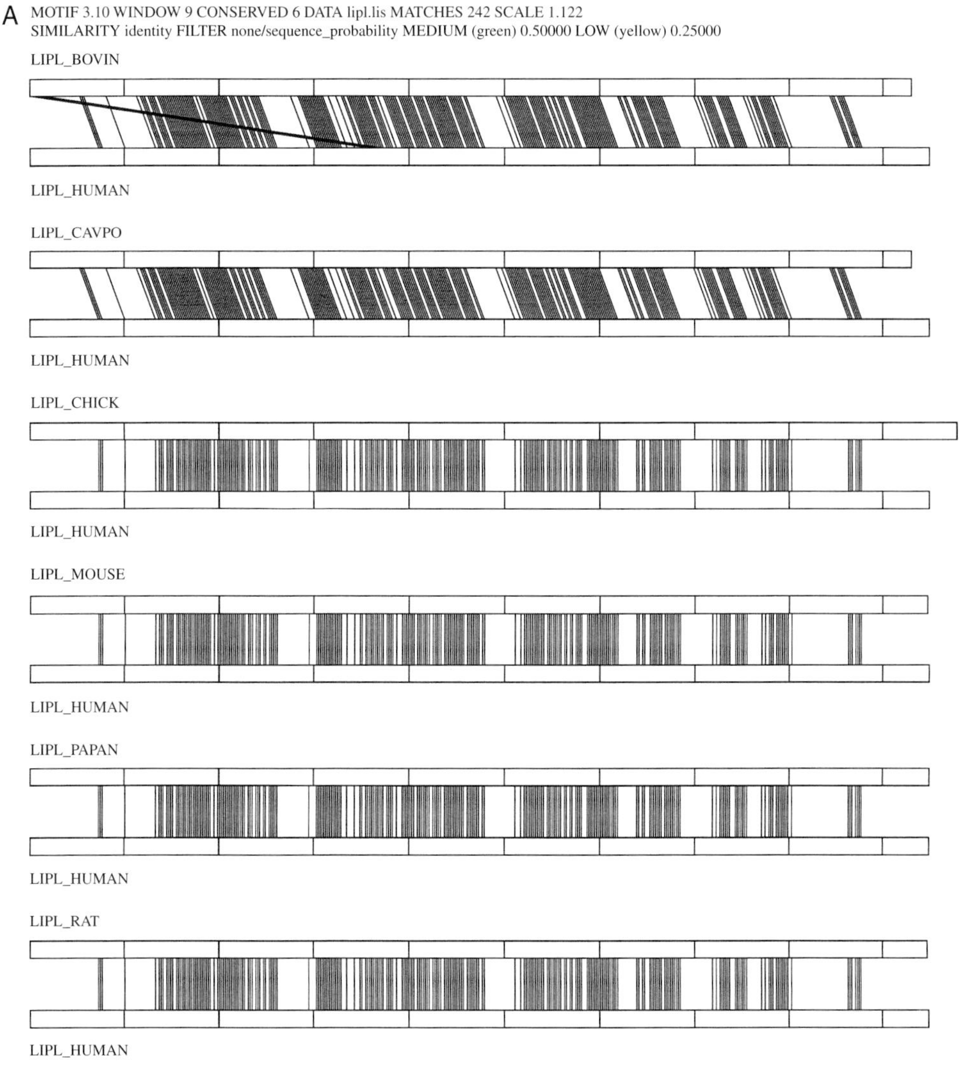

FIG. 3. (A) Subset of 7 sequences from the MULTIM analysis of 10 lipoprotein lipase precursors (LIPLs). A window size of nine residues has been used, with six residues conserved. No additional MOTIF options were implemented. All sequences were extracted from the Swiss-Prot database and the Swiss-Prot codes have been used. The species involved are as follows: *Bos taurus* (LIPL_BOVIN), *Homo sapiens* (LIPL_HUMAN), *Cavia porcellus* (LIPL_CAVPO), *Gallus gallus* (LIPL_CHICK), *Mus musculus* (LIPL_MOUSE), Papio anubis (LIPL_PAPAN), *Rattus norvegicus* (LIPL_RAT), *Sus scrofa domestica* (LIPL_PIG), *Ovis orientalis aries* (LIPL_SHEEP), and *Felis cattus* (LIPL_FELIS). (B) MULTIM analysis of six evolutionarily related lipases. A window size of six residues has been used with three residues conserved. No additional MOTIF options were implemented. All sequences were extracted from the Swiss-Prot database and the Swiss-Prot codes have been used. The proteins shown are human lipoprotein lipase (LIPL_HUMAN), human pancreatic lipase (LIPP_HUMAN), human hepatic lipase (LIPH_HUMAN), rat lipoprotein lipase (LIPL_RAT), Rat pancreatic lipase (LIPP_RAT), and rat hepatic lipase (LIPH_RAT). The LIPL_HUMAN and LIPH_HUMAN sequences containing the known disease-related mutations have been annotated with an asterisk (*). The LIPP_HUMAN sequence has been annotated with secondary structure features.[7]

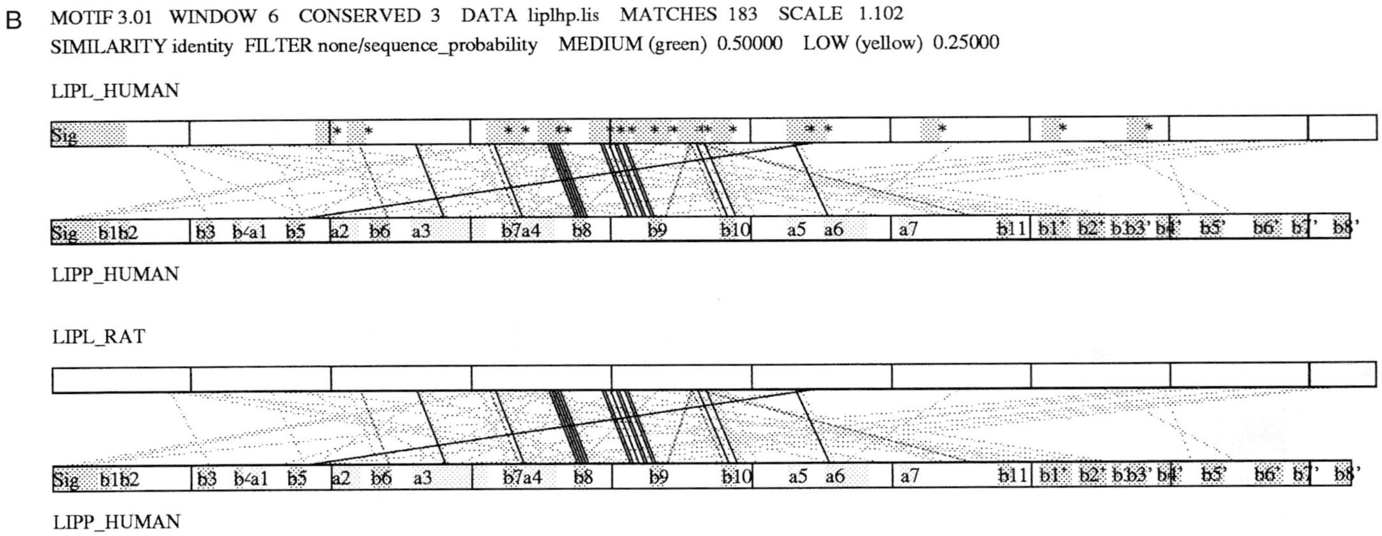

FIG. 3. (*continued*)

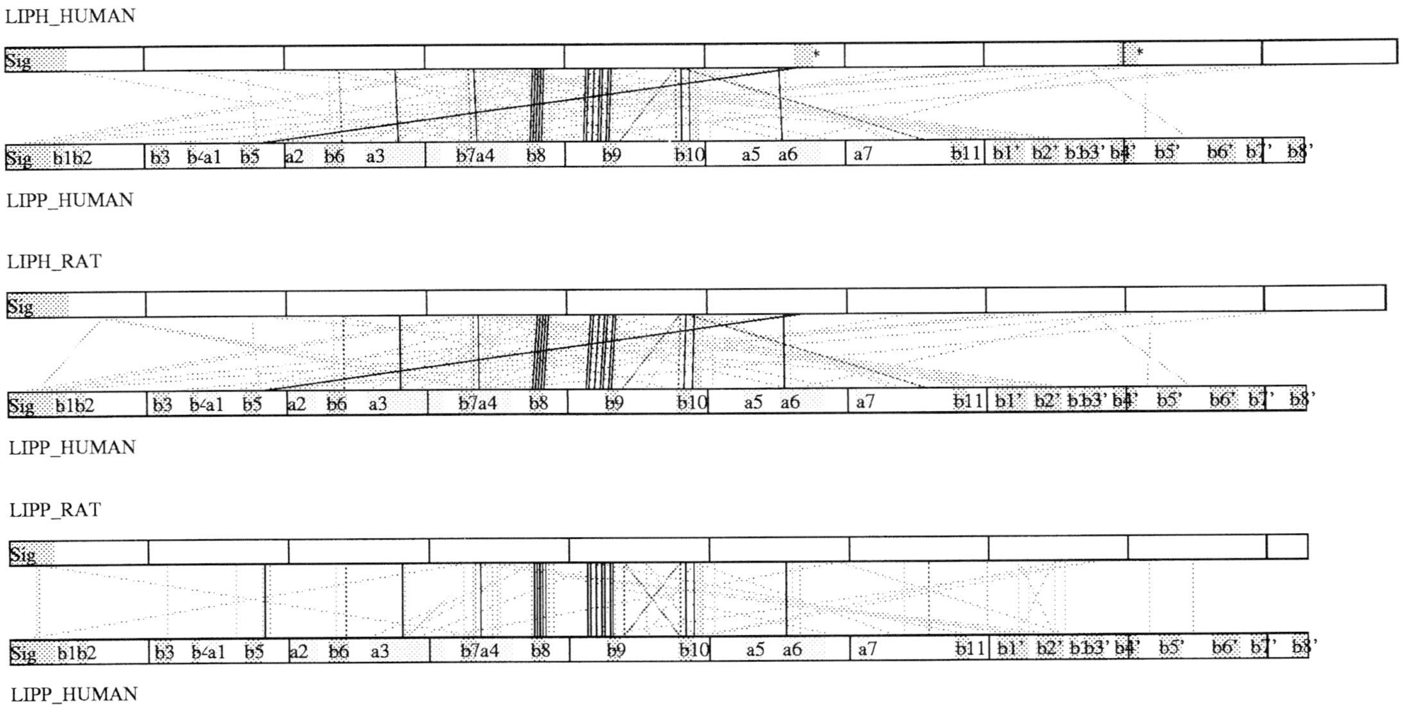

FIG. 3. (*continued*)

TABLE II
DISEASE-RELATED MUTATIONS IN HUMAN LIPOPROTEIN AND HEPATIC LIPASE PRECURSORS[a]

Protein	Mutation
LIPL_HUMAN	R102S, S199C, P234L, W113R, A203T, C243S, H163R, D107E, R270H, G169E, G215E, S271T, G181S, I221T, D277N, D183G, G222E, N318S, D183N, D231E, A361T, P184R, I232S, L392V
LIPH_HUMAN	S289F, T405M

[a] The sequence numbers refer to the precursor sequence. The sizes of the LIPL_HUMAN and LIPH_HUMAN precursors are 475 and 499 amino acids, respectively. The disease-related mutations are the ones listed in the Swiss-Prot database.

sequences. The sequences differ in origin, from human to *Drosophila melanogaster,* yet a consistent pattern of conservation emerges in this analysis. The *Drosophila* sequence displays an approximate 20-amino acid gap in the range ~135–150. The presence of the gap can be deduced from the line patterns in that range, which display groups of parallel lines on both sides of the insertion, but the two groups of lines are not parallel with each other. The chicken sequence reveals a highly surprising feature: a 155-amino acid insertion. Considering that the chicken must be regarded as a closer relative to humans than *Drosophila* or the nematode is (*Caenorhabditis elegans,* ACE1_CAEEL), it is remarkable that neither *Drosophila* nor the nematode sequence has any indication of any insertion in this range. They are in fact both similar to human acetylcholinesterase.

Another type of enzyme that carries the lipase descriptor in its name is bile salt-activated lipase. From its sequence relationship to the esterase family it should be named an esterase.[8] It presents a beautiful example of sequence repeats in the C-terminal region. In the bovine sequence no repeats of this motif are found; in rat 4 repeats are found, and in human 16 repeats of length 11 amino acids are found. In Fig. 5A, these three sequences have been studied with MULTIM and they align well, displaying high-sequence homology as judged by the density of connecting lines. However, in the C-terminal part (after 550 amino acids) a new line pattern appears: A single range in the bovine sequence displays connecting lines to multiple locations in the human sequence, thus revealing the location of the many repeats in the human sequence. In the rat sequence, in which four repeats are present, the connecting line pattern becomes more complex and multiple lines criss-cross between repeat locations in the two sequences.

[8] H. W. Anthonsen, A. Baptista, F. Drablos, P. Martel, S. B. Petersen, M. Sebastiao, and L. Vaz, *in* "Biotechnology Annual Review," vol. 1 (M. R. El-Gewely, ed.), pp. 315–371. Elsevier Science, New York, 1995.

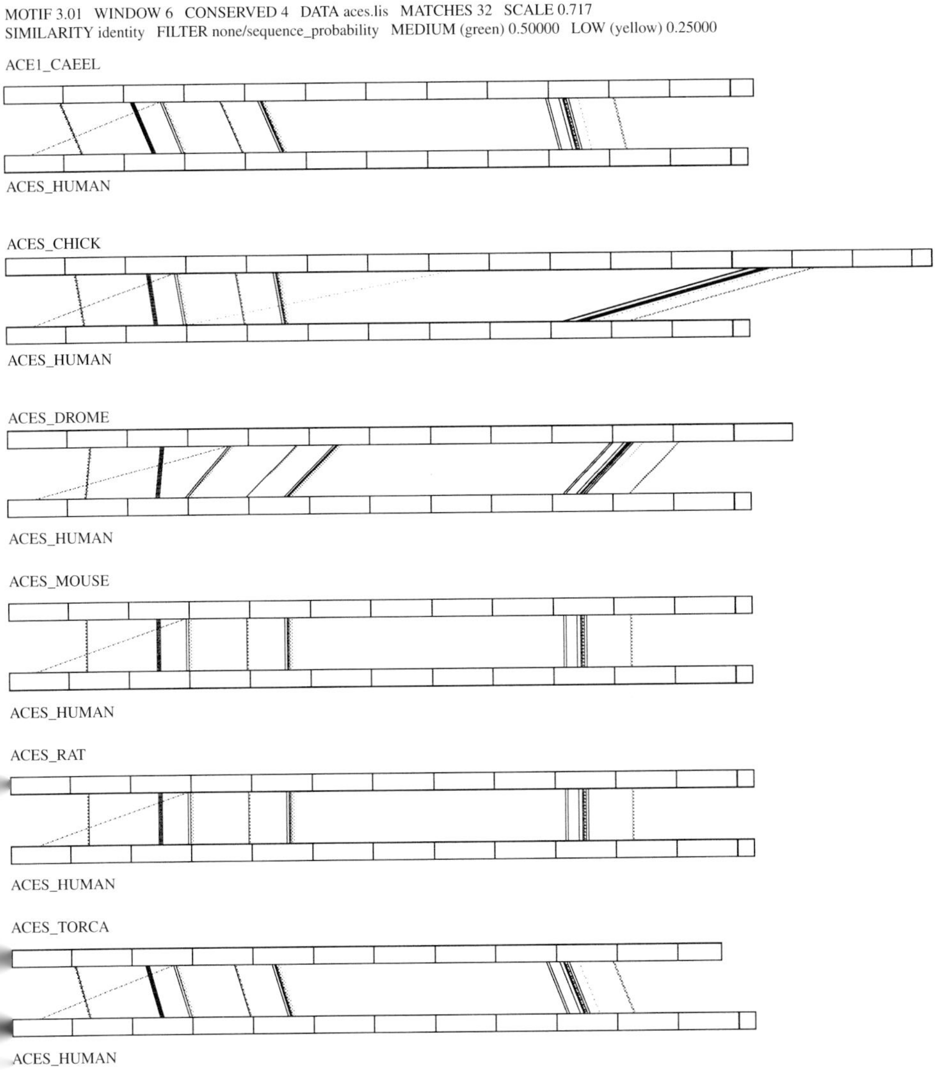

FIG. 4. Subset of seven sequences from the MULTIM alignment of nine acetylcholinesterase precursors. A window size of six has been used, with four residues conserved. No additional MOTIF options were implemented. All sequences were extracted from the Swiss-Prot database and the Swiss-Prot codes have been used. The species involved are *Caenorhabditis elegans* (ACE1_CAEEL), *Homo sapiens* (ACES_HUMAN), *Gallus gallus* (ACES_CHICK), *Drosophila melanogaster* (ACES_DROME), *Mus musculus* (ACES_MOUSE), *Rattus norvegicus* (ACES_RAT), *Torpedo californica* (ACES_TORCA), *Torpedo marmorata* (ACES_TORMA, data not shown), and *Bos taurus* (ACES_BOVIN, data not shown).

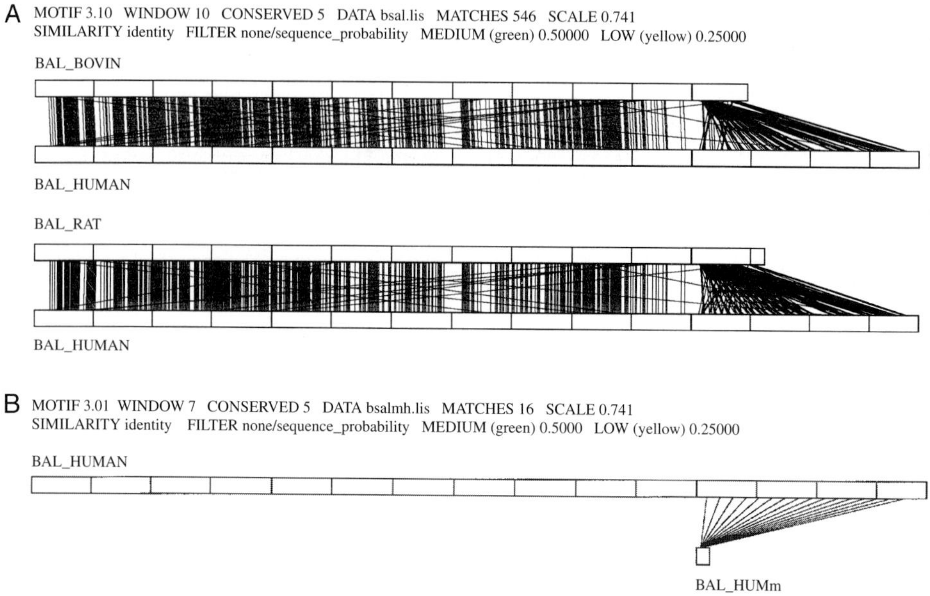

FIG. 5. (A) MULTIM analysis of three bile salt-activated lipase precursors. A window size of 10 has been used, with 6 residues conserved. No additional MOTIF options were implemented. All sequences were extracted from the Swiss-Prot database and the Swiss-Prot codes have been used. The protein precursors shown are as follows: bovine bile salt-activated lipase (BAL_BOVIN, from *Bos taurus*), human bile salt-activated lipase (BAL_HUMAN), and the rat bile salt-activated lipase (BAL_RAT, from *Rattus norvegicus*). Four and 16 sequence repeats are reported in the C-terminal parts of the BAL_RAT and BAL_HUMAN sequences, respectively. (B) MULTIM analysis of the human bile salt-activated lipase (BAL_HUMAN) against a consensus motif (xxxPVxPTxDx, where *x* represents any amino acid) found in the C terminus. A window size of seven has been used, with five residues conserved. No additional MOTIF options were implemented. The BAL_HUMAN sequence was extracted from the Swiss-Prot database and the Swiss-Prot code has been used.

In Fig. 5B, we have extracted a single such 11-amino acid repeat and searched with this repeat against all possible positions in the BAL_HUMAN sequence. To allow for some evolutionarily induced variations in the sequence repeats, five residues were conserved using a window size of seven residues. It is clear from Fig. 5B that 16 repeats have been identified using this approach.

In Fig. 6, another MULTIM example for the visualization of internal sequence repeats is shown. The case shown is the nodule-25 protein from *Medicago sativa,* which in this case has been plotted against itself. The vertical grid pattern of lines represent trivial identity. This trivial identity

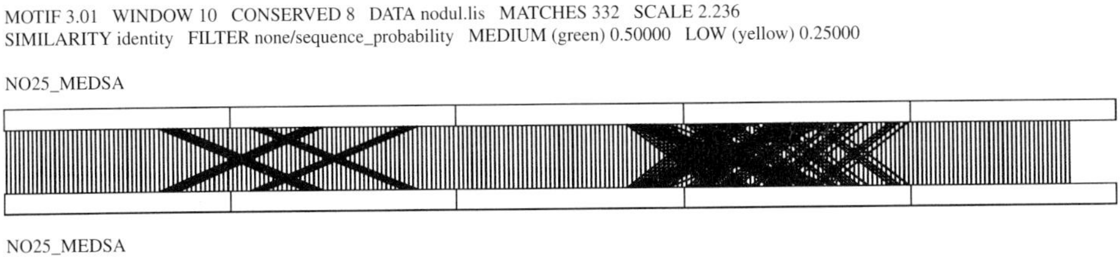

FIG. 6. Identification of internal repeats in a protein precursor sequence (nodulin-25 precursor) using MULTIM. A window size of 10 residues has been used, with 8 residues conserved. No additional MOTIF options were implemented. All sequences were extracted from the Swiss-Prot database and the Swiss-Prot codes have been used. The investigated species is *Medicago sativa* (NO25_MEDSA).

can be suppressed by using the -i option when running MOTIF. Strong X patterns appear in the region 140–200; most connecting lines display either a forward slope or a backward slope of the same magnitude. From the anchor points these lines have on the two sequence bars one can deduce that the length of the sequence repeat corresponds to 18 amino acids. The amino acid composition of this repeat can subsequently be identified by manual inspection of the ⟨*.pro⟩ file.

In Fig. 7, seven sequences belonging to the cutinase family of enzymes have been aligned. Despite the fact that they pairwise display 49–55% identity for the mature sequences, indicating that they constitute a close family of enzymes, only a relatively small number of identical motifs are retained when the whole family is studied with MULTIM. In the present run, a window size of six residues was used, with three residues conserved. The secondary structure for the *Fusarium solani pisi* cutinase has been used for annotation of the CUTI_FUSSO sequence. It is seen from Fig. 7 that most highly significant connecting lines join positions at or close to the edges of secondary structural features such as helices and β strands. This is typical of conserved residues. At lower gray-tone intensity one can observe more than 15 additional connecting lines, parallel to the more intense lines. On the basis of this visual inspection, the positions that these lines are connecting should be included in the cutinase consensus sequence. The sequence location for the active site residues histidine, serine, and aspartate have been annotated in all the sequences. It is clear from Fig. 7 that the location of the active site residues closely follows the connecting line pattern produced by MULTIM, except in the case of the active site histidine. Note also that two of the essential oxyanion hole residues[9] have been identified.

[9] C. Martinez, A. Nicolas, H. van Tilbeurgh, M. P. Egloff, C. Cudrey, R. Verger, and C. Cambillau, *Biochemistry* **33,** 83 (1994).

MOTIF 3.01 WINDOW 6 CONSERVED 3 DATA pcut.lis MATCHES 140 SCALE 2.391
SIMILARITY identity FILTER none/sequence_probability MEDIUM (green) 0.50000 LOW (yellow) 0.25000

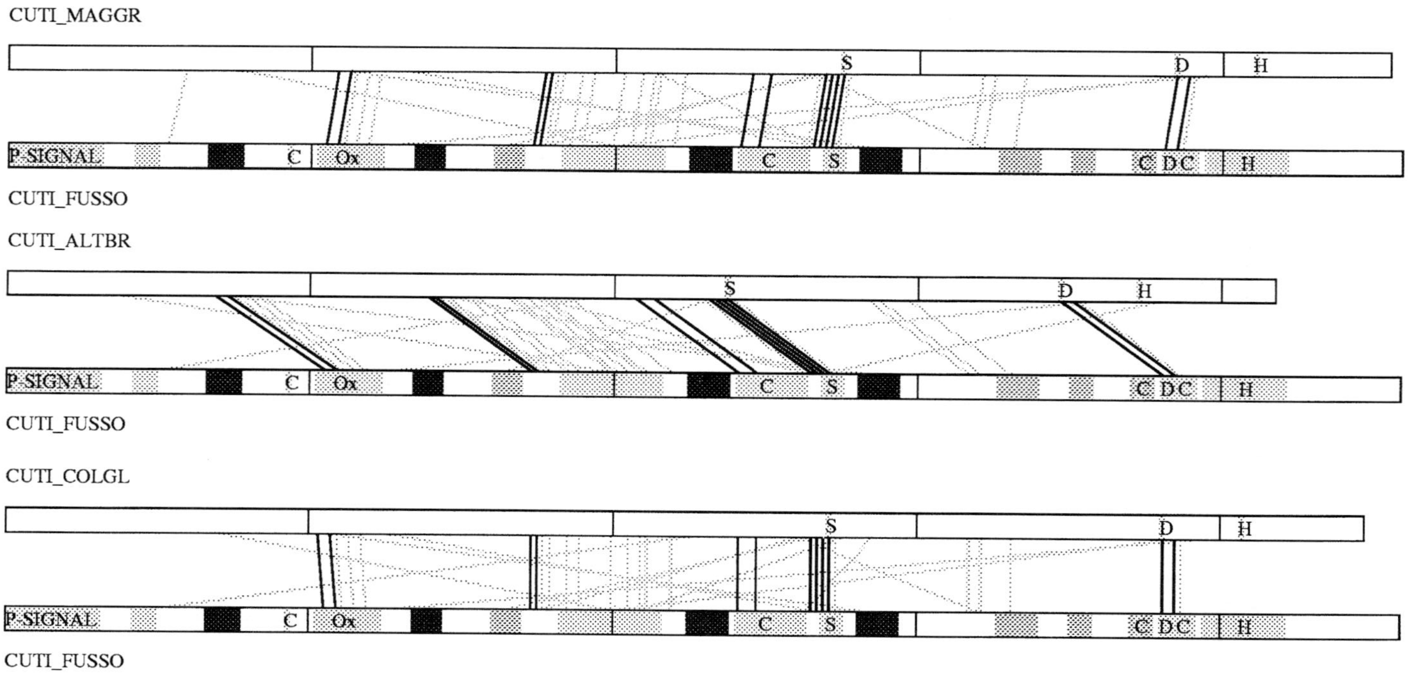

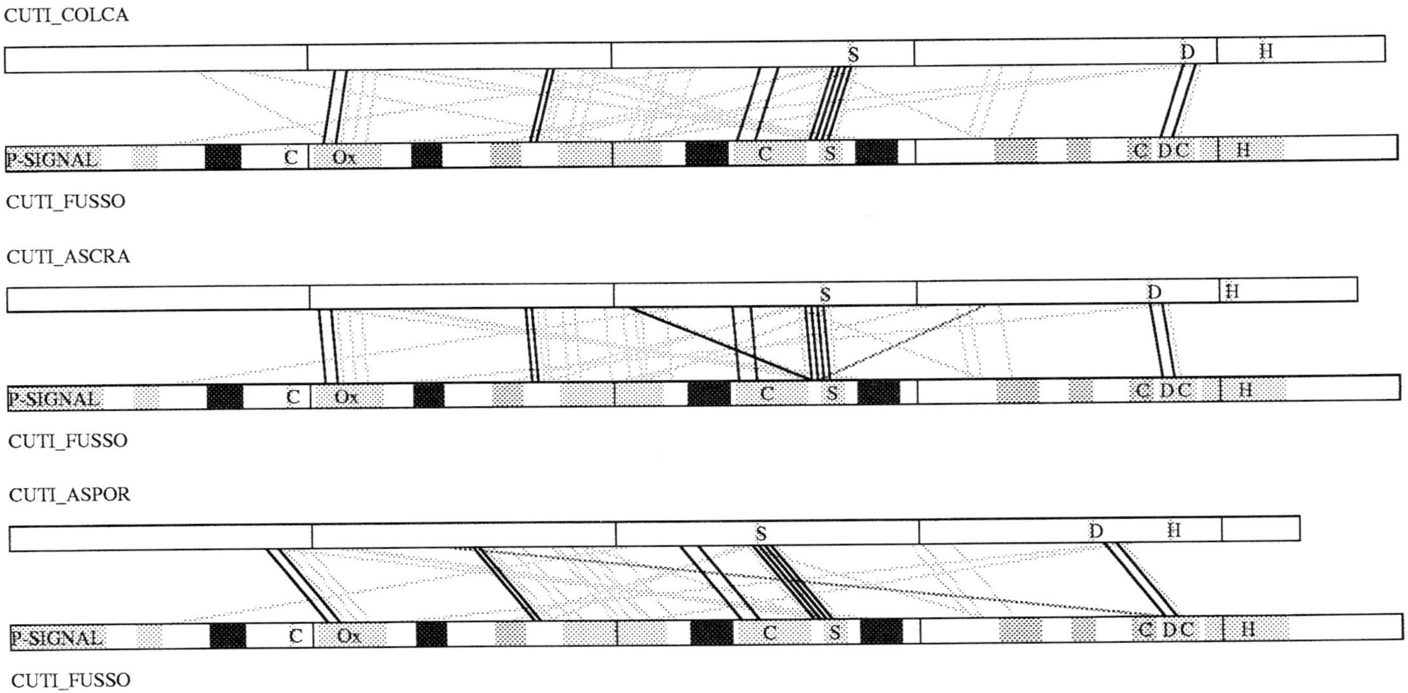

FIG. 7. MULTIM analysis of seven fungal cutinase precursors. A window size of six has been used, with three residues conserved. No additional MOTIF options were implemented. All sequences were extracted from the Swiss-Prot database and the Swiss-Prot codes have been used. The species involved are *Magnaporthe grisea* (CUTI_MAGGR), *Alternaria brassicicola* (CUTI_ALTBR), *Colletotrichum gloeosporioides* (CUTI_COLGL), *Colletotrichum capsici* (CUTI_COLCA), *Ascochyta rabiei* (CUTI_ASCRA), *Aspergillus oryzae* (CUTI_ASPOR), and *Fusarium solani pisi* (CUTI–FUSSO). The CUTI_FUSSO sequence is annotated with its secondary structure features, using the data in the 1cus.hssp file (hssp, homology-derived secondary structure of proteins) available at the World Wide Web (http://swift. embl-heidelberg.de/hssp/). In all comparisons the signal sequence of CUTI_FUSSO is highlighted with "P-Signal." The location of the β strands and helices are shown in black and gray, respectively, in the annotation for CUTI_FUSSO. For all sequences, the location of the active site residues S, D, and H have been highlighted. In the CUTI_FUSSO sequence the location of four cysteine residues involved in S–S bonds are displayed as well. Ox, Oxyanion hole serine.

MOTIF identified both the oxyanion hole serine as well as the glutamine (annotation not shown in Fig. 7), which is located next to the active site serine.

In Fig. 8, six sequences from enzymes of widely different size and origin have been studied using MULTIM. The enzymes belong to two classes of proteins: two citrate lyases and the α chain of four succinate-CoA ligases. The α chain from the succinate-CoA ligases are known to be conserved among different species (about 70% identity has been found between the *Escherichia coli* and rat liver succinate-CoA ligases[10]). This is clearly revealed in the MULTIM output as well. The connecting parallel line pattern is constant for all of the organisms studied. Whereas the latter sequences all have a length of about 300–350 amino acids, both citrate lyases have a length of about 1100 amino acids. The MULTIM output gives unambiguous evidence of a significant homology between the two groups of proteins. It also clearly identifies the local region (~570–720) of the citrate lyase sequences that contain a succinyl-CoA ligase-type domain. Interestingly, by homology the active site, as well as CoA-binding and ATP-binding sites, have been assigned to His-765.[11] Note that the relative location of the active site histidine is conserved in the human citrate lyase and in the *E. coli* succinate-CoA ligase. It is remarkable that apparently no mutations involving larger insertions or deletions have taken place over such a large evolutionary span. As far as we know this homology between the two Krebs cycle-associated enzymes has not been reported elsewhere.

In Fig. 9, a MULTIM alignment of five envelope proteins from different viruses is shown. The alignment reveals a variety of different, interesting features all involving insertions and/or deletions. It is apparent from Fig. 9 that all sequences are closely related, but two of the sequences are significantly shorter. Nevertheless, the C termini of the shorter sequences display a large number of parallel connecting lines to the C terminus of the longer ENV_RMC sequence used as a reference in Fig. 9. The likely interpretation of this observation is that a large, almost 200-amino acid insertion has appeared in ENV_RMC, ENV_MC, and ENV_FL, or alternatively, that an identically sized deletion has been introduced into the two shorter envelope sequences. When observing the last pair of aligned sequences (ENV_FL and ENV_RMC) it is obvious that at least two other insertions have appeared in ENV_FL when compared to ENV_RMC. The first is located in the sequence range ~150–165 and the second in the range

[10] W. D. Henning, C. Upton, G. McFadden, R. Majumdar, and W. A. Bridger, *Proc. Natl. Acad. Sci. U.S.A.* **85,** 1432 (1988).

[11] N. A. Elshourbagy, J. C. Near, P. J. Kmetz, T. N. Wells, P. H. Groot, B. A. Saxty, S. A. Hughes, M. Franklin, and I. S. Gloger, *Eur. J. Biochem.* **204,** 491 (1992).

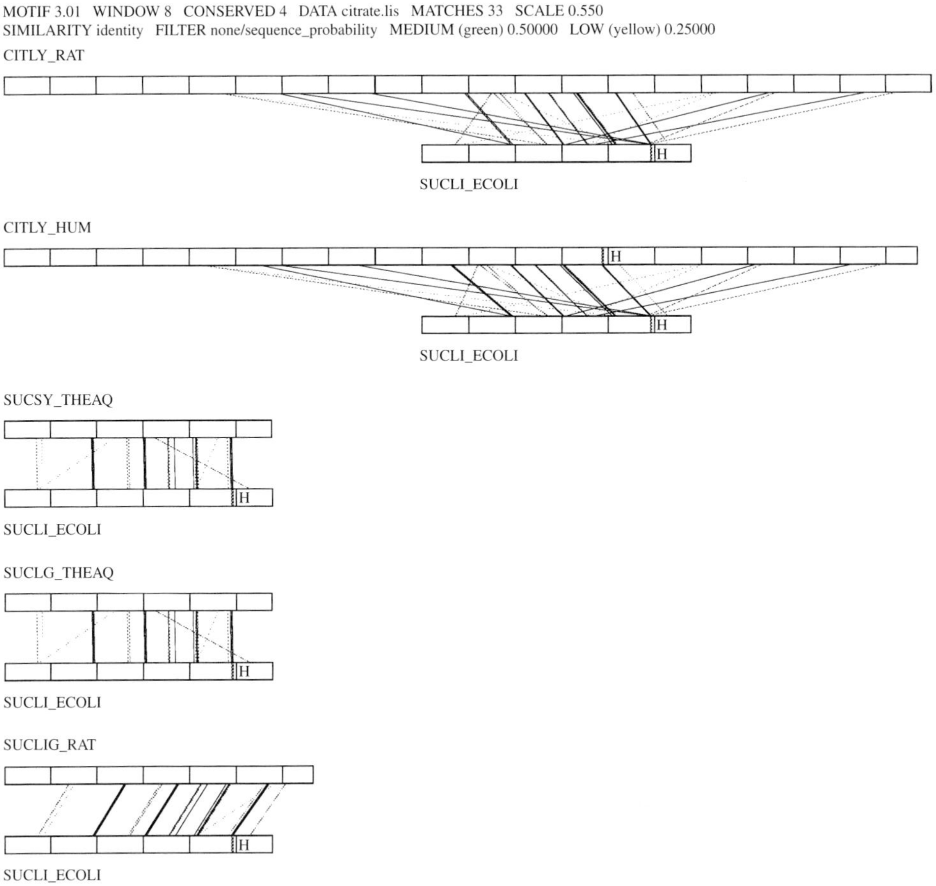

FIG. 8. MULTIM analysis of six evolutionarily related proteins. A window size of eight residues has been used, with four residues conserved. No additional MOTIF options were implemented. The proteins shown are ATP citrate (pro-*S*)-lyase from rat (CITLY_RAT, *PIR ID* A35007) truncated at position 1 to 100, succinate-CoA ligase (ADP-forming) α chain from *Escherichia coli* (SUCLI_ECOLI, *PIR ID* SYECSA), ATP citrate (pro-*S*)-lyase from human (CITLY_HUMA, *PIR ID* S21173) truncated at position 1 to 120, succinyl-CoA synthetase from *Thermus aquaticus* (SUCSY_THEAQ, *PIR ID* S12138), succinate-CoA ligase (ADP-forming) α chain from *Thermus aquaticus* (SUCLG_THEAQ, *PIR ID* S15951), and succinate-CoA ligase (GDP-forming) α chain from rat (SUCLIG_RAT, *PIR ID* A28962). The annotated histidine (H) in the SUCLI_ECOLI sequence is a catalytic negatively charged phosphorylated histidine.

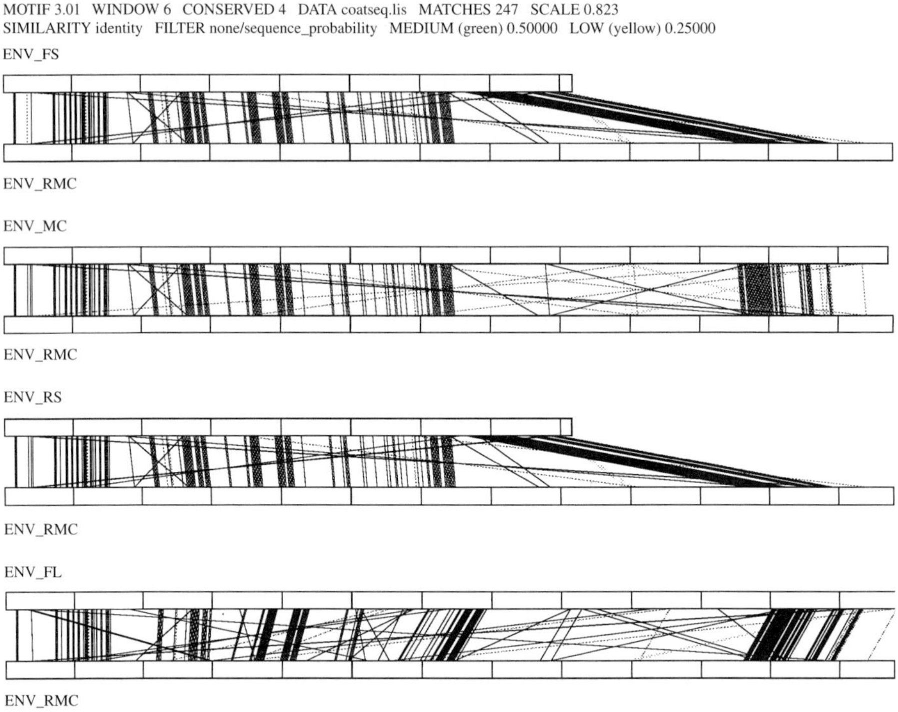

FIG. 9. MULTIM analysis of five retrovirus envelope proteins. A window size of six residues has been used, with four residues conserved. No additional MOTIF options were implemented. The envelope proteins shown belong to the following virus: Friend spleen focus-forming virus (ENV_FS, *PIR ID* A21170), mink cell focus-forming virus (ENV_MC, *PIR ID* VCVWFS), Rauscher mink cell focus-forming virus (ENV_RMC, *PIR ID* VCMVRV), Rauscher spleen focus-forming virus (ENV_RS, *PIR ID* VCMVSR), and feline leukemia virus (ENV_FL, *PIR ID* VCMVFP).

~250–280. By manual inspection of the ⟨*.pro⟩ file it is found that the regions identified by MULTIM are extremely well conserved, thus it is no surprise that other multialignment programs, such as CLUSTALW, perform well in this case, despite the large gap in the C-terminal half of the sequence.

In Fig. 10, the alignment between four different phosphatidylinositol phosphatases from rat is shown. It is apparent from Fig. 10 that a ~400-amino acid insertion of Sarc[12] homology domains 2 and 3 has taken place in PIP4_RAT and PIP5_RAT sequences when comparing to the other

[12] C. Schumacher, B. S. Knudsen, T. Ohuchi, P. P. Di Fiore, R. H. Glassman, and H. Hanafusa, *J. Biol. Chem.* **270,** 15341 (1995).

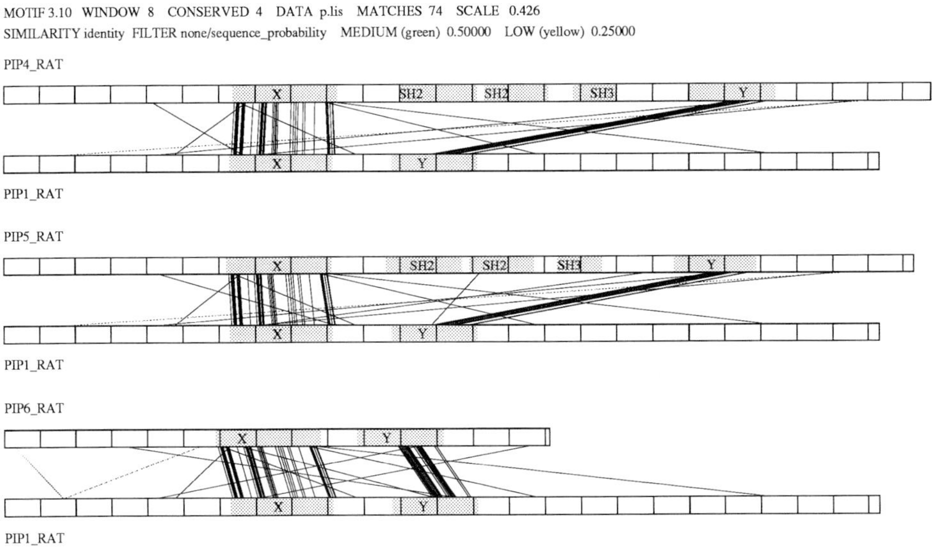

FIG. 10. MULTIM analysis of four 1-phosphatidylinositol-4,5-bisphophate phosphodiesterase from rat. A window size of eight residues has been used, with four residues conserved. No additional MOTIF options were implemented. All sequences were extracted from the Swiss-Prot database and the Swiss-Prot codes have been used. The protein precursors shown are 1-phosphatidylinositol-4,5-bisphophate phosphodiesterase γ_1 (PIP4_RAT, 1290 amino acids), 1-phosphatidylinositol-4,5-bisphophate phosphodiesterase β_1 (PIP1_RAT, 1216 amino acids), 1-phosphatidylinositol-4,5-bisphophate phosphodiesterase γ_2 (PIP5_RAT, 1265 amino acids), and 1-phosphatidylinositol-4,5-bisphosphate phosphodiesterase δ_1 (PIP6_RAT, 756 amino acids). All four sequences are annotated with *X* and *Y*. These annotations refer to domains essential for catalytic activity. In addition, the sequences PIP4_RAT and PIP5_RAT are annotated with SH2 and SH3, which refer to the Sarc homology domains present in these proteins (phosphorylatable domains).

sequences. MULTIM clearly identifies the homology between distant regions in the sequences as judged by the two sets of parallel connecting lines joining the sequences pairwise. If the same sequences are analyzed with CLUSTALW, the large insertion is not found because the homology between the different C termini is much less pronounced than in the case of the viral envelope sequences in Fig. 9.

Limitations and Future Developments of MULTIM

In the preceding sections we have presented the many positive features of MULTIM. But as with so many other tools, it is not perfect for all types of problems. We point out here some of the most significant limitations that we know of.

Motif Alignment by MULTIM

In its present version MULTIM presents graphically the alignment of motifs found by MULTIM. Because motifs contain at the least two residues each, and most often three or more, the graphic alignment is intrinsically coarse grained. If it is essential that the alignment be completed at the single-residue level, the user will have to do it manually. We are seeing various possibilities for improving this through a multistage process, wherein MULTIM first completes the graphically displayed alignment, and then the user selects the most trustworthy connecting lines, following which MULTIM proceeds to align at the single-residue level all other positions.

Sequence Repeat Problems for MULTIM

If a sequence contain sequence repeats or recurrent motifs, a given motif in such a repeat will be found multiple times by MULTIM. This may not cause problems unless a sufficient number of sequences contain such repeats, because the exhaustive nature of the MULTIM search algorithm will result in a combinatorial explosion of potential motifs that MULTIM will try to test. In Fig. 11, this is illustrated. All sequences in this example contain three recurrent motifs. The first motif is tested against all possible locations in the second sequence. Three hits are found. For each of these three motifs are found in the third sequence. Thus for every motif found in the first sequence 12 linked motifs will be evaluated. The larger the total number of such recurrent motifs is in a given batch of sequences, the more significant this problem becomes. At some stage the embedded maximum number of motifs that MULTIM can handle will be exceeded. Currently this is around 1000. This problem can be reduced by increasing the number

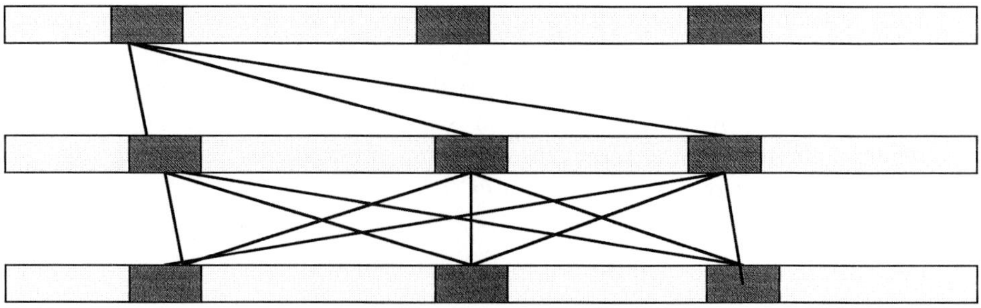

FIG. 11. Combinatorial explosion. Shown here are three sequences, each of which has three recurrent motifs (in dark gray). When MULTIM selects a possible motif in the first sequence, it will find (at least) three matching motifs in sequence 2, and for each of these three in sequence 3.

of conserved residues in the search. This will work if the size of the motif exceeds the size of the recurrent motif, and therefore includes some of the neighboring residues that are not conserved. A special case of recurrent motifs is simple amino acid repeats that, e.g., occur in storage proteins such as gliadin from wheat.

Acknowledgments

The work described in this chapter was supported in part by the European Commission (BIO2-CT94-3016) and the Norwegian Research Council (29345/213). E. I. P. was supported by the Austrian Bundesministerium für Wissenschaft Forchnung and Kunst, Grant Number GZ 558.009/141-IV/A/5a/95. M. T. N. P. acknowledges support from the Portuguese JNICT (Junta Nacional de Investigação Cientifica e Tecnológica), Program PRAXIS XXI, Project PRAXIS/2/2.1/BIO/34/95.

[4] Lipases and α/β Hydrolase Fold

By Joseph D. Schrag and Miroslaw Cygler*

Introduction

Since the pioneering work of Anfinsen (reviewed in Ref. 1) demonstrated that the amino acid sequence of a protein contains all of the necessary information to fold the protein chain into a compact molecule, the quest to decipher the "protein-folding code" has attracted the attention of biologists, chemists, and physicists alike. Much has been learned about the forces and influences that govern the process of protein folding, but the key to understanding the folding process and the ability to predict accurately the tertiary structure of a protein solely from its amino acid sequence have remained elusive. The advent of modern molecular biology techniques has resulted in an explosion of available protein sequence information. Despite the exponential increase in the number of three-dimensional protein structures determined by X-ray crystallography and, more recently, by nuclear magnetic resonance (NMR) methods, the number of known sequences far exceeds the number of known structures. This excess of known amino acid sequences over known three-dimensional structures promises to continue

[1] C. B. Anfinsen, *Science* **181,** 223 (1973).

to increase in parallel with the advances in the efforts to sequence entire genomes, underscoring the importance of developing methods for identifying structural motifs from sequence information.

The number of unique protein folds that should be expected from the ~50,000 protein sequences in the Swiss-Prot database is difficult to estimate, but the ~4200 current entries in the Protein Data Bank[2] have been estimated to represent only a few percent of the total number of unique folds.[3] Comparison and classification of known three-dimensional structures into unique protein folds have been the subject of numerous studies and a number of different classification schemes have resulted.[4–9] Several different protein classification databases based on superposition of three-dimensional structures have been established. The structural classification of proteins developed by Murzin *et al.*[10] categorizes proteins primarily on the basis of visual inspection of the structure, aided by several automated search methods. This classification divides the known three-dimensional structures in the Protein Data Bank (PDB) into five major classes of protein: all α, all β, α/β with mostly parallel β strands, α/β with mostly antiparallel β strands, and multidomain proteins. Proteins having greater than 30% sequence identity, implying a common evolutionary origin, are grouped together in a protein family. Families are also identified by similar structures and functions despite lower sequence identity. This classification of proteins can be accessed on the Internet at http://scop.mrc-lmb.cam.ac.uk/scop. Another structure-based classification of proteins is the FSSP database established by Holm *et al.*[6] This database was produced by automated comparison of two-dimensional distance matrices calculated from the three-dimensional structures.[11] All of the entries in the Protein Data Bank have been compared in an all-against-all pairwise manner. The proteins are then clustered on the basis of a specific scoring function. The FSSP database can be accessed at http://www.sander.embl-heidelberg.de.

Comparison and analysis of the known three-dimensional protein structures demonstrate two points clearly. First, proteins with 30% sequence

[2] F. C. Bernstein, T. F. Koetzl, G. J. B. Williams, E. F. Meyer, Jr., M. D. Brice, J. R. Rodgers, O. Kennard, T. Shimaouchi, and M. Tasumi, *J. Mol. Biol.* **112,** 535 (1977).

[3] C. A. Orengo, D. T. Jones, and J. M. Thornton, *Nature (London)* **372,** 631 (1994).

[4] M. Levitt and C. Chothia, *Nature (London)* **261,** 552 (1976).

[5] J. Richardson, *Adv. Protein Chem.* **34,** 167 (1981).

[6] L. Holm, C. Ouzounis, C. Sander, G. Tuparev, and G. Vriend, *Protein Sci.* **1,** 1691 (1992).

[7] C. A. Orengo, T. P. Flores, W. R. Taylor, and J. M. Thornton, *Protein Eng.* **6,** 485 (1993).

[8] J. P. Overington, Z. Y. Zhu, A. Sali, M. S. Johnson, R. Sowdhamini, C. Louie, and T. L. Blundell, *Biochem. Soc. Trans.* **21,** 597 (1993).

[9] D. P. Yee and K. A. Dill, *Protein Sci.* **2,** 884 (1993).

[10] A. G. Murzin, S. E. Brenner, T. Hubbard, and C. Chothia, *J. Mol. Biol.* **247,** 536 (1995).

[11] L. Holm and C. Sander, *J. Mol. Biol.* **233,** 123 (1993).

identity or more always adopt similar tertiary folds. This means that the general features of three-dimensional structures of some of the 50,000 protein sequences known can be deduced by comparison with homologs whose three-dimensional structures have been determined. This homology modeling has been practiced for many years. Sander and Schneider[12] have created the HSSP database, which is a compilation of sequence database searches for all of the known three-dimensional structures in the PDB. For each known structure one can find a sequence alignment with other proteins with similar sequences and, therefore, similar folds. The second observation arising from comparison of the known three-dimensional protein structures is that similar folds are sometimes adopted by proteins with no obvious sequence similarity. Some examples indicate that even in the difficult situation of a sequence with little similarity to any protein whose structure is known, comparison of 3D–1D profiles (a one-dimensional representation of a protein that contains some information about the three-dimensional structure) can provide indications of similarities in tertiary structures.[13,14]

The α/β hydrolase fold is one example of a tertiary fold that is adopted by proteins displaying no sequence similarity. The first description of this fold was based on comparison of the three-dimensional structures of five different enzymes, one of which is a lipase. A number of lipase structures have now been determined and, despite a lack of sequence similarity, all have proved to be members of this fold family. Numerous other studies on lipases and related enzymes provide examples of the identification of other likely α/β hydrolase proteins, using a variety of methods including crystallographic structure determination, sequence comparisons, secondary structure predictions, and 3D–1D profile comparisons. With emphasis on lipases, we review here the features of this fold and the resources used to identify similarities in the rapidly growing number of enzymes and proteins that share this fold.

Features of Canonical α/β Hydrolase Fold

The α/β hydrolase fold belongs to the doubly wound α/β superfold[3,5] and is 1 of 55 different folds in the SCOP classification of mostly parallel α/β structures.[10] This fold was first described after visual inspection of the three-dimensional structures of five enzymes, dienelactone hydrolase,[15] haloalkane dehalogenase (1ede[16]), wheat serine carboxypeptidase II

[12] C. Sander and R. Schneider, *Proteins* **9,** 56 (1991).

[13] J. U. Bowie, R. Lüthy, and D. Eisenberg, *Science* **253,** 164 (1991).

[14] R. N. Alon, L. Mirny, J. L. Sussman, and D. L. Gutnick, *FEBS Lett.* **371,** 231 (1995).

[15] D. Pathak and D. Ollis, *J. Mol. Biol.* **214,** 497 (1990).

[16] S. M. Franken, H. J. Rozeboom, K. H. Kalk, and B. W. Dijkstra, *EMBO J.* **10,** 1297 (1991).

(1wht[17]), acetylcholinesterase (1ace[18]), and the lipase from *Geotrichum candidum* (1thg[19]), revealed a common fold with unique features.[20] Although all five of the enzymes catalyze hydrolysis reactions, the substrates hydrolyzed are quite diverse. Only two of the five enzymes, acetylcholinesterase and the *G. candidum* lipase, showed enough sequence similarity to suggest that their structures were similar.[21] On the basis of superposition and comparison of these five structures, the conserved elements of this new fold were first delineated.[20]

The active site of each of the five enzymes is composed of a catalytic triad similar to those first observed in serine proteases. However, α/β hydrolase fold enzymes can be distinguished from the trypsin, papain, and subtilisin families of proteases by the linear order of catalytic triad residues. In α/β hydrolase fold proteins the linear sequence is always nucleophile, acid, histidine compared to His–Asp–Ser for the trypsin family, Cys–His–Asn for the papain family, and Asp–His–Ser for the subtilisin family.[20] A viral cysteine protease was shown to have a His–Glu–Cys triad and may represent a fifth enzyme family containing a stable active site triad.[22] The identities of the triad residues are, however, more variable in α/β hydrolase fold enzymes than in the protease families. Serine, aspartate, and cysteine have all been identified as catalytic nucleophiles in α/β hydrolase fold enzymes, whereas serine is required in serine proteases[23] and cysteine is required in the papain family if significant activity is to be maintained.[24] In addition, α/β hydrolase fold enzymes have provided the first examples of glutamate rather than aspartate taking the role of the catalytic acid.[18,19]

The α/β hydrolase fold has a core composed of a central, mostly parallel β sheet (Fig. 1). Using the nomenclature of Richardson,[5] the canonical connectivity of the eight β strands follows the topology $+1$, $+2$, $-1x$, $+2x$, $(+1x)_3$. The β sheet has a left-handed superhelical twist and the first and last strands cross at an approximate angle of 90° to one another.[20] The twist of the β sheet differs significantly among the various enzymes. Helices pack on either side of the β sheet. Helices A and F pack against the concave

[17] D.-I. Liao and S. J. Remington, *J. Biol. Chem.* **265,** 6528 (1990).

[18] J. L. Sussman, M. Harel, F. Frolow, C. Oefner, A. Goldman, L. Toker, and I. Silman, *Science* **253,** 872 (1991).

[19] J. D. Schrag, Y. Li, S. Wu, and M. Cygler, *Nature (London)* **351,** 761 (1991).

[20] D. L. Ollis, E. Cheah, M. Cygler, B. Dijkstra, F. Frolow, S. M. Franken, M. Harel, S. J. Remington, I. Silman, J. D. Schrag, J. L. Sussman, K. H. G. Verschueren, and A. Goldman, *Protein Eng.* **5,** 197 (1992).

[21] A. R. Slabas, J. Windust, and C. M. Sidebottom, *Biochem. J.* **269,** 279 (1990).

[22] J. Ding, W. J. McGrath, R. M. Sweet, and W. F. Mangel, *EMBO J.* **15,** 1778 (1996).

[23] J. N. Higaki, L. B. Evnin, and C. S. Craik, *Biochemistry* **28,** 9256 (1989).

[24] P. I. Clark and G. Lowe, *Eur. J. Biochem.* **84,** 293 (1978).

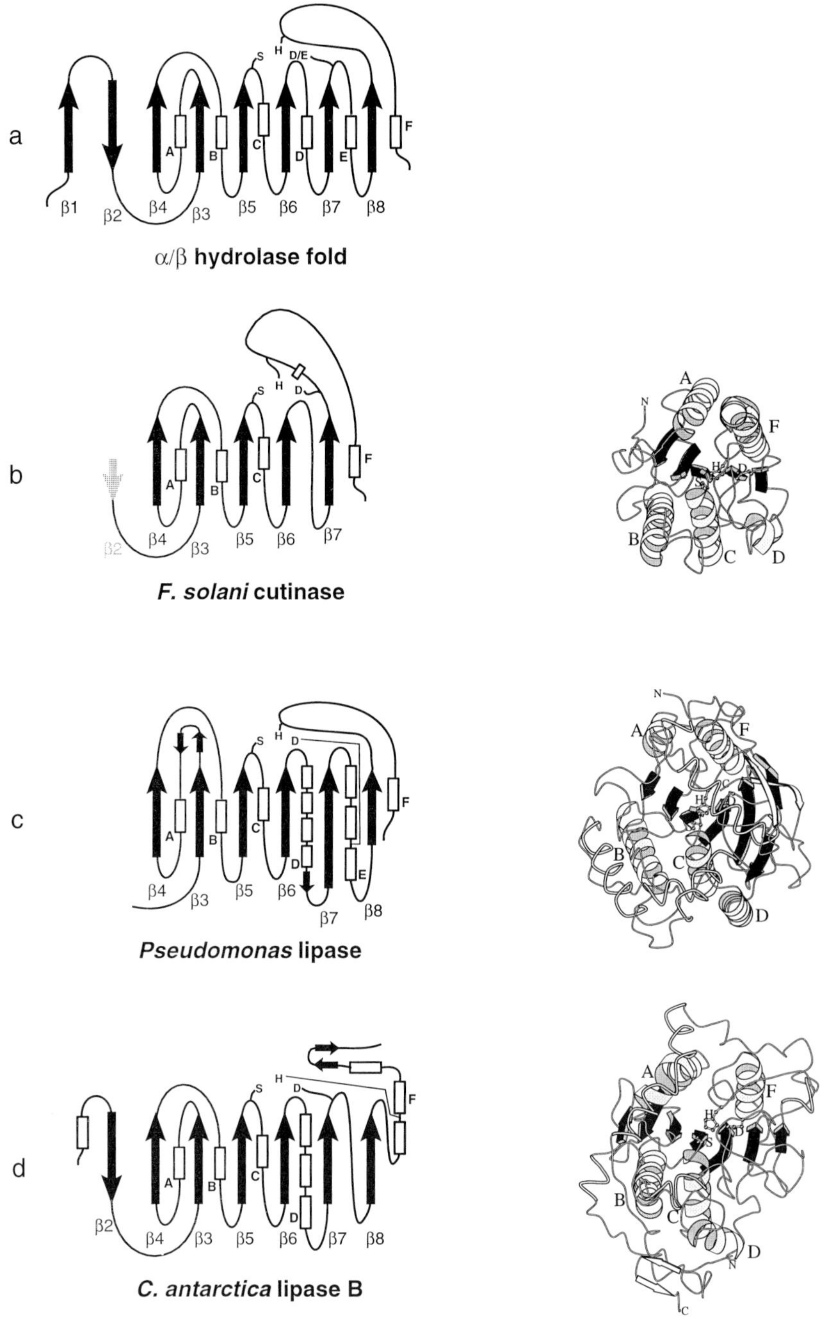
a
β1
β2
β4
β3
β5
β6
β7
β8
α/β hydrolase fold
b
F. solani cutinase
c
Pseudomonas lipase
d
C. antarctica lipase B

Fig. 1

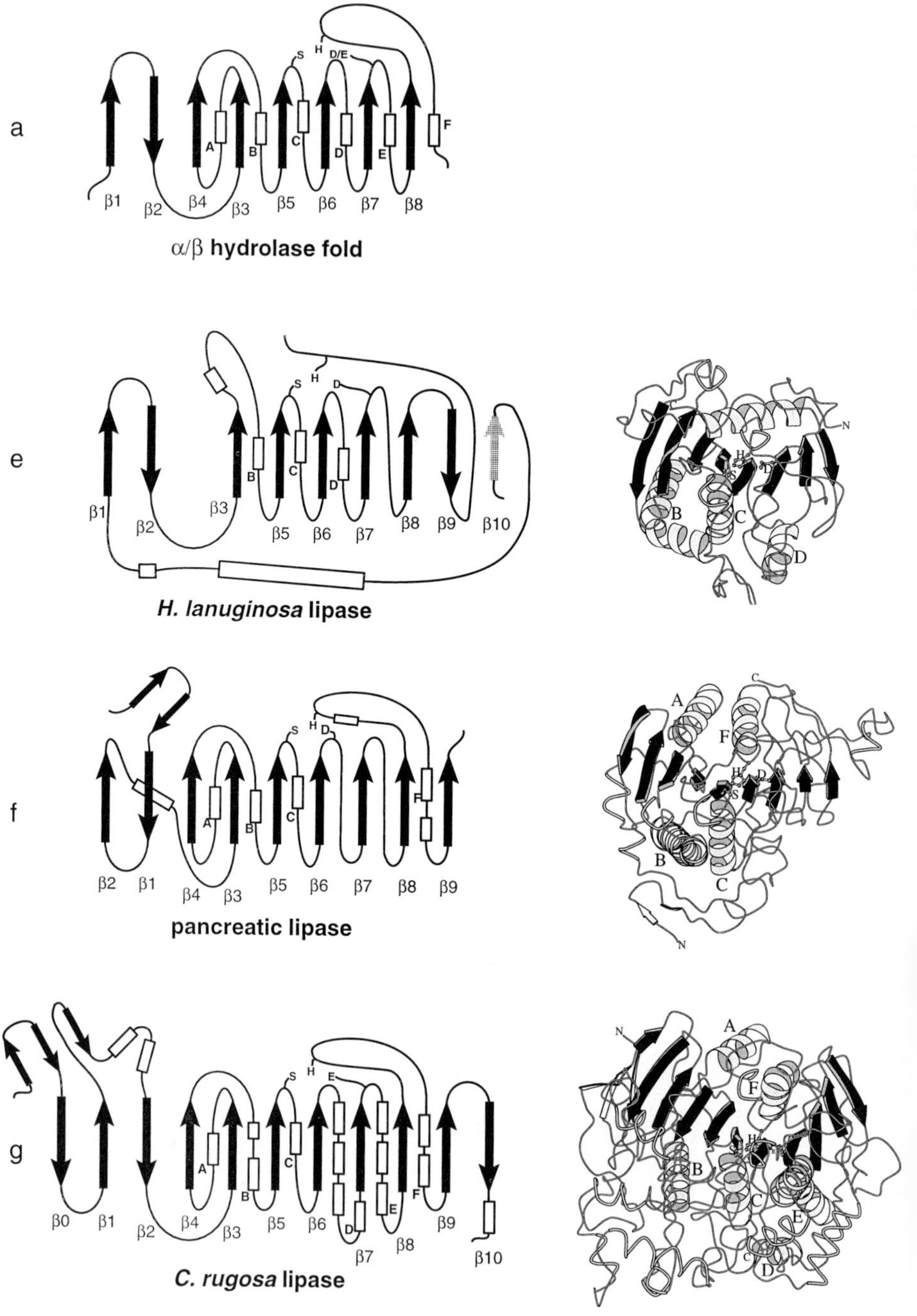

FIG. 1 (*continued*)

side of the β sheet and helices B–E pack against the convex face of the β sheet. The positions of the helices in three dimensions vary considerably from one enzyme to another despite their topological equivalence. The superposition of the α/β hydrolase fold enzymes based on the central β sheet in general gives poor alignment of the corresponding α helices. Helix C, which forms part of the nucleophile elbow, is the exception to this generality. The central position of this helix and its importance in positioning the catalytic nucleophile require the position of this helix to be highly conserved among all of the α/β hydrolase fold enzymes.

Nucleophile Elbow

The catalytic site is located at the C-terminal ends of the β strands. The catalytic nucleophile is located in a highly conserved pentapeptide, G–X–S–X–G. This motif has often been referred to as a consensus sequence, but some enzymes have sequences around their catalytic nucleophiles that do not conform entirely to this pentapeptide sequence.[20,25] As mentioned above, the nucleophile is serine, aspartate, or cysteine and the glycine residues at Nu−2 and Nu+2 are sometimes substituted for other small amino acids including alanine and serine.[26] A true consensus might better be described as Sm–*X*–Nu–*X*–Sm. This pentapeptide forms a tight turn in a strand–turn–helix motif, which forces the nucleophile to adopt unusual main-chain ϕ, ψ torsion angles of $-50°$, $-130°$. The strand involved in this motif is generally near the center of the β sheet and is strand $\beta 5$ in the canonical fold. For consistency with this terminology, the

[25] S. Brenner, *Nature* (*London*) **334,** 528 (1988).

[26] D. M. Lawson, U. Derewenda, L. Serre, S. Ferri, R. Szittner, Y. Wei, E. A. Meighen, and Z. S. Derewenda, *Biochemistry* **33,** 9382 (1994).

FIG. 1. Canonical α/β hydrolase fold, and the folds of six lipases representative of all of the known three-dimensional structures of lipases. On the left is the schematic representation of the fold. Numbering of β strands and naming of α helices are consistent with Ollis *et al.*[20] Locations of the catalytic residues are marked with small-size letters S, H, and D/E. To the right of each fold is a schematic drawing of a representative lipase. Only β strands from the main sheet are colored black and only the helices common to the α/β hydrolase fold are shown in ribbon representation with letters next to them. The catalytic triad residues are shown in larger letters. All of the structures are shown on the same scale to allow for an easy visual comparison of the relative sizes of lipases from various families. (a) The canonical α/β hydrolase fold. (b) Cutinase; the N terminus is in an extended conformation and can be considered an equivalent of strand $\beta 2$. (c) Group 1 *Pseudomonas* lipases. (d) *Candida antarctica* lipases B. (e) *Humicola lanuginosa* lipase (note that strand $\beta 10$, marked in gray, does not occur in all lipases in this family). (f) Pancreatic lipase. (g) *Candida rugosa* lipase.

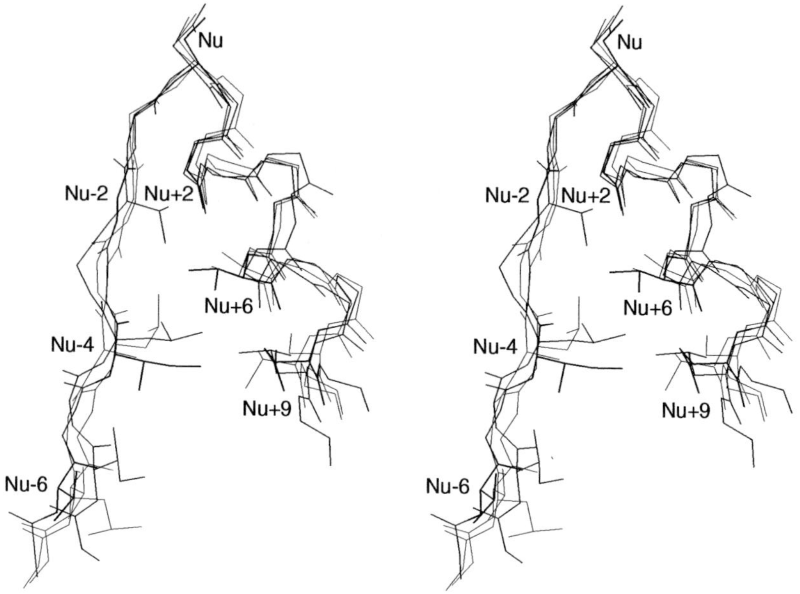

FIG. 2. Superposition of the nucleophile elbows of four representatives of α/β hydrolase fold enzymes. The four enzymes are dienelactone hydrolase, haloalkane dehalogenase, cutinase, and *G. candidum* lipase. The side chains involved in packing of the β strand and α helix are shown in full, numbered with regard to the position of the catalytic nucleophile Nu.

strand involved in this motif is numbered $\beta5$ throughout this text for all of the structures discussed. Helix C is the helix of the strand–turn–helix motif. This "nucleophile elbow" is the most conserved feature of the α/β hydrolase fold. Steric restrictions in forming this turn require a small side chain or no side chain at positions Nu−2 and Nu+2 and in most cases these residues are glycine (Fig. 2). The Nu+3 residue must also be small to prevent steric conflicts with strand $\beta4$ and usually is either glycine or alanine. The structural constraints of this motif were further analyzed by Derewenda and Derewenda,[27] using sequence comparison of various lipases and esterases with the three-dimensional structure of the *Rhizomucor miehei* lipase. This analysis showed that residues Nu−4 and Nu−6 of the β strand are generally hydrophobic residues and generally have small side chains. The Nu+6 residue is also a small hydrophobic residue, often valine. These residues are structurally constrained because they are involved in hydrophobic interactions and packing of the β strand against the α helix (Fig. 2).

[27] Z. S. Derewenda and U. Derewenda, *Biochem. Cell. Biol.* **69,** 842 (1991).

At the C-terminal end of strand $\beta6$ the main chain takes a sharp bend of approximately 90°, which moves the main chain away from the catalytic nucleophile at the end of strand $\beta5$. The residue located at this bend is either glycine or a residue with a small side chain. The C-terminal end of strand $\beta7$ extends beyond the end of strand $\beta6$ by at least one residue, and this extension combined with the sharp change in the main-chain direction at the end of $\beta6$ allows approach of the triad histidine and acid side chains to assemble the catalytic triad and allow attack on the substrate by the nucleophile.[20]

Catalytic Acid and Histidine

The catalytic acid is located in a loop following strand $\beta7$ of the β sheet and is found in a reverse turn. The hydrogen-bonding stabilization of the acid side chain differs depending on whether the acid is aspartate or glutamate.[20,28] In both cases, one oxygen atom of the carboxylate group adopts a similar position and hydrogen bonds to the histidine imidazole nitrogen. If aspartate is the catalytic acid, the stabilizing hydrogen bonds between main-chain or side-chain atoms to the other carboxylate oxygen of the catalytic acid are on one side of the plane of the histidine imidazole ring containing enzymes, and when glutamate is the catalytic acid these stabilizing hydrogen bonds are on the opposite side. The catalytic histidine is located in a loop following strand $\beta8$. The length and conformation of the histidine-containing loop are variable.

Handedness of Catalytic Triad

The triad handedness of the α/β hydrolase fold enzymes is opposite to that of the serine proteases of the trypsin and subtilisin families. The handedness of the triad defines the stereoselectivity of the hydrolysis reaction catalyzed and again will be opposite to that of the serine proteases. The configurations of the catalytic triads of a α/β hydrolase fold enzyme, the lipase from *Candida rugosa* (PDB code 1lpm), and a serine protease, chymotrypsin (1acb), are compared in Fig. 3. The phosphonyl inhibitor covalently bound to the lipase provides an analog of the ester bond hydrolyzed by the lipase and the noncovalently bound chymotrypsin inhibitor, eglin *c*, provides a view of the amide bond hydrolyzed by the protease. These two complexes are shown after superposition of the three atoms of the scissile bonds: the carbonyl oxygen, the carbonyl carbon (a phosphorus atom in the covalent complex with the lipase), and the ester oxygen or

[28] J. D. Schrag, T. Vernet, L. Laramеé, D. Y. Thomas, A. Recktenwald, M. Okoniewska, E. Ziomek, and M. Cygler, *Protein Eng.* **8,** 835 (1995).

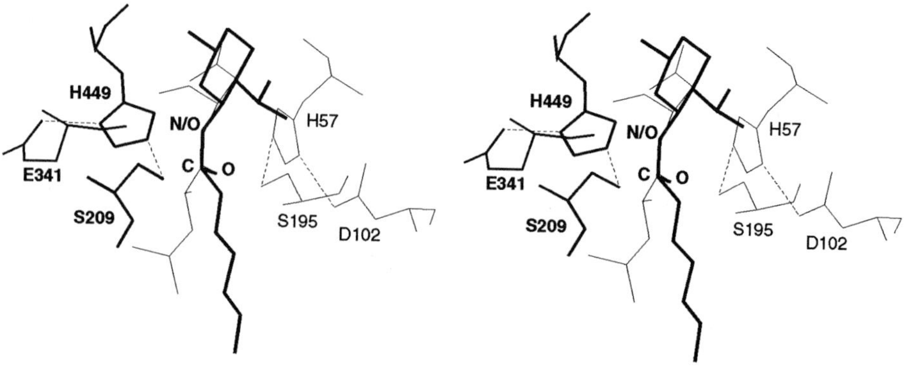

FIG. 3. Superposition of enzyme–inhibitor complexes of *C. rugosa* lipase and chymotrypsin, demonstrating the opposite-handedness of the triad. The superposition was based on the three atoms defining the scissile bond: the carbonyl C and O and the N/O of the amide/ester bond hydrolyzed. The catalytic triads of these enzymes are approximately mirror images, with the mirror plane being the plane of the scissile bond.

amide nitrogen. The triads are approximate mirror images, with the scissile bond lying on a pseudo-mirror plane. It is apparent that the nucleophilic attack of lipases on the ester bond of the substrate occurs on the opposite face of the scissile bond relative to the attack on the amide by the serine proteinases (Fig. 3).

Lipases and α/β Hydrolase Fold

Since the first description of the α/β hydrolase fold based on five enzymes, the structures of additional members of this fold family, many of them lipases, have been determined. All lipases whose three-dimensional structures are known are α/β proteins and have a central, mostly parallel β sheet. The conserved pentapeptide G–*X*–S–*X*–G containing the active site serine is located in a "nucleophile elbow." Among these lipases, the linear sequences of the catalytic residues is invariably Ser–acid–His. The triad handedness is the same in all lipases and corresponds to that of the canonical α/β hydrolase fold. These features clearly establish these lipases as members of the α/β hydrolase fold family of enzymes. All of the lipases whose three-dimensional structures have been solved conform in large part to the α/β hydrolase fold features, but some variations from the fold as originally characterized do occur. Their folds are described in detail below and are shown schematically in Fig. 1.

Cutinase

Cutinase is capable of hydrolyzing lipids and is, thus, considered to be a lipase, although it does not show the interfacial activation that was long believed to be characteristic of true lipases. This enzyme from *Fusarium solani* is the smallest of the lipolytic enzymes whose structures have been determined and its central β sheet has only five parallel strands.[29] Maintaining a numbering scheme consistent with that used by Ollis *et al.*,[20] these strands are the equivalents of strands $\beta3$–$\beta7$ of the α/β hydrolase fold and the connectivity is $-1x$, $+2x$, $+1x$, $+1x$ (Fig. 1). The antiparallel $\beta2$ equivalent in cutinase is distorted and makes only two hydrogen bonds typical of antiparallel β structures. Cutinase has topological equivalents to helices A, B, C, D, and F of the α/β hydrolase fold. The position of helix D is shifted to a position in three-dimensional space between that usually observed for helices D and E. Following strand $\beta7$ is a distorted single helical turn that lies in the expected position of helix E. Because there is no strand $\beta8$, both the catalytic aspartate and histidine follow strand $\beta7$. Helix F, which in most α/β hydrolase fold enzymes follows $\beta8$, is found near the C terminus of the protein at the end of a long loop following $\beta7$.

Pseudomonas Lipases

Two different gene families of lipases have been identified in *Pseudomonas* (referred to as group 1 and group 2 lipases[30]). Amino acid sequence comparisons show significant similarity only in the region of the consensus pentapeptide containing the active site serine. The structures of several members of the group 1 lipases have been determined.[31–33] The central β sheets of the *Pseudomonas glumae, Pseudomonas cepacia,* and *Chromobacterium viscosum* lipases are one strand larger than that of cutinase and contain strands equivalent to strands $\beta3$–$\beta8$ of the α/β hydrolase fold. All of these strands are parallel and the connectivity is $-1x$, $+2x$, $(+1x)_3$, identical to that originally described for the equivalent strands of the canonical α/β hydrolase fold enzymes. All of the expected helices are present, except that helix E is represented by a single helical turn. The crystal

[29] C. Martinez, P. DeGues, M. Lauwereys, G. Matthyssens, and C. Cambillau, *Nature (London)* **356,** 615 (1992).

[30] J. Gilbert, *Enzyme Microb. Technol.* **15,** 634 (1993).

[31] M. E. M. Noble, A. Cleaseby, L. N. Johnson, M. R. Egmond, and L. G. J. Frenken, *FEBS Lett.* **331,** 123 (1993).

[32] D. Lang, B. Hoffmann, L. Haalck, H.-J. Hecht, F. Spener, R. D. Schmid, and D. Schomburg, *J. Mol. Biol.* **259,** 704 (1996).

[33] J. D. Schrag, Y. Li, M. Cygler, D. Lang, T. Burgdorf, H.-J. Hecht, D. Schomburg, T. Rydel, J. Oliver, L. Strickland, M. Dunaway, S. Larson, and A. McPherson, *Structure* **5,** 187 (1997).

structures of these lipases reveal an aspartate residue that, as expected, is located after strand $\beta 7$ and makes a hydrogen bond to the histidine residue of the catalytic triad. On the basis of these structures one would conclude that this is the catalytic acid. Surprisingly, however, mutation of this residue in *P. glumae* has shown that this aspartate is not essential for catalytic activity.[34] The three-dimensional structures of these enzymes suggest a possible explanation for this observation. A glutamate residue, located after strand $\beta 8$ and after the histidine in the linear sequence, may also be able to hydrogen bond to the catalytic histidine and, in the absence of the aspartate residue, provide the negative charge required for activity.

Candida antarctica Lipase B

Another fungal lipase whose three-dimensional structure has been reported is that of *Candida antarctica*.[35] The central β sheet of this lipase is two strands larger than that of cutinase and contains strands equivalent to strands $\beta 2$–$\beta 8$ of the canonical α/β hydrolase fold (Fig. 1). The first of the strands is antiparallel to the other six. The connectivity of these strands is identical to that of the canonical α/β hydrolase fold, $+2, -1x, +2x, (+1x)_3$. Helices equivalent to helices A, B, C, D, and F are present in *C. antarctica* lipase, but helix E is missing. The amino acids of the catalytic triad are located after strands $\beta 5$, $\beta 7$, and $\beta 8$ as observed in the canonical fold.

Rhizomucor miehei and Related Lipases

The *Rhizomucor miehei* lipase is representative of a group of related fungal lipases, which includes those of *Rhizopus delemar, Humicola lanuginosa,* the mono- and diglyceride lipase from *Penicillium camembertii,*[36–38] and the triglyceride lipase from *Penicillium* sp. strain UZLM 4 (Ref. 39; and J. D. Schrag and M. Cygler, unpublished results, 1997). The topologies

[34] L. G. Frenken, M. R. Egmond, A. M. Batenburg, J. W. Bos, C. Visser, and C. T. Verrips, *Appl. Environ. Microbiol.* **52,** 3787 (1992).

[35] J. Uppenberg, M. T. Hansen, S. Patkar, and T. A. Jones, *Structure* **2,** 293 (1994).

[36] L. Brady, A. Brzozowski, Z. S. Derewenda, E. Dodson, G. Dodson, S. Tolley, J. P. Turkenburg, L. Christiansen, B. Huge-Jensen, L. Norskov, L. Thim, and U. Menge, *Nature (London)* **343,** 767 (1990).

[37] U. Derewenda, L. Swenson, R. Green, Y. Wei, G. Dodson, S. Yamaguchi, M. J. Haas, and Z. S. Derewenda, *Nat. Struct. Biol.* **1,** 36 (1994).

[38] U. Derewenda, L. Swenson, Y. Wei, R. Green, P. M. Kobos, R. Joerger, M. J. Haas, and Z. S. Derewenda, *J. Lipid Res.* **35,** 524 (1994).

[39] K. Gulomova, E. Ziomek, J. D. Schrag, K. Davranov, and M. Cygler, *Lipids* **31,** 379 (1996).

of these lipases are identical. The central β sheets of these lipases contain nine strands. The seven strands centered around $\beta5$ are equivalent in position and direction to strands $\beta1$–$\beta7$ of the canonical α/β hydrolase fold. However, the connectivity between these strands in Richardson terminology is $+1$, $+1$, $(+1x)_4$, compared to the usual connectivity of $+1$, $+2$, $-1x$, $+2x$, $(+1x)_3$. This topology suggests that strand $\beta4$ was lost in these lipases (Fig. 1). The chain following the positional equivalent of strand $\beta8$ of the α/β hydrolase fold carries the catalytic histidine of this lipase, but this segment includes an antiparallel β strand that is added to the central β sheet. Following this additional strand, the loop that holds the catalytic histidine has no regular secondary structure. The lipases in this group have helices equivalent to helices B, C, and D,[20] but lack equivalents for helices A, E, and F. Helices A and F of the canonical fold both pack against the same side of the central β sheet, with the helix axis roughly parallel to the strands of the sheet. In the *R. miehei* and related lipases, helices A and F are replaced by one long helix that packs against this side of the β sheet with the helix axis roughly perpendicular to the β strands.

Pancreatic Lipases

The pancreatic lipases are two-domain proteins.[40] The N-terminal domain of the pancreatic lipases is the catalytic domain and the C-terminal domain binds the colipase, a cofactor required for activity in the presence of bile salts. The central β sheet of the catalytic domain is nine stranded.[40,41] The first eight strands of this sheet occupy positions and have directions equivalent to those of the canonical fold, but the connectivities of the first two strands differ. In pancreatic lipases the connectivity is -1, $+3x$, $-1x$, $+2x$, $(+1x)_3$ (Fig. 1). The connection between strands $\beta8$ and the additional strand $\beta9$ is a rarely observed left-handed crossover. Helices equivalent to helices A, B, C, and F are present in pancreatic lipases.

An important deviation from the canonical fold in the pancreatic lipases is the location of the catalytic acid residue. This acid in pancreatic lipase is located in a loop following strand $\beta6$, rather than after strand $\beta7$ as in other $\alpha\beta$ hydrolase fold enzymes. This change in position does not affect the handedness of the catalytic triad, which remains the same as in other lipases.

[40] F. K. Winkler, A. D'Arcy, and W. Hunziker, *Nature* (*London*) **343,** 771 (1990).

[41] Y. Bourne, C. Martinez, B. Kerfelec, D. Lombardo, C. Chapus, and C. Cambillau, *J. Mol. Biol.* **238,** 709 (1994).

Geotrichum candidum and Candida rugosa Lipases

Multiple lipase genes have been identified in both *Geotrichum candidum*[42–44] and *Candida rugosa* (formerly *cylindracea*).[45] The isozymes in each of these species show at least 80% sequence identity and the *G. candidum* and *C. rugosa* lipases are about 45% identical in sequence. The three-dimensional structures of one *G. candidum* isozyme[19,46] and two *C. rugosa* isozymes[47,48] have been reported. These lipases are the largest lipases whose structures are known. Their central β sheets have 11 strands, of which the central 8 strands constitute the α/β hydrolase fold with the canonical topology. There are numerous helices in addition to helices A–F expected in the α/β hydrolase fold. Despite the additional strands and helices, these lipases maintain all of the features of the canonical fold. In addition, the connection between strand $\beta 8$, the last strand of the canonical fold, and strand $\beta 9$ is left handed in these enzymes, as in the pancreatic lipases. Unique among the known lipases, the catalytic acid in these enzymes is glutamate rather than aspartate.

Fold Versatility

Superposition of the catalytic domain of human pancreatic lipase (HPL) with the lipase from *G. candidum* lipase (GCL) showed that HPL has an aspartate residue located after strand $\beta 7$ whose C_α atom superimposes on that of the catalytic Glu-354 of GCL.[49] The side chain of this aspartate residue (Asp-205) is directed toward the surface of the protein and in the crystal structure is hydrogen bonded to a water molecule. Rotation of the side chain around the χ_1 angle places the carboxyl group of the aspartate into a position where it could participate in the catalytic triad, although the geometry and distance are not ideal for a hydrogen bond to the histidine residue. Conversely, the residue in GCL that occupies the position equiva-

[42] Y. Shimada, A. Sugihara, Y. Tominaga, T. Iizumi, and S. Tsunasawa, *J. Biochem.* **106,** 383 (1989).

[43] Y. Shimada, A. Sugihara, T. Iizumi, and Y. Tominaga, *J. Biochem.* **107,** 703 (1990).

[44] M. C. Bertolini, L. Laramée, D. Y. Thomas, M. Cygler, J. D. Schrag, and T. Vernet, *Eur. J. Biochem.* **219,** 119 (1994).

[45] M. Lotti, R. Grandori, F. Fusetti, S. Longhi, S. Brocca, A. Tramontano, and L. Alberghina, *Gene* **124,** 45 (1993).

[46] J. D. Schrag and M. Cygler, *J. Mol. Biol.* **230,** 575 (1993).

[47] P. Grouchulski, Y. Li, J. D. Schrag, F. Bouthillier, P. Smith, D. Harrison, B. Rubin, and M. Cygler, *J. Biol. Chem.* **268,** 12843 (1993).

[48] D. Ghosh, Z. Wawrzak, V. Z. Pletnev, N. Li, R. Kaiser, W. Pangborn, H. Jornvall, M. Erman, and W. L. Duax, *Structure* **3,** 279 (1995).

[49] J. D. Schrag, F. K. Winkler, and M. Cygler, *J. Biol. Chem.* **267,** 4300 (1992).

lent to that of the catalytic aspartate of HPL is a serine residue (Ser-249). Modeling studies indicated that an aspartate in this position and replacement of the catalytic glutamate with serine could produce a stable triad with a geometry similar to that of HPL.[49]

A double mutant *G. candidum* lipase, S249D/E354S, in which the catalytic acid was moved to a position after strand $\beta6$, retained at least 10% of the wild-type enzymatic activity.[28] The crystal structure was also determined and showed that the only structural differences between the wild-type enzyme and the double mutant were localized to the active site region where the mutations were introduced. The geometry of the mutant catalytic site was, indeed, similar to that of HPL and the positions of the side chains were almost exactly as the modeling studies predicted. Sequence comparisons show that hepatic and lipoprotein lipases are close relatives of the pancreatic lipase.[50] In these lipases the catalytic aspartate is in the same position as observed in the pancreatic lipases. However, these lipases do not have an acid that is equivalent to Asp-205 of the pancreatic lipases. These sequence comparisons and the modeling studies of HPL and GCL were interpreted as evidence that the pancreatic lipases represent an evolutionary intermediate in a move in the position of the catalytic acid. The crystallographic analysis of the double mutant in combination with its catalytic competency lends solid support to the notion that HPL is an intermediate in an evolutionary switch in the position of the catalytic aspartate in the pancreatic, hepatic, and lipoprotein lipase family of lipases. It also underscores the ability of the α/β hydrolase fold to accommodate catalytic sites that vary in terms of the identity and positions of catalytic residues.

Structural Features Common to Lipases

The structural features that are common to all of the lipases represent a subset of the canonical α/β hydrolase fold. They are as follows: (1) a Nu–acid–His linear sequence of catalytic triad residues is present; (2) the nucleophile is located in a strand–turn–helix nucleophile elbow; (3) the central β sheet has at least five consecutive parallel strands; (4) strand $\beta6$ is shorter than strand $\beta7$ and is followed by a sharp change in the direction of the main chain; (5) the triad handedness is opposite to that of the trypsin and subtilisin protease enzyme families; (6) each lipase has an α helix that packs against the convex side of the central β sheet in the crossover between strand $\beta5$ and a preceding β strand. In most cases, the connectivity between strands is conserved, exceptions being the *Rhizomucor* and related lipases and the pancreatic lipases. The catalytic acid usually follows strand $\beta7$. The

[50] M. C. Komaromy and M. C. Schotz, *Proc. Natl. Acad. Sci. U.S.A.* **84,** 1526 (1987).

only known exceptions to this rule are the pancreatic and related lipases and the engineered mutant of *G. candidum* lipase. Helices are generally found in each of the crossover connections between the β strands, but various lipases are lacking one or more of these helices.

Interestingly, the SCOP classification[10] groups cutinase with the flavodoxin-like fold rather than with the α/β hydrolase fold, indicative of the ambiguity that can arise in attempting to determine evolutionary relationships among proteins. The flavodoxins most similar to cutinase have a central β sheet with the same connectivity between the strands but, in stark contrast to cutinase, do not have a supersecondary structure comparable to the "nucleophile elbow." The automated method used in the FSSP database[6,11] clusters cutinase with the other enzymes identified as α/β hydrolase fold enzymes. Although the structural similarity does not necessarily imply an evolutionary relationship between the proteins, similarity in both structure and function is generally taken to imply an evolutionary relationship among proteins. On the basis of these criteria, we presume an evolutionary relationship among the lipases. This method of comparing three-dimensional structures consistently shows that *R. miehei* and related lipases give the poorest overlap with other lipases. Together with the deviations from the canonical fold, this suggests that *R. miehei* and the related lipases of *Rhizopus delemar, H. lanuginosa, P. camembertii,* and *Penicillium* sp. strain UZLM 4 are the most distantly related to other lipases.

Three-Dimensional Structures of Other α/β Hydrolase Fold Enzymes

In the SCOP classification,[10] the α/β hydrolase fold enzymes have been separated into nine families. In addition to the fungal lipases, bacterial lipases, and pancreatic lipases are acetylcholinesterase, serine carboxypeptidase, haloalkane dehalogenase, dienelactone hydrolase, bromoperoxidase A_2, and thioesterases. Three-dimensional structures of α/β hydrolase fold proteins include the bromoperoxidase A_2 from *Streptomyces aureofaciens,*[51] the thioesterase from *Vibrio harveyi,*[26] and several serine carboxypeptidases including carboxypeptidase Y,[52] human protective protein,[53] and the yeast Kex1 carboxypeptidase.[52a]

Around the active site serine the thioesterase has a sequence A–A–S–L–S.[26] Although the residues at positions Nu–2 and Nu+2 are not glycine, but have small side chains, the nucleophile elbow supersecondary structure

[51] H. J. Hecht, H. Sobek, T. Haag, O. Pfeifer, and K.-H. van Pée, *Nat. Struct. Biol.* **1,** 532 (1994).
[52] J. A. Endrizzi, K. Breddam, and S. J. Remington, *Biochemistry* **33,** 11106 (1994).
[52a] B. Shilton, D. V. Thomas, and M. Cygler, *Biochemistry* in press (1997).
[53] G. Rudenko, E. Bonten, A. d'Azzo, and W. G. J. Hol, *Structure* **3,** 1249 (1995).

is maintained. The larger side chains at these positions are accommodated by adjustment of a single main-chain torsion angle, the ϕ angle of residue Nu–1.

The bromoperoxidase A_2 from *S. aureofasciens* shares some sequence similarity with haloalkane dehydrogenase.[51] A number of other bacterial enzymes involved in various detoxification pathways share sequence similarity with the bromoperoxidase A_2 and haloalkane dehydrogenase and are predicted to have the same fold.[54–56]

The three-dimensional structures of four serine carboxypeptidases have now been determined, second only to lipases among the various groups of α/β hydrolase fold enzymes. These enzymes have all of the hallmarks of the α/β hydrolase fold with the caveat that, like the wheat serine carboxypeptidase II, the central β sheet is two strands larger than the canonical fold. The two outermost strands of the central β sheet can be viewed as parts of a long loop between strands that are equivalents of $\beta7$ and $\beta8$ of the α/β hydrolase fold.

Identification of α/β Hydrolase Fold Enzymes by Sequence Comparisons

Since the α/β hydrolase fold was first described on the basis of the three-dimensional structures of five enzymes, the number of proteins and enzymes that have been identified as members of this fold family has increased rapidly and continues to grow. In the SCOP classification, proteins with sequence identity of at least 30% are identified as members of the same fold family. The amino acid sequences in the GenBank and Swiss-Prot databases can readily be scanned for matches by automated methods and multiple sequence alignments of related proteins are easily produced. Many of the enzymes and proteins assigned as members of the α/β hydrolase fold family were identified by sequence similarity to other α/β hydrolase fold enzymes.

Cygler *et al.*[57] used a combination of three-dimensional structural alignment of acetylcholinesterase (AChE) and *G. candidum* lipase, sequence database searches, and automated sequence alignment to identify a large family of lipases and esterases. The sequence alignments obtained by automated sequence search methods were improved by comparison with the

[54] K. Kimbara, T. Hashimoto, M. Fukuda, T. Koana, M. Takagi, M. Oishi, and K. Yano, *J. Bacteriol.* **171,** 2740 (1989).

[55] E. Díaz and K. N. Timmis, *J. Biol. Chem.* **270,** 6403 (1995).

[56] P. C. K. Lau, J. Garnon, D. Labbé, and Y. Wang, *Gene* **171,** 53 (1996).

[57] M. Cygler, J. D. Schrag, J. L. Sussman, M. Harel, I. Silman, M. K. Gentry, and B. P. Doctor, *Protein Sci.* **2,** 366 (1993).

superimposed three-dimensional structures of the two enzymes, providing a more accurate identification of structurally equivalent residues. Among the proteins identified as α/β hydrolase fold proteins by similarity to AChE and GCL are the bile salt-stimuated lipase, cholesterol esterases, and carboxylesterases. This combination of sequence and structural alignments facilitated distinguishing structural versus functional conservation of sequences within this group of enzymes. Highly conserved residues were concentrated in the core of these molecules at positions, often at the ends of strands and helices, important for maintaining the fold and direction of the polypeptide chain. A number of conserved salt bridges that stabilize the positions of loops in substrate-binding regions were also identified. Highly variable positions were primarily on the protein surface and regions that may be involved in substrate recognition. Although most of the related proteins are enzymes whose catalytic triad residues could easily be identified in the sequence alignment, two protein domains lacking catalytic activity were also identified as members of this group. Subsequently, amino acid sequences of other members of this lipase/esterase group have been reported.[45,58]

Sequence homology to carboxypeptidases with known three-dimensional structures identified yet other related carboxypeptidases and allowed analysis of the sequence conservation pattern among them.[59] Included in this group of serine carboxypeptidases are enzymes from both prokaryotes and eukaryotes including plants, microbes, invertebrates, and mammals. Alignments of these sequences on the basis of superposition of the four three-dimensional structures should provide a wealth of information with regard to sequence conservation due to structural constraints in contrast to variable regions related to substrate specificity differences.

Haloalkane dehalogenase and bromoperoxidase A_2 are both involved in bacterial detoxification pathways. Amino acid sequence comparisons of these enzymes with other enzymes involved in these and other detoxification pathways suggest strongly that many enzymes in various detoxification pathways are α/β hydrolase fold enzymes. Examples include 2-hydroxymuconic semialdehyde hydrolase,[54,55] the *Rhodococcus* sp. M5 *bpdF* gene product,[56] and the epoxide hydrolases.[60,61] The secondary structure of haloalkane dehalogenase was compared to the secondary structure prediction of the 2-hydroxymuconic semialdehyde hydrolase and the agreement between

[58] D. R. Gjellesvik, J. B. Lorens, and R. Male, *Eur. J. Biochem.* **226,** 603 (1994).

[59] K. Olesen and K. Breddam, *Biochemistry* **34,** 15689 (1995).

[60] D. B. Jansen, F. Pries, J. van der Ploeg, B. Kazemier, P. Terpstra, and B. Witholt, *J. Bacteriol.* **171,** 6791 (1989).

[61] M. Arand, D. F. Grant, J. K. Beetham, T. Freidberg, F. Oesch, and B. D. Hammock, *FEBS Lett.* **338,** 251 (1994).

prediction and the observed secondary structure reinforces the expectation of a similar fold.[55]

Identification of α/β Hydrolase Fold Features in Lipases with Limited Sequence Similarity

Proteins with common functions sometimes have similar tertiary structures despite little overall sequence similarity or with sequence similarity limited to short stretches around essential catalytic residues. Sequence conservation around essential residues in proteins with common functions suggests structural conservation as well. In the case of lipases, the G–*X*–S–*X*–G pentapeptide has frequently been presumed to contain the catalytic serine residue. In many cases this has been confirmed by either chemical labeling with inhibitors or site-directed mutagenesis or both. Comparison of the linear sequence of triad residues identified by local sequence comparisons combined with site-directed mutagenesis has been used to infer that many other lipases are α/β hydrolase fold enzymes.

The catalytic residues of *Staphylococcus hyicus* lipase have been identified by site-directed mutagenesis as Ser-369, Asp-559, and His-600[62] and the lipase is inactivated by sulfonyl fluoride inhibitors.[63] The linear sequence Ser–Asp–His strongly suggests that this lipase also conforms to the α/β hydrolase fold. Additional support for this possibility is that the sequence around the catalytic serine is consistent with the packing of residues in a nucleophile elbow.[27]

Lingual lipase, gastric lipase, and lysosomal acid lipase have similar amino acid sequences and are expected to have similar structures, but they show little sequence similarity to other lipases.[64] Although these lipases have been shown to be serine hydrolases,[64,65] further characterization is necessary to establish whether or not these enzymes conform to the α/β hydrolase fold. The sequences of these lipases on either side of the G–*X*–S–*X*–G motif expected to carry the catalytic serine fit the extended pattern described for the nucleophile elbow,[27] suggesting that this supersecondary structure is present in these lipases as well. This group of lipases also shares sequence similarity with egg-specific protein from *Bombyx* and

[62] S. Jäger, G. Demleitner, and F. Götz, *FEMS Microbiol. Lett.* **100,** 249 (1992).

[63] M. L. Tjeenk, Y. B. M. Bulsink, A. J. Slotboom, H. M. Verheij, G. H. de Haas, G. Demleitner, and F. Götz, *Protein Eng.* **7,** 579 (1994).

[64] R. A. Anderson and G. N. Sando, *J. Biol. Chem.* **266,** 22479 (1991).

[65] H. Moreau, A. Moulin, Y. Gargouri, J.-P. Noël, and R. Verger, *Biochemistry* **30,** 1037 (1991).

yolk polypeptides from *Drosophila melanogaster*.[66] The functions of these proteins are unknown, but may involve lipid binding.

Hormone-sensitive lipase is the only lipase whose activity is known to be regulated by reversible phosphorylation. The active site serine has been identified as Ser-423 by site-directed mutagenesis,[67] but no further structural characterization has been done. Some sequence similarity to microbial lipases has been identified.[68] The region of similarity includes the G–*X*–S–*X*–G motif and a HG (His–Gly) dipeptide located about 70 residues N terminal to the catalytic serine. This HG dipeptide is found in many, but not all, lipases. When present, it is located at the margins of the lipid-binding pocket.

The platelet-activating factor acetylhydrolase is an intracellular phospholipase A_2 that, unlike the secretory phospholipase A_2, is independent of calcium. Inhibition of the enzyme with diisopropyl fluorophosphate suggested the importance of a serine residue in catalysis.[69] The amino acid sequence of the protein shows a typical G–*X*–S–*X*–G motif. Subsequently a triad of Ser-273, Asp-296, and His-351 was identified by site-directed mutagenesis.[70] The linear sequence of triad residues is that expected of a α/β hydrolase fold enzyme. The spacing between the serine and aspartate is similar to that in pancreatic lipase and on the basis of this spacing it was suggested that the secondary structure of this enzyme in this region might be similar to that of pancreatic lipases.[70] The acetylhydrolase was inferred to be another α/β hydrolase fold enzyme and, like pancreatic lipases, the catalytic acid may follow strand $\beta 6$ rather than $\beta 7$. Structural characterization will be required to confirm these inferences.

The observation that all of the known lipase three-dimensional structures have similar folds leads to the obvious question of whether all lipases have the same fold. There are still, however, many lipases with unrelated sequences that have not been structurally characterized, e.g., the group 2 *Pseudomonas* lipases.[30] The identity of the catalytic serine has been inferred from the presence of the consensus pentapeptide, but the other members of the presumed catalytic triad have not yet been identified. Sequence comparisons have revealed another group of lipases that might differ considerably from the usual α/β hydrolase fold.[71] This group of enzymes includes

[66] Y. Sato and O. Yamashita, *Insect Biochem.* **21,** 495 (1991).

[67] C. Holm, R. C. Davis, T. Østerland, M. C. Schotz, and G. Fredrikson, *FEBS Lett.* **344,** 234 (1994).

[68] D. Langin and C. Holm, *Trends Biochem. Sci.* **18,** 466 (1993).

[69] D. M. Stafforini, S. M. Prescott, and T. M. McIntyre, *J. Biol. Chem.* **262,** 4223 (1987).

[70] L. W. Tjoelker, C. Eberhardt, J. Unger, H. L. Trong, G. A. Zimmerman, T. M. McIntyre, D. M. Stafforini, S. M. Prescott, and P. W. Gray, *J. Biol. Chem.* **270,** 25481 (1995).

[71] C. Upton and J. T. Buckley, *Trends Biochem. Sci.* **20,** 178 (1995).

lipases from *Xenorhabdus luminescens* and *Aeromonas hydrophila,* thioesterase from *Escherichia coli,* and arylesterase from *Vibrio mimicus* in addition to proline-rich proteins from *Brassica napus* and *Arabidopsis thaliana.* The G–*X*–S–*X*–G consensus that contains the catalytic serine and is instrumental in the formation of the strand–turn–helix nucleophile elbow is changed to G–*X*–S–*X*–S and in some of these enzymes is found near the N terminus of the protein rather than near the middle of the protein. Furthermore, the mutation of the second serine of this pentapeptide to glycine to produce the usual G–*X*–S–*X*–G consensus sequence inactivates the enzymes.[72] On the basis of these observations it has been suggested that these lipases may differ considerably from those neutral lipases conforming to the α/β hydrolase fold. Sequence comparisons suggest the existence of aspartate and histidine residues, which could complete a catalytic triad, but the presence of a triad has not been experimentally confirmed. Interestingly, however, the linear sequence of the candidate aspartate and histidine residues suggests a Ser–Asp–His linear sequence as found in α/β hydrolase fold enzymes. Further characterization of these enzymes is required to distinguish these enzymes from or confirm their identity as members of the α/β hydrolase fold family.

Enzymes Other Than Lipases

The conserved pentapeptide G–*X*–S–*X*–G containing the catalytic serine is common not only to lipases and esterases, but also the thioesterases and proteases.[25] Amino acid sequence comparisons identified a group of peptidases that are distinct from the trypsin, subtilisin, and chymotrypsin families of proteases. This group of enzymes includes dipeptidyl-peptidase IV, acylaminoacyl peptidase, and prolyl oligopeptidase.[73,74] The catalytic serine was identified first because of its occurrence in the conserved pentapeptide common to proteases, esterases, and lipases. Searches for conserved histidine and aspartate residues to complete the suspected triad showed some sequence similarity to regions surrounding the triad acids of some lipases, suggesting the identity of the catalytic acid in the peptidases. On the basis of these sequence similarities, a triad similar to that of lipases was suggested. The role of a serine and a histidine in catalysis was confirmed by chemical modification and subsequently the identities of all of the triad

[72] D. L. Robertson, S. Hilton, K. R. Wong, A. Koepke, and J. T. Buckley, *J. Biol. Chem.* **269,** 2146 (1994).

[73] N. D. Rawlings, L. Polgár, and A. J. Barrett, *Biochem. J.* **279,** 907 (1991).

[74] L. Polgár, *FEBS Lett.* **311,** 281 (1992).

residues were confirmed in dipeptidyl-peptidase IV by site-directed mutagenesis.[75] The linear sequence of triad residues, Ser–Asp–His, is consistent with these peptidases being members of the α/β hydrolase fold family. Secondary structure predictions were also suggestive of a fold similar to that of lipases.[76]

Sequence similarity between haloalkane dehalogenase and the epoxide hydrolases is limited to the regions around the catalytic nucleophile and an essential histidine residue of haloalkane dehalogenase.[61] The equivalent residues in epoxide hydrolases were shown to be essential for activity by both chemical modification and site-directed mutagenesis.[77] The reaction mechanism of haloalkane dehalogenase was elegantly visualized by X-ray crystallography[78] and a similar mechanism was proposed for the epoxide hydrolases.[77] On the basis of this limited sequence similarity around residues essential for activity in both enzymes and the proposed similarity in reaction mechanism, the epoxide hydrolases were presumed to be α/β hydrolase fold enzymes.

With the exception of GCL and AChE, the original five enzymes from which the α/β hydrolase fold was first described showed no sequence similarity beyond the G–*X*–S–*X*–G motif.[20] One might expect, then, that other enzymes and proteins that show no sequence similarity to known α/β hydrolase fold proteins may also have this conserved tertiary structure. The amino acid sequences of esterases from the bacterium *Acinetobacter* suggested that these proteins were serine esterases, but no structures of closely related proteins are known.[79] These sequences were compared to known three-dimensional structures using a 3D–1D algorithm.[14] The algorithm developed by Bowie *et al.*[13] produces a one-dimensional representation of the three-dimensional structures on the basis of the polarity, solvent accessibility, and secondary structural environment of each residue. The amino acid sequences of the *Acinetobacter* esterases were compared to 200 such linearized structures from the Protein Data Bank. The highest scores in this comparison consistently were with the known structures of α/β hydrolase fold enzymes, suggesting that these esterases are also likely to have a three-dimensional fold comparable to the α/β hydrolase fold.[14]

[75] F. David, A.-M. Bernard, M. Pierres, and D. Marguet, *J. Biol. Chem.* **268,** 17247 (1993).

[76] F. Goosens, I. DeMeester, G. Vanhoof, D. Hendriks, G. Vriend, and S. Scharpé, *Eur. J. Biochem.* **233,** 432 (1995).

[77] F. Pinot, D. F. Grant, J. K. Beetham, A. G. Parker, B. Borhan, S. Landt, A. D. Jones, and B. D. Hammock, *J. Biol. Chem.* **270,** 7968 (1995).

[78] K. H. G. Verschueren, F. Seljée, H. J. Rozeboom, K. H. Kalk, and B. Dijkstra, *Nature (London)* **363,** 693 (1993).

[79] R. N. Alon and D. L. Gutnick, *FEMS Microbiol. Lett.* **112,** 275 (1993).

Conclusion

The three-dimensional structures of more than 20 representatives of the α/β hydrolase fold are now known and many more members have been identified by sequence and secondary structure comparisons. The fold is proving to be a common and stable way to assemble a wide variety of catalytic activities. The enzymes in this fold family include peroxidases, proteases, lipases, esterases, dehalogenases, and epoxide hydrolases. This fold is versatile in terms of the identities of catalytic residues and in their locations. The amino acids thus far observed as catalytic nucleophiles are serine, cysteine, and aspartate and both glutamate and aspartate have been observed as the catalytic acid. Although the acid is generally located after strand $\beta7$, functional triads can also be constructed with the acid located after strand $\beta6$. This fold family is also known to include proteins with no catalytic activity. Despite its only recent recognition, the rapidly growing list of enzymes and proteins sharing this fold shows that it is, indeed, one of the most common protein folds known.

Acknowledgments

The authors thank Mr. René Coulombe for his help in preparation of the figures. This chapter adapted by permission of the National Research Council of Canada.

[5] Pancreatic Lipases and Their Complexes with Colipases and Inhibitors: Crystallization and Crystal Packing

By CHRISTIAN CAMBILLAU, YVES BOURNE, MARIE PIERRE EGLOFF, CHRISLAINE MARTINEZ and HERMAN VAN TILBEURGH

Introduction

The pancreatic lipase family has been divided into three subgroups based on primary structure comparisons and according to the nomenclature of Giller *et al.*[1]: (1) classic pancreatic lipases, (2) RP1 pancreatic lipases, and (3) RP2 pancreatic lipases. Classic pancreatic lipases are enzymes of 50 kDa capable of hydrolyzing fat triglycerides *in vitro*. Under physiological conditions, however, dietary fat digestion is possible only if a small ancillary

[1] T. Giller, P. Buchwald, D. B. Kaelin, and W. Hunziker, *J. Biol. Chem.* **267,** 16509 (1992).

0076-6879/97 $25.00

protein, colipase (M_r 11,000), is present.[2] Colipase prevents lipase from denaturation at the water–lipid interface and reverses the inhibitory effect of bile salts on lipase binding at this interface. The human pancreatic lipase (HPL) was the first mammalian lipolytic enzyme to be solved structurally,[3] followed later by structures of horse pancreatic lipase (HoPL).[4] Winkler *et al.*[3] demonstrated the presence of a surface loop called the lid domain (residues 237 to 261) covering the active site in the closed form and stabilized via van der Waals contacts by two others loops called $\beta 5$ (residues 76 to 85) and $\beta 9$ (residues 206 to 216) according to the Winkler numbering system.[3] These three structural elements were hypothesized to undergo a spatial reorganization to facilitate the entrance of the substrate in the active site.

Colipases are homologous proteins, which have the capability of cross-activating pancreatic lipases from different organisms.[5] We used this characteristic to obtain well-diffracting crystals of a heterologous complex between human pancreatic lipase and porcine (pro)colipase.[6] A first form of a binary lipase–colipase complex had the active site covered by a lid, as found for the pancreatic lipase alone. When crystallized in the presence of a surfactant mixture including a phospholipid[7] or of a C11 alkyl phosphonate,[8] the ternary complex lipase–colipase–inhibitor was found to exhibit large conformational changes. In fact, only the lid domain and the $\beta 5$ surface loop were shown to adopt a totally different conformation in the presence of lipids, while the $\beta 9$ loop remained unchanged.[7,8] The $\beta 5$ loop folds back on the core of the protein and the lid domain moves away from its closed position with a maximal amplitude of 29 Å, interacting with colipase and forming the lipid–water interface binding site (Fig. 1, see color insert). This "open" conformation has been shown to give free access to the catalytic triad.[7,8] These events probably mimic in the crystal those that in solution are associated to "the interfacial activation," namely a burst of activity of lipase in the presence of a water–lipid interface.[9]

[2] B. Borgström and C. Erlanson, *Biochim. Biophys. Acta* **242,** 509 (1971).
[3] F. K. Winkler, A. d'Arcy, and W. Hunziker, *Nature* (*London*) **343,** 771 (1990).
[4] Y. Bourne, C. Martinez, B. Kerfelec, D. Lombardo, C. Chapus, and C. Cambillau, *J. Mol. Biol.* **238,** 709 (1994).
[5] J. Rathelot, R. Julien, I. Bosc-Bierne, Y. Gargouri, P. Canioni, and L. Sarda, *Biochimie* **63,** 227 (1981).
[6] H. van Tilbeurgh, L. Sarda, R. Verger, and C. Cambillau, *Nature* (*London*) **359,** 159 (1992).
[7] H. van Tilbeurgh, M. P. Egloff, C. Martinez, N. Rugani, R. Verger, and C. Cambillau, *Nature* (*London*) **362,** 814 (1993).
[8] M. P. Egloff, F. Marguet, G. Buono, R. Verger, C. Cambillau, and H. van Tilbeurgh, *Biochemistry* **34,** 2751 (1995).
[9] L. Sarda and P. Desnuelle, *Biochem. Biophys. Acta* **30,** 513 (1958).

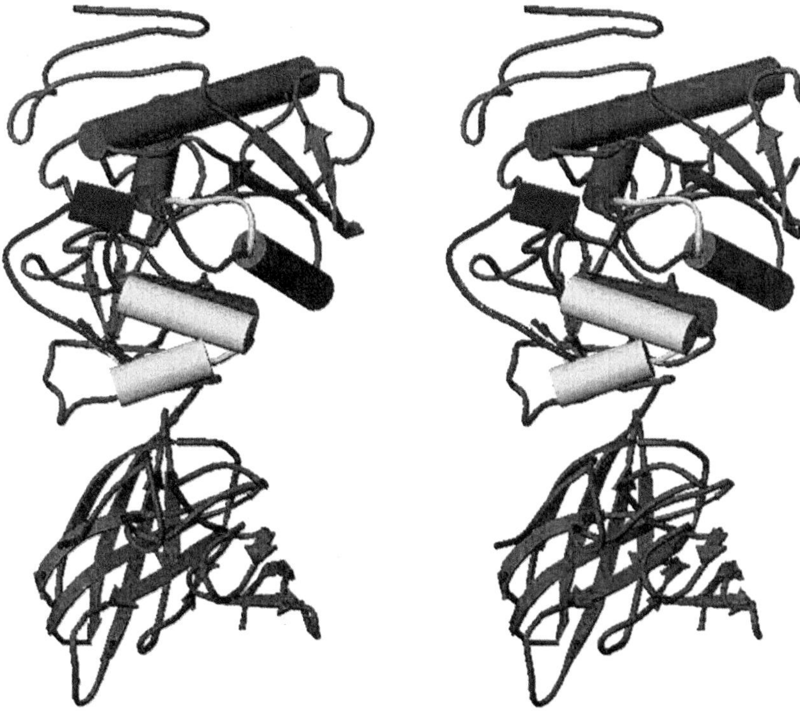

FIG. 1. Stereoview of the binary complex of human pancreatic lipase and porcine colipase (blue)[6]; the lid and the $\beta 5$ loop are colored in red. The lid and the $\beta 5$ loop of the open lipase (ternary complex) have been superimposed (yellow).[7,8]

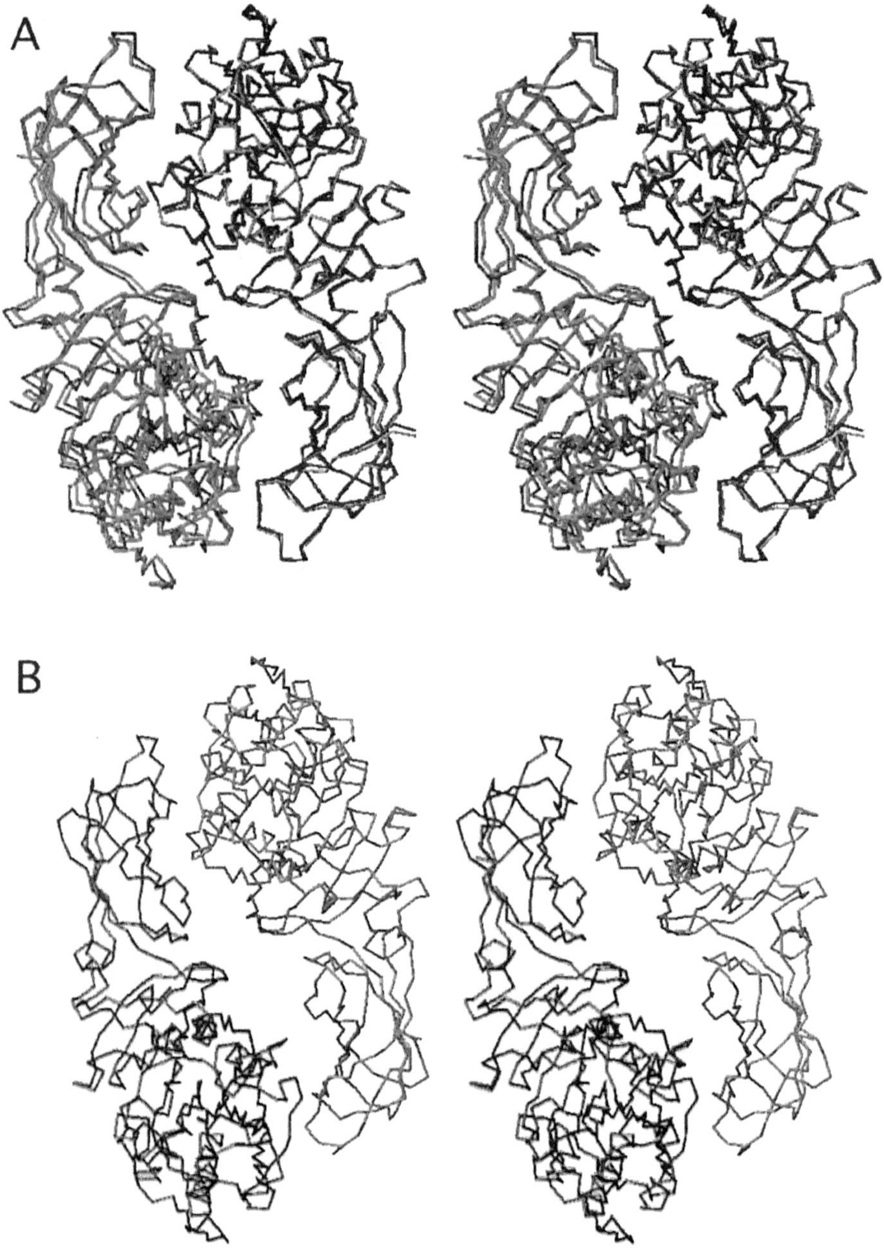

FIG. 2. Stereoviews of the pancreatic lipase dimers contained in the crystal asymmetric units (Table I). (A) Human enzyme form 1[3] (yellow and blue) and human enzyme form 3[15] (green and red). (B) Horse enzyme.[4]

Hjorth and co-workers[10] sequenced and expressed a cDNA encoding a guinea pig pancreatic (phospho)lipase belonging to the RP2 pancreatic subfamily (GPLRP2). Although the amino acid sequence is highly homologous to that of other known pancreatic lipases, this enzyme possesses a large deletion of 18 residues in the lid domain and has been shown to differ kinetically from the classic mammalian pancreatic lipases through the following: (1) no interfacial activation, (2) no effect of colipase, and (3) unusually high phospholipase A_1 activity associated with a classic lipase activity.[9,11,12] A chimeric GPL was obtained using the GPLRP2 N-terminus catalytic domain and C-terminus HPL domain.[10,13] This GPLRP2 chimera has been crystallized, and its structure determined.[14]

In this chapter, we describe the crystallization of different pancreatic lipases and of their complexes with colipases and inhibitors. We analyze their crystal packing in light of the crystallization experiments.

Crystallization

Pancreatic Lipases Alone

Pancreatic lipases are proteins soluble in water at concentrations up to 20 mg/ml; they crystallize in a narrow range of conditions, although with some liberty in the addition of detergent. Their crystallization conditions are also similar, in terms of pH and precipitant, to those used for the binary and the ternary complexes with colipase and inhibitors. Guinea pig lipase (PGLRP2), instead, crystallizes at basic pH and in the presence of a high salt concentration.

Human Pancreatic Lipase. The crystal forms of human pancreatic lipase are highly variable, a fact that cannot be related only to the protein source. Winkler *et al.*[3] obtained two different crystal forms (forms 1 and 2; Table I) of HPL, which was purified from pancreas powder. We obtained a third crystal form with lipase extracted from pancreas juice, although under the

[10] A. Hjorth, F. Carrière, C. Cudrey, H. Wöldike, E. Boel, D. M. Lawson, F. Ferrato, C. Cambillau, G. G. Dodson, L. Thim, and R. Verger, *Biochemistry* **32,** 4202 (1993).

[11] J. Fauvel, M. J. Bonnefis, L. Sarda, H. Chap, J. P. Thouvenot, and L. Douste-Blazy, *Biochim. Biophys. Acta* **663,** 446 (1981).

[12] K. Thirstrup, R. Verger, and F. Carrière, *Biochemistry* **33,** 2748 (1994).

[13] K. Thurstrup, F. Carrière, S. A. Hjorth, P. B. Rasmussen, H. Wöldike, P. F. Nielsen, and L. Thim, *FEBS Lett.* **327,** 79 (1993).

[14] C. Withers-Martinez, F. Carrière, R. Verger, D. Bourgeois, and C. Cambillau, *Structure* **4,** 1363 (1997).

TABLE I
CRYSTAL DATA OF PANCREATIC LIPASES AND THEIR COMPLEXES[a]

Source	Space group	Cell dimensions (Å)	Resolution limit (Å)	Molecule(s) per asym. unit.	V_m (Å^3/Da)	Solvent (%)	Ref.
HPL							
Form 1	$P2_1$	68.5, 83.8, 91.9 (β = 92.2°)	—	2	2.65	54	3
Form 2	$P2_1$	47.8, 112.8, 91.0 (β = 99.3°)	2.3	2	2.5	51	3
Form 3	$P2_1$	68.0, 88.2, 91.9 (β = 92.2°)	3.0	2	2.7	55	15
HoPL	$P2_12_12_1$	79.7, 97.8, 145.3	2.3	2	2.8	56	4, 18
GPLRP2	C_2	62.0, 55.9, 144.0 (β = 93.2°)	2.1	1	2.65	54	14
HPL–PCL	$P3_221$	80.3, 80.3, 251.0	3.0	1	3.8	68	6, 23
HPL–PCL–PC	$P4_22_12$	133.4, 133.4, 92.6	3.0	1	3.4	64	7, 23
HPL–PCL–C11P	$P4_22_12$	134.0, 134.0, 94.0	2.46	1	3.4	64	7, 8
HPL–HCL–C11P	$P4_22_12$	133.7, 133.7, 93.3	3.40	1	3.4	64	26

[a] HPL, Human pancreatic lipase; HoPL, horse pancreatic lipase; GPLRP2, guinea pig-related protein 2 lipase; PCL, porcine colipase; PC, phosphatidylcholine; C11P, *O*-(2-methoxyethyl)-*O*-(p-nitrophenyl)-*n*-undecylphosphonate.

same crystallization conditions.[15] Crystals were obtained by vapor diffusion and the hanging drop technique[16] at 20°, by adding 3 μl of a protein solution at 8 mg/ml to 3 μl of a well solution containing 2-(*N*-morpholino)ethanesulfonic acid (MES) buffer (0.1 *M*, pH 5.5), polyethylene glycol (PEG) 8000 (between 5 and 10%, w/v), LiCl (150 m*M*), and β-octylglucoside (β-OG; 1%, v/v). Birefringent crystals appeared in a few days and grew as thin plates of size 0.3 × 0.3 × 0.1 mm^3. All three forms contain two molecules in the asymmetric unit, and belong to the monoclinic space group $P2_1$ (Table I). V_m values range from 2.5 to 2.65 Å^3/Da, indicating solvent contents between 51 and 54%.[17]

Horse Pancreatic Lipase. Crystals of HoPL were grown[18] at 20° using the vapor diffusion technique from hanging drops. Best crystals were obtained by mixing 4 μl of an 8-mg/ml protein solution in a 0.1 *M* MES buffer (pH 5.6) with 4 μl of well solution containing the same buffer, 15 to 20%

[15] C. Cambillau, Y. Bourne, C. Chapus, D. Lombardo, and M. Rovery, *in* "Lipases: Structure, Mechanism and Genetic Engineering," Vol. 6, pp. 31–34. GBF Monographs. VCH Weinheim, New York, 1991.

[16] A. McPherson, *Eur. J. Biochem.* **189,** 1 (1990).

[17] B. W. Matthews, *J. Mol. Biol.* **33,** 491 (1968).

[18] D. Lombardo, C. Chapus, Y. Bourne, and C. Cambillau, *J. Mol. Biol.* **205,** 259 (1989).

(w/v) PEG 8000, 10 mM $MgCl_2$, 0.1 M NaCl. Crystals reached a maximum size of 0.3 × 0.3 × 0.5 mm^3 after a few weeks. Crystals of HoPL belong to space group $P2_12_12_1$, with cell dimensions a = 79.8 Å, b = 97.2 Å, c = 145.3 Å (Table I). With two molecules of 50 kDa each in the asymmetric unit, these cell dimensions correspond to a specific volume, V_m, of 2.8 $Å^3$/Da, and to a solvent content of 56%.

Guinea Pig Pancreatic Lipase Chimera. Samples of pure GPLRP2 chimera in 10 mM MES, 50 mM NaCl (pH 6.5) were concentrated up to 12 mg of protein per milliliter. Crystals were grown at 20° using the vapor diffusion technique with hanging drops. Typically, 2 μl of a 12-mg/ml protein solution in 50 mM NaCl, 10 mM MES (pH 6.5) is mixed with 2 μl of a reservoir solution containing 100 mM N,N-bis(2-hydroxyethyl)glycine (bicine, pH 9), 1.0 M LiCl and 10 to 15% (w/v) PEG 6000. Crystals appeared in 2 to 4 days as clusters of thin plates next to cubelike crystals of final dimensions 0.2 × 0.2 × 0.2 mm^3. Because of the presence of twinned crystals, microseeding techniques were used with a cat whisker[19] plunged into a solution of crushed crystals and passed through drops containing 8 to 10% (w/v) PEG 6000. The best monocrystals were obtained at pH 9.25 with 12% (w/v) PEG 6000, using the fine sampling pH method,[20] and yielded one large crystal (0.7 × 0.7 × 0.7 mm^3) per drop. Crystals belong to space group $C2$, with unit cell dimensions a = 62.0 Å, b = 55.9 Å, c = 144. Å, β = 93.2 Å, indicating the presence of one molecule in the asymmetric unit and a V_m value of 2.65 $Å^3$/Da (53% estimated solvent volume).

Complexes of Pancreatic Lipases

As mentioned above, colipases from one species are able to activate pancreatic lipases from a different species. The possibility of combining colipases and lipases from different species increases the chances for obtaining crystalline material, using a lipase–colipase crystallization matrix (Table II). An important question concerns whether the structures of a heterologous complex have physiological significance. We therefore determined and compared the structures of human lipase–porcine colipase and human lipase–human colipase complexes. Both complexes crystallize in the same crystal form and their three-dimensional structures are virtually the same, indicating that no fundamental difference exists between inter- and intraspecies complexes. These structural results are thus in agreement with the preceding biochemical observations.

Unlike pancreatic lipases alone, binary and ternary complexes (lipase–colipase–inhibitor) can be crystallized only when a detergent (β-octylgluco-

[19] E. A. Stura and I. A. Wilson, *J. Crystal Growth* **110,** 270 (1991).
[20] A. McPherson, *J. Appl. Crystallogr.* **28,** 362 (1995).

TABLE II
CRYSTALLIZATION MATRIX OF LIPASE–COLIPASE COMPLEXES[a]

	Lipase		
Colipase	Human	Equine	Porcine
None	+[b]	+[c]	—
Human	+[d]	—	—
Equine	—	—	—
Porcine	+[e]	—	+[f]

[a] The plus sign (+) indicates that crystals were obtained, whereas the minus sign (−) indicates no attempt at crystallization or no crystal obtained.
[b] See Refs. 3 and 15.
[c] See Ref. 4.
[d] See Ref. 26.
[e] See Refs. 6–8, 23, and 26.
[f] See Ref. 15.

side) is added. In fact, this detergent is not used as an additive, but as a precipitating agent.

Binary Human Lipase–Porcine Colipase Complex. The purification of human pancreatic lipase was carried out according to the procedure of DeCaro *et al.*[21] and that of porcine-activated colipase according to Canioni *et al.*[22] The binary complex between HPL and porcine colipase (PCL) was prepared by mixing these components in a 1 : 1.3 ratio (8 mg of HPL and 2.5 mg of PCL per milliliter), both being dissolved in distilled water. Sitting drops were set up by mixing 4 μl of the complex and 4 μl of a solution containing NaCl (ranging from 0.2 to 0.5 *M*), MES (0.1 *M*, pH 6.0), and 2% (w/v) PEG 8000 at 18°. Crystals were obtained, after 5 days of equilibration, on addition of 1 μl of a 250 m*M* solution of β-OG. Crystals appeared 5 min after addition of the detergent, and reached their full size ($0.3 \times 0.3 \times 0.5$ mm^3) after a few hours.[23] Both HPL and PCL were present in the crystal, as shown by sodium dodecyl sulfate (SDS) gel or high-performance liquid chromatography (HPLC). The space group was $P3_221$, with cell edges

[21] H. DeCaro, C. Figarella, J. Amic, R. Michel, and O. Guy, *Biochim. Biophys. Acta* **490,** 411 (1977).
[22] P. Canioni, R. Julien, J. Rathelot, H. Rochat, and L. Sarda, *Biochimie* **59,** 919 (1979).
[23] H. van Tilbeurgh, Y. Gargouri, C. Dezan, M. P. Egloff, M.-P. Nésa, N. Rugani, L. Sarda, R. Verger, and C. Cambillau, *J. Mol. Biol.* **229,** 552 (1993).

$a = b = 80.3$ Å and $c = 251$ Å. The V_m value was 3.8 Å^3/Da for one complex in the asymmetric unit (68% solvent) (Table I).

Ternary Human Lipase–Porcine Colipase–Phosphatidylcholine Complex. Cocrystallization of the HPL–PCL complex with mixed micelles of 1,2-didodecanoyl-*sn*-3-glycerophosphorylcholine and sodium taurodeoxycholate resulted in crystals of a different space group, compared to the preceding conditions. First, a concentrated solution of mixed micelles was prepared in water (30 m*M* phosphatidylcholine and 60 m*M* sodium taurodeoxycholate). One microliter of this micellar solution was added to a preequilibrated sitting drop of 4 μl of a solution of the complex in water (8 mg of HPL and 2.5 mg of PCL per milliliter) and to 4 μl of the well solution containing NaCl (0.3 *M*), MES (0.1 *M*, pH 6.0), and 2% (w/v) PEG 8000 at 20°. After 2 days of equilibration, crystallization was started by adding 1 μl of β-OG to the drop. β-Octylgucoside actually induced the formation of crystals. The crystals remained small ($0.05 \times 0.05 \times 0.1$ mm^3) but diffract at 3.0-Å resolution (Table I). These crystals belong to the space group $P4_22_12$ with cell dimensions $a = b = 134.0$ Å and $c = 94.0$ Å. With one complex in the asymmetric unit the V_m value is 3.4 Å^3/Da (64% solvent).

Ternary Human Lipase–Porcine Colipase–C11 Complexes. Synthesis of *O*-(2-methoxyethyl)-*O*-(p-nitrophenyl)-*n*-undecylphosphonate (C11P) was carried out as previously described.[24] Crystallization experiments were carried out with a racemic mixture of the inhibitor C11P. The ternary C11P–lipase–colipase complex was crystallized under conditions similar to those described for the ternary phospholipid–lipase–procolipase complex (see preceding section). At room temperature, C11P powder is soluble only in organic solvents. Two milligrams of this inhibitor was dissolved in 10 μl of THF (tetrahydrofuran). Two microliters of this solution was added to a solution containing lipase (8 mg/ml) and colipase (2.5 mg/ml), which corresponds to a molar excess of inhibitor vs lipase of 100. After 30 min of incubation, 4 μl of this solution was mixed with 4 μl of the mother liquor containing 2% (w/v) PEG 8000, 0.1 *M* MES (pH 6.0), and 0.4 *M* NaCl, and deposited as sitting drops. A few days later, 1 μl of a β-octylglucoside solution in water (250 m*M*) was added to the drops. Tetragonal crystals developed overnight. They grew within about 2 days to a maximum size of $0.5 \times 0.5 \times 0.5$ mm^3. Alternatively, β-octylglucoside was added without any previous equilibration period; in this case, much more numerous but smaller crystals developed in each drop. These crystals belong to the space group $P4_22_12$ and are isomorphous ($a = b = 134.0$ Å and $c = 94.0$ Å) with those obtained in the case of the ternary phospholipid–lipase–colipase complex.

[24] F. Marguet, C. Cudrey, R. Verger, and G. Buonu, *Biochim. Biophys. Acta* **1210,** 157 (1994).

Ternary Human Lipase–Human Colipase–C11 Complexes. Recombinant human procolipase was prepared according to Lowe *et al.*[25] Lyophilized procolipase was dissolved to 25 mg/ml in water and mixed with a solution of wild-type human pancreatic lipase to obtain a complex solution containing lipase (8 mg/ml) and colipase (2.5 mg/ml). A 100-fold excess of C11 alkyl phosphonate was then added to the mixture. This solution was subjected to crystallization conditions similar to those described for the heterologous complex. Briefly, 4 μl of the mixture and 4 μl of a reservoir solution containing PEG 8000 (2%, w/v), MES (pH 6.0), and NaCl (0.1–0.5 *M*) were mixed and equilibrated over the reservoir solution for 2 days. One microliter of β-octylglucoside was then added, which provokes the rapid appearance of small crystals. Several macroseeding cycles yielded crystals isostructural to the heterologous crystals and usable for data collection: space group $P4_22_12$ and cell dimensions $a = b = 133.7$ Å, $c = 93.3$ Å. However, the quality of the homologous crystals was inferior to that of the heterologous crystals, and diffraction extended only to 3.4-Å resolution.[26]

Porcine Lipase–Porcine Colipase Complex. Crystals of a complex between porcine lipase and porcine colipase have been obtained. These crystals were not of sufficient quality to determine the crystal structure: They diffracted at best to 5-Å resolution and were often disordered. They were obtained at 20° by the hanging drop method. Four microliters of a 10-mg/ml protein solution (8 mg of lipase and 2 mg of colipase per milliliter) in MES (0.1 *M*, pH 6.0) were mixed with 4 μl of a well solution containing MES (0.1 *M*, pH 6.0), PEG 8000 (8–12%, w/v), and NaCl (0.1 *M*). The crystals grew as bunches of elongated ovoid leaves, and were of dimensions $0.1 \times 0.2 \times 0.7$ mm^3.

Crystal Packing

Classic Pancreatic Lipases

Classic pancreatic lipases tend to form well-defined dimers in crystal packing; dimers, however, were not detected in solution. The asymmetric unit of human and horse lipase crystals contains two independent molecules arranged in a head-to-tail fashion and related by a twofold axis. In both cases, molecule A and molecule B have similar orientations, and their lids, covering the active site, are facing each other (Fig. 2, see color insert). However, the HoPL dimer is less compact than that of HPL, and the interactions between the monomers in both molecules are different. It is

[25] M. E. Lowe, J. L. Rosenblum, P. McEven, and A. W. Strauss, *Biochemistry* **29,** 823 (1990).
[26] M.-P. Egloff, Ph.D. thesis. Université de Paris XI Orsay, Paris, 1995.

remarkable that the HPL dimer association is preserved in two different crystal packing environments: that of Winkler *et al.* (form 1) and ours (form 3; Table I). It was clear that this crystal would impede all efforts to obtain open forms of HPL or HoPL by diffusing inhibitors in the active site, because the opening of the lid was blocked by face-to-face packing.

Guinea Pig Related Protein 2 Lipase–Phospholipase Chimera

The crystal packing of GPLRP2 lipase chimera is made of layers of molecules having the same orientation, with the molecules of one layer being head to tail with those of the next layer (Fig. 3). Therefore, the active site of GPLRP2 chimera, at the front face of the molecule, is packed against the back face of a symmetry-related lipase (Fig. 3). The crystal packing is such that the active site is facing a solvent channel and is partly accessible to solvent (Fig. 3B). Because this enzyme has a reduced lid, the active site is accessible to substrates or inhibitors. This could be used advantageously to diffuse such molecules in the crystal.

Binary Lipase–Colipase Complex

The binary lipase–colipase complex exhibits the loosest crystal packing among the crystals reported here, with 68% solvent. The position of the complex is again head to tail, but in a side-by-side fashion. Consequently, the lid covering the active site is only partly involved in crystal contacts (Fig. 4). Furthermore, superimposing the structure of the open lipase (as found in the ternary complex) on the binary complex lipase structure results in a fully accessible active site. The lid, however, does not open under the crystallization conditions of the ternary complex, and addition of inhibitors in the crystallization medium leads to another crystal form.

Ternary Lipase–Colipase–Inhibitor Complexes

When the structure of a ternary lipase–colipase–phospholipid (LCP) complex was determined, some spectacular differences were observed in comparing it with the binary complex involving only lipase and colipase. However, the rather low resolution of this ternary complex (3 Å) did not make it possible to carry out more detailed studies on the ligand binding and on the interactions of lipase with the detergent molecules used in crystallization. The crystals of a ternary complex involving lipase, colipase, and a covalently bound C11P alkyl phosphonate inhibitor (LCC) belong to the same space group as the LCP crystals, but diffract to 2.46-Å resolution. This structure confirmed the previous findings about the lid and the β5 loop opening (Fig. 1) and revealed the presence of five detergent molecules (β-octylglucoside) per complex.

A

B

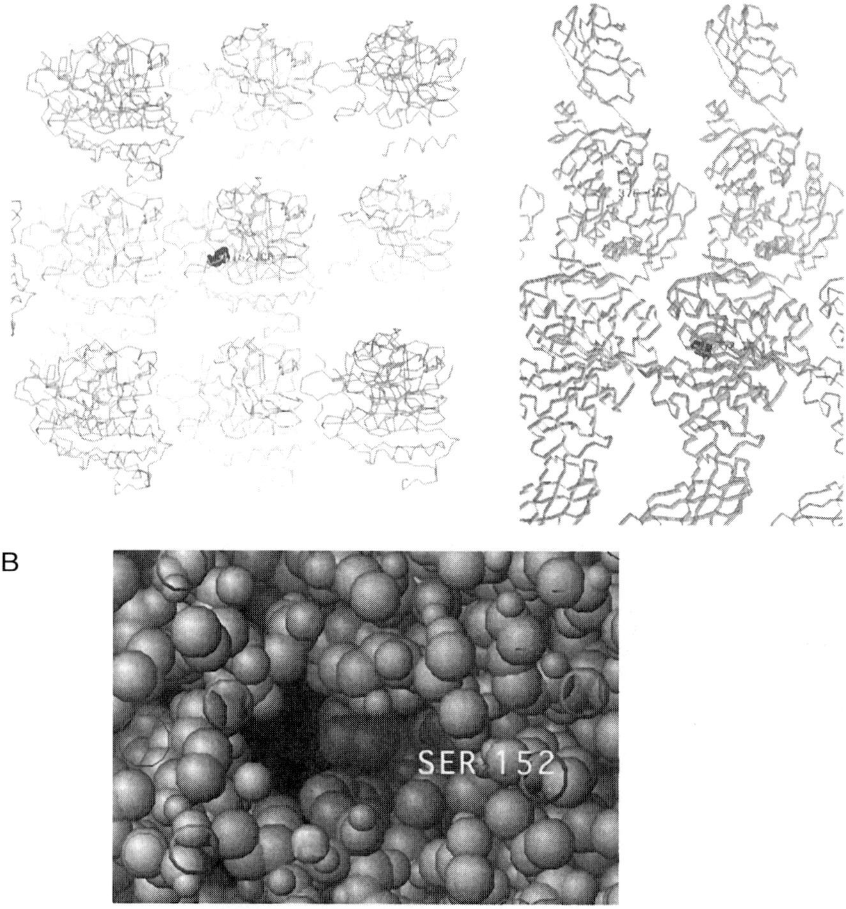

FIG. 3. (A) Views at 90° of the GPLRP2 chimera, showing the solvent channels. The surface of the catalytic Ser-152 has been added in one of the molecules. (B) A CPK view of a solvent channel facing the GPLRP2 chimera catalytic Ser-152.

The racemic C11 alkyl phosphonate was found to consist of a mixture of bound phosphorus enantiomers in the active site, each having an occupancy of about 50%. The position of the alkyl chains of the phosphonate enantiomers coincides with that observed in the case of the alkyl chains of the phospholipid in the LCP structure. Both chains interact with the same patch of hydrophobic/aromatic residues in the active site. The five β-OG detergent molecules have been carefully refined and are located on the

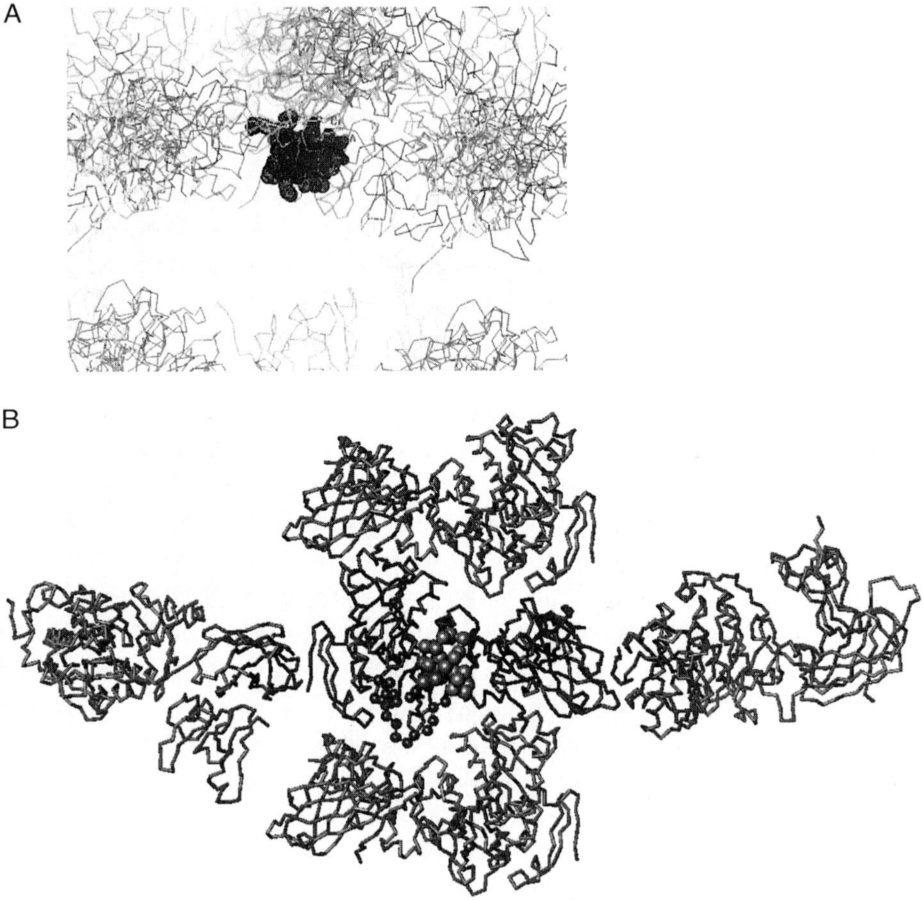

FIG. 4. (A) View of the C_α tracing of the crystal packing of the binary lipase–colipase complex. A large solvent channel is facing the lid, which has been surfaced. (B) View of the C_α tracing of the four nearest symmetry-related binary complexes. The closed lid has been represented in ball-and-stick representation, whereas the superimposed open lid is shown as clear CPK spheres, and is not involved in any clash.

same side of the lipase active site, along the putative surface interacting with lipids.

It has been suggested that the lid-opening mechanism might possibly have to do with the low solubility of pancreatic lipase in its open form, which would mean that it would present to solvent an extensive hydrophobic interface. A closed lipase form is obviously much more hydrophilic. Which mechanism would therefore yield solutions containing high concentrations of the open complex lipase–colipase–phosphonate in the crystallization medium? A possible answer to this question arises from the observation of the molecules in the crystal lattice. A tetramer of complexes can be identified in the crystal packing, which can best be described as a dimer of dimers (Fig. 5). Here two lipase–colipase complex molecules form a side-to-side, head-to-tail dimer. This dimer is packed against another dimer, which is rotated by 90°. The extensive hydrophobic surface of this dimer, formed jointly by the active site region of lipase and by the hydrophobic extremities of the colipase fingers, lies opposite to the equivalent hydrophobic surface of the dimer packing partner in the crystal. The resulting structure presents a classic protein surface to the solvent, because the two hydrophobic faces are hidden inside the tetramer. A total of 8 half-occupied C11P and 20 β-OG molecules are buried at the interface, filling the "holes" of the molecular assemblage (Fig. 5). It has been assumed that each hydrophobic surface plays the role of the lipid surface for the other dimer,

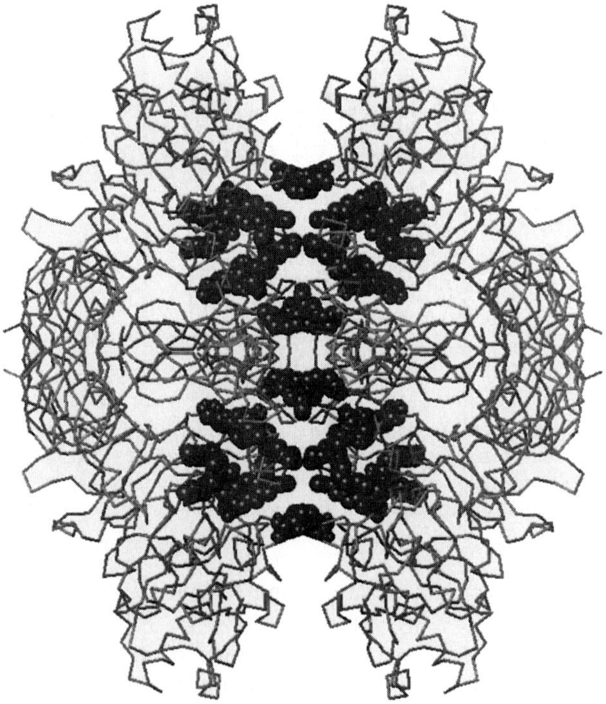

FIG. 5. C_α Tracing of the crystallographic tetramer of LCC[8] formed by four symmetry-related molecules of the ternary complex. β-Octylglucoside and inhibitor molecules C11P (shown in CPK) are located at the hydrophobic interface formed by this crystal packing.

thus mimicking the interaction between the lipase–colipase dimer and the lipid interface.

Note Added in Proof

Since the manuscript has been accepted, the homologous porcine lipase–colipase complex has been solved at 2.8 Å resolution.[27]

Acknowledgments

We thank Dr. Robert Verger and Dr. Frederic Carrière for their enthusiastic collaboration and discussions. These studies were supported in part by the EC BRIDGE BIOT-CT91-0274 and the BIOTECH BIO2-CT94-3041 programs, by the program IMABIO of the Centre National de la Recherche Scientifique, and by the PACA region.

[27] J. Hermode, D. Pignol, B. Kerfelec, I. Crenon, C. Chapus, and J. C. Fontecilla, *Comps. J. Biol. Chem.* **271(30),** 18007 (1990).

[6] Impact of Structural Information on Understanding Lipolytic Function

By MAARTEN R. EGMOND and CARLA J. VAN BEMMEL

Introduction

The elucidation of several lipase structures by crystallographic methods has given us the opportunity to improve our understanding of lipase structure–function relationships considerably. Lipases (triacylglycerol hydrolases, E.C. 3.1.1.3) are versatile enzymes showing many interesting properties for application in industrial products and processes. Among these are, e.g., the hydrolysis of wide ranges of poorly water-soluble esters. At low water activities transesterification of fats and oils or (stereo)selective esterification of alcohols can be achieved. By far the largest application area is the addition of lipases to laundry detergents, e.g., aiming at improved removal of fatty stains from fabrics.

Industrially applied lipases are mainly obtained from microbial sources. Application of protein-engineering techniques has widened the range of lipases to choose from considerably. Lipases from eukaryotic species are produced on a large scale using, e.g., *Aspergillus* species. However, some bacterial lipases require the original host for production, either because production levels are low, or secretion of these enzymes is complex and insufficiently understood. An example of this class of lipases includes those obtained from *Pseudomonas* species.

0076-6879/97 $25.00

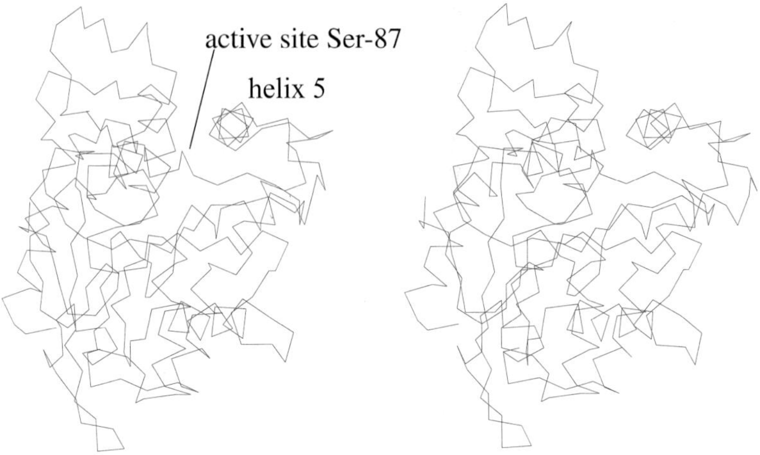

FIG. 1. Stereo view of the backbone tracing of *Pseudomonas glumae* lipase.

A central question that can now be addressed using structural data is whether or not—and if so to what extent—lipases bind to, react with, or are activated by lipid–water interfaces. The fact that water-soluble lipases are able to hydrolyze poorly water-soluble substrates requires carefully controlled assay conditions. These are essential before detailed understanding of structure–function relations can be obtained from such kinetic and mechanistic studies. In this chapter, methods are described that can be applied to achieve this goal.

The inaccessible active site in many lipases, including that of *Pseudomonas glumae* lipase,[1,2] adds further complexity. In contrast, lipolytic enzymes do exist, e.g., *Fusarium solani pisi* cutinase,[3] that have an exposed active site, but still show behavior also noted for the more complex lipases. The structure and function of these two lipases are discussed in further detail.

Structural and Functional Aspects of Lipases

Pseudomonas Lipases

As an example of this class of lipases, Fig. 1 shows a stereo view of the backbone tracing of *P. glumae* lipase (PGL). The active site Ser-87 of this lipase is buried deeply in the interior of the protein. One possibility for

[1] M. E. M. Noble, A. Cleasby, L. N. Johnson, M. R. Egmond, and L. G. J. Frenken, *FEBS Lett.* **331,** 123 (1993).

[2] M. E. M. Noble, A. Cleasby, L. N. Johnson, M. R. Egmond, and L. G. J. Frenken, *Protein Eng.* **7,** 559 (1994).

[3] C. Martinez, P. Degeus, M. Lauwereys, G. Matthyssens, and C. Cambillau, *Nature* (*London*) **356,** 615 (1992).

allowing access of substrate molecules into the active site is a conformational change involving movement of helix 5, which may occur after binding of the enzyme to lipid–water interfaces. Helix 5 is flanked by two flexible peptide regions that will allow such a conformational change. When the active site is exposed to solvent a large hydrophobic area is exposed as well, allowing interaction with a number of substrate molecules.

Principle of Kinetic Assays

A systematic investigation of the functional properties of this lipase should be carried out under carefully controlled conditions. This is done ideally by incubation of lipase with an inert micellar detergent containing a small and fixed molar percentage of substrate relative to detergent. Substrates should contain a single ester bond and should preferably be pure enantiomers, yielding straightforward hydrolysis kinetics assayed spectrophotometrically or using a pH-stat device. In this way multiple modes of lipolytic attack are avoided as far as possible. From such turnover data apparent Michaelis–Menten constants (K_m and k_{cat}) are obtained by varying the amount of detergent. Further mechanistic aspects may be obtained when kinetic assays are repeated for different amounts of substrate present in the detergent micelles. Replots of the apparent K_m and k_{cat} values versus the amount of substrate can provide information regarding interaction between single substrate molecules and the enzyme when bound to the micellar lipid–water interface.

In previous models of lipolytic action,[4] it is assumed that the enzyme will remain bound to the interface for several turnovers at least. A model describing this situation is represented in Fig. 2 (model I). A consequence of this assumption is that apparent K_m values will not be perturbed by turnover rates that are fast compared to the rate of equilibration of enzyme with detergent micelles.

In contrast, when a substantial amount of enzyme will dissociate from the micellar interface after product formation, it can easily be derived that apparent K_m values will increase when k_{cat} considerably exceeds the k_{off} rate constant.

Reaction Conditions Using pH-Stat Equipment

Tris (5 mM), $CaCl_2$ (10 mM), NaCl (50 mM) at pH 8 and 25° under N_2 atmosphere; all substrates at 8 mM, dissolved in 200 mM Triton X-100; titration with 10 mM NaOH

Substrates used are shown in Fig. 3. These compounds mimic diacylglycerol rather than triacylglycerol substrates as only two longer (C_9) chains are present.

[4] R. Verger and G. H. de Haas, *Annu. Rev. Biophys. Bioeng.* **5,** 77 (1976).

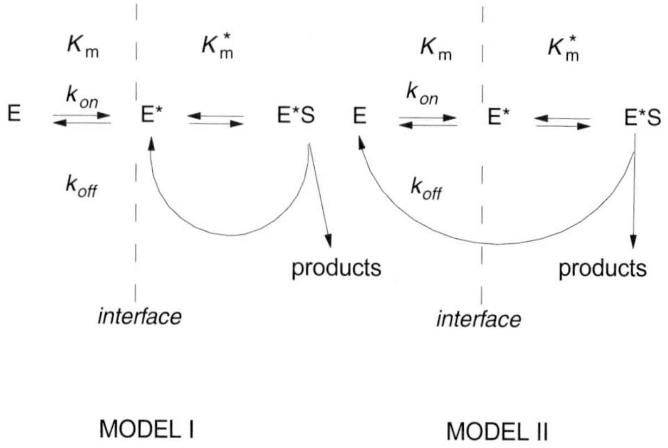

FIG. 2. Models for interfacial catalysis displayed by lipases in which the enzyme stays in the interface during catalysis (model I), or returns to the bulk water phase when products are formed (model II). The dashed lines represent the lipid–water interface. Products are liberated in the bulk water phase, and/or remain in the lipid phase.

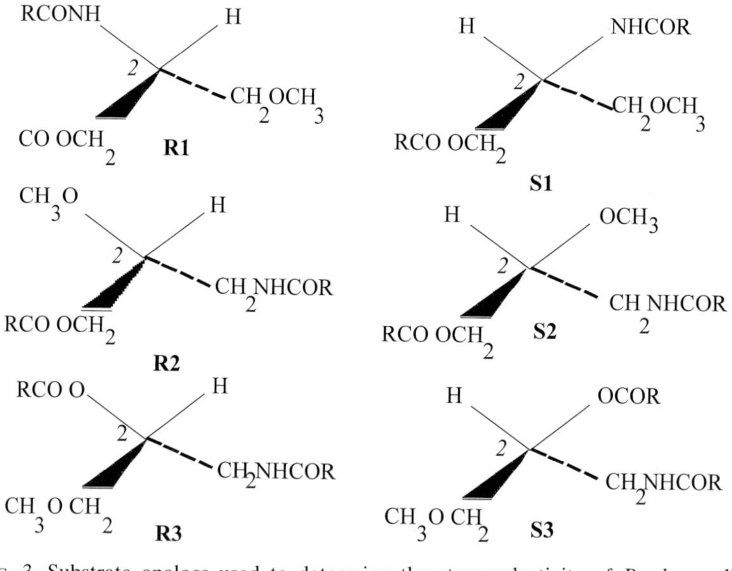

FIG. 3. Substrate analogs used to determine the stereoselectivity of *P. glumae* lipase. R groups = C_9H_{19}. These compounds were dissolved in Triton X-100 at 4% molar ratio to detergent.

TABLE I
SUMMARY OF RESULTS FOR SIX ISOMERS

Substrate	Apparent K_m (mM)	k_{cat} (sec^{-1})
R1	0.3	40
S1	0.4	16
R2	0.3	44
S2	0.3	90
R3	0.9	405
S3	0.1	11

The results of these experiments for all six isomers, but only at a 4% molar ratio of substrate to detergent, are summarized in Table I. The data shown in Table I for compound **R3** do provide some support for the mechanism according to model II, because the apparent K_m is large relative to the other compounds. However, it will only be possible to conclude whether the enzyme must leave the lipid–water interface after each cycle of product formation when independent measurements of k_{on} and k_{off} rates are obtained for the micellar system used. In principle these rates may be obtained, e.g., by spectroscopic means, but were not measured for native PGL owing to lack of reporter groups that are perturbed on binding to a micellar interface.

The turnover data (k_{cat}) listed in Table I are rather unexpected as the highest turnover was found for splitting an ester bond at the 2-position (compound **R3**). As the compounds investigated resemble diacylglycerols rather than triacylglycerols, these data do not contradict previous work.[5] In fact, splitting primary ester bonds as in compounds **R1** and **S1** confirms the previous finding that PGL prefers *sn*-3 (i.e., **R1** resembling 2,3-diacylglycerol) over *sn*-1 (**S1** resembling 1,2-diacylglycerol) hydrolysis. The data in Table I also predict that 1,3-diacylglycerols (represented by compounds **R2** and **S2**) are hydrolyzed at rates comparable to the 1,2- or 2,3-diacylglycerols. However, in contrast with **R1**/**S1**, now the **S2** isomer is hydrolyzed more rapidly than the **R2** isomer.

It should be realized that the data presented in Table I cannot be used to derive the enantioselectivity constant E (equaling $[k_{cat}/K_m]^R/[k_{cat}/K_m]^S$). This is because the apparent K_m values describe interfacial binding rather than binding of single substrate molecules to the enzyme. The latter process is described by the constant K_m^* shown in Fig. 2. This constant (or the ratio k_{cat}/K_m^*) may be separately obtained from experiments in which the molar fraction of substrate is changed relative to detergent. The consequences of

[5] E. Rogalska, C. Cudrey, F. Ferrato, and R. Verger, *Chirality* **5,** 24 (1993).

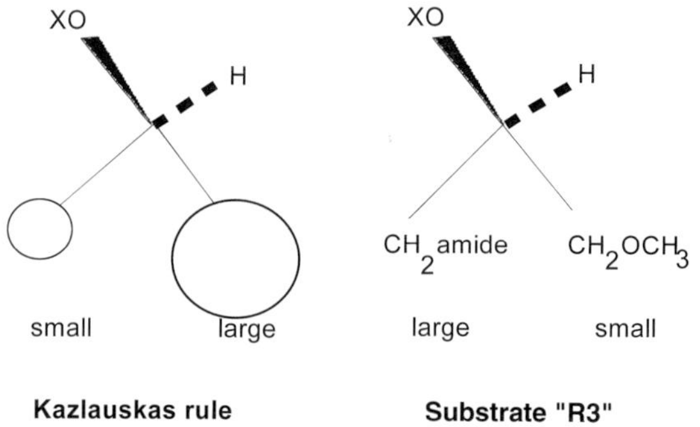

FIG. 4. Comparison of model compound **R3** and model of preferred bulkiness of groups bound to the carbon at which ester is cleaved or formed according to Kazlauskas.

the kinetic studies using single ester substrate analogs may be summarized as follows. Starting from triacylglycerol substrates it will be most likely that the *sn*-3 acyl chain is preferentially split. The resulting 1,2-diacylglycerol would be an ideal substrate for *sn*-2 hydrolysis, as predicted by the high turnover of compound **R3**. Via this sequence of events 1-monoacylglycerol is formed most rapidly. The less preferred pathway following formation of 2,3-diacylglycerol would yield 2-monoacylglycerol as product, because *sn*-2 hydrolysis of this diacylglycerol compound is slow. This can be predicted from the hydrolysis rate of compound **S3**. In conclusion, efficient breakdown of diacylglycerol substrates does not seem to require acyl migration from the 2-position to the primary positions for hydrolysis. The observed stereo preference for compound **R3** does not conform to the selection rule according to Kazlauskas *et al.*[6] From Fig. 4 it can be derived that the bulkiness of the two groups bound to the C-2 carbon is opposite to the preferred orientation noted. Apparently, apart from bulkiness other factors also play a part in positioning substrates, leading to hydrolysis or esterification at secondary alcohols.

The stereo preference noted above for PGL cannot easily be reconciled with the structural data obtained for this enzyme. Several uncertainties remain. One is the extent of conformational change involved in allowing access to the active site. Another is the large area of hydrophobic surface of PGL that is exposed once the enzyme opens up. This allows many binding modes of substrates, making predictive modeling studies an impossible task. The conclusion therefore seems justified that simpler structural models are required to investigate lipolytic action in detail.

[6] R. J. Kazlauskas, A. N. E. Weissfloch, A. T. Rappaport, and L. A. Cuccia, *J. Org. Chem.* **56,** 2656 (1991).

Cutinase Structure and Function

The full-atom tracing of cutinase from *F. solani pisi* shown in Fig. 5 demonstrates that for this enzyme the active site is accessible to solvent. The active site can be approached via a rather narrow cleft that is flanked by two flexible loops containing a number of hydrophobic residues.

The first problem to solve is to investigate whether cutinase can be considered a member of the family of lipases or rather an esterase. When a lipase is defined as an enzyme that binds to (and reacts with) substrates present in a lipid–water interface, such binding behavior needs to be demonstrated for cutinase. In principle, there are several ways to observe directly the formation of enzyme–lipid aggregates. An example using time-resolved fluorescence is shown in Fig. 6. Figure 6 shows an increase in rotational correlation time with increasing concentrations of lithium dodecyl sulfate

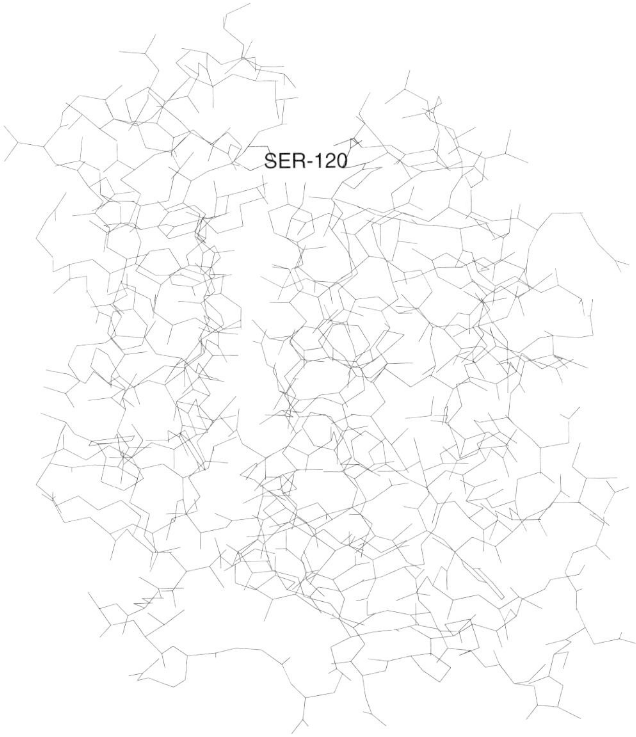

FIG. 5. Full-atom X-ray model of cutinase. The active site Ser-120 is indicated, showing its solvent accessibility.

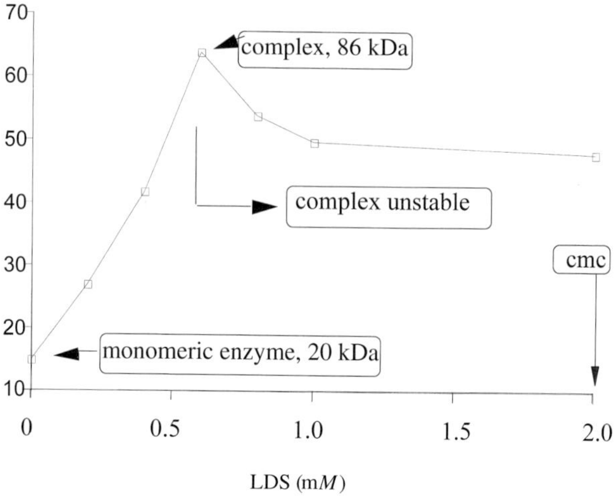

FIG. 6. Time-resolved fluorescence data for cutinase as a function of total LDS concentration, below the cmc of LDS (2 m*M* under the conditions used). A complex is formed between enzyme and detergent about four times the size of the free enzyme. Above 0.5 m*M* LDS the enzyme becomes unstable.

(LDS). The lithium salt is preferred over the sodium salt for solubility reasons.

Reagents

Tris buffer (10 m*M*, pH 8.5), 40 m*M* NaCl, 1 μM cutinase

From these data the size of the complex formed can be calculated, yielding 86 kDa, which is about four times the size of free cutinase (20.6 kDa). At LDS concentrations above 0.5 m*M* the size of complexes formed apparently decreases. It was found that under these conditions cutinase becomes unstable, losing activity. At, e.g., 2 m*M* LDS enzyme activity is lost in approximately 30 min. This loss in activity was found to be reversible, as full activity could be recovered after removal of LDS from the enzyme preparation by column chromatography. Enzyme activity is also restored by addition of detergents that are inert toward cutinase. For example, with the addition of 10 m*M* taurodeoxycholate (TDOC) enzyme activity is completely recovered in approximately 30 min. The kinetics of these changes can be monitored spectrophotometrically at 288 nm, as is shown in Fig. 7. The optical density (OD) changes brought about by LDS are most likely due to unfolding of cutinase involving tyrosine residues and the single tryptophan residue.

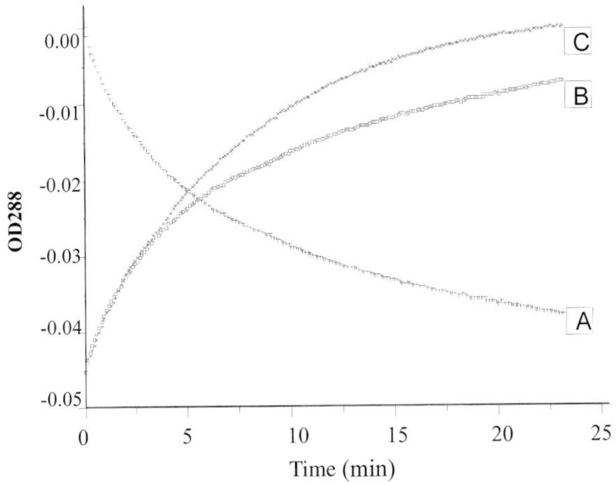

FIG. 7. Unfolding of cutinase (at 24 μM) due to 2 mM lithium dodecyl sulfate (trace A) and refolding after dissolution of LDS in taurodeoxycholate micelles (10 mM, trace C). The inverted trace of the unfolding process is also shown for comparison (trace B).

Reagents

Cutinase (24 μM), lithium dodecyl sulfate (2 mM), NaCl (40 mM), Tris (10 mM, pH 8.5) and 25°

Whereas refolding of cutinase in the presence of mixed TDOC–LDS micelles is simply first order, the unfolding process in the presence of LDS is more complex and can be described by at least two first-order rate constants.

These studies demonstrate that cutinase behaves like a lipase rather than an esterase. Even below the critical micellar concentration (cmc) of micellar substrates kinetically relevant enzyme complexes may be formed. This same phenomenon can also be demonstrated by time-resolved fluorescence studies for other lipases (data not shown). It explains why interfacial activation around the cmc of substrates can remain unnoticed, because the enzyme is already activated by premicellar complex formation below the cmc. It also tells us that enzymes like cutinase, having an exposed active site serine, can behave like lipases having a buried active site.

Given the fact that cutinase can be considered a lipase, the same type of kinetic studies were carried out as described above for PGL. Because of a different substrate preference, other (triacylglycerol) analogs were used.[7] Furthermore, it was found[5] that cutinase is highly specific for hydroly-

[7] M. L. M. Mannesse, R. C. Cox, B. C. Koops, H. M. Verheij, G. H. de Haas, M. R. Egmond, H. T. W. M. van der Hijden, and J. de Vlieg, *Biochemistry* **34,** 6400 (1995).

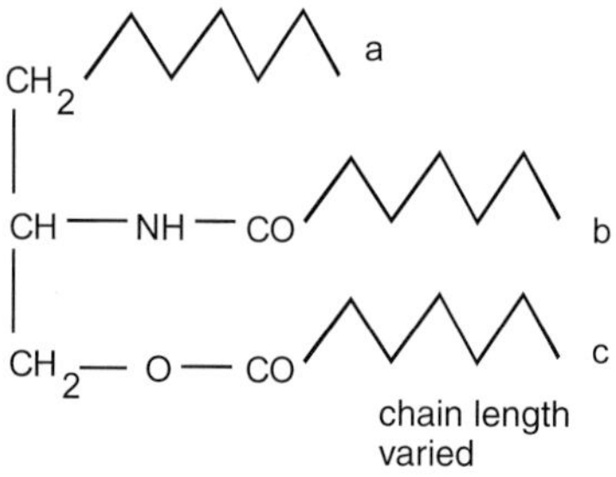

FIG. 8. Example of triacylglycerol analogs for which only one ester bond is available for cleavage by the enzyme. A series of compounds (both R and S isomers) was made in which the chain lengths a, b, and c were varied separately.

sis of the *sn*-3 acyl chain of triacylglycerols. Therefore, the general structure of compounds shown in Fig. 8 was chosen. Substrates were dissolved in a detergent mixture consisting of 150 m*M* Triton X-100 and 17 m*M* deoxycholate. Rates were measured using the pH-stat technique at several constant molar ratios of substrate relative to detergent. From these experiments apparent maximal rates were derived. The maximal rates were replotted versus the amount of substrate present in the detergent mixtures. These replots essentially yield linear dependencies owing to the low levels of substrate present. From the slopes of these plots the catalytic power (k_{cat}/

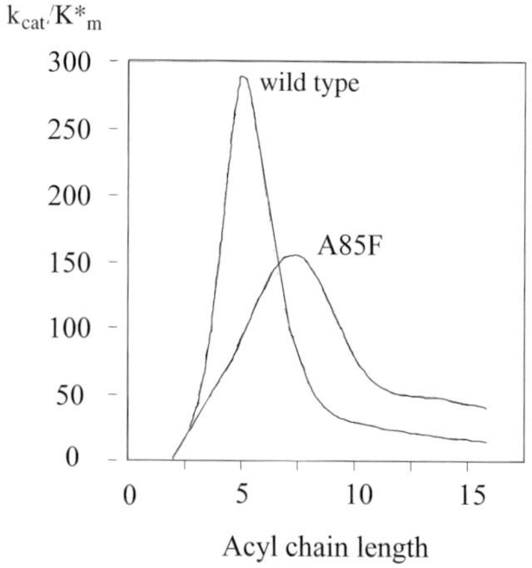

FIG. 9. Catalytic power obtained for the series of (R) substrate analogs with varying acyl chain lengths. The number of carbon atoms of the chain that is split off (chain c) is indicated on the *x* axis. The other two chains (a and b) have a fixed length of three carbons only.

K_m^*) can be derived. Although this method does not allow us to obtain information about binding affinities and turnover as such for single substrate molecules bound to cutinase in the detergent complexes formed, we can obtain further insight into the mechanism of action by changing the chain length, e.g., of the fatty acid that is cleaved off by cutinase. In addition we can modify cutinase by protein-engineering techniques[8] in those areas that are likely to interact with these substrate analogs during turnover. The results of such studies are summarized in Fig. 9.

For wild-type cutinase it is shown that maximum catalytic power is obtained for chain lengths (c) of about five carbon atoms, when the other two chains (a and b) have a fixed length of three carbons. This tells us that the binding cleft of cutinase that accommodates the chain to be split off must be relatively short indeed.

From inspection of the molecular model of cutinase (Fig. 5) it can be predicted that amino acid residues 80–89, present in a short helical region and in one of the loops close to the active site serine, should be in close contact with the bound acyl chain. This was verified by construction of the cutinase variant Ala85Phe. The results of the kinetic experiments for this variant are also summarized in Fig. 9. Owing to the presence of the more bulky phenylalanine residue, the maximum chain length was found to increase by approximately two carbon atoms. This confirms the functional importance of this loop and also shows a way to modify the chain length preference of cutinase. The changed kinetic properties of the Ala85Phe variant could also be shown using, e.g., the standard triolein emulsion pH-stat assay [one emulsified volume of 10% (w/w) triolein, 10% (w/w) gum arabic, and 80% (w/w) water, admixed with 2 vol of a buffer solution containing 5 m*M* Tris, 20 m*M* $CaCl_2$ and 50 m*M* NaCl at pH 9.0 and 40°]. The specific activity increased about fivefold relative to wild-type cutinase for this type of substrate.

In summary, this work on cutinase is starting to provide some insight into the mechanism of action of lipolysis, but requires further investigation to gather detailed understanding of the mechanism of action of lipolytic enzymes.

[8] I. A. van Gemeren, W. Musters, C. A. M. J. J. van Hondel, and C. T. Verrips, *J. Biotechnol.* **40,** 155 (1995).

[7] Surface and Electrostatics of Cutinases

By Maria Teresa Neves Petersen, Paulo Martel, Evamaria I. Petersen, Finn Drabløs, and Steffen B. Petersen

Introduction

The surface of a protein constitutes the interface through which the protein senses the environment surrounding it. Therefore, it should be central to any study aiming at a better understanding of the molecular basis for the interaction between an enzyme and its substrate or inhibitor, a receptor and its ligand, or any other type of molecular recognition. Thus, in a broader sense, protein interactions with aqueous and nonaqueous solvents are also a consequence of the kind of surface composition the protein has. Lipases constitute a special challenge for protein electrostatic studies because they operate at or close to a molecular interface. In proteins for which we have three-dimensional structural insight, the majority of the residues have one or more atoms in contact with the solvent, whereas the majority of the atoms are buried. The surface-positioned residues are (on average) 70% hydrophilic, including charged residues, and the rest are hydrophobic residues (see Figs. 1 and 2[1,1a]). The surface is a complex steric arrangement of residues, where any residue type can be found. Yet, the type and exact position of the different residues have a crucial impact on functional parameters, such as thermal stability, substrate specificity, activity, and pH optimum. What may at first sight appear to be a random distribution of various amino acid residues on the protein surface is, in reality, nature's outstanding achievement with respect to the 3.7 billion-year evolution of biological sequences and their associated three-dimensional structures. We are presented with many variations on the same theme, i.e., we may have many members of a given protein family, each member of which has an amino acid sequence that is different but shares distinct similarity with the other members of its family.

All long-range intermolecular forces are believed to be essentially electrostatic in origin. Therefore, the molecular understanding of the initial interaction between a protein and, e.g., its substrate or inhibitor, is essentially an undestanding of the role of electrostatics in intermolecular interactions, such as molecular recognition. Short-range interactions, such as hydrophobic interactions, may not be explained solely on the basis of current

[1] P. H. Jonson and S. B. Petersen, unpublished results (1995).
[1a] S. Miller, J. Janin, A. M. Lesk, and C. Chothia, *J. Mol. Biol.* **196,** 641 (1987).

0076-6879/97 $25.00

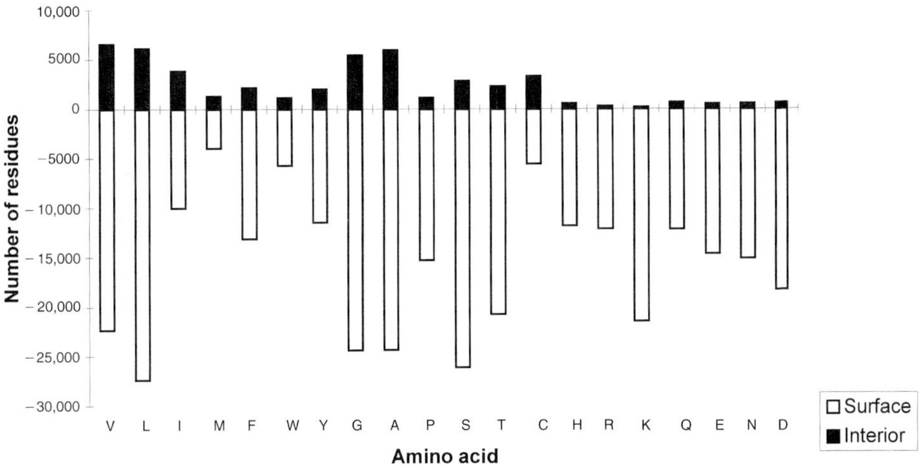

FIG. 1. Number of amino acids (aa) present in the 10 following proteins[1]: 1ace (acetylcholinesterase, 526 aa), 1cgt (CGTase, 684 aa), 1lyb (cathepsin D, 343 aa), 1lz3 (lysozyme, 129 aa), 1pis (phospholipase A_2, 124 aa), 1piv (poliovirus, 846 aa), 1sgt (trypsin, 223 aa), 2hbf (hemoglobin, 287 aa), 2hpq (hydrolase, 363 aa), and 8fab (immunoglobulin, 642 aa). Positive values indicate number of interior residues, negative values indicate surface residues. The expected low number of charged residues in the interior of the proteins can be observed. A cutoff of zero was used, meaning that residues with a solvent accessibility greater than zero were considered surface residues.

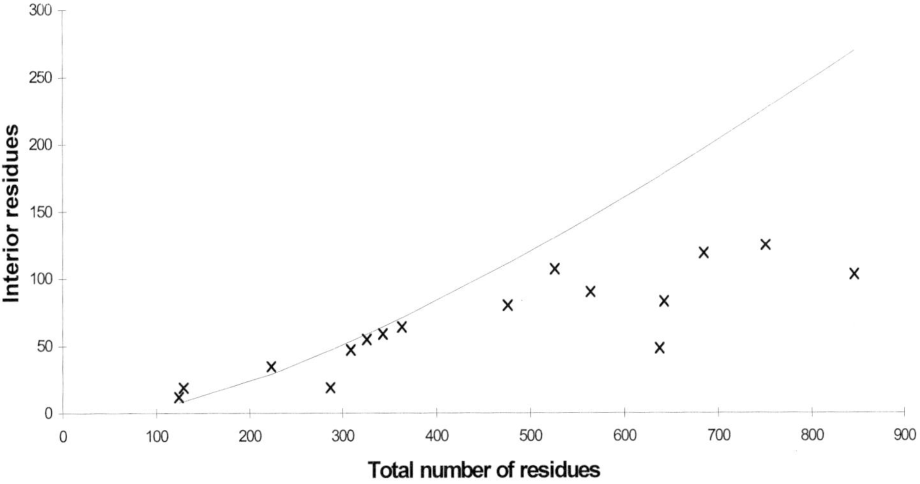

FIG. 2. Number of interior residues (×) as a function of the total number of amino acids.[1] The exponential curve is given by the Miller *et al.* (1987)[1a] function.

classic electrostatic models. The force fields that protein molecules generate are predominantly electrostatic in nature. Typically such fields may extend 10–15 Å away from the protein, depending on solvent, temperature, and the net charge of the protein. Beyond this loosely defined threshold, the random thermal fluctuations of the solvent dominate any role that the electrostatic field may have. The distance dependence of the electrostatic potential is modulated by the medium through which the electrostatic forces act. It reaches further in a low-dielectric medium as compared to a high-dielectric medium. Lowering the temperature can also be expected to expand the sphere inside which the electrostatic forces are dominant.

The fundamentals of classic electrostatics are well known and can be stated concisely with a few elegant equations. The apparent simplicity of these equations, however, can hide the difficulties of applying them to complex systems. The problem is particularly important in studies of proteins and nucleic acids, owing to the vast amount of structural information about these macromolecules now available. In contrast to the traditional models, in which proteins were treated as low-dielectric spheres and DNA as a charged cylinder,[1b] most current questions of interest are asked at the atomic level. The challenging task of representing atomic and molecular properties within the framework of electrostatic theory poses new conceptual as well as numerical difficulties. It is not uncommon to encounter the opinion that models based on classic electrostatics have been superseded, or even invalidated, by the advent of computer simulations of atomic motions.[1] However, it is possible to apply the theory in a physically meaningful way to the system being studied and classic electrostatics still provides a satisfying and intuitive approach to a wide range of microscopic phenomena.

The total electrostatic energy of a macromolecular system can be partitioned into contributions from interactions between pairs of fully or partially charged atoms on the macromolecule and from the interactions of individual charges with the solvent.[2] Charge–solvent interactions play an important role, for example, in the energetics of protein folding and denaturation and the strong favorable interactions between charged amino acids and the aqueous solvent may explain the tendency of these groups to be found at or near the protein surface.[3,4] However, these residues can generate electrical potentials that, depending on solvent and temperature, may extend several angstroms into the solution. Such electrical potentials have an important role in catalysis because they can enhance substrate binding, stabilize transi-

[1b] K. A. Sharp and B. Honig, *Annu. Rev. Biophys. Biophys. Chem.* **19,** 301 (1990).
[2] M. K. Gilson, A. Rashin, R. Fine, and B. Honig, *J. Mol. Biol.* **183,** 503 (1985).
[3] C. H. Paul, *J. Mol. Biol.* **155,** 53 (1982).
[4] A. A. Rashin and B. H. Honig, *J. Mol. Biol.* **173,** 515 (1984).

tion states, and assist in efficient product release. We are convinced that the shape and intensity of the electrostatic potentials as a function of pH determine, to a large extent, the pH activity profile of the enzyme.[5]

In the study of biological macromolecular systems it is becoming increasingly evident that electrostatic interactions contribute significantly to folding, conformational stability, enzyme activity, and binding energies as well as to protein–protein interactions. In this chapter, we present approaches to modeling electrostatic interactions in biomolecular systems. We describe an approach for the calculation of the pK_a values of titratable groups in proteins. We also present some methods that can be used to map and study the amino acid distribution on the molecular surface of proteins. The combination of graphic visualization of the electrostatic fields with the knowledge about the location of key residues on the protein surface allows us to envision atomic models for enzyme function.

Last, we apply some of these methods to the enzymes of the cutinase family.

Modeling Electrostatic Interactions

Electrostatic interactions in macromolecular systems arise from the following sources: the presence of local charges, the polarization stemming from the nonspherical distribution of electron density around atoms, the redistribution of electrons caused by local electrical fields (electronic polarization), and the reorientation of polar groups in the solute and solvent molecules in response to the electric field (orientation polarization).[6] As was alluded to above, electrostatic interactions are believed to be essential for the understanding or prediction of the behavior of native or mutated proteins.

The approaches to model electrostatic interactions in chemical and biochemical systems, either from a purely theoretical or a computational point of view, can be divided in two broad types. The earlier models obviated the atomic-level description by treating solute and solvent as homogeneous dielectric media in which charges were distributed in a discrete or continuous fashion.[7–10] In this way, the treatment of atomic electrostatic interactions was reduced to a problem of classic continuum electrostatics (CE). These models were justifiable given the lack of atomic-level information about

[5] M. T. Neves Petersen, E. I. Petersen, and S. B. Petersen, in preparation (1996).

[6] J. Moult, *Curr. Opin. Struct. Biol.* **2,** 223 (1992).

[7] M. Born, *Z. Phys.* **1,** 45 (1920).

[8] V. P. Debye and E. Hueckel, *Phys. Z.* **24,** 185 (1923).

[9] L. Onsager, *J. Am. Chem. Soc.* **58,** 1486 (1936).

[10] J. G. Kirkwood, *J. Chem. Phys.* **2,** 351 (1934).

biological molecules at the time they were developed, and the limited computational facilities. With the advent of computers and high-resolution molecular structure techniques, new methods were introduced for calculations based on simulations at the atomic level, namely Monte Carlo[11] and molecular dynamics.[12,13] These atomic-level methods became a common practice in chemical physical studies and were later extended to a wide range of systems of chemical and biological interest.[14–17] The atomic detail of these methods led to a neglect of CE-based methods, whose less detailed nature became regarded as a crude approximation. However, the development of fast numerical and computational methods made it possible to achieve a quantitative level in CE calculations and caused a revival in the use of CE methods.[18,19] The work presented in this chapter is primarily based on CE methods.

Continuum Electrostatics

In the most simplistic approach to the modeling of electrostatic interactions in proteins, one can assume that charges on a protein interact through a medium characterized by a single dielectric constant and that all interactions can be described by Coulomb's law. However, this approach fails because the protein and the solvent have different dielectric properties.

In a system with different dielectric constant regions, the apparent forces between charges do not have to occur along the straight line connecting them. Instead, electric field lines travel in curved paths over all space, and apparent paradoxical situations may result, e.g., two charged residues may interact with an electrostatic energy characteristic of a dielectric constant even higher than the one from the solvent.[20] The use of a single dielectric constant cannot account for this kind of effect. A more realistic approach

[11] N. Metropolis, A. W. Rosenbluth, M. N. Rosenbluth, A. H. Teller, and E. Teller, *J. Chem. Phys.* **21,** 1087 (1953).

[12] B. J. Alder and T. E. Wainwright, *J. Chem. Phys.* **31,** 459 (1959).

[13] A. Rahaman, *Phys. Rev.* **136A,** 405 (1964).

[14] M. P. Allen and D. J. Tildesley, "Computer Simulation of Liquids." Clarendon, Oxford, 1987.

[15] G. Ciccotti, D. Frenkel, and I. R. McDonald, "Simulation of Liquids and Solids." North-Holland, Amsterdam, 1987.

[16] J. A. McCammon and S. C. Harvey, "Dynamics of Proteins and Nucleic Acids." Cambridge University Press, Cambridge, 1987.

[17] W. F. van Gunsteren and P. K. Weiner, "Computer Simulation of Biomolecular Systems." Escom, Leiden, 1989.

[18] A. A. Rashin and M. A. Bukatin, *Biophys. Chem.* **51,** 167 (1994).

[19] B. Honig and A. Nicholls, *Science* **268,** 1144 (1995).

[20] A. Baptista, T. Brautaset, F. Drablos, P. Martel, S. Valla, and S. B. Petersen, *in* "Carbohydrate Bioengineering" (S. B. Petersen, B. Svensson, and S. Pedersen, eds.), pp. 181–204. Elsevier Science B. V., Amsterdam, 1995.

is to model the protein and solvent as regions with different dielectric constants. This means that the interactions can no longer be computed using Coulomb's law. Instead, the Poisson equation of the system of charges and dielectrics must be solved.[21,22] In addition, it is reasonable to assume that the protein surrounds itself with an atmosphere of counterions, as described by the Debye–Hückel theory of electrolytes.[21,23,24] In this case the Poisson–Boltzmann equation, usually in its linear form, is solved. When the system has some symmetry it is often possible to express the solution of either the Poisson or the Poisson–Boltzmann equation in an analytic form. A simple approximation is to consider the protein to be a sphere with the charges placed at a small distance beneath the surface and surrounded by an ionic atmosphere.[20] Although proteins are never perfectly spherical, this model was shown to give satisfactory results in many cases,[25] especially when the interactions are corrected according to the solvent accessibility of the residues.[26] Although these simple spherical models do not include atomic detail to any substantial extent, they have the advantage of being analytically solvable and computationally accessible with present-day computers.

The dielectric properties of a system are described by the dielectric constant, which reflects the reorientation of dipoles under the local electric field. These dipoles are essentially of two types: permanent and induced. Permanent dipoles occur when the distribution of charge over neighboring atoms is not symmetric. Typical examples are the peptide bond and the water molecule. Induced dipoles arise from electronic polarization, i.e., from the distortion of electron clouds immersed in an electric field. In liquid water, the relative freedom of the molecules allows a high dipolar rotation and consequently a high dielectric constant (78.5 at 298 K). The contribution of electronic polarization to this overall value is very small, ~4. In contrast, permanent dipoles in the protein interior are virtually fixed and the orientation of the induced dipoles leads to a much smaller dielectric constant. Both experiment and theory point to 2–4 for the protein dielectric constant, with electronic polarization considered to be the most important contribution.[27]

The resulting dielectric regions can be seen as a cavity (see Fig. 3) with a low dielectric constant ε_p (the protein) immersed in a continuous medium

[21] P. W. Atkins, "Physical Chemistry." Oxford University Press, Oxford, 1994.

[22] J. D. Jackson, "Classical Electrodynamics." Wiley & Sons, New York, 1975.

[23] J. O. Bockris and A. K. N. Reddy, "Modern Electrochemistry." MacDonald, London, 1970.

[24] D. A. McQuarrie, "Statistical Mechanics." Harper, New York, 1976.

[25] J. B. Matthew, *Annu. Rev. Biophys. Biophys. Chem.* **14,** 387 (1985).

[26] S. J. Shire, G. I. H. Hanania, and F. R. N. Gurd, *Biochemistry* **13,** 2967 (1974).

[27] P. Martel, Ph.D. thesis. Universidade Nova de Lisboa, Lisbon, Portugal, 1996.

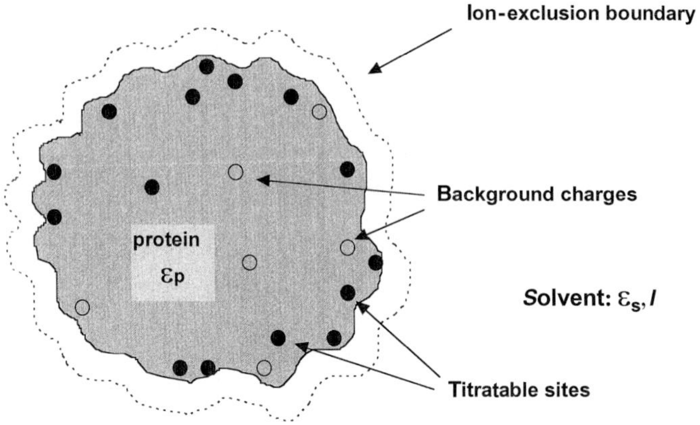

FIG. 3. Continuum electrostatic model of a protein in solution. ε_p, Protein dielectric constant; ε_s, solvent dielectric constant; *I*, solvent ionic strength.

with a high dielectric constant ε_s (the solvent). In this model, formal charges are assigned to all titratable residues, depending on pH and pK_a, and bound ions can be included. In more detailed models, partial charges on all atoms can be included. The charge sources have been divided in two groups: the background charges and the titratable charges. Although some of the atomic charges are independent of the protonation state of the molecule (background charges, e.g., partial charges carried by the peptide bond atoms and partial charges carried by nontitratable polar groups, such as the hydroxyl group of serine and threonine), partial atomic charges in the vicinity of the titratable protons of ionic (titratable) residues (aspartate, glutamate, lysine, arginine, histidine, tyrosine, free cysteine, N- and C-termini) are generally pH dependent, as a consequence of the protonation/deprotonation reactions. In some cases, the contribution of the background charges is not included and the formal charges of the titratable residues are taken as the only electrostatic field sources in the protein. The spatial location of the titratable moieties on the protein derives from the coordinate information obtained from X-ray or nuclear magnetic resonance (NMR) studies. In the absence of such information one may have to rely on homology-based modeling. The charges on the solvent molecules, however, are assumed to be averaged out in the dielectric-based continuum description. However, the polarization of the molecular surface reflects the orientation of the water molecules throughout the solvent. If there are ions present in the aqueous phase, their distribution will be affected by the protein charges, and in the CE model this effect is normally accounted for through the use of a counterionic charge. The counterions cannot approach the protein more closely than is allowed by their ionic radii, which define an ionic exclusion boundary. The counterion distribution is usually assumed to be determined simply by the electrostatic potential and the solution ionic strength, I, as in the Debye–Hückel theory of ionic solutes.[23,24]

The boundary between the two regions can be obtained by using one of the commonly used definitions of molecular surface,[28] such as the Connolly molecular surface[29] equivalent to the Richards contact and reentrant surfaces.[30] In most cases, the surface is determined by rolling a spherical probe, with the radius of a solvent molecule, e.g., water, on the surface of a molecule.

Modeling Effects of pH on Proteins

Enzymes require that the catalytic residues have the appropriate protonation state in the active pH range. Thus, pH is of key importance for enzyme activity. Usually, proteins become unstable at extreme pH values, not only because of acid- and base-catalyzed reactions but also because of changes in the formal charge states of the titratable groups. Ever since these principles were recognized, there has been great interest in the physical basis of the pH-dependent phenomena in proteins. It is clear that a successful structure-based model for the prediction of such phenomena would contribute significantly to our understanding of enzyme mechanisms, protein stability, and molecular recognition.

Electrostatic Model

The direct result of a pH change is a modification in the equilibrium concentrations of the protonated and deprotonated forms of the titratable sites. The most pronounced consequence of this modification is a corresponding change in the average charge of the titratable sites. Therefore, electrostatic interactions are widely believed to be the primary forces controlling pH-dependent phenomena. As a consequence, the development of the Poisson–Boltzmann method for computing detailed electrostatic fields in and around macromolecules has led to a burst of new activity in the theory of pH-dependent phenomena.[31]

At first glance, the theoretical task of explaining and predicting these pH-dependent electrostatic changes may seem straightforward—given the

[28] B. Lee and F. M. Richards, *J. Mol. Biol.* **55,** 379 (1971).
[29] M. L. Connolly, *Science* **221,** 709 (1983).
[30] F. M. Richards, *Annu. Rev. Biophys. Bioeng.* **6,** 151 (1977).
[31] J. Antosiewics, J. A. McCammon, and M. K. Gilson, *J. Mol. Biol.* **238,** 415 (1994).

TABLE I
pK_a VALUES OF TITRATABLE GROUPS[a,b]

Group	pK_a Model compounds (pK^{model})	pK_a Usual range in proteins
Amino acid α-COOH	3.6	
Aspartate (COOH)	4.0	2–5.5
Glutamate (COOH)	4.5	
Histidine (imidazole)	6.4	5–8
Amino acid α-NH_2	7.8	~8
Lysine (ε-NH_2)	10.4	~10
Arginine (guanidine)	~12	—
Tyrosine (OH)	9.7	9–12
Cysteine (free SH)	9.1	8–11

[a] T. E. Creighton, "Proteins: Structure and Molecular Properties." W. H. Freeman & Company, New York, 1993.

[b] A. Fersht, "Enzyme Structure and Mechanism." W. H. Freeman & Company, New York, 1985.

pK_a of the titratable residue (available in any biochemistry handbook; see Table I) it would be a trivial matter to tell whether a given group is charged or not at a particular pH value. However, the situation is far more complicated because the other charged sites and the local environment in the protein may shift the pK_a of a given site from its typical value by several pH units[27] (see Tables I and II). In fact, as shown below, even the usual concept of pK_a becomes, to some extent, inappropriate.

Following the nomenclature of Bashford and Karplus,[32] we use the following terms.

pK^{model}: The pK_a of a titrating group in a small model compound, supposedly free from the action of other titrating groups. It can be measured by NMR or other titration methods

pK^{int}: The pK_a of a titrating site with all other groups in the protein neutralized. This quantity depends not only on the residue type but also on its location in the protein. It is pH independent

pK^{eff}: The pK_a displayed by a given group at a given pH by the fully charged protein. This quantity changes with pH throughout the titration owing to the mutual interactions between groups

$pK_{1/2}$: The pH at which the residue is half-protonated

[32] D. Bashford and M. Karplus, *Biochemistry* **29,** 10219 (1990).

TABLE II
HIGHLY PERTURBED pK_a VALUES IN PROTEINS[a,b]

Enzyme	Residue	pK_a
Lysozyme	Glu-35	6.5
Lysozyme–glycolchitin complex	Glu-35	~8.2
Carboxypeptidase A	Glu-27	7.0
Acetoacetate decarboxylase	Lys (ε-NH_2)	5.9
Chymotrypsin	Ile-16 (α-NH_2)	10.0
α-Lactalbumin	COOH	7.5
Rhodanese	Cys-247	6.5
Papain	His-159	3.4
	Cys-25	3.3
Pepsin	Asp-32	1.5

[a] T. E. Creighton, "Proteins: Structure and Molecular Properties." W. H. Freeman & Company, New York, 1993.
[b] A. Fersht, "Enzyme Structure and Mechanism." W. H. Freeman & Company, New York, 1985.

Let us start by considering a single titratable site. In this case, the protonation equilibrium is fully described by the pK_a of the site through the familiar Henderson–Hasselbalch equation of the acid-base equilibrium[21,27]:

$$pK_a = pH + \log \frac{f}{1-f} \tag{1}$$

where f is the degree of protonation, i.e., the fraction of molecules that has the site protonated. From Eq. (1), it can be seen that the pK_a is the pH value at which the site is half-protonated.

The pK_a value measured in solution for the model compound (pK^{model}), typically Gly–X–Gly, where X is the residue in question, reflects an aqueous environment for the residue, considered completely solvent accessible. However, when the titratable residue is transferred from the model compound into a specific site in the protein (see Fig. 4), new terms contribute to the energetics of its titration[28,31]: The *born*, or desolvation, term represents the free energy change in the protonation reaction resulting from the burial of the residue in the protein low dielectric; and the *background* term describes the free energy change resulting from the interaction of the residue with the other nontitratrable charges in the protein (e.g., peptide bond dipoles and polar atoms). Together, these two terms account for the difference between pK^{int} and pK^{model}.

A third energetic term describes the interaction of the residue with all other titratable residues in the protein. The magnitude of this interaction

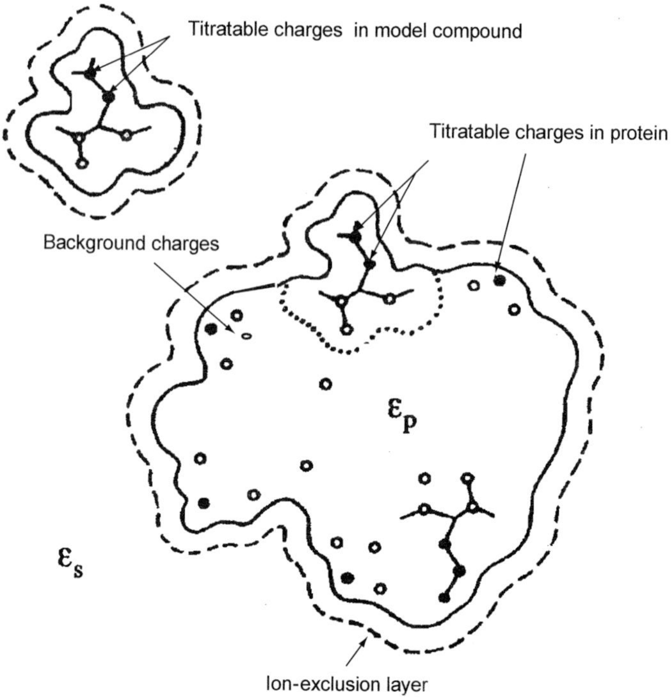

FIG. 4. Calculation of pK^{int} and pK^{eff} values.[27] See text for details. ε_p, Protein dielectric constant; ε_s, solvent dielectric constant.

term is pH dependent. All terms can be computed by the use of the linearized Poisson–Boltzmann equation.

The pK_a value resulting from the insertion of the amino acid residue into a neutral protein (see Fig. 4) is usually referred to as the *intrinsic* pK_a, pK^{int}, and may be written as

$$pK^{int} = pK^{model} + 1/(2.3k_BT)\Delta\,\Delta G^{env} \tag{2}$$

where ΔG^{env} is the free energy change due to moving the residue from water into the neutral form of the protein.

If we had only one titratable residue in the protein molecule, the protonation equilibrium would be given by Eqs. (1) and (2), with $pK_a = pK^{int}$. However, when other titratable or permanently charged sites exist, the electrostatic interaction between them needs to be considered as well. Thus, the way in which the pK^{int} of a given site is affected by a nearby group depends on whether the latter is charged or not. But, conversely, the protonation state of the second group will also depend on the protonation of the first. Another way of stating the problem is to say that a protein with s titratable sites has 2^s possible protonation states, and to characterize the protonation equilibrium of a single titratable site we must specify the populations of each of its two forms at each of the 2^s forms of the protein at each pH value. The probability of each protonation state can be computed,[32–36] and this task is sometimes referred to as the multiple state titration problem. Thus, to account for the additional interactions that an amino acid residue displays with other charged sites in the protein, an effective pK_a is defined (see Fig. 4):

$$pK^{\text{eff}} = pK^{\text{int}} + 1/(2.3k_BT)\,\Delta G^{\text{inter}} \quad (3)$$

where ΔG^{inter} is the electrostatic contribution due to the interaction with other charged residues. Because the interaction term is a pH-dependent quantity, the pK^{eff} itself becomes pH dependent and it can no longer be equated with the pH corresponding to half-protonation.

Methods: A Practical Approach

After presenting a view of electrostatic interactions from the point of view of CE and how to model the effect of pH on proteins it seems appropriate to present and discuss some methods from the point of view of their implementation and use, introducing the software tools that we use in this work. The programs TITRA, ACC_RUN, DelPhi, and Grasp are therefore briefly described. Some examples of the use of Coulombic and pH-dependent electrostatic potential mapping is presented in this section, using cutinases as model proteins.

Program TITRA: A Program for pK_a and Titration Curve Calculations in Proteins

The TITRA program, written by Martel and Petersen,[27] is a protein titration program implementing the Tanford–Kirkwood (TK) sphere model for site–site interactions (see Fig. 5)[27,33] and the Tanford–Roxby iterative mean field approximation[37] for calculation of the average protonation state of the titratable sites. Here we outline the general workings of the program.

The general flow of the TITRA program is shown in Fig. 6. First, files containing atomic accessibilities (AA) or side-chain accessibilities (SA) for

[33] C. Tanford and J. G. Kirkwood, *J. Am. Chem. Soc.* **79,** 5333 (1957).
[34] D. Bashford and M. Karplus, *J. Phys. Chem.* **95,** 9956 (1991).
[35] A. S. Yang, M. R. Gunner, R. Sampogna, K. Sharp, and B. Honig, *Proteins* **15,** 252 (1993).
[36] M. K. Gilson, *Proteins* **15,** 266 (1993).
[37] C. Tanford and R. Roxby, *Biochemistry* **11,** 2192 (1972).

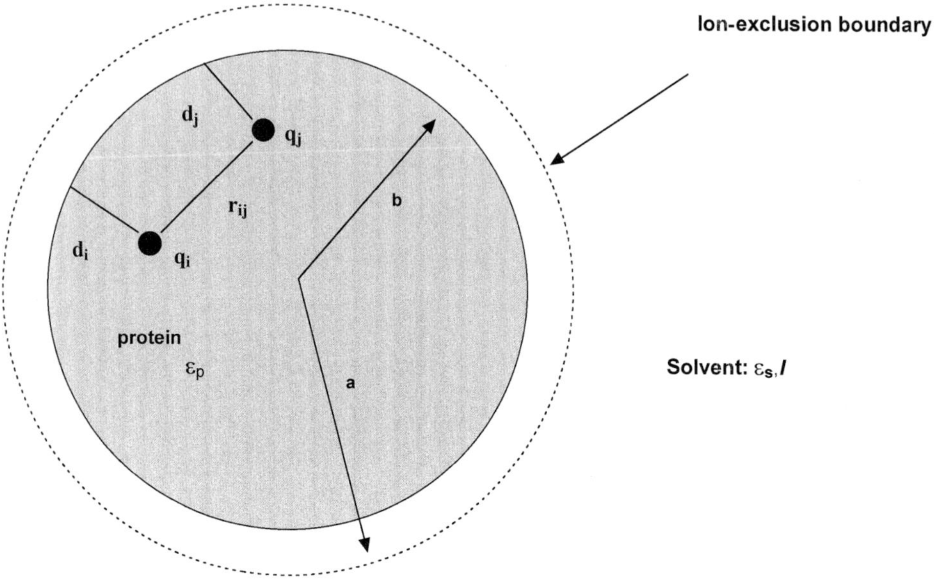

FIG. 5. The Tanford–Kirkwood model. The protein is considered spherical. Usually, charges (e.g., q_i and q_j) are considered to be at the same depth from the surface of the sphere (i.e., the same d_i) and separated by the experimental (e.g., crystallographic) r_{ij} distances. ε_p, Protein dielectric constant; ε_s, solvent dielectric constant; *I*, solvent ionic strength; *b*, protein radius; *a*, protein radius plus ion-exclusion boundary radius.

all the atoms, pK^{int} values for each of the titratable sites, and Tanford–Kirkwood model parameters are read, and user options and arguments processed. A set of titratable residues and atomic locations for charge placement is selected according to default internal rules and/or information specified in user input files.

Values for the site–site coupling function W_{ij} (described below) are then computed, using the Tanford–Kirkwood formula [Eq. (4)], for a range of distances specified by the cutoff values, and stored in a table for later use. The fractional charge of each site is computed at the starting pH value, using the pK^{int} value for that group and Eq. (1). The total electrostatic potential at each group, generated by the remaining groups, is determined using the previously calculated partial charges and W_{ij} coupling terms, which may be adjusted by the modified Tanford–Kirkwood correction factor $(1 - SA_{ij})$,[38] depending on the user input options. The total potential at the site is then used to calculate its ΔpK^{eff} value, and the site pK^{eff} is updated.

[38] S. J. Shire, G. I. H. Hanania, and R. F. N. Gurd, *Biochemistry* **13,** 2967 (1974).

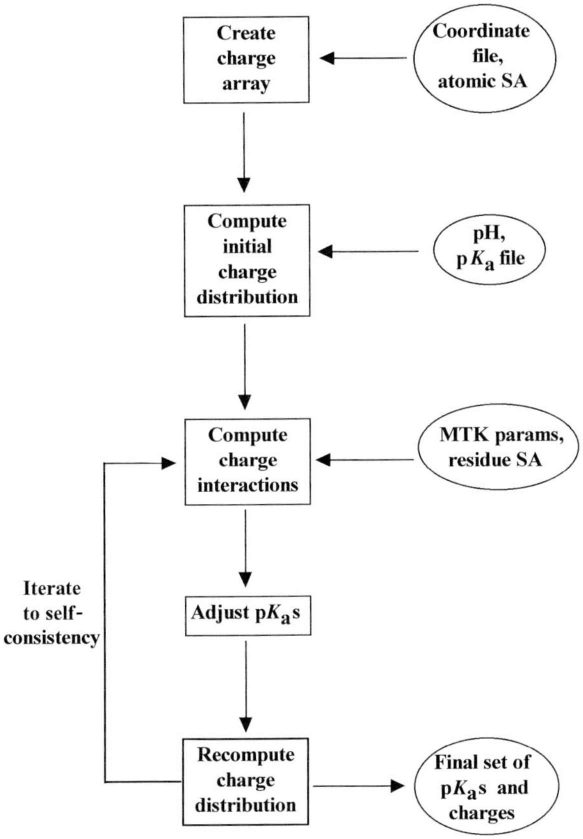

FIG. 6. Flow chart describing the steps within the program TITRA. SA, Side-chain accessibility.

After all site pK^{eff} values have been updated, a new cycle starts. Iteration proceeds until the change in calculated mean charges $\overline{z}_i$ and pK^{eff} values goes below a predefined error tolerance or the maximum allowed number of iterations is reached. The whole iterative procedure is repeated for the range of pH values requested by the user, and the total charge and pK^{eff} values at each pH are stored in separate output files. A number of user options may change details of this procedure. Energy values may be read from a precomputed table stored in disk, or a set of site–site coupling constants W_{ij} may be read from a file. The format of the pK^{int} input file[27] allows the values of selected residues to be preset or fixed at given pK_a or charge values (fixing the charge value of a site creates a background charge, with a pH-independent value).

Site–Site Coupling Factors. The pairwise interaction energies W_{ij} between positive unit charges i and j, placed at a certain depth under the surface of a sphere of radius b and ion-exclusion radius a, and at a distance r_{ij} from each other (see Fig. 5), are calculated using the Tanford–Kirkwood formula[10,33]:

$$W_{ij} = R[(A_{ij} - B_{ij})/b - C_{ij}/a] \tag{4}$$

where R is a constant dependent on the choice of units.[27]

The terms A_{ij}, B_{ij}, and C_{ij} in Eq. (4) can be understood in physical terms as follows[27]: A_{ij} represents the direct Coulombic interaction of charges i and j in an unbounded medium with dielectric constant ε_p, B_{ij} is the interaction between charge i and the surface polarization induced by charge j on the spherical dielectric boundary, and C_{ij} describes the interaction of charge i with the counterionic atmosphere of charge j. If $\varepsilon_p = \varepsilon_s$ and $I = 0$, then the terms B_{ij} and C_{ij} vanish and Eq. (4) becomes the familiar Coulomb formula for a medium with dielectric ε_p.

Mean Field Titration. After the coupling factors W_{ij} have been evaluated, using Eq. (4), for all site pairs (i, j), the calculation of pK_i^{eff} and $\overline{z_i}$ takes place using the preceding self-consistent algorithm, based on the iteration of the Eqs. (5) and (6)[27]:

$$\overline{z_i} = z_i^0 + \frac{10^{(pK_i^{eff}-pH)}}{1 + 10^{(pK_i^{eff}-pH)}} \tag{5}$$

$$pK_i^{eff} = pK_i^{int} - \frac{1}{2.303k_BT}\sum_{i \neq j} W_{ij}\overline{z_j} \tag{6}$$

where z_i^0 is the charge of site i in the deprotonated state, and $\overline{z_i}$ the average charge of the same site. As previously described, iteration of Eqs. (5) and (6) proceeds until the change in z_i and pK_i^{eff} falls below the error threshold, for all sites.

Bashford and Karplus[34] have shown that the iterative scheme based on Eqs. (5) and (6) corresponds to a mean field approximation, where the probability $p(\mathbf{n})$ of a protonation state can be written as

$$p(\mathbf{n}) = \prod_{i=1} p(n_i) \tag{7}$$

where $p(n_i)$ is the probability of finding site i in state n_i. The mean field approximation can lead to erroneous results only in the case of strongly coupled sites ($W_{ij} > 1$ pH unit), and if the interacting residues titrate in the same pH range. Also, the errors are much smaller for integral properties such as the total protein charge.[35]

The titration curve, i.e., the total charge of the protein ($\overline{Z}$) versus pH, is computed by TITRA according to

$$\overline{Z} = \sum_{i} \overline{z_i} \tag{8}$$

and the total electrostatic energy (relative to the discharged state) is

$$G_{\text{Elec}} = \frac{1}{2} \sum_{i \neq j} W_{ij} \overline{z_i} \overline{z_j} \tag{9}$$

TITRA also computes pK_i^{app} values, defined as the pH value at which the site is half-protonated, by using a bisection search algorithm, by which an initially large pH interval is progressively narrowed around the point at which $n_i = 0.5$.[27]

Program ACC_RUN

ACC_RUN is a simple program that computes contact solvent accessibilities.[30] Each atom is modeled as a collection of evenly distributed points on the surface of a sphere. The atom is considered solvent accessible if a water probe tangent to one or more of these points does not overlap any other protein atoms (the water probe is usually modeled as a 1.4-Å sphere). The solvent accessibility is calculated from the fraction of exposed dots on the surface of each atomic sphere.

The program takes as input a Protein Data Bank (PDB) file and a water probe radius value (default value 1.4 Å), and outputs solvent accessibility files for all individual atoms as well as the side chains. The side-chain file contains accessibilities for all side chains, normalized with the standard areas for tripeptides Gly–*X*–Gly in extended conformation,[38] while the atomic accessibility file contains absolute solvent-exposed atomic areas in angstroms squared. The program is written in C and runs under SGI IRIX and Linux. On a typical Unix workstation it takes about 5 sec to compute accessibilities for a 250-residue protein.

The two accessibility files produced by ACC_RUN are required as input for a TITRA calcultion.

Program DelPhi

DelPhi is a software package that calculates the electrostatic potential in and around macromolecules, using a finite difference solution to the nonlinear Poisson–Boltzmann equation.[39] It was developed by Honig and

[39] G. D. Rose, A. R. Geselowitz, G. J. Lesser, R. H. Lee, and M. H. Zehfus, *Science* **229,** 834 (1985).

co-workers at Columbia University (New York, NY)[40–43] and marketed by Biosym Technologies, Inc.[44]

Typical uses for DelPhi include calculating electrostatic potential in and around a protein and displaying isopotential contour maps to gain qualitative information on protein–substrate interactions, and to determine the effects of site-directed mutagenesis on the pK_a values of important residues, on binding energies, and on catalytic rates.

Poisson–Boltzmann Equation: Finite Difference Poisson–Boltzmann Method. A molecule such as a protein has a low dielectric constant because its dipolar groups are frozen into a hydrogen-bonded lattice and cannot reorient in an external electrostatic field. A value near 2 measures its electronic polarization response whereas a value near 4 includes some additional contributions from dipole reorientation. Water, however, has a very high dielectric constant (78.5 at 298 K) because its dipoles reorient more freely. Therefore, a protein molecule in aqueous solution yields a system with two very different dielectric media. The effects of this large difference can be considerable and should be accounted for by the electrostatic model. The Poisson–Boltzmann equation is considered to be a good model for describing protein molecules in water solutions because it accounts for both the effect of dielectrics and ionic strength. Unfortunately, this equation can be solved analytically only for systems with simple dielectric boundary shapes, such as spheres and planar surface. However, most molecules of interest have complex shapes, and their convoluted conformations may have a significant effect on the resulting electrostatic properties. The alternative to analytical solutions is to use numerical techniques to find an approximate solution. DelPhi uses the finite difference Poisson–Boltzmann method (FDPB),[40,41,45] which involves mapping of the molecule onto a three-dimensional cubic grid. The Poisson–Boltzmann equation must be satisfied at each grid point.

The input files for DelPhi include a coordinate file (in PDB format), an atomic radii file, an atomic charge file, and a parameter file containing various parameters and options that control the behavior of the program. These include the grid step, its extent and placement relative to the protein molecule, as well as the ionic strength, the dielectric constants for protein and solvent, and the maximum number of iterations and boundary conditions. Specification of both charged and noncharged atoms is required

[40] B. Honig, K. Sharp, and A. S. Yang, *J. Phys. Chem.* **97,** 1101 (1993).

[41] I. Klapper, R. Hagstrom, R. Fine, K. Sharp, and B. Honig, *Proteins* **1,** 47 (1986).

[42] M. K. Gilson and B. H. Honig, *Nature* (*London*) **330,** 84 (1987).

[43] M. K. Gilson and B. Honig, *Proteins* **3,** 32 (1988).

[44] M. K. Gilson and B. Honig, *Proteins* **4,** 7 (1988).

[45] Biosym/MSI, "DelPhi and Solvation," version 95.0. Biosym/MSI, San Diego, CA 1995.

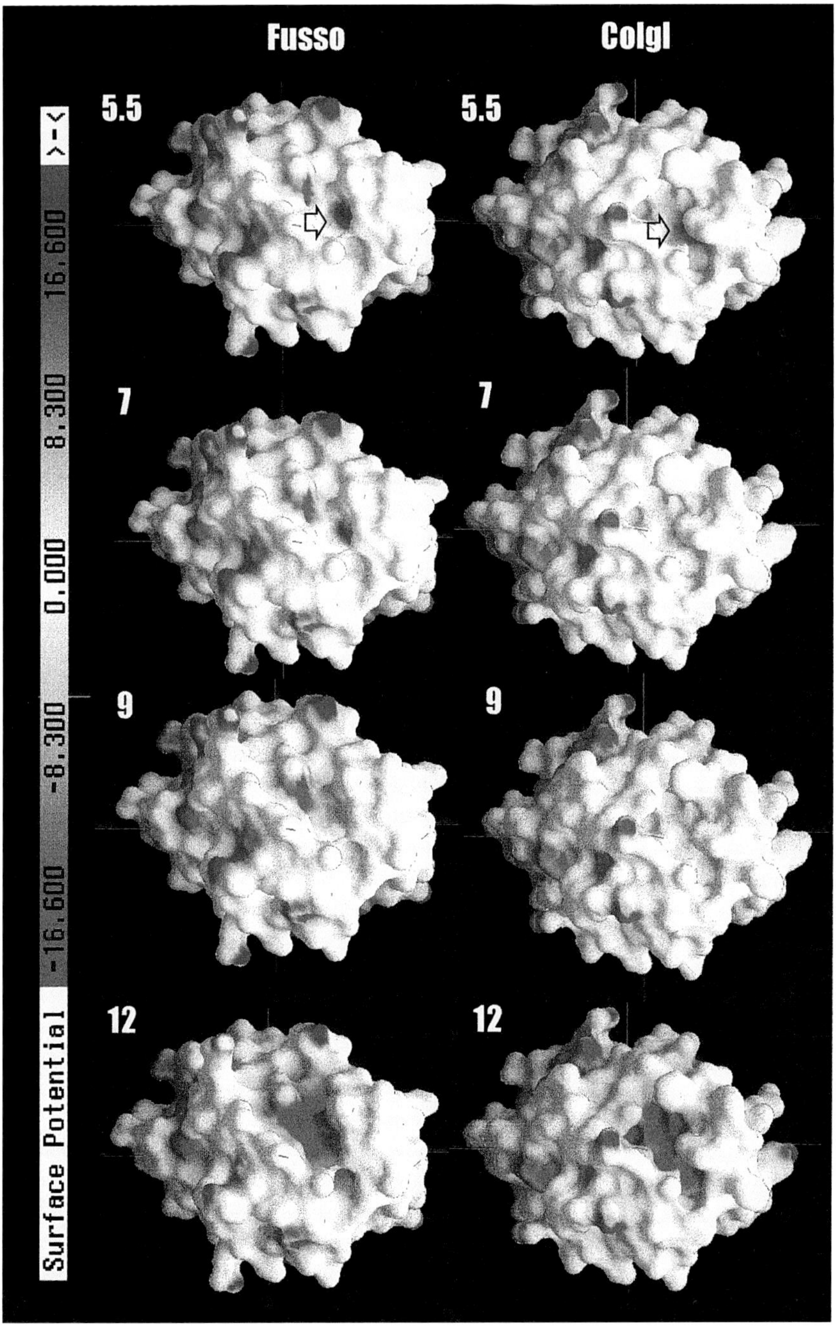

FIG. 7. Electrostatic potential maps displayed on the molecular surface of *Fusarium solani pisi* (Fusso) and *Colletotrichum gloeosporioides* (Colgl) cutinases at different pH values: 5.5, 7, 9, and 12. Blue color represents positive potential and red color negative potential. The white arrow points to the catalytic cleft.

because both contribute to the overall protein surface and, in particular, to the definition of the protein–solvent interface. The program outputs a grid file containing potentials for every grid point and a file containing the potential and electrostatic field vectors at the location of each atom in the system. The grid file can be read by the Biosym viewer program, InsightII,[46] and color-coded equipotential surfaces can be displayed at defined kT/e values. DelPhi charge files can be generated by another application, e.g., TITRA, to allow the display of equipotential surfaces. DelPhi calculation times depend primarily on the total number of grid nodes, but also on the chosen ionic strength and number of point charges of the system, the first two having an effect on the rate of convergence of the iteration. Setting up the molecular surface dielectric boundary takes very little time, owing to the use of an efficient algorithm.[47] The DelPhi computations are not time consuming when compared with, e.g., protein molecular dynamics calculation.

Program Grasp

The program Grasp[48,49] was developed as a consequence of the need for visualizing electrostatic potentials at surfaces, in particular the surface of biological molecules, where the surface is modeled as a solid surface. The program DelPhi, which calculates electrostatic potentials from the Poisson–Boltzmann equation, can be used to obtain quantitative numbers for a variety of biochemical phenomena but visualization has been limited to qualitative isopotential contouring. The limitation of this approach is that the contours do not highlight local topology or shape. They often extend significant distances away from the surface of the molecule, whereas one would expect most of the interactions to be close to the molecule, in fact at the surface of molecules. Although DelPhi can give detailed information about the molecular electrostatic signature or shape, it does not permit concurrent viewing of the electrostatic potential and the molecular surface. On the other hand, Grasp allows for the production of a solid surface, color coded with the local electrostatic potential.

As is shown in the next section, Grasp has proved to be an ideal tool for the study of electrostatics in the active site cleft of cutinases; it allows details of the molecular surface to be viewed simultaneously with the electrostatic potential features [see Fig. 7 (color insert)].

[46] Biosym/MSI, "InsightII," version 95.0. Biosym/MSI, San Diego, CA, 1995.
[47] A. Nicholls and B. Honig, *J. Comput. Chem.* **12,** 435 (1991).
[48] A. Nicholls, K. Sharp, and B. Honig, *Proteins* **11,** 281 (1991).
[49] A. Nicholls, R. Bharadwaj, and B. Honig, *Biophys. J.* **64,** Part 2, A166 (1993).

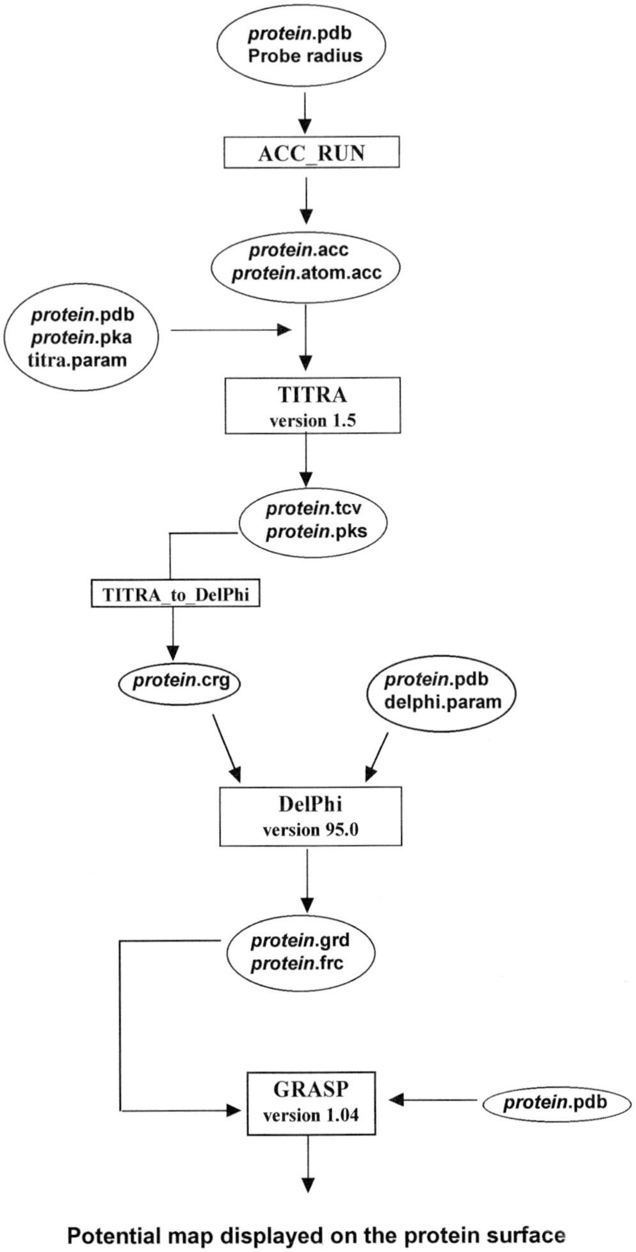
protein.pdb
Probe radius
ACC_RUN
protein.acc
protein.atom.acc
protein.pdb
protein.pka
titra.param
TITRA
version 1.5
protein.tcv
protein.pks
TITRA_to_DelPhi
protein.crg
protein.pdb
delphi.param
DelPhi
version 95.0
protein.grd
protein.frc
GRASP
version 1.04
protein.pdb
Potential map displayed on the protein surface

A flow chart representing the necessary steps for displaying electrostatic potential maps on the molecular surface of a protein is displayed in Fig. 8.

Electrostatic Properties of Cutinases

Cutinases catalyze the hydrolysis of the water-insoluble biopolyester cutin, which covers the surface of plants. They also display a hydrolytic activity toward a broad variety of esters ranging from soluble *p*-nitrophenyl esters to insoluble long-chain triglycerides. Because cutinases hydrolyze soluble esters and emulsified triacylglycerols as efficiently as esterases and lipases, these enzymes can be considered as a link between esterases and lipases.[50]

Cutinases have been isolated from several different sources—fungi,[51–55] bacteria,[56,57] and pollen[58,59]—and at the time of writing this chapter, 13 protein sequences were available in the sequence databases. Fungal cutinases are the best studied cutinases; they exhibit a molecular mass in the 22- to 25-kDa range. The cutinase from the fungus *Fusarium solani* has been cloned and expressed in a heterologous host and its three-dimensional structure has been determined at an exceptionally high resolution using

[50] M. L. M. Mannesse, R. C. Cox, B. C. Koops, H. M. Verheij, G. H. de Haas, M. R. Egmond, H. T. W. M. van der Hijden, and J. de Vlieg, *Biochemistry* **34,** 6400 (1995).

[51] C. J. Baker and D. F. Bateman, *Phytopathology* **68,** 1577 (1978).

[52] C. Yao and W. Koller, *Physiol. Mol. Plant Pathol.* **44,** 81 (1994).

[53] J. A. Sweigard, F. G. Chumley, and B. Valent, *Mol. Gen. Genet.* **232,** 174 (1992).

[54] M. B. Dickman, S. S. Patil, and P. E. Kolattukudy, *Physiol. Plant Pathol.* **20,** 333 (1982).

[55] G. T. Cole and H. C. Hoch, "The Fungal Spore and Disease Initiation in Plants and Animals." Plenum, New York, 1991.

[56] W. F. Fett, H. C. Gerard, R. A. Moreau, S. F. Osman, and L. E. Jones, *Appl. Environ. Microbiol.* **58,** 2123 (1992).

[57] W. F. Fett, H. C. Gerard, R. A. Moreau, S. F. Osman, and L. E. Jones, *Curr. Microbiol.* **25,** 165 (1992).

[58] M. Shayk and P. E. Kolattukudy, *Plant Physiol.* **60,** 907 (1977).

[59] I. B. Maiti, P. E. Kolattukudy, and M. Shaykh, *Arch. Biochem. Biophys.* **196,** 412 (1979).

FIG. 8. Flow chart representing the steps necessary for displaying electrostatic potential maps on the protein molecular surface. The files needed for the different programs are *protein*.acc (side-chain static accessibility file); *protein*.atom.acc (atomic accessibility file); *protein*.tcv (titration curve data); *protein*.pks (information about each titratable site: Residue_name, Residue_number, pH, pK^{int}, pK^{eff}, Partial_charge); *protein*.crg (information about each titratable site: pH, Residue_name, Residue_number, Partial_charge); *protein*.pdb (coordinates of the residues); delphi.param (solvent, solute, and grid parameters); *protein*.grd (potential map file); *protein*.frc (optional file: lists the coordinates, charges, potential, and field components for a specified set of atoms).

X-ray diffraction methods (1.25-Å resolution and an R factor of 0.158 Å.[60] Biochemical properties, such as the pH optimum or substrate specificity, have been reported for some of these enzymes.[54,61] The cutinase of *F. solani,* for instance, shows its highest activity at about pH 8.5 (e.g., toward tributyrin and triolein),[61] whereas the cutinase from *Colletotrichum gloeosporioides* is most active at pH 10 (toward *p*-nitrophenyl butyrate).[54]

To understand the different pH optima of the highly homologous fungal cutinases, sequence alignment, homology modeling, and protein electrostatics studies were carried out.

The distribution of titratable charges at various pH values was computed using the TITRA program[27] (see Fig. 6). Solid molecular surface representations of the enzymes were generated using the Grasp program (see Fig. 7). Electrostatic potential maps of the surfaces of each cutinase at different pH values were computed using the charge distribution obtained from the TITRA program as input for the DelPhi program. The surface electrostatic maps were displayed using the Grasp program. The residues belonging to the active site region were identified, defining spherical subsets around the hydroxyl oxygen of the catalytic serine.

The solvent was treated as a continuum with a dielectric constant of 78.5. Theoretical and experimental arguments suggest a value of 2–4 for the dielectric constant of the protein interior. In the present study, a value of 4 was used.[1b]

Results

The simulated titration curves of two fungal cutinases are shown in Fig. 9. The curves appear remarkably similar and can by no means explain the experimental pH optimum difference of 1.5 pH units. However, if we look at the electrostatic potentials at the molecular surfaces displayed on each of the cutinase molecules (see Fig. 7) clear differences can be observed. In the pH ranges where highest activity is reported for these two cutinases, the molecular surface at the active site entrance displays a slightly negative potential. When changing pH to more acidic conditions we observe an increasing polarization of the active site pocket, which presents a more and more positive potential, and when changing pH to more basic conditions the active site potential becomes more and more negative. The same is observed on the active site-flanking regions. If we postulate that the charge

[60] C. Martinez, P. de Geus, M. Lauwereys, G. Matthyssens, and C. Cambillau, *Nature* (*London*) **356,** 615 (1992).

[61] M. Lauwereys, P. de Geus, J. de Meutter, P. Stanssens, and G. Matthyssens, *in* "Lipases: Structure, Mechanism and Genetic Engineering" (L. Alberghina, R. D. Schmid, and R. Verger, eds.). Verlag Chemie, Weinheim, Germany, 1990.

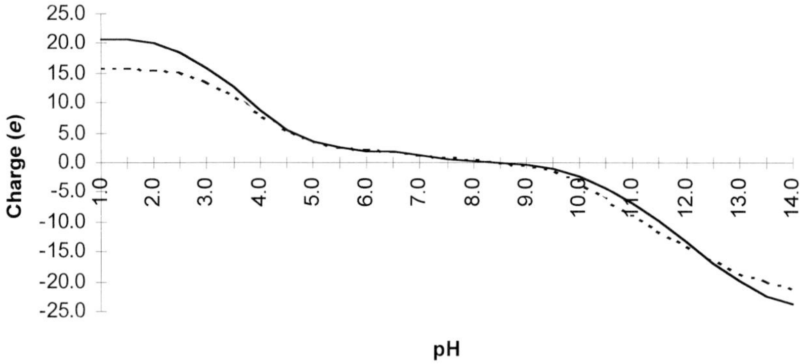

FIG. 9. Titration curves for *Fusarium solani pisi* (Fusso, —) and *Colletotrichum gloeosporioides* (Colgl, ---) cutinases. All titratable residues in the protein are taken into account.

polarity is essential, then the different potential properties observed in the active site clefts correlate with the different pH optima for the enzymes.

To investigate whether charged residues close to the active site could be responsible for the observed pH optima differences, a subset consisting of only the charged residues that have atoms located closer than 10 Å to the hydroxyl oxygen of the active site serine was identified. The "titration curves" for the 10-Å cutoff (see Fig. 10) describe the local charge properties for the active site environment of the enzymes. From Fig. 10 it can be seen that, at the presumed pH optima for the two enzymes, they exhibit very

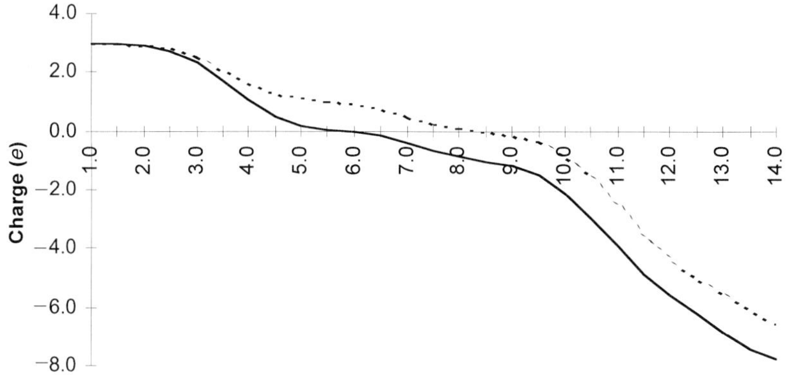

FIG. 10. Titration curves for *Fusarium solani pisi* (Fusso, —) and *Colletotrichum gloeosporioides* (Colgl, ---) cutinases. Only titratable residues within a 10-Å distance of the hydroxyl oxygen of the active site serine are taken into account.

similar net charge (~−1) if the 10-Å cutoff is applied. Although the choice of the 10-Å cutoff is somewhat arbitrary, it does tend to highlight the presence of local charges. This subset effectively highlights the qualitative importance and influence of the local electrostatic environment, which from a physical–chemical perspective is something we could expect.

These results have provided novel insight into the relationship between enzymatic activity and the electrostatic profile of cutinase as a function of pH. This approach may therefore also become of use when protein-engineered cutinases (or other enzymes for which the three-dimensional structure is known or can be modeled) are being evaluated using theoretical methods. If successful, such methods may save significant amounts of time-consuming laboratory work.

Conclusion

The methodologies described in this chapter allow us to predict the pK_a values for each of the amino acid residues of a protein with reasonable accuracy for all solvent-accessible and titratable residues and, consequently, to compute and project onto the surface of the protein the electrostatic potential map at a given pH value. With the program TITRA it is possible to compute, as an example, the theoretical titration curve for *F. solani pisi* cutinase and a good match exists between the observed theoretical isoelectric point (see Fig. 9) and the experimentally determined one (pI_{exp} = 7.8[62]). Using TITRA results as input for DelPhi, the potential maps at different pH values for *F. solani pisi* and *C. gloeosporioides* cutinases were calculated. With the program Grasp, these potential maps were projected and displayed on the molecular surface of each cutinase (see Fig. 7). The described methodologies were applied to these two cutinases to investigate whether electrostatic features could explain the different pH optima for two proteins. After displaying the potential maps on the surface of the proteins, the distribution of electrostatic potentials in the active cleft of both enzymes could be compared. At the active pH optimum, the bottom of the active cleft in both enzymes displays a similar negative potential value. Further studies have been done in this laboratory on other cutinases and similar electrostatic signatures have been identified in these proteins as well.[5]

The potential map around the catalytic cleft allows us to explore the interactions between a particular protein and its substrate as well as with the products of the reaction at different pH values and ionic strength,

[62] M. R. Egmond, J. de Vlieg, H. M. Verhey, and G. H. de Haas, *in* "Engineering of/with Lipases" (F. X. Malcata, ed.), pp. 193–202. Kluwer Academic Publishers, Dordrecht, 1995.

and in the presence of different solvents. The electrostatic potentials are presumed to have an important role in catalysis because they can affect the substrate association rate as well as influence the product release rate. If the substrates or the products are charged compounds, the electrostatic attractive or repulsive interactions are even more important for the molecular basis for catalysis. Another application of these methodologies is the study of electrostatic potential distribution in the metal-binding site(s) and/or allosteric region of an enzyme, when present. These studies may help us to unravel the electrostatic requirements for metal binding or allosteric regulation.

However, the study of the charge distribution at the surface of a protein helps us to predict electrostatic effects on catalytic activity and is an essential component in protein design. It is possible then to predict the pK_a shifts of particular residues as a consequence of charge mutations and thus improve our understanding of the electrostatic requirements for the catalytic mechanism, pH optimum, substrate specificity, substrate-binding rates, and product release rates. We see a clear potential for engineering the surface of proteins to obtain more hydrophobic or hydrophilic proteins while still maintaining the essential electrostatic features in the active cleft and, thereby, enzyme activity. Other applications may include the prediction of mutations that enhance, for example, metal-binding rates or decrease the inactivation rates of enzymes when exposed to denaturants (e.g., it is known that the presence of surfactants may inactivate proteins[62]). Finally, the consequence of the introduction of new salt bridges with the aim of increasing the protein thermal stability may also be evaluated. As in the case of all theoretical approaches, their true value can emerge only when their predictive qualities have been tested against experimental observables. One route, in the case of protein electrostatics, is to test it against a family of related proteins, such as we have demonstrated briefly here in the case of cutinases. Another route is to do rational protein engineering on the charged residues and predict the outcome of such mutations. Such work is currently in progress in our laboratory, as well as in other laboratories.[63,64]

Ultimately, protein electrostatics holds the promise of being capable of contributing to a better understanding of the role of charged residues in protein catalysis, folding and denaturation, as well as protein stabilization or destabilization effects stemming from the presence of substances such as polyols, sugars, and metal ions.

[63] A. Fersht, "Enzyme Structure and Mechanism." W. H. Freeman & Company, New York, 1985.

[64] M. J. E. Sternberg, F. R. F. Hayes, A. J. Russell, P. G. Thomas, and A. R. Fersht, *Nature* (*London*) **330,** 86 (1987).

One challenge is to develop more complex dielectric models that could take into account the presence of multiple dielectric regions, each with their own dielectric constant. A case that would be of specific interest to treat would be the interaction between a lipase and aggregated (emulsified) lipid substrate, where at least three regions with different dielectric constants should be considered.

Although appealing as a methodology, protein electrostatics is not developed to the stage at which all relevant phenomena can be treated in a rigorous manner. It is currently difficult to model all of the effects of metal ion binding to appropriate sites on the protein surface. Of particular importance in the present context is the fact that the treatment of multiple dielectric environments is still in its infancy. Thus, modeling the enzyme while attached to the lipid interface, but still in contact with solvent water, is not possible yet.

In this laboratory, work has more recently been aimed at establishing a link between molecular dynamics and electrostatics. It is our belief that significant new methodologies have been developed, but they still await proper testing in model systems. Finally, a proper treatment of molecular recognition and the docking process that this involves, has obvious electrostatic components embedded. Whereas the initial trajectory can be modeled in a plausible way using long-range effects of protein electrostatics, a proper treatment of the short-range effects, including the final stages of docking and subsequent binding, is not possible using protein electrostatics alone, because hydrophobic effects are also involved and possibly quantum mechanical effects as well.

Acknowledgments

The work described in this chapter has been supported in part by the European Commission (BIO2-CT94-3016) and the Norwegian Research Council (29345/213). E. I. P. was supported by the Austrian Bundesministerium für Wissenschaft, Forchnung and Kunst, Grant GZ 558.009/141-IV/A/5a/95. M. T. N. P. acknowledges the support from the Portuguese JNICT (Junta Nacional de Investigação Científica e Tecnológica), program PRAXIS XXI, project PRAXIS/2/2.1/BIO/34/95. P. M. acknowledges PRAXIS XXI for grant PRAXIS XXI/BPD/9967/96.

Section II

Isolation, Cloning, Expression, and Engineering

[8] Site-Specific Mutagenesis of Human Pancreatic Lipase

By MARK E. LOWE

Introduction

The three-dimensional structures of fungal and human pancreatic lipases have provided great insight into the mechanism of lipolysis (see [4], [5], and [37] in this volume[1]). The tertiary structures identified a potential serine–histidine–acidic acid catalytic triad supporting earlier suspicions that the lipolytic mechanism was similar to that of serine proteases. In addition, a potential mechanism for interfacial activation, the marked increase in lipolysis at an oil–water interface, was proposed from these X-ray crystal studies. A surface loop, named the lid domain, covered the catalytic site and closed the active site to substrate. It was clear that the lid would have to change conformation for lipolysis to proceed and several investigators suggested that this movement represented the physical correlate of interfacial activation.[2,3] This hypothesis was supported by subsequent crystal structures of lipases obtained with inhibitors or substrates in the crystallization solution.[4–7] In these crystals, the lid adopted a markedly different position, exposing the catalytic site and creating an open substrate-binding site.

One of the first lipases to be crystallized was human pancreatic lipase.[2] In addition to crystals of pancreatic lipase alone, it was cocrystallized with colipase in two different conformations, with the lid closed and with the

[1] J. D. Schrag and M. Cygler, *Methods Enzymol.* **284,** [4], 1997 (this volume); C. Cambillau, Y. Bourne, M. P. Egloff, C. Martinez, and H. van Tilbeurgh, *Methods Enzymol.* **284,** [5], 1997 (this volume); F. Ferrato, F. Carriere, L. Sarda, and R. Verger, *Methods Enzymol.* **284,** [37], 1997 (this volume).

[2] F. K. Winkler, A. D'Arcy, and W. Hunziker, *Nature* (*London*) **343,** 771 (1990).

[3] L. Brady, A. M. Brzozowski, Z. S. Derewenda, E. Dodson, G. Dodson, S. Tolley, J. P. Turkenburg, L. Christiansen, B. Huge-Jensen, L. Norskov, L. Thim, and U. Menge, *Nature* **343,** 767–770 (1990).

[4] H. van Tilbeurgh, M. P. Egloff, C. Martinez, N. Rugani, R. Verger, and C. Cambillau, *Nature* **362,** 814–820 (1993).

[5] M. P. Egloff, F. Marguet, G. Buono, R. Verger, C. Cambillau, and H. van Tilbeurgh, *Biochemistry* **24,** 2751–2762 (1995).

[6] A. M. Brzozowski, U. Derewenda, Z. S. Derewenda, G. G. Dodson, D. M. Lawson, J. P. Turkenburg, F. Bjorkling, B. Huge-Jensen, S. A. Patkar, and L. Thim, *Nature* **251,** 491–494 (1991).

[7] U. Derewenda, A. M. Brzozowski, D. M. Lawson, and Z. S. Derewenda, *Biochemistry* **31,** 1532–1541 (1992).

0076-6879/97 $25.00

lid open.[4,8] These studies supported the hypothesis that interfacial activation involved a conformational change in the lid. They also revealed information about the interaction of lipase with colipase. In particular, the movement of the lid into the open position created new contacts between colipase and residues in the lid. As a result, an extensive hydrophobic area, which could bind water-insoluble substrates, was formed by the exposed surfaces of the lid and colipase.

Despite the important role of crystal structures in defining relationships between structure and function, they provided only a static view of a dynamic molecule crystallized at high concentrations. The concepts derived from crystal structures must be supported by additional kinetic and biophysical data before firm conclusions can be drawn about structure–function relationships. A powerful tool that can complement crystal data is site-specific mutagenesis. The application of this technology to the study of pancreatic triglyceride lipase is described in this chapter. The shortfalls and overall approach to site-specific mutagenesis as well as the details of the techniques of mutagenesis and recombinant DNA technology have been presented previously in this series and are not covered in this chapter except as they apply to pancreatic lipase specifically.[9,10]

Construction of Mutants

Mutagenesis

Many methods are available to mutate either single-stranded or double-stranded cDNA. The most versatile and efficient of these methods are based on the polymerase chain reaction (PCR). Point mutations, deletions, and truncations can be introduced anywhere in the cDNA encoding pancreatic lipase by PCR overlap extension.[11,12] This method employs two primary PCR reactions to produce two overlapping cDNA fragments that together comprise the entire length of the desired product. The first set of reactions is primed with complementary oligonucleotides, each bearing the same mutation, and with the appropriate flanking primers that hybridize at each end of the target sequence (Fig. 1). Double-stranded wild-type cDNA is the template for both reactions. The conditions for the initial PCR reactions vary depending on the primers, the template, and the product length. Thus,

[8] H. van Tilbeurgh, L. Sarda, R. Verger, and C. Cambillau, *Nature* **359,** 159–162 (1992).
[9] B. V. Plapp, *Methods Enzymol.* **249,** 91 (1995).
[10] *Methods Enzymol.* **68, 100, 101, 153, 154, 155, 216,** and **218.**
[11] S. N. Ho, H. D. Hunt, R. M. Horton, J. K. Pullen, and L. Pease, *Gene* **77,** 51–59 (1989).
[12] R. Higuchi, B. Krummel, and R. K. Saiki, *Nucleic Acids Res.* **16,** 7351 (1988).

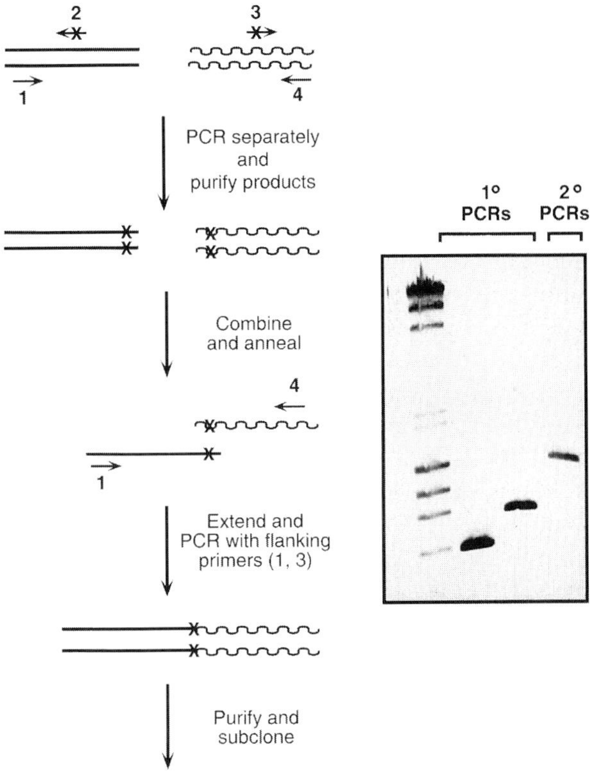

FIG. 1. *Left:* Schematic of overlap PCR. With this method mutations can be introduced anywhere in a cDNA. Two separate reactions are done as the first step. The template for both reactions is the cDNA to be mutated. The first round of PCR is done with two sets of primers, the internal primers (2 and 3) containing the mutation, which can be substitution, insertion, or deletion mutations, and the flanking primers (1 and 4), which are complementary to the wild-type sequence. Although an aliquot of these reactions can be used to perform the second PCR step, consistent results are reliably obtained if the primary PCR products are purified by agarose gel electrophoresis. In the second PCR reaction, the products of the primary PCR are combined with the flanking primers (1 and 4) to make the full-length product containing the desired mutation. *Right:* Results of the primary and secondary PCR reactions after analysis by agarose gel electrophoresis and ethidium bromide staining.

the temperature and extension times for each step of the PCR reaction need to be determined empirically. Typically, one cycle at 94° for 2 min followed by 20–30 cycles of 94° for 1 min to denature the cDNA, 50–55° for 1 min to anneal the primers, and 72° for 1–2 min to extend the reaction work well with a human pancreatic lipase template and primers of 24–27 nucleotides. The first-round products are isolated by agarose gel electrophoresis, using one of several available gel extraction kits [QIAquick (Qiagen, Chatsworth, CA); GeneClean (Bio 101, La Jolla, CA)]. The two products are then mixed and the full-length product made by PCR with the flanking primers. As before, the product is isolated from an agarose gel. The isolated product is then digested with restriction enzymes and subcloned into a suitable vector (see below).

The major advantages of site-specific mutagenesis by overlap extension are its simplicity and virtual 100% efficiency. The method eliminates the need for single-stranded DNA and the need to screen colonies for clones containing mutant cDNAs. The major disadvantage is the erorr rate of the *Taq* polymerases. The precise error rate varies for available polymerases and probably varies with different PCR conditions, templates, and primers. For this reason, it is essential to sequence the entire PCR product both to determine if it contains the desired mutations and to ensure that no undesired mistakes are present. One strategy that decreases the length of cDNA to sequence is to subclone a portion of the PCR product containing the mutation into unique restriction sites in the wild-type cDNA. In this way, only the region between the restriction sites needs to be sequenced and the product is conveniently introduced into the desired vector.

Expression of Recombinant Proteins

Expression System

The choice of expression system is critical to successful production of recombinant proteins. Human pancreatic lipase can be expressed in bacterial systems, but it forms insoluble inclusions that must be renatured prior to analysis.[13] The need to renature the recombinant protein adds complexity to the expression of recombinant proteins because folding *in vitro* may not reproduce the folding pattern of the native proteins, particularly for mutants. As a result, each refolded recombinant protein must be carefully characterized to ensure that it is not grossly misfolded. This requirement greatly slows the screening of multiple mutants.

Because of the difficulties in expressing soluble protein in bacteria, lipases have been expressed in mammalian cells.[14] Mutants of human pancreatic lipase have been screened by transient expression in COS cells. These cells are readily transfected and will robustly express cDNAs driven by simian virus 40 (SV40) promoters, making them suitable for the rapid and efficient screening of multiple mutant proteins. Furthermore, these

[13] M. E. Lowe, unpublished observations (1996).

[14] M. E. Lowe, *J. Biol. Chem.* **267,** 17069 (1992).

cells will secrete human pancreatic lipase into the medium, providing an initial, rapid assessment of folding integrity.

For transient expression in COS cells, the cDNA encoding human pancreatic lipase or mutants of lipase is subcloned into pSVL, a vector containing the SV40 late promoter (Pharmacia Biotech, Piscataway, NJ). Typically, transfections are done in 100-mm tissue culture dishes by the DEAE-dextran method previously described in detail.[15] After transfection, the cells are incubated in Dulbecco's modified Eagle's medium with 10% fetal calf serum overnight and changed to serum-free medium the following morning. The medium and cell pellet are harvested 60–72 hr after transfection and an aliquot of each is analyzed by sodium dodecyl sulfate–polyacrylamide gel electrophoresis (SDS–PAGE) and immunoblot. The separated proteins are electrophoretically transferred to an Immobilon polyvinylfluoride membrane (Millipore, Bedford, MA) (see [17] in this volume[16]). Lipase is detected with a rabbit polyclonal antibody raised against recombinant human pancreatic lipase, followed by an alkaline phosphatase-conjugated second antibody against rabbit IgG, and stained with nitroblue tetrazolium and 5-bromo-4-chloro-3-indolyl phosphate. The lower limit of detection with this method is 20 to 50 ng of human pancreatic lipase.

The major limitation of mammalian expression systems is the cost to express milligram quantities of recombinant proteins, which are necessary for enzymatic characterization and many biophysical techniques. For this reason, wild-type human pancreatic lipase and interesting mutants are expressed in baculovirus-infected insect cells. Recombinant baculovirus has several features that distinguish it from mammalian cell expression systems and provide potential advantages over other expression systems. Baculovirus has the potential to express high levels of protein in an environment in which the expressed protein is likely to fold properly and to have post-translational modifications correctly processed. In general, the recombinant protein is targeted to the appropriate cellular compartment. These properties increase the probability that measurements on mutant lipases reflect the contribution of the mutated residue and not changes in the protein secondary to expression in a heterologous cell.

The construction of a recombinant baculovirus is accomplished in two steps. First, the cDNA encoding the desired protein is subcloned into a plasmid vector downstream from a baculovirus promoter. Next, the plasmid is transfected into insect cells together with circular, genomic viral DNA. The cDNA is incorporated into the viral DNA by homologous recombination at a frequency of 0.1–1.0%. The recombinant virus is purified by

[15] B. Cullen, *Methods Enzymol.* **152,** 684 (1987).
[16] M. E. Lowe, *Methods Enzymol.* **277,** [17], 1997 (this volume).

repeated infections and plaque purification. Alternatively, several systems are available that select for recombinants over wild type and eliminate the requirement for repeated plaque isolations [i.e., BaculoGold (Pharmingen, San Diego, CA) and Bac-to-Bac (GIBCO-BRL, Gaithersburg, MD)].

Once the recombinant baculovirus is produced, it is amplified to produce a high-titer viral stock. This stock is used to infect cells adapted to growth in spinner culture and serum-free medium (EX-CELL 400; JRH Biosciences, Lenexa, KS). These techniques have been described elsewhere.[17] Lipase appears in the medium by 48 hr postinfection and continues to increase until 96 hr postinfection. Because cell lysis occurs between days 3 and 4 after infection, the amount of contaminating insect cell protein increases in the medium, and the medium is routinely harvested between 72 and 96 hr.

Purification of Recombinant Lipase

The following purification scheme has been used to purify wild-type recombinant human pancreatic lipase and multiple mutants of human pancreatic lipase.[18,19] Another procedure has been published for purifying human pancreatic lipase from insect cell medium in a single column step.[20] In our hands, lipase prepared by this method includes significant insect cell protein contaminants. The medium, 1 liter, is harvested by centrifugation at 3000 *g* for 5 min to remove cells and debris. The clarified medium is desalted with a hollow-fiber system (Spectrum, Houston, TX) using 10 m*M* Tris-HCl, pH 8.0, buffer. Generally, 1–2 liters of buffer is passed through the fibers over 2–3 hr. The desalted medium is adjusted to pH 8.0 with NaOH if necessary and centrifuged at 10,000 *g* for 10 min to remove any precipitate. The supernatant is applied to a DEAE-Blue Sepharose column (200-ml bed volume) equilibrated in 10 m*M* Tris-HCl, pH 8.0, and washed through with 1 bed volume of equilibration buffer. Human pancreatic lipase and all mutants made to date have been passed through this column. The passthrough is brought to pH 6.2 with 0.5 *M* succinate and is applied to a CM-Sepharose column (40-ml bed volume) equilibrated in 30 m*M* Tris–succinate, pH 6.2. The column is washed with 2–3 bed volumes of equilibration buffer and eluted with a linear NaCl gradient (0 to 300 m*M*). Lipase or mutant lipases are located by activity assay against tributyrin if the mutants have activity or by immunoblot of individual fractions if the mu-

[17] M. D. Summers and G. E. Smith, "A Manual of Methods for Baculovirus and Insect Cell Culture Procedures." Texas Agriculture Experiment Station, College Station, Texas, 1987.

[18] M. E. Lowe, *Biochim. Biophys. Acta* in press (1996).

[19] M. L. Jennens and M. E. Lowe, *J. Biol. Chem.* **269,** 25470 (1994).

[20] K. Thirstrup, F. Carriere, S. Hjorth, P. B. Rasmussen, H. Wldike, P. F. Nielsen, and L. Thim, *FEBS Lett.* **327,** 79–84 (1993).

tants have low activity. The lipase-containing fractions are pooled and concentrated by ultrafiltration over an Amicon (Danvers, MA) membrane (YM30). Lipase and all mutants have bound to CM-Sepharose, but the salt concentration for elution has varied for some mutants containing charge differences. At this stage the preparation is usually homogeneous by SDS–PAGE. If minor impurities are present, they can be removed by gel filtration over an S-200 column (200-ml bed volume) eluting with 30 m*M* Tris-HCl (pH 8.0), 0.15 *M* NaCl, 21 m*M* $CaCl_2$, and 1.0 m*M* EDTA. Yields vary, from 1 to 10 mg of purified lipase or mutant from 1 liter of culture medium.

Evaluation of Tertiary Structure

Site-specific mutagenesis has provided much information about multiple enzymes including pancreatic lipase, but, like all methods, this one has limitations. An important concern when making mutants is the effect of that mutation on the structure of the protein. Altered activity can occur because critical residues are modified or because protein folding is disrupted. Distinguishing between these possibilities is crucial to the interpretation of results from any mutagenesis study.

Several criteria for assessing proper folding of mutants have been applied in published studies on pancreatic lipase. The initial screen for proper folding of lipase was secretion of the recombinant protein into the medium. Improperly folded proteins will remain inside the cells and will not be secreted. The ability to isolate the secreted lipase mutants by usual methods also suggested that the mutant was not grossly misfolded. Further evidence for the structural integrity of the mutants was provided by determining stability to pH, storage, urea, and proteases. Finally, several properties of pancreatic lipase were measured including activity and interactions with colipase or substrate. The preservation of one or more of these properties provided confirmatory evidence that folding was not dramatically altered by the mutation.

Examples of Pancreatic Lipase Mutagenesis

Mutagenesis of Catalytic Residues

The crystal structure combined with chemical modification data implicated Ser-153, Asp-177, and His-264 (the numbering is based on the human sequence) in the catalytic triad of human pancreatic lipase. Site-specific mutagenesis at these positions provided additional evidence of the role of each residue in the hydrolysis of triglycerides.[19] First, multiple mutations were introduced in the Ser-153 codon by using a mixture of synthetic oligonucleotides containing all combinations of bases at the first two positions of the codon as the internal primers in the overlap PCR reaction. The PCR product was ligated into a pGEM vector containing the wild-type cDNA for human pancreatic lipase by exchanging a 60-bp restriction fragment containing the mutations with the corresponding wild-type sequence. The newly created vector was transformed into bacterial cells and individual colonies analyzed for the presence of vector containing insert. Each clone was sequenced to identify the mutation in the Ser-153 codon. In this way, nine different mutations were isolated.

Each mutant cDNA was subcloned into pSVL for expression in COS cells. Analysis of the medium from the transfected cells revealed that all mutants except S153R were secreted into the medium. This suggested that the mutations did not greatly disrupt the tertiary structure of the protein. Each mutant had no detectable activity against triolein, indicating that the replacement of Ser-153 decreased the specific activity of the mutants more than 400-fold.

Despite the decreased activity of the mutants, they bound to an interface to the same extent as wild-type lipase. One mutant, S153A, was examined in detail for its ability to bind to tributyrin. The binding assays were done with metabolically labeled recombinant proteins produced by incubating transfected COS cells with [^{3}H]leucine and isolating the labeled lipases by antibody affinity chromatography. Binding of the purified lipases to tributyrin particles was done in 100-μl volumes of 20 m*M* Tris-HCl (pH 8.0), 1 m*M* $CaCl_2$, 0.15 *M* NaCl and 3 μl of tributyrin. The mixture was sonicated to disperse the tributyrin. Varying amounts of the purified, labeled lipase were added and the mixture was incubated with constant shaking for 1–2 min. Following the incubation, the mixture was centrifuged at high speed in a microcentrifuge for 5 min and an aliquot of the upper, aqueous phase was analyzed in a liquid scintillation counter. This analysis showed that the binding of wild type and the mutant lipase was virtually identical, indicating that the mutation of Ser-153 did not adversely affect this important property of pancreatic lipase.

Two other residues, Asp-177 and His-264, were also mutated in this study to determine their role in catalysis. The introduction of a leucine residue at position 264 or of an asparagine at position 177 produced mutant lipases that were secreted, but were inactive. In contrast, a D177E mutant had a specific activity for triolein that was 80% that of wild type. Additional characterization of the purified D177E mutant demonstrated near normal activity toward tributyrin and trioctanoin and showed that it possessed interfacial activation and normal interaction with colipase.[18] In the context of available data from crystal structures and chemical modification data, the most reasonable explanation of the site-specific mutagenesis studies was that human pancreatic lipase possesses a Ser-153–His-264–Asp-177 catalytic triad.

Although the available data were consistent with this interpretation, the preserved activity of the D177E mutant raised the possibility that the assignment of this residue to the triad was incorrect. Two other observations supported this possibility. First, a similar exchange of one acidic residue for the other in two other lipases, *Geotrichum candidum* lipase and rat pancreatic carboxyl ester lipase, greatly altered their activity, suggesting that there was a strong preference for the naturally occurring acidic residue.[21,22] Second, examination of the human pancreatic lipase crystal structure revealed that a simple dihydral angle adjustment brought Asp-206 into position to form a hydrogen bond with His-264 and create a triad with the same *syn*-carboxylate interaction that was observed in other serine hydrolases instead of the *trans*-carboxylate interaction formed with Asp-177.

Site-specific mutagenesis was employed to determine the relative roles of Glu-177 and Glu-206 in the lipolytic mechanism of human pancreatic lipase.[18] An alanine was introduced at each site singly and in combination and the activity of the expressed proteins determined against tributyrin. The D206A mutant possessed full activity whereas the D177A mutant had 20% activity compared to wild type. The double-alanine mutant, D177A/D206A, had about 2% of wild-type activity. In addition to these mutants, a D177E/D206A mutant was tested and found to have the same activity as the D177E mutant. The best explanation for these results and the previous study showing that D177N was inactive is that Asp-177 is the acidic group in the catalytic triad, as predicted from the X-ray crystal structure, and that glutamic acid can effectively replace Asp-177, in contrast to results with other lipases.

A potential explanation for the preference of aspartate over glutamate in the catalytic triad of pancreatic lipase became apparent during the characterization of D177E lipase.[18] It was noted that the mutant was much more susceptible to degradation by endogenous insect cell proteases than was wild-type lipase. This was confirmed by incubating purified mutant and wild-type lipase with trypsin and showing that the D177E mutant was degraded at a much faster rate than wild-type lipase. These results suggested that the substitution of Asp-177 with glutamate led to conformational changes in the protein that did not greatly alter its enzymatic properties, but markedly increased its susceptibility to proteases. The increased susceptibility of the D177E mutant to proteases would decrease the ability of the

[21] T. Vernet, E. Ziomek, A. Recktenwald, J. D. Schrag, C. de Montigny, D. C. Tessier, D. Y. Thomas, and M. Cygler, *Biol. Chem.* **268,** 26212–26219 (1993).

[22] L. P. DiPersio and D. Y. Hui, *J. Biol. Chem.* **268,** 300 (1993).

mutant to function in an environment that was rich in proteases, such as the mileau in which human pancreatic lipase functions.

These studies of the catalytic site residues provided examples of several problems associated with site-specific mutagenesis of enzymes. The different activities of the D177L and D177A mutants demonstrated that the observed effect is dependent on the nature of the substitution. Multiple substitutions at a single site can greatly increase the information gained from mutagenesis. The conclusion that Asp-177 participates in the catalytic triad was based on multiple mutagenesis studies of both Asp-177 and Asp-206 and the X-ray crystal data emphasizing the complementary roles of these techniques in relating protein structure to function. Finally, the D177E mutant provided an excellent example of a mutation that did not significantly affect catalysis, but caused a marked effect on the sensitivity to proteases, presumably by affecting the conformation of residues distant from the mutation. It must always be kept in mind that mutations may affect other residues in the protein and the measured outcome may be related to distant effects on conformation or other properties of the protein.

C-Terminal Domain Truncations

Another feature of human pancreatic lipase structure revealed by the crystal data was the presence of two domains.[2] A globular N-terminal domain was connected through an unstructured region to a β-sheet C-terminal domain. No interactions were observed between residues of the two domains. The C-terminal domain formed all of the bonds with colipase identified in the colipase–lipase crystal structure when the lid domain was closed and contributed the majority of interactions between colipase and lipase when the lid was open. From these studies, it was suggested that the primary role of the C-terminal domain was to interact with colipase.

The importance of the C-terminal domain was investigated by deleting various lengths of the domain, including the entire C-terminal domain.[23] These truncated mutants were constructed by PCR using flanking primers that introduced a stop codon at the desired location. The flanking primers also contained sequence encoding a restriction site that permitted the PCR product to be subcloned into pSVL. After confirming the presence of the stop codon by sequence analysis, the mutant cDNAs were expressed in COS cells and analyzed for secretion and activity.

Two mutants with stop codons after Asp-388 and Phe-432 were not secreted, suggesting that the truncations within the C-terminal domain

[23] M. L. Jennens and M. E. Lowe, *J. Lipid Res.* **36,** 1029 (1995).

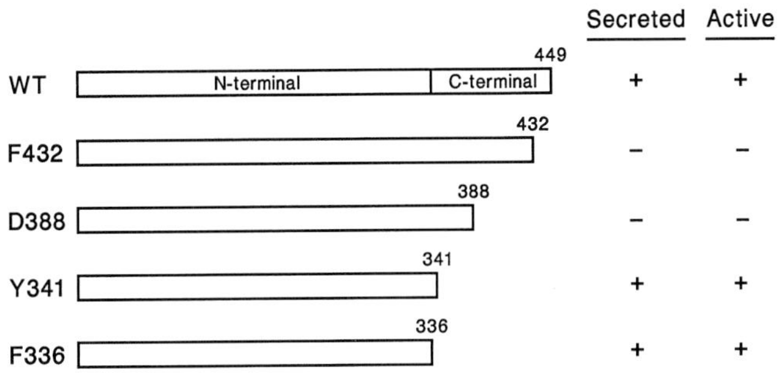

		Secreted	Active
WT	N-terminal / C-terminal 449	+	+
F432	[──────────] 432	−	−
D388	[──────────] 388	−	−
Y341	[──────────] 341	+	+
F336	[──────────] 336	+	+

FIG. 2. Truncation mutants of human pancreatic lipase. Schematics of the four truncated mutants of human pancreatic lipase are given. Also, the results of analyzing the mutants in transfected COS cells are summarized. *Secreted:* +, the mutant was secreted into the medium; −, the mutant was present only in the cells and was not secreted into the medium. *Active:* +, the mutant had lipolytic activity with triolein; −, the mutant was inactive whether the medium or cell extracts were tested.

affected folding of that region (Fig. 2). Two other mutants with stop codons at Phe-336 and Tyr-341 in the region between the two domains were secreted and shown to have activity against triolein. Importantly, this activity, although only 20–25% of the wild-type activity, required colipase. This study provided evidence that colipase may form important functional interactions with the N-terminal domain of human pancreatic lipase.

Lid Domain Mutations

An important observation from the tertiary structures of lipases was the identification of a surface loop that changes its conformation to uncover the active site of these lipases. This observation provided an explanation for interfacial activation that could be tested by site-specific mutagenesis. If the lid movement correlates with interfacial activation, then removing the lid should permanently open the active site and abolish interfacial activation. Removal of the lid could be accomplished by deleting the portion of the cDNA encoding the lid domain.

Two mutants with deletions in the lid domain were created (Fig. 3).[19] The first deleted amino acids 248–257, composing the α helix overlying the catalytic site, and the second deleted the entire lid domain (amino acids 240–260) except for one maino acid. Both mutants were created by the PCR overlap extension method described above. The mutant cDNA was subcloned into baculovirus vectors, expressed in insect cells, and purified from the insect cell medium over DEAE-Blue Sepharose and CM-Sepharose. The mutants behaved like wild type on these columns, passing through

Wild Type
C K K N I L S Q I V D I D G I W E G T R D F A A C
248–257
C K K N I L S Q I V D - - - - - - - - - - D F A A C
240–260
C K - C

FIG. 3. The amino acid sequence of the lid domain mutants. The predicted amino acid sequence of the lid domain between Cys-239 and Cys-261 is given for wild-type human pancreatic lipase and for the two lid domain mutants. The single-letter amino acid code is used. Dashes indicate a deletion at that position. The numbering is based on the human pancreatic lipase sequence.

the DEAE-Blue Sepharose and eluting at the same position from the CM-Sepharose column.

Evaluation of the mutants revealed several interesting properties. Both mutants had activity against tributyrin and triolein, but the activity compared to wild type was decreased 8-fold for the 248–257 mutant and 17-fold for the 240–260 mutant. The relative activities of the mutants against the two substrates were similar to wild-type lipase, suggesting that the lid domain did not influence substrate specificity in contrast to findings with lipoprotein lipase.[24] Despite the similar substrate preferences, there was an important difference in the reaction course between the mutants and wild-type lipase. With triolein but not tributyrin, the mutants had a long lag phase before hydrolysis that was not seen with wild-type lipase. This example demonstrates the need to do multiple point assays of all mutants because the activity of the deletion mutants may have been missed in a single point assay with a short incubation.

Several possible reasons for the lag phase were investigated. First, the ability of the mutants to interact with colipase was tested. Impaired interactions with colipase would decrease the binding of the mutants to mixed emulsions. The requirement of colipase for maximal activity of both mutants suggested that the mutants could interact with colipase and that the lid domain residues were not essential for this interaction. Because activity measurements do not measure the strength of the interaction between the mutants and colipase, the interactions between colipase and the mutants could still be defective. In fact, the decreased activity of the mutants could be a result of poorer interactions with colipase.

The ability of the mutants to interact with colipase was determined in a competition assay that measured the mutant's capacity to inhibit the colipase-dependent activity of wild-type lipase. The assay conditions in-

[24] K. A. Dugi, H. L. Dichek, and S. Santamarina-Fojo, *J. Biol. Chem.* **270,** 25396 (1995).

cluded a large excess of substrate, suboptimal concentrations of colipase, a short incubation period within the lag time of the mutants, a constant amount of lipase, and increasing amounts of mutant. Because of the short incubation time, all of the tributyrin hydrolysis could be attributed to the wild-type enzyme. Activity would be inhibited by molar excesses of mutant if the affinity of the mutant for colipase was close to that of the wild-type lipase. If the mutant had decreased affinity for colipase, then increasing concentrations of mutant would not inhibit the activity. The 240–260 mutant completely inhibited the wild-type lipase in a dose-dependent fashion over a 10-fold molar excess range. A twofold molar excess of colipase over the total concentration of mutant plus wild-type lipase, overcame the inhibition. This result suggested that the 240–260 mutant had an affinity for colipase that was similar to that of wild-type lipase and that the interaction between colipase and the mutant occurred quickly. These results make it unlikely that a decreased affinity of the mutant for colipase could account for the lag period seen with the mutants.

Next, binding to tributyrin was measured directly with the centrifugation assay described above, except that radiolabeled lipases were not used and activity was assayed in the aqueous phase. In addition, the buffer was 0.1 *M* Tris-HCl, pH 8.0, to prevent the decrease in pH that might occur from hydrolysis of tributyrin during the incubations. The incubation was for 1 min to limit hydrolysis. Under these conditions, wild-type lipase bound completely to tributyrin, but did not bind to mixed micelles of tributyrin and taurocholate unless colipase was included in the assay. In contrast, neither mutant bound well to tributyrin, with greater than 90% remaining in the aqueous phase compared to 2% for wild type. The 248–257 mutant bound slight better to mixed micelles of tributyrin and taurodeoxycholate when colipase was present, about 60% in the aqueous phase compared to 20% for wild type. These results suggested that the lid domain influenced the interfacial binding site and provided an explanation for the decreased activity and lag time of the mutants.

Finally, the mutants were tested for interfacial activation by measuring activity against increasing concentrations of tributyrin in the absence of bile salt. The activity of wild-type lipase was low (below 0.5 m*M* tributyrin) and increased sharply above that concentration. In contrast, both mutants showed near maximal activity below 0.5 m*M* tributyrin, indicating that they could hydrolyze monomeric substrate and were not activated by an interface. These findings were consistent with the model that required movement of the lid domain to open the active site to substrate and were consistent with the hypothesis that the conformational change of the lid domain correlates with the kinetic phenomena of interfacial activation.

Measuring interfacial activation with tributyrin has been reported by several different laboratories and the data obtained validate its use, but this method of measuring interfacial activation can be difficult to perform. The primary difficulty is in obtaining monomeric solutions of tributyrin because it has low solubility in water and its physical state is influenced by salt concentrations. As a result, aggregates of tributyrin may form even at low concentrations and the activity may be high enough to obscure the increase in activity at saturation that is typical of interfacial activation. In the course of our work, we have obtained ambiguous results with wild-type lipase when testing interfacial activation by this method. For this reason, we have also determined interfacial activation with another substrate, *p*-nitrophenylbutyrate.[18,25] The substrate has water solubility but forms aggregates at higher concentrations and the reaction can be followed spectrophotometrically. The primary difficulty is the high rate of spontaneous hydrolysis, thus appropriate blanks must be performed at each concentration. Even so, we have been consistently able to obtain unambiguous interfacial activation with wild-type lipase using this substrate. The lid deletion mutants do not show interfacial activation with this assay.[26]

Conclusion

Site-specific mutagenesis is a powerful technique that has provided useful insights into the function of human pancreatic lipase and other lipases. It can provide information about residues that contribute to catalysis, that mediate conformational changes, that interact with interfaces, and that bind to colipase. Examples of these studies have been presented here to illustrate techniques, approaches, and the utility of site-specific mutagenesis in the study of pancreatic lipase. Several of the caveats and difficulties of this approach are also illustrated by these examples. Despite the potential utility of site-specific mutagenesis, the results must be correlated with data from other techniques and with the constant remembrance that enzymes are dynamic, cooperative macromolecules.

[25] M. Martinelle, M. Holmquist, and K. Hult, *Biochim. Biophys. Acta* **1258,** 272 (1995).
[26] M. E. Lowe, unpublished observations (1996).

[9] Lipase Engineering: A Window into Structure–Function Relationships

By HOWARD WONG, RICHARD C. DAVIS, JOHN S. HILL, DAWN YANG, and MICHAEL C. SCHOTZ

Lipase Engineering Strategy

Lipoprotein lipase (LPL), hepatic lipase (HL), and pancreatic lipase form a family of lipolytic enzymes related to an ancestral gene and by a high degree of sequence homology.[1–3] Pancreatic lipase participates in the hydrolysis of dietary lipids and subsequent intestinal absorption, whereas the other two enzymes are plasma proteins involved in the metabolism of different lipoprotein classes. Collectively, they share the same catalytic-triad reaction mechanism[4–7] and similar monomer molecular weights. The pancreatic lipase crystal structure has been solved,[8] indicating that the enzyme is composed of two globular domains separated by a short spanning region. Lipoprotein and hepatic lipases are presumed to have a similar structure, although substantial functional differences demarcate each enzyme. These differences form the underlying basis and rationale for several strategies described in this chapter to better understand the structure–function relationships within this lipase gene family.

Functional LPL is bound to the capillary endothelium of numerous tissues and hydrolyzes chylomicrons and very low-density lipoprotein (VLDL) triglycerides to free fatty acids for energy utilization and triglycer-

[1] T. G. Kirchgessner, J.-C. Chuat, C. Heinzmann, J. Etienne, S. Guilhot, K. Svenson, D. Ameis, C. Pilon, L. D'Auriol, A. Andalibi, M. C. Schotz, F. Galibert, and A. J. Lusis, *Proc. Natl. Acad. Sci. U.S.A.* **86,** 9647 (1989).

[2] W. A. Hide, L. Chan, and W.-H. Li, *J. Lipid Res.* **33,** 167 (1992).

[3] C. F. Semenkovich, S.-H. Chen, M. Wims, C.-C. Luo, W.-H. Li, and L. Chan, *J. Lipid Res.* **30,** 423 (1989).

[4] R. L. Jackson, Lipoprotein lipase and hepatic lipase. *In* "The Enzymes" (P. D. Boyer, ed.), Vol. XVI, pp. 141–180. Academic Press, New York, 1983.

[5] R. C. Davis, G. Stahnke, H. Wong, M. H. Doolittle, D. Ameis, H. Will, and M. C. Schotz, *J. Biol. Chem.* **265,** 6291 (1990).

[6] T. Olivecrona and G. Bengtsson-Olivecrona, *Curr. Opin. Lipidol.* **1,** 222 (1990).

[7] S. Santamarina-Fojo and K. A. Dugi, *Curr. Opin. Lipidol.* **5,** 117 (1994).

[8] L. Brady, A. M. Brzozowski, Z. S. Derewenda, E. Dodson, G. Dodson, S. Tolley, J. P. Turkenburg, L. Christiansen, B. Huge-Jensen, L. Norskov, L. Thim, and U. Menge, *Nature (London)* **343,** 767 (1990).

0076-6879/97 $25.00

ide synthesis.[4,6,7,9] In a complementary role, HL is found primarily in the liver, with lesser amounts present in the adrenal gland and ovary, where lipoprotein triglycerides are processed.[10,11] A number of studies have also postulated a remnant catabolic role for both LPL and HL via a noncatalytic mechanism, whereby lipoproteins are bridged by the lipase to cell surface components and intracellularly metabolized.[12–19] The critical role of HL and LPL in plasma lipoprotein metabolism and implications for the development of atherosclerosis have generated considerable interest in determining specific functional mechanisms. This chapter describes a molecular biology-based approach toward the elucidation of lipase structure–function relationships.

Exploration of HL and LPL structure–function relationships by the use of genetic engineering techniques is motivated and underscored by the lack of crystal structure information about these important enzymes of lipid metabolism. In the absence of crystal data, alternative methods are employed to better understand the molecular mechanisms underlying lipase function. Site-directed mutagenesis is one example, and has been extensively used to examine lipase functional groups. However, because this method generally results in the production of inactive protein, it is not known whether the loss of function is a direct result of the change, or should be attributed to a global effect at a remote site. This primary shortcoming prompted the development of alternative approaches to site-direct mutagenesis for lipase structure–function determinations.

[9] C. F. Semenkovich, C.-C. Luo, M. K. Nakanishi, S.-H. Chen, L. C. Smith, and L. Chan, *J. Biol. Chem.* **265,** 5429 (1990).

[10] E. A. Hixenbaugh, T. R. Sullivan, Jr., J. F. Strauss III, E. A. Laposata, M. Komaromy, and L. G. Paavola, *J. Biol. Chem.* **264,** 4222 (1989).

[11] M. H. Doolittle, H. Wong, R. C. Davis, and M. C. Schotz, *J. Lipid Res.* **28,** 1326 (1987).

[12] U. Beisiegel, W. Weber, and G. Bengtsson-Olivecrona, *Proc. Natl. Acad. Sci. U.S.A.* **88,** 8342 (1991).

[13] D. A. Chappell, G. L. Fry, M. A. Waknitz, P.-H. Iverius, S. E. Williams, and D. K. Strickland, *J. Biol. Chem.* **267,** 25764 (1992).

[14] A. Nykjaer, G. Bengtsson-Olivecrona, A. Lookene, S. K. Moestrup, C. M. Peterson, W. Weber, U. Beisiegel, and J. Gliemann, *J. Biol. Chem.* **268,** 15048 (1993).

[15] P. Diard, M.-I. Malewiak, D. Lagrange, and S. Griglio, *Biochem. J.* **299,** 889 (1994).

[16] S. E. Williams, I. Inoue, H. Tran, G. L. Fry, M. W. Pladet, P.-H. Iverius, J.-M. Lalouel, D. A. Chappell, and D. K. Strickland, *J. Biol. Chem.* **269,** 8653 (1994).

[17] A. Nykjaer, M. Nielson, A. Lookene, N. Meyer, H. Roigaard, M. Etzerodt, U. Beisiegel, G. Olivecrona, and J. Gliemann, *J. Biol. Chem.* **269,** 31747 (1994).

[18] M. Z. Kounnas, D. A. Chappell, H. Wong, W. S. Argraves, and D. K. Stickland, *J. Biol. Chem.* **270,** 9307 (1995).

[19] D. A. Chappell, G. L. Fry, M. A. Waknitz, L. E. Muhonen, M. W. Pladet, P. H. Iverius, and D. K. Strickland, *J. Biol. Chem.* **268,** 14168 (1993).

The approach described here differs fundamentally from other methods because the goal is to produce structurally altered, but active, lipase molecules. Preservation of catalytic activity following the engineered change ensures that the tertiary structure of the molecule is not significantly altered. Retention of lipolytic activity indicates that folding of regions into domains, interdomain interactions, and crucial intersubunit associations are maintained in a native, or near-native, state. The engineered change can then be evaluated within the context of a functional lipase and subsequent conclusions reached are substantially strengthened. We review several lipase molecules, developed utilizing this strategy, that have extended our understanding of the structural requirements for lipase heparin binding and subunit orientation.

Description of Engineered Lipases

Domain-Exchanged Chimeric Lipases

HL–LPL chimera: Contains the first 312 amino acids of HL linked to the last 136 amino acids of LPL (Fig. 1, top)

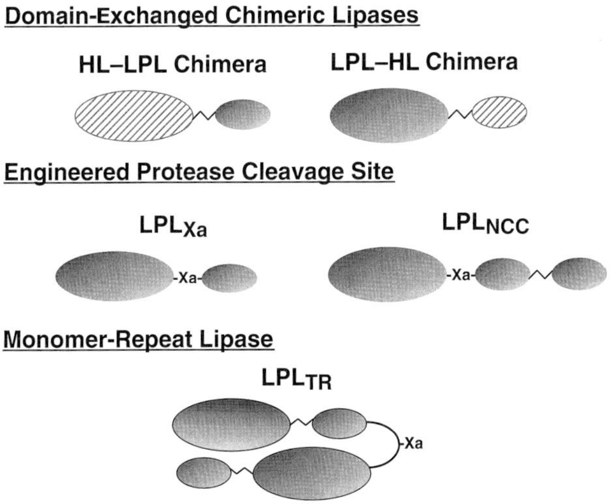

FIG. 1. Schematic diagrams of engineered lipases. Large and small ovals depict the lipase N- and C-terminal domains that constitute a single monomer unit. *Top:* Domain-exchanged lipases created from human LPL and rat HL. The HL N-terminal domain is hatched; the LPL C-terminal domain is shaded. The jagged line represents a shorter linker region joining the domains. *Middle:* LPL monomer showing the relative location of the engineered factor Xa site and the attached second C-terminal domain (right). *Bottom:* Tandem repeat of the LPL monomer, with the eight-amino acid tether shown in boldface. See text for additional details.

LPL–HL chimera: Contains the first 329 amino acids of LPL joined to the last 121 amino acids of HL

Engineered Protease Cleavage Site

LPL_{Xa}: Corresponds to native LPL amino acid sequence, except residues 330–333 have been removed and replaced with the factor Xa recognition sequence, IEGR (Fig. 1, middle)

LPL_{NCC}: The starting point begins as LPL_{Xa}, but this molecule has an additional C-terminal domain such that the linear order from N- to C-terminus is as follows: N-terminal domain (residues 1–329), factor Xa site (IEGR, plus four other amino acids), C-terminal domain (residues 330–448), an alternating 11-residue Gly–Ser linker, and another C-terminal domain (residues 330–448)

Monomer-Repeat Lipase

LPL_{TR}: Tandem repeats of the LPL monomer joined by an eight-amino acid linker region containing a factor Xa cleavage site (IEGR) (Fig. 1, bottom)

Construction of Engineered Lipases

General Considerations

The basic strategy used to generate the constructs for these experiments is polymerase chain reaction (PCR) cloning of the appropriate lipase cDNA subregions. Primer sequences are chosen without regard to melting temperature or GC content, because the primer must define the cDNA region to be amplified. Thus, the 3′-terminal 18 to 24 nucleotides of each primer are a direct copy of the appropriate cDNA sequence or, in the case of reverse primers, a copy of the reverse complement. Preservation of the reading frame requires that this primer segment begin with the first full codon of the desired cDNA region. However, primer length can be varied to some extent to avoid problems such as terminal loops or long stretches of a single nucleotide. Additional upstream (5′) sequences of the primer region vary depending on the desired function. In the case of splicing primers designed to join two cDNA fragments, the upstream sequence is simply another 18 to 24 nucleotides representing the in-frame complement or reverse complement of the coding sequence to be joined. For primers defining the 5′ and 3′ termini of the construct, restriction endonuclease sites are added to allow directional cloning. In these instances, a 5′ cap of three or four nucleotides is added to allow efficient restriction enzyme cleavage of the final PCR product.

The completed expression vector carries a strong promoter capable of driving high-level synthesis of transcripts in transfected cells. For lipoprotein lipase, a short portion of the 3′ untranslated (UT) region, extending about 40 nucleotides from the stop codon, is included in the construct, although the remaining 1–2 kb of 3′ untranslated region has little effect on expressed activity and is deleted in these experiments. For HL, the 3′ UT is shorter and is usually included in the constructs.

To achieve the expression of a functional protein, it is vital to avoid introducing unwanted mutations during PCR steps, thus all constructs are made using VENT polymerase (New England Biolabs, Beverly, MA). In conjunction with relatively high nucleotide concentrations, the VENT polymerase 3′ → 5′ exonuclease activity effectively protects against nucleotide misincorporation, but necessitates several adjustments in the PCR procedures. For example, hot-start PCR is crucial to avoid primer and template degradation; yields are improved by adjusting the Mg^{2+} concentration and placing the products on ice immediately on completion of the amplification cycle.

Construction of Chimeric Lipases from Domains of Lipoprotein Lipase and Hepatic Lipase

Although closely related proteins, lipoprotein lipase and hepatic lipase show strong differences in a variety of properties, including substrate specificity, enzymatic parameters V_{max} and K_m, heparin affinity, cofactor dependence, and sensitivity to salt. The possibility of constructing chimeras of these enzymes promised to help identify the domains responsible for these various attributes. Placement of a splice site between the enzymes to initiate domain exchange was an enigma in the absence of any three-dimensional structure information. However, once the PL three-dimensional structure was solved in 1990,[8] it seemed likely that all members of this lipase gene family were composed of two well-resolved structural domains: The N- and C-terminal domains constituted about two-thirds and one-third, respectively, of the lipase monomer mass. The boundary region separating domains coincided with the junction between LPL and HL exons 6 and 7. The presence of this natural division in the lipase molecule led us and others to construct lipase chimeras with splice sites at or near this junction.

Chimera Construction. The LPL–HL chimera containing the LPL N-terminal domain and the rat HL C-terminal domain is constructed from two PCR products (Fig. 2). The N-terminal product is amplified from a human LPL cDNA template with an upstream primer designed to amplify the 5′ UT sequences and to add a selected restriction site (R1). The down-

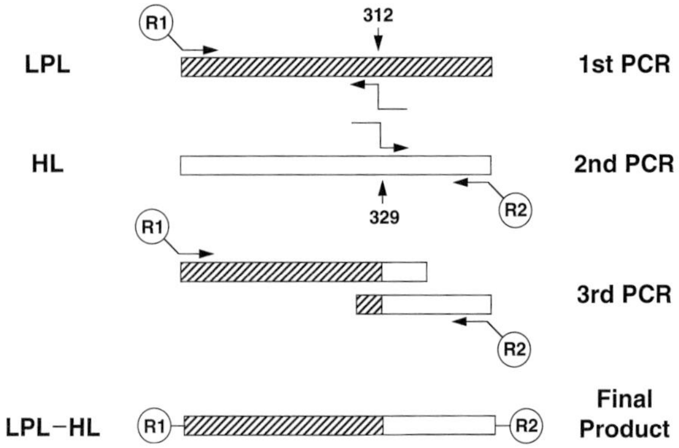

FIG. 2. PCR strategy for chimeric lipase construction. LPL cDNA (hatched) and HL cDNA (open) were templates for separate PCR reactions that utilized primers complementary to both cDNAs (lines with arrowheads) and that introduced restriction sites (R1 and R2). These steps created products with sequences from both lipases, which overlapped by 36 bp. The final product was generated by priming off the ends of the first two PCR products to form the full-length chimera monomer. The junction point in the final product was the N-terminal 312 LPL residues followed by HL residues 329 to 440.

stream primer consists of the reverse complement to cDNA encoding LPL residues 306 to 312, plus a splicing sequence corresponding to the reverse complement for recombinant HL (rHL) residues 329 to 335. For the C-terminal product, rHL cDNA template is used with an upstream primer encoding residues 329 to 335 of rHL with splicing sequences derived from the codons for residues 306 to 312 of human LPL. The downstream primer consists of reverse complement to the last 18 residues of the rHL 3′ UT, plus sequences for a second restriction site (R2) and a 4-base cap.

The resultant two PCR products contain a 36-base overlap between the 3′ end of the N-terminal product and the 5′ end of the C-terminal product. The appropriately joined final construct is made by purifying the two products and combining them in a third PCR reaction along with the flanking primers. The joined product formed by mutual priming between the two overlapping fragments is then amplified by the flanking primers.

The HL–LPL chimera containing the HL N-terminal domain and the LPL C-terminal domain is constructed with primers amplifying the appropriate regions of the rHL and human LPL cDNAs in a manner analogous to that described for the LPL–HL chimera.

Polymerase Chain Reaction Conditions. Each reaction (50 ml) contains 800 ng of each primer, 50 ng of plasmid DNA template, a 200 m*M* concentration of each deoxynucleotide, 1× reaction buffer (New England BioLabs, Beverly, MA) supplemented with $MgSO_4$ (2 m*M* final concentration), bovine serum albumin (10 mg/ml), 1% (v/v) dimethyl sulfoxide (DMSO), and 1 unit of VENT polymerase. The reactions are allowed to proceed for 40 sec at 94°, 45 sec at 58°, and 96 sec at 72° for a total of 30 cycles. Intermediate PCR products are purified on 1% (w/v) low melting temperature agarose gels before being mixed together in a final PCR reaction with flanking primers.

Expression. The chimeric and parental lipase cDNAs are cloned in the expression vector pcDNA3 (Invitrogen, San Diego, CA) and confirmed by complete nucleotide sequencing. Using electroporation for transfection of the plasmid DNAs, the lipases are transiently expressed in COS-7 cells.[20] Subconfluent cells are harvested by trypsinization and suspended in phosphate-buffered saline with 20 m*M* HEPES, pH 7.5. An 800-ml sample containing 1.7×10^7 cells and 40 mg of supercoiled plasmid DNA is placed in a 0.4-cm electroporation cuvette (Bio-Rad, Hercules, CA). After 5 min on ice, the cuvette is pulsed at 960 mfd, 0.33 kV with a time constant of about 21 msec. After another 10 min on ice, the suspended cells are mixed with 12 ml of culture medium [Dulbecco's modified Eagle's medium, high glucose supplemented with glutamine, sodium pyruvate, nonessential amino acids, antibiotics, and 10% (v/v) fetal bovine serum] and plated in a 75-cm^2 culture flask at 37° in a 5% CO_2 atmosphere. After 3 hr the medium is replaced, and after 24 hr the cells are washed five times with phosphate-buffered saline and then incubated in serum-free medium supplemented with heparin (10 units/ml; Sigma, St. Louis, MO). The medium is replaced at 24-hr intervals, and the harvested medium stored at −80°.

Construction of LPL_{Xa}: A Construct Containing a Factor Xa Cleavage Site between Lipoprotein Lipase N-Terminal and C-Terminal Domains

To investigate the physical properties of the isolated LPL N- and C-terminal domains, a mutant LPL protein has been expressed containing a factor Xa cleavage site (IEGR) in place of Ser^{328}–Glu^{329}–Thr^{330}–His^{331}. The N-terminal LPL_{Xa} product is amplified from the human LPL cDNA template with an upstream primer designed to include 5′ UT sequences and a selected restriction site (R1) (Fig. 3). The downstream primer consists of the reverse complement cDNA encoding LPL residues 322 to 327, plus a splicing sequence corresponding to reverse complement for the factor Xa cleavage site. For the C-terminal product, the LPL cDNA template is used with an upstream primer encoding residues 332 to 337 of LPL, again with splicing sequences derived from the factor Xa cleavage site. The downstream primer consists of the reverse complement to 18 residues of the

[20] R. C. Davis, H. Wong, J. Nikazy, K. Wang, Q. Han, and M. C. Schotz. *J. Biol. Chem.* **267,** 21499 (1992).

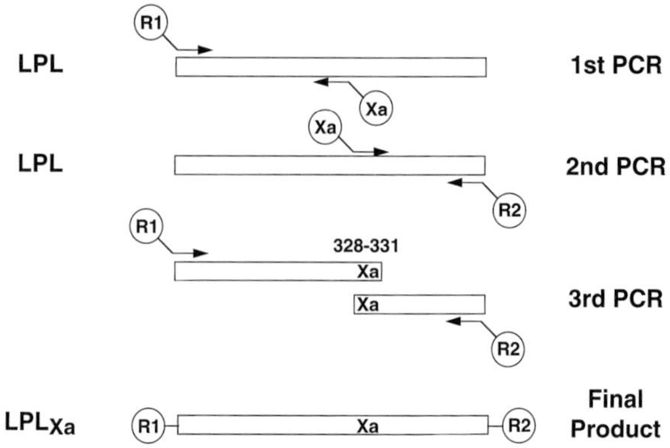

FIG. 3. PCR strategy for the introduction of a factor Xa cleavage site into LPL. The LPL cDNA was the template for two separate PCR reactions. The first reaction amplified a fragment that corresponded to the N-terminal LPL_{Xa} product and the second reaction created the coding region for the C-terminal LPL_{Xa} product. The middle primers for each reaction contained regions encoding the native LPL cDNA sequence and the engineered sequence encoding the factor Xa cleavage site (328–331) whereas the flanking primers contained restriction sites (R1 and R2) to facilitate cloning. The final PCR reaction utilized the flanking primers in combination with the two overlapping PCR products to produce the full-length cDNA product encoding the LPL_{Xa} molecule.

LPL 3′ UT region plus sequences for a second restriction site (R2) and a 4-base cap. The resulting two PCR products contain a 16-base overlap between the 3′ end of the N-terminal product and the 5′ end of the C-terminal product. The appropriately joined construct is made by purifying the two PCR products and combining them in a third PCR reaction along with the flanking primers. The joined product formed by mutant priming between the two overlapping fragments is then amplified by the flanking primers. The PCR conditions and expression are carried out as described above.

Construction of LPL_{NCC}: A Lipoprotein Lipase Construct Containing Two C-Terminal Domains per Monomer and a Factor Xa Cleavage Site

Experimental results suggest that all of the heparin affinity of an LPL homodimer can be attributed to the C-terminal domains. To test this hypothesis, a construct has been made to express a modified LPL_{Xa} molecule containing a second C-terminal domain covalently attached to the normal C terminus.

As with other engineered lipases, two PCR fragments are joined to make this particular construct (Fig. 4). The N-terminal product is amplified from the LPL_{Xa} template with the same upstream primer as for LPL_{Xa},

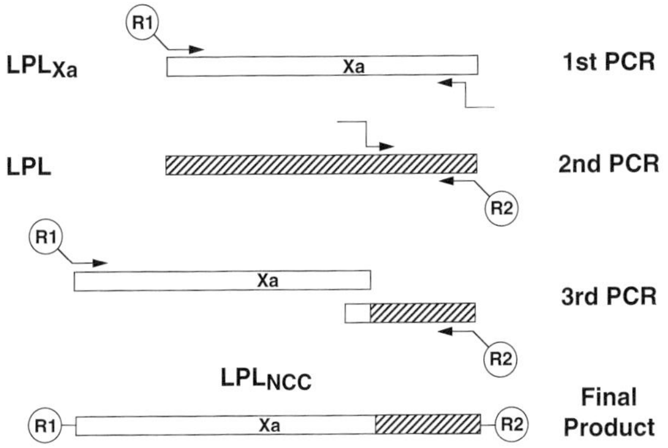

FIG. 4. PCR strategy for the creation of an LPL construct containing two C-terminal domains joined in tandem. The LPL_{Xa} cDNA was used as a template for the first PCR reaction, in which the upstream primer contained a restriction site (R1) and the downstream primer encoded the final four residues of the C-terminal domain as well as the first four residues of a second C-terminal domain. A second PCR reaction utilized the LPL cDNA to amplify a C-terminal product in which the upstream primer encoded residues 331–334 of LPL and the last four coding residues of the C-terminal domain. The downstream primer encoded for the LPL 3′ UT region plus sequences for a second restriction site (R2). The overlapping PCR products were used in a final PCR reaction with the flanking primers to create the full-length LPL_{NCC} cDNA.

but containing a restriction site (R1). The downstream primer consists of the reverse complement to the cDNA encoding LPL residues 445–448 (the final four residues of the C-terminal domain) and the reverse complement for LPL residues 331–334 (the first four residues of the second C-terminal domain). For the C-terminal product, the LPL cDNA template is used with an upstream primer encoding residues 331–334 of LPL and the last four coding residues of the first C-terminal domain. The downstream primer consists of the reverse complement to 18 residues of the LPL 3′ UT region plus sequences for a second restriction site (R2) and a 4-base cap. The resulting two PCR products contain an overlap corresponding to the last four residues of the first C-terminal domain and the first four residues of the second C-terminal domain. The joined PCR product is directionally cloned in the pcDNA3 vector and transiently transfected into COS-7 cells.

Construction of LPL_{TR}: A Construct Expressing a Tandem Repeat of Lipoprotein Lipase Monomer Joined by Eight-Amino Acid Linker

To test the hypothesis that the LPL homodimer is arranged in a head-to-tail configuration, a construct has been designed to express two LPL monomers joined by a short peptide linker. This construct is judged too long to join by VENT polymerase PCR; therefore, the two halves are separately fabricated by PCR, transferred to separate plasmids and later joined by conventional cloning techniques (Fig. 5). The N-terminal monomer is amplified from the LPL cDNA template with an N-terminal primer that encodes the signal peptide and a C-terminal primer consisting of reverse complement to the last six codons of LPL (residues 443–448), plus sequences for the restriction site R2 and a four-base cap. In the resulting construct, the R2 site replaces the stop codon and all 3′ UT sequence. The second LPL monomer construct is also amplified from the LPL cDNA template, but in this case the upstream primer consists of sequence encoding residues 1–7 (lacking the signal peptide), plus sequence encoding a factor Xa cleavage site, an R2 restriction site, and a four-base cap. The downstream primer consists of the reverse complement to six C-terminal residues of LPL, plus sequences for a third restriction site (R3) and a four-base cap. This PCR product lacks signal peptide sequences, and thus the R2 site is followed immediately by sequence for a factor Xa site joined directly to the first residue of the mature enzyme. For the final expression construct, the two constructs are joined at the R2 site via the linker between the C terminus of the first monomer and the N terminus of the second monomer. This construct is subsequently transferred to the pcDNA3 expression vector and transiently transfected in COS-7 cells.

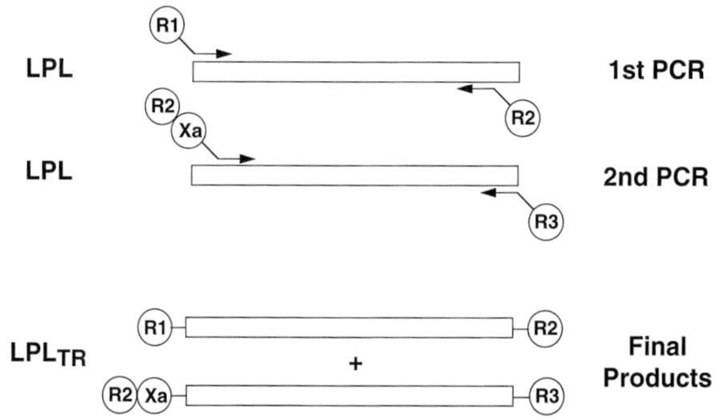

FIG. 5. PCR strategy for the construction of a tandem repeat of LPL monomers. Each PCR reaction utilized the LPL cDNA as template to create full-length products encoding the first and second LPL monomer sequences. The first monomer was created by an upstream primer coding for the signal peptide and a restriction site (R1) whereas the sequence of the downstream primer corresponded to the last six codons of LPL and a second restriction site (R2). To create the second monomer, the upstream primer encoded residues 1–7 (lacking the signal peptide) plus sequence encoding a factor Xa cleavage site (Xa) and a restriction site (R2). The downstream primer encoded the fianl six residues of LPL plus sequence in the 3′ UT region and a third restriction site (R3). Each product was sequentially subcloned and joined at the common R2 site to produce the full-length LPL_{TR} molecule.

Results of Investigations with Engineered Lipases

Domain-Exchanged Lipases

The initial test of this approach was the construction and expression of a hybrid between rat HL and human LPL. The first 312 amino acids of HL were joined to the last 136 amino acids of LPL, which correspond to the domains revealed by the crystal structure of the related enzyme, pancreatic lipase. The high degree of amino acid identity among members of the lipase gene family suggested that the two-domain structure may also be present in LPL and HL. The constructed hybrid (HL–LPL chimera) was expressed in COS cell culture and determined to be lipolytically active. Interestingly, the lipase activity behaved like that of HL rather than LPL, suggesting that the catalytic character of the chimera arose from elements within the N-terminal domain. In contrast, the HL–LPL chimera heparin-binding properties were comparable to those of LPL, not HL, inferring that elements within the C-terminal domain controlled heparin binding. These initial conclusions were reinforced with the construction and characterization of the reverse chimera, designated the LPL–HL chimera, which contained the N-terminal domain of LPL and the C-terminal domain of HL. For this domain-exchanged lipase, the catalytic character was like that of LPL, not HL, verifying that the N-terminal domain controlled catalysis, whereas heparin-binding affinity was like that of HL, substantiating that the C-terminal domain controlled that function.

These studies hinged on the retention of lipolytic activity for a clear interpretation of results. Because both chimeras were lipolytically active, subsequent conclusions regarding heparin binding and catalytic character were strengthened. Critical peptide folding and side group interactions required for lipolytic activity was preserved, strongly suggesting folding of regions important for other functions were also retained. The investigation of lipase chimeras has led to the identification of several domain-specific functions, including catalytic character, cofactor site, and heparin binding, and a better view of how these separate functions are integrated to form a lipase molecule.[20–23]

[21] H. Wong, R. C. Davis, J. Nikazy, K. E. Seebart, and M. C. Schotz. *Proc. Natl. Acad. Sci. U.S.A.* **88,** 11290 (1991).

[22] H. Wong, R. C. Davis, T. Thuren, J. W. Goers, J. Nikazy, M. Waite, and M. C. Schotz, *J. Biol. Chem.* **269,** 10319 (1994).

[23] H. L. Dichek, C. Parrott, R. Ronan, J. D. Brunzell, H. B. Brewer, and S. Santamarina-Fojo, *J. Lipid Res.* **34,** 1394 (1993).

Engineered Protease Cleavage Site

Proteins engage in interdomain interactions to accomplish overall function. On the basis of the presence of a putative heparin-binding motif, it was proposed that the LPL N-terminal domain participated in anchoring the lipase to heparan sulfate proteoglycans. Our chimera studies showed the C-terminal domain of the molecule controlled heparin binding, but did not rule out N-terminal domain participation, thereby raising the possibility of two binding sites per monomer. If multiple sites did contribute to heparin binding, then interaction between them was certainly possible. Multiple heparin-binding sites per monomer was further complicated by the obligate dimer structure of the enzyme, implying four binding sites per dimer. To clarify the number of LPL heparin-binding regions and to investigate domain and subunit interactions involved with heparin binding, two engineered molecules were constructed.

The first molecule incorporated a factor Xa cleavage site at residues 330–333, between the borders of the two domains that compose the molecule. The LPL_{Xa} molecule was designed to determine whether multiple heparin-binding sites existed, the domain location of the site(s), and whether those sites interacted during the binding process. LPL_{Xa} was expressed in COS cells and was lipolytically active. The activity was similar to that of native LPL, i.e., stimulated by serum, and inhibited by high ionic strength and monospecific antiserum. Heparin binding of LPL_{Xa} was nearly identical to that of the native enzyme; both eluted from heparin–Sepharose at ~1.1 *M* NaCl. Purified LPL_{Xa} was cleaved with factor Xa and the mixture rechromatographed on heparin–Sepharose. Only the C-terminal domain bound and eluted from the matrix, and at a salt concentration of ~0.75 *M*. The N-terminal domain did not bind to the column, suggesting that LPL heparin-binding elements were present solely on the C-terminal domain. That suggestion was supported by the amount of salt required to elute the isolated C-terminal domain (0.75 *M*), the same concentration necessary to elute monomeric LPL from heparin–Sepharose. Hence, total heparin-binding affinity of the LPL monomer could be accounted for solely by elements within the C-terminal domain of the molecule. These findings indicated a single binding site per monomer, operating without apparent cooperativity or interaction.

The second molecule constructed to examine the heparin-binding issue was a variation of LPL_{Xa}. Instead of the factor Xa site separating one N- and one C-terminal domain, the site separated one N-terminal domain from two C-terminal domains joined in tandem. It was anticipated that the molecule (LPL_{NCC}) would bind heparin with more avidity than either native LPL or LPL_{Xa}, because of four C-terminal domains per dimer rather than two per dimer. That was the result observed: LPL_{NCC} bound to heparin–Sepharose and required a higher salt concentration for elution. Typically, LPL_{NCC} eluted from the column at a 0.2–0.3 *M* higher salt concentration than LPL, necessitating 1.3–1.4 *M* NaCl for elution. The presence of the additional C-terminal domain altered the heparin-binding characteristics compared to native enzyme and confirmed that this domain contained heparin-binding elements.

A factor that suggests the heparin-binding properties exhibited by LPL_{NCC} were indeed significantly different from that of native enzyme was the preservation of lipolytic activity. The LPL_{NCC} molecule folded properly for catalytic activity, in spite of the presence of the extra domain, strengthening the interpretation that heparin-binding sites were also correctly folded. In addition, sucrose gradient centrifugation studies showed LPL_{NCC} subunit structure to be a dimer, indicating normal quaternary interactions. These findings, considered together, indicated the LPL monomer has a single heparin-binding site, located within the C-terminal domain.

Lipoprotein Lipase Monomer Repeat

On the basis of the known obligate dimer structure for LPL activity and chimeric lipase results, a model for the enzyme was proposed.[22] The model placed monomer subunits in a head-to-tail orientation such that the C-terminal domain from one subunit was in close proximity to the N-terminal domain from the other subunit. This subunit arrangement was proposed to explain how elements in the C-terminal domain participated in lipolysis, while conforming to the rules of protein symmetry. To test LPL subunit organization directly, an engineered molecule was designed and constructed.[24] The molecule consisted of two complete monomer-coding regions, interrupted by an eight-amino acid linker that contained a factor Xa cleavage site (LPL_{TR}). The rationale for design was that if LPL monomers were arranged in a head-to-tail conformation, then covalent linkage via a short spacer region (too short to permit a head-to-head arrangement) may allow proper domain folding and subunit interaction, and result in the expression of a functional lipase. Retention of LPL-like functions in LPL_{TR} would provide persuasive evidence for such a subunit arrangement in the native enzyme.

LPL_{TR} was transiently expressed in COS cells and the medium assayed for lipolytic activity. Observed activity was dependent on protein added and demonstrated stimulation by serum and inhibition by a monospecific antibody to LPL, all characteristics of authentic LPL. Western blot analyses

[24] H. Wong, D. Yang, J. S. Hill, R. C. Davis, J. Nikazy, and M. C. Schotz, *Proc. Natl. Acad. Sci. U.S.A.* in press (1997).

indicated a specific protein of molecular mass ~120 kDa, which was not present in COS cells transfected with vector lacking insert. Heparin–Sepharose chromatography of LPL_{TR} demonstrated an affinity for the matrix nearly identical to that of native LPL, indicating no gross misfolding of the heparin-binding region had occurred. Normal lipolytic activity and heparin binding suggested LPL_{TR} was the functional equivalent of the native enzyme. Furthermore, sucrose gradient centrifugation of medium from cells transfected with either LPL_{TR} or LPL demonstrated that the active form of both enzymes had a molecular weight of ~120,000, indicating LPL_{TR} was active as a monomer (or tethered dimer), and LPL as a dimer.

These findings showed that tandem linkage of LPL monomers into a head-to-tail subunit arrangement resulted in the production of active enzyme with properties virtually indistinguishable from those of native LPL. The lipolytically active, tethered-dimer form of LPL_{TR} suggested one half of the molecule folded back on the other to mimic the subunit arrangement of active native dimeric enzyme. A reasonable conclusion is that the subunit arrangement in the native enzyme is also in a head-to-tail organization.

Summary

Utilization of genetic engineering techniques to create novel functional lipases has increased knowledge of structure–function relationships in this important class of enzymes. The examples of engineered lipases presented in this chapter addressed the investigation of domain-specific properties, heparin binding, and subunit orientation. Conclusions reached are credible because the designed lipases retained catalytic activity, implying native, or near-native, conformation. This approach has demonstrated vigor by determining the domain location of several important enzyme functions and by providing the first evidence that LPL subunits are arranged in a head-to-tail orientation. In conjunction with physical techniques, such as crystallography and nuclear magnetic resonance spectroscopy, the engineered lipase approach could reveal new insights into the mechanism by which lipolysis is accomplished. The studies described here represent only the first attempts to explore that subject; more sophisticated lipase engineering will be used in future as a window into structure–function relationships.

Acknowledgment

This study was supported by VA Merit Review and NIH funds.

[10] Purification of Carboxyl Ester Lipase (Bile Salt-Stimulated Lipase) from Human Milk and Pancreas

By LARS BLÄCKBERG, RUI-DONG DUAN, and BERIT STERNBY

Background

Purification of carboxyl ester lipase is a story of purification of several enzymes using different substrates. Thus several strategies and methods have been used to obtain a pure enzyme. It has been determined that all these enzymes are the same and are found in both human milk and pancreas. The exact role and importance of this enzyme for lipid digestion and absorption, both in health and disease, are still under investigation. The results obtained so far suggest great importance, which has previously been underestimated. We therefore begin with a brief summary of the physiological motivation for the interest in the purification of the human enzyme.

Digestion of lipids starts in the stomach, where in the adult human mainly gastric lipase is responsible for the enzymatic activity. In the breast-fed infant a contribution from bile salt-stimulated lipase cannot be neglected.[1] The gastric lipase originates from the chief cells in the fundic part of the human stomach,[2,3] and is responsible for about 20% of the lipid hydrolysis.[4,5] By changing the surface properties of dietary lipids it prepares ingested fat for more efficient digestion in the small intestine.

In the duodenum the gastric contents are mixed with pancreatic juice (approximately 1 liter/day with 20 g of protein) and bile. The mammalian exocrine pancreas synthesizes, stores, and secretes about 20 different proteins, of which the majority take part in the intestinal digestion.[6] Among these digestive proteins are three lipolytic enzymes and one cofactor, namely, carboxyl ester lipase, phospholipase A_2, and the "classic" pancreatic lipase with its cofactor colipase.

[1] B. Fredrikzon, O. Hernell, L. Bläckberg, and T. Olivecrona, *Pediatr. Res.* **12,** 1048 (1978).
[2] H. Moreau, R. Laugier, Y. Gargouri, F. Ferrato, and R. Verger, *Gastroenterology* **95,** 1221 (1988).
[3] C. K. Abrams, M. Hamosh, T. C. Lee, M. J. Collen, J. H. Lewis, S. B. Benjamin, and P. Hamosh, *Gastroenterology* **95,** 1460 (1988).
[4] B. Borgström, A. Dahlqvist, G. Lundh, and J. Sjövall, *J. Clin. Invest.* **36,** 1521 (1957).
[5] F. Carriere, J. A. Barrowman, R. Verger, and R. Laugier, *Gastroenterology* **105,** 876 (1993).
[6] J. C. Kukral, A. P. Adams, and F. W. Preston, *Ann. Surg.* **162,** 63 (1965).

0076-6879/97 $25.00

Substrate Spectrum and Nomenclature

Carboxyl ester lipase is an enzyme with many names, owing to its action on a great variety of substrates. The first substrate used in 1970 to identify the human enzyme as carboxyl-ester hydrolase[7] was the synthetic *p*-nitrophenylacetate, which differentiated the activity of this enzyme from the classic pancreatic lipase.[8] High lipase and esterase activity from a milk enzyme against milk lipids (especially retinol esters) in the presence of certain bile salts was demonstrated in 1978,[1] and the enzyme was called bile salt-stimulated lipase. In 1979, additional names (nonspecific carboxylesterase and cholesterol ester hydrolase) appeared after the substrates vinyl 8-phenyloctanoate and cholesterol esters had been tested.[9,10]

A broad substrate specificity of the enzyme was shown in 1980, when it was reported to hydrolyze triacetin, methyl butyrate, and glycerides solubilized by bile salts, and to deacylate phospholipids and lysophospholipids at different rates.[11] These data indicated a correspondence to bovine lysophospholipase. Carboxyl-ester hydrolase was also found to hydrolyze lipid-soluble vitamins A, D_3, and E esters,[12] as well as to catalyze the esterification of cholesterol and the lipid-soluble vitamins A, E, and D_3 with oleic acid.[13]

Lipoamidase activity was found in normal and mutagenized pancreatic cholesterol esterase (bile salt-stimulated lipase) using lipoyl-4-aminobenzoate as reported in 1993, and the milk lipoamidase was confirmed to be the bile salt-stimulated lipase in milk.[14]

Questions concerning the existence of a human pancreatic lysophospholipase have been answered: No evidence was found for such an enzyme.[15] The substrate spectrum of the enzyme, although already broad, is still extending, and we have found that it also has a galactolipidase activity.[16]

There are thus a large number of substrates available for the quantification of the enzyme activity during the purification process. Some of these substrates have generated the names bile salt-stimulated lipase, carboxylesterase, carboxyl-ester hydrolase, carboxyl ester lipase, cholesterol esterase, lipoamidase, lysophospholipase, and nonspecific carboxylesterase for

[7] C. Erlanson, *Scand. J. Gastroenterol.* **5,** 333 (1970).

[8] C. Erlanson and B. Borgström, *Scand. J. Gastroenterol.* **5,** 395 (1970).

[9] W. Junge, K. Leybold, and B. Philipp, *Clin. Chim. Acta* **94,** 109 (1979).

[10] D. Lombardo and O. Guy, *Biochimie* **61,** 415 (1979).

[11] D. Lombardo, J. Fauvel, and O. Guy, *Biochim. Biophys. Acta* **611,** 136 (1980).

[12] D. Lombardo and O. Guy, *Biochim. Biophys. Acta* **611,** 147 (1980).

[13] D. Lombardo, P. Deprez, and O. Guy, *Biochimie* **62,** 427 (1980).

[14] D. Y. Hui, K. Hayakawa, and J. Oizumi, *Biochem. J.* **291,** 65 (1993).

[15] R.-D. Duan and B. Borgström, *Biochim. Biophys. Acta* **1167,** 326 (1993).

[16] L. Andersson, C. Bratt, K. C. Arnoldsson, B. Herslöf, N. U. Usson, B. Sternby, and Å. Nilsson, *J. Lipid Res.* **36,** 1392 (1995).

the enzyme. We believe it is most easily described by the name carboxyl ester lipase, given by Brockman in 1984.[16a]

Origin

The presence of some kind of lipolytic activity in human milk has been known since 1901.[17,18] Not unitl 1970 was a carboxyl-ester hydrolase, which differed from the classic lipase, identified in human pancreatic juice and intestinal contents by using a water-soluble substrate.[8] The milk bile salt-stimulated lipase was first purified around 1980,[19,20] and the first published procedure for purifying the human enzyme from pancreatic juice came from France in 1978.[21] A corresponding report from the United States concerning the enzyme from human pancreatic gland appeared 10 years later.[22]

The idea that the enzyme present in human milk was the same as the one in exocrine pancreas was suggested in 1981.[23] The characteristics found for either one of them (milk or pancreas) are mostly applicable on the other. The suggestions of a common origin was confirmed when the complete primary structure became available through cloning and sequencing of cDNA from the respective sources.[24,25] The carboxyl ester lipase gene was found in a highly polymorphic locus on human chromosome 9q-34ter,[26,27] the same for both the milk and pancreatic enzymes.

Structure, Function, and Properties

The first publication on the purification of the enzyme soon resulted in information about structure and function, which has been useful for further

[16a] E. A. Rudd and H. L. Brockman, *in* "Lipases" (B. Borgström and H. L. Brockman, eds.), pp. 185–204. Elsevier Science B. V., Amsterdam, 1984.

[17] P. Marfan, *Presse Med.* **9,** 13 (1901).

[18] E. Freudenberg, "Die Frauenmilch-Lipase." S. Karger, Basel, 1953.

[19] C.-S. Wang, *Anal. Biochem.* **105,** 398 (1980).

[20] L. Bläckberg and O. Hernell, *Eur. J. Biochem.* **116,** 221 (1981).

[21] D. Lombardo, O. Guy, and C. Figarella, *Biochim. Biophys. Acta* **527,** 142 (1978).

[22] C.-S. Wang, *Biochem. Biophys. Res. Commun.* **155,** 950 (1988).

[23] L. Bläckberg, D. Lombardo, O. Hernell, O. Guy, and T. Olivecrona, *FEBS Lett.* **136,** 284 (1981).

[24] J. Nilsson, L. Bläckberg, P. Carlsson, S. Enerbäck, O. Hernell, and G. Bjursell, *Eur. J. Biochem.* **192,** 443 (1990).

[25] K. Reue, J. Zambaux, H. Wong, G. Lee, T. H. Leete, M. Ronk, J. E. Shively, B. Sternby, B. Borgström, D. Ameis, and M. Schotz, *J. Lipid. Res.* **32,** 267 (1991).

[26] A. K. Taylor, J. L. Zambaux, I. Klisak, T. Mohandas, R. S. Sparkes, M. C. Schotz, and A. J. Lusis, *Genomics* **10,** 425 (1991).

[27] U. Lidberg, J. Nilsson, K. Strömberg, G. Stenmark, P. Sahlin, S. Enerbäck, and G. Bjursell, *Genomics* **13,** 630 (1992).

purification procedures: It is a glycoprotein (9% neutral and amino carbohydrates) with a molecular weight of 100,000, whose amino acid composition is characterized by a high content of proline residues (12.7%).[21] Its activity increases in the presence of bile salts; the rate (V) is not modified, but the Michaelis–Menten constant (K_m) is decreased 10 times by the addition of bile salts.[21] Gel filtration of the enzyme showed an apparent molecular weight of 300,000, indicating a peculiar configuration.[21]

In 1980, the existence of two sites of bile salt recognition was suggested. One site, specific to the 3α,7α-hydroxyl group of cholanic acid, would induce dimerization and activation of the enzyme. The other site, unspecific toward bile salt hydroxylation, would be located at the active center, and would be implicated in substrate recognition.[12]

More information about the structure was reported in 1981.[28] The carbohydrate content was determined to be 20%. Sedimentation studies indicated an ellipsoid form of the enzyme. The three-dimensional structure was estimated to belong to the "all-β" class of proteins; about 54–60% of its residues are in β sheets and β turns, most probably forming the surface of an ellipsoid.

In 1990–1991 the total amino acid sequence of the carboxyl ester lipase was completed by cDNA cloning, and earlier observations on the covalent structure were confirmed.[24,25] In 1993, the structure of N-linked oligosaccharides of human pancreatic bile salt-dependent lipase was published, and it was suggested that the carbohydrates in the enzyme possess a blood group-related antigenic determinant.[29] It was also seen that glycosylation in the enzyme varies among individuals, appears in the N linkages, and in some way seems to reflect disease. The enzyme is the only one in the pancreas that has both N- and O-linked oligosaccharides.[30]

This information on structure and function provides the basis for how the different purification methods work. When the enzyme is pure its activity is stable. It can be kept for 20 years in a glycerol solution in a freezer,[31] and is also quite stable at acidic pH.[1]

Lipolytic activity by carboxyl ester lipase requires binding of the enzyme to the lipid surface.[32] All three positions in the triglycerides are hydrolyzed with similar potency.[33] The rate of hydrolysis of triglycerides is faster with

[28] O. Guy, D. Lombardo, and J. G. Brahms, *Eur. J. Biochem.* **127,** 457 (1981).

[29] T. Sugo, E. Mas, N. Abouakil, T. Endo, M.-J. Escribano, A. Kobata, and D. Lombardo, *Eur. J. Biochem.* **216,** 799 (1993).

[30] E. Mas, N. Abouakil, J.-L. Franc, J. Montreuil, and D. Lombardo, *Eur. J. Biochem.* **216,** 807 (1993).

[31] B. Sternby, unpublished observations (1995).

[32] D. Lombardo and O. Guy, *Biochim. Biophys. Acta* **659,** 401 (1981).

[33] O. Hernell and L. Bläckberg, *Pediatr. Res.* **16,** 882 (1982).

short-chain fatty acids than with long-chain fatty acids.[34] The same pattern was obtained with monoglycerides.[35] Diglycerides are hydrolyzed faster than triglycerides and the lowest activity is obtained with monoglycerides.[36] Stimulation by bile salts is needed for the hydrolysis of fatty acids longer than eight carbons.[37] Bile salt-stimulated lipase provides the essential long fatty acids to the breast milk-fed infant.[38] The importance of carboxyl ester lipase is further emphasized by its concerted action with all other known gastrointestinal lipases.[39–43]

Purification Procedures

Before going into the details on recommended purification procedures some previous achievements should be mentioned. Using Sephadex gel filtration and HTP (hydroxyapatite; Biochemika, Fluka, Lund, Sweden) chromatography at pH 6.2, in the presence of mercaptoethanol and glycerol, a pure carboxylesterase from human pancreatic juice was obtained in 1975.[31] The pure enzyme produced by this procedure was used to obtain the amino acid sequence.[25]

The first published procedure for purifying the enzyme from pancreatic juice came from France in 1978.[21] This group used ion-exchange chromatography with CM-Sephadex gel and Sephadex gel filtration in the presence of benzamidine to prevent proteolytic activity.

Concanavalin A– and cholate–Sepharose 4B gels were used to purify the milk enzyme in a procedure first published in 1980.[19] In 1981, the milk enzyme was purified by another two-step chromatography procedure using heparin–Sepharose and Affi-Gel blue gels.[20] A combination of cholate–Sepharose and Bio-Rex 5 anion-exchange chromatographies was used for a large-scale purification of the milk enzyme, as published in 1983.[44]

Chromatography of pancreatic juice with immobilized antibodies directed against pure human pancreatic carboxyl-ester hydrolase gave pure

[34] C.-S. Wang, A. Kuksis, F. Manganaro, J. J. Myher, D. Downs, and H. B. Bass, *J. Biol. Chem.* **258,** 9197 (1983).

[35] C.-S. Wang, *Biochem. J.* **279,** 297 (1991).

[36] C.-S. Wang, J. A. Hartsuck, and D. Downs, *Biochemistry* **27,** 4834 (1988).

[37] C.-S. Wang and D. H. Lee, *J. Lipid Res.* **26,** 824 (1985).

[38] C.-S. Wang, M. E. Martindale, M. M. King, and J. Tang, *Am. J. Clin. Nutr.* **49,** 457 (1989).

[39] S. Bernbäck, L. Bläckberg, and O. Hernell, *J. Clin. Invest.* **85,** 1221 (1990).

[40] Q. Chen, L. Bläckberg, Å. Nilsson, B. Sternby, and O. Hernell, *Biochim. Biophys. Acta* **1210,** 239 (1994).

[41] Q. Chen, B. Sternby, and Å. Nilsson, *Biochim. Biophys. Acta* **1004,** 372 (1989).

[42] Q. Chen, B. Sternby, and Å. Nilsson, *Biochim. Biophys. Acta* **1044,** 111 (1990).

[43] M. Lindström, B. Sternby, and B. Borgström, *Biochim. Biophys. Acta* **959,** 178 (1988).

[44] C.-S. Wang and K. Johnson, *Anal. Biochem.* **133,** 457 (1983).

enzyme in a single step with up to 90% yield.[45] This purification procedure, which in one step gives a pure enzyme from either milk or pancreas, is of course recommended because it is a simple procedure. However, to have this purification option a certain amount of the enzyme must already be available. Furthermore, to obtain pure antibodies for immobilization a great deal of work must be done. Large-scale purification by this method may also be more difficult.

Purification of the enzyme from pancreatic gland is more cumbersome, which may explain why there is just one published study from 1988 available for the human enzyme.[22] This purification scheme includes homogenization of the pancreatic tissue in the presence of trypsin inhibitor and bile salts, centrifugations, ammonium sulfate precipitation, filtration, cholate–Sepharose chromatography, and Bio-Rex and heparin–Sepharose chromatography.

Purification of the milk enzyme is easier, owing to the presence of fewer proteases. We present different strategies for purification of the human enzyme from milk and pancreas, owing to the significant difference in the source of the enzyme—milk or pancreatic juice/pancreas.

Milk Enzyme

Bile salt-stimulated lipase from milk is a whey protein.[19,20] To obtain the enzyme the initial two purification steps involve removal of the cream and caseins.

1. Skimmed milk is obtained simply by centrifugation of breast milk (e.g., 10,000 *g* for 30 min at 4°) and removal of the top lipid layer.
2. To eliminate the caseins, precipitation at acidic pH and increased temperature are used. The pH of the skimmed milk is lowered to pH 4.7 by addition of HCl, and heated to 40° for 30 min.
3. After a second centrifugation (10,000 *g*, 30 min, 4°) the whey can be collected. The decanted whey is dialyzed overnight against the appropriate starting buffer (see the following sections under the respective chromatographies) for the subsequent chromatographic step.

The most commonly used chromatography step for purification of bile salt-stimulated lipase from milk is chromatography on immobilized heparin.[20,44] Immobilized heparin can either be prepared by the procedure of Iverius[46] or obtained premade. It is believed that the lipase interaction with heparin is not just a useful tool for purification purposes, but also an

[45] N. Abouakil, E. Rogalska, J. Bonicel, and D. Lombardo, *Biochim. Biophys. Acta* **961,** 299 (1988).

[46] P.-H. Iverius, *Biochem. J.* **124,** 677 (1971).

important physiological event, being a part of binding of the lipase with an intestinal receptor.[47]

The chromatography of human whey on heparin–Sepharose can be performed in a number of buffers. One that is frequently used is 5 m*M* sodium barbital, pH 7.4. Typically 50 ml of human whey is applied to 10 ml of settled heparin–Sepharose gel in the barbital buffer in absence of, or in the presence of, up to 0.05 *M* sodium chloride. After washing with the same buffer the bound proteins are eluted with a linear gradient of sodium chloride (0.05–1.0 *M*). The lipase elutes at approximately 0.4 *M* NaCl. This step has several advantages. It can easily be scaled up so that large quantities of the enzyme can be obtained. The recovery of enzyme is generally good (80% or more), and a high degree of purification is obtained.

The resulting lipase preparation contains one principal contaminant, i.e., lactoferrin,[48] which is one of the main whey proteins, and elutes at a slightly higher salt concentration. To obtain a homogeneous preparation several strategies can be employed. The easiest procedure is to repeat the described heparin–Sepharose chromatography after dialysis of the pooled lipase-containing fractions. This will give a homogeneous preparation if the lactoferrin-containing fractions of this second chromatography are excluded. The disadvantage may be that the recovery is decreased by excluding some lipase-containing fractions.

A second approach is to perform a gel filtration. Bile salt-stimulated lipase has a rather high molecular mass, i.e., 110–120 kDa, when determined by sodium dodecyl sulfate-polyacrylamide gel electrophoresis (SDS–PAGE). Moreover, when applied to a gel-filtration column the enzyme will elute at a position corresponding to a molecular mass of more than 300 kDa.[49] This behavior has been explained by the highly glycosylated C-terminal repetitive part of the enzyme.

A suitable gel-filtration matrix to use is Superdex 200 (Pharmacia-Biotechnology, Piscataway, NJ) run in 10 m*M* sodium phosphate (pH 7.2), 0.5 *M* NaCl. The disadvantage of this procedure is that for large quantities of enzyme a very large column is required.

A third procedure is an immunosorbent technique.[45] This procedure will give a pure preparation but has the limitation that it may be difficult to scale up. The procedures described here have been used not only for purification of bile salt-stimulated lipase from milk but also for the purification of recombinant lipase, including variants obtained by site-directed

[47] M. S. Bosner, T. Gulick, D. J. S. Riley, C. A. Spilburg, and L. G. Lange III, *Proc. Natl. Acad. Sci. U.S.A.* **85,** 7438 (1988).

[48] L. Bläckberg and O. Hernell, *FEBS Lett.* **109,** 180 (1980).

[49] L. Bläckberg and O. Hernell, *FEBS Lett.* **157,** 337 (1983).

mutagenesis, from conditioned media containing cultures of (for example) C-127 cells.[50,51]

Pancreatic Enzyme

To purify the enzyme from pancreatic juice or gland is always more difficult than from milk, owing to the presence of proteases, which always seem to be activated when it is most undesirable. Although purification by hydroxyapatite chromatography gives a homogeneous enzyme,[31] it is a somewhat more tricky procedure compared to the one described below, which was used for the latest purifications of the pancreatic enzyme in our laboratory. This is the reason why we choose to present it as the preferred purification procedure. It is based on the stimulation of carboxyl ester lipase by bile salts, particularly trihydroxy bile salts, which bind to the enzyme and stimulate its activity. A method using cholate as a ligand to prepare an affinity column to purify the enzyme has been developed.[15,19,52] The following procedure has been used successfully to purify the two forms from human pancreatic juice.[15] *p*-Nitrophenyl butyrate was used as substrate[53] to monitor the enzyme activity.

Materials

Sephacryl S-200 gel for a column with 56-mm diameter and 200-cm height
Cholate–Sepharose 4B gel for a column with 20-mm diameter and 20-cm height
Sephadex G-100 gel for a column with 26-mm diameter and 40-cm height
Sephadex C-50 gel for a column with 26-mm diameter and 19-cm height
Human pancreatic juice and buffers

If the cholate–Sepharose 4B gel is not available premade it can be prepared as follows.

1. Pour 50 ml of epoxy-activated AH Sepharose 4B (EAH–Sepharose 4B) into a sintered glass filter (porosity G3, Pyrex, England) funnel and wash it with 4000 ml of 0.5 *M* NaCl.
2. Wash the gel with 250 ml of 50% (v/v) dioxane.

[50] L. Hansson, L. Bläckberg, M. Edlund, L. Lundberg, M. Strömqvist, and O. Hernell, *J. Biol. Chem.* **268,** 26692 (1993).
[51] L. Bläckberg, M. Strömqvist, M. Edlund, K. Juneblad, L. Lundberg, L. Hansson, and O. Hernell, *Eur. J. Biochem.* **228,** 817 (1995).
[52] D. R. Gjellesvik, D. Lombardo, and B. T. Walther, *Biochim. Biophys. Acta* **1124,** 123 (1992).
[53] K. Shirai and R. L. Jackson, *J. Biol. Chem.* **257,** 1253 (1982).

3. Transfer the Sepharose to a beaker and add 50 ml of 50% (v/v) dioxane containing 2 g of sodium cholate. Keep within pH 4.7–5.0.

4. Add 12.5 ml of 50% (v/v) dioxane containing 600 mg of 1-ethyl-3-(dimethylaminopropyl)carbodiimide dropwise over 4 hr, with gentle shaking throughout (avoid stirring by magnetic bar).

5. Incubate overnight in a shaker at room temperature.

6. Once again pour the Sepharose suspension onto a glass filter funnel and wash with 1000 ml of 50% (v/v) dioxane and then with 0.1 *M* NaOH. During this wash, check the absorbance at 280 nm (A_{280}), and wash until the absorbance becomes stable (about 1000 ml of the NaOH solution is required).

7. Transfer the washed gel from the filter and resuspended it in 50 m*M* NH_4OH–HCl buffer at pH 8.5.

The cholate-coupled EAH–Sepharose can be stored at 4° until use. Deaerate the gel suspension by vacuum suction before packing a column.

Method

Step 1: Sephacryl S-200 Gel Chromatography

1. Filter 100 ml of human pancreatic juice through a nylon cloth.

2. Apply the filtered solution to the Sephacryl S-200 column equilibrated in 30 m*M* Tris buffer, 1 m*M* benzamidine to prevent activation of proteolytic zymogens at pH 8.5.

3. Elute the column with the same buffer at a rate of 4 ml/min; collect 4-min fractions.

4. Under these conditions the enzyme activity peak appears at about 600 ml of elution volume.

5. Pool the fractions containing activity above 0.5 μmol/min · ml.

Step 2: Cholate–Sepharose 4B Affinity Chromatography

1. The pooled fractions from the Sephacryl column are applied directly to a cholate–Sepharose column equilibrated in NH_4OH–HCl at pH 8.0.

2. Wash the column with the buffer at a rate of 2 ml/min and collect 12-ml fractions.

3. Wash the column with 150 ml of NH_4OH–HCl buffer containing 1% (w/v) sodium deoxycholate at the preceding rate, which serves to remove unspecifically bound proteins.

4. The enzyme is then eluted with 200 ml of 2.5% (w/v) sodium cholate in the same buffer.

5. Fractions with activity above 0.5 μmol/min · ml are pooled.

Step 3: Removal of Bile Salts

1. The pooled peak is concentrated to 10–15 ml by ultrafiltration through a PM10 (Diaflo, Amicon Inc., Bejerly, Ireland) membrane with a molecular cutoff of 10 kDa.
2. The enzyme sample is then applied to a Sephadex G-100 column and eluted with 0.01 *M* Tris buffer containing 0.15 *M* NaCl at pH 7.0 at a rate of 20 ml/hr. Fractions of 20 min are collected and the pooled enzyme activity peak is found at 60–85 ml.

Step 4: Isolation of Two Forms of Enzyme by CM-Sephadex C-50 Ion-Exchange Chromatography

In some human pancreatic juice, two forms of the enzyme with the same specific activity are seen, which can be separated by ion-exchange chromatography.

1. Apply the pool from the last gel-filtration step to a CM-Sephadex C-50 column, equilibrated in 10 m*M* morpholineethanesulfonic acid (MES) buffer containing 0.075 *M* NaCl, pH 6.5.
2. Wash with 150 ml of the same buffer.
3. Elute the two enzyme peaks with a gradient (0.075–0.25 *M* NaCl) of the preceding buffer (two 260-ml volumes) at a rate of 18 ml/hr and collect fractions of 3 ml. The two forms will appear at 0.14 and 0.20 *M*, respectively.

Step 5: Electrophoresis to Check Purity. The purity of the enzyme can be checked by SDS–7.5% (w/v) polyacrylamide gel electrophoresis. The protein is visualized with Coomassie Blue R-250. Following these steps we have repeatedly obtained homogeneous enzymes.

[11] Two Novel Lipases from Thermophile *Bacillus thermocatenulatus:* Screening, Purification, Cloning, Overexpression, and Properties

By Claudia Schmidt-Dannert, M. Luisa Rúa, and Rolf D. Schmid

Introduction

Proteins from thermophiles generally exhibit high thermodynamic stability, both at elevated temperatures and in organic solvents,[1–4] important

[1] R. Jaenicke and P. Závsdosky, *FEBS Lett.* **268,** 344 (1990).
[2] V. V. Mozhaev, *Trends Biotechnol.* **11,** 88 (1993).

0076-6879/97 $25.00

properties for the industrial application of enzymes.[5–7] Although many lipases from mesophiles are stable at elevated temperatures,[8–11] we have focused our attention on lipases from thermophiles to enhance thermostability further and to gain insight into the structural requirements for thermodynamic stability.

The thermophile *Bacillus thermocatenulatus* produces two lipases of different sizes: 16 and 43 kDa. The 16-kDa lipase (BTL-1) was purified directly from the culture broth of *B. thermocatenulatus,*[12] whereas the 43-kDa lipase (BTL-2) was cloned from an expression library of genomic DNA.[13] This novel lipase was weakly expressed in *Escherichia coli* under the control of its native promoter and was purified for a first characterization. For structural investigations and further characterization of lipase BTL-2, we developed a system for the overexpression of the enzyme in *E. coli,* which allows us to produce large amounts of lipase.

In the first part of this chapter, we describe the screening process for a thermophilic microbial lipase producer and the purification and characterization of the lipase BTL-1. The second part of this chapter deals with the cloning and overexpression of an additional lipase of *B. thermocatenulatus,* BTL-2. The large-scale production, purification, and characterization of this recombinant lipase BTL-2 are also described.

Lipase BTL-1

Screening for a Thermophilic Microbial Lipase Producer

Fifteen different thermophilic *Bacillus* strains were acquired from the German Collection of Microorganisms in Braunschweig (DSM), Germany: *B. stearothermophilus* (DSM 22, 2027, 2349), *B. caldolyticus* (DSM 405),

---

[3] R. K. Owusu and D. A. Cowan, *Enzyme Microb. Technol.* **11,** 568 (1989).
[4] T. Brock, "Thermophiles: General, Molecular and Applied Microbiology" (T. Brock, ed.). John Wiley & Sons, New York, 1986.
[5] R. A. Herbert, *Trends Biotechnol.* **10,** 395 (1992).
[6] T. D. Brock, *Science* **230,** 132 (1985).
[7] B. Sonnleitner, *Adv. Biochem. Eng. Biotechnol.* **28,** 69 (1983).
[8] A. Sugihara, T. Tani, and Y. Tominaga, *J. Biochem.* **109,** 211 (1991).
[9] E. J. Gilbert, A. Cornish, and C. W. Jones, *J. Gen. Microbiol.* **137,** 2223 (1991).
[10] Y. Shabati and N. Daya-Mishne, *Appl. Environ. Microbiol.* **58,** 147 (1992).
[11] A. Makhzoum, R. K. Owusu, and J. S. Knapp, *Food Chem.* **46,** 355 (1993).
[12] C. Schmidt-Dannert, H. Sztajer, W. Stöcklein, U. Menge, and R. D. Schmid, *Biochim. Biophys. Acta* **1214,** 43 (1994).
[13] C. Schmidt-Dannert, M. L. Rúa, H. Atomi, and R. D. Schmid, *Biochim. Biophys. Acta* **1301,** 105–114 (1997).

B. caldotenax (DSM 406), *B. caldovelox* (DSM 411), *B. coagulans* (DSM 459, 2321, 2319), *B. sphaericus* (DSM 461, 462), *B. thermodenitrificans* (DSM 465), *B. thermocatenulatus* (DSM 730), *B. thermoglucosidasius* (DSM 2542), and *B. flavothermus* (DSM 2641). These strains were screened for lipase activity on rhodamine B–olive oil agar plates according to Kouker and Jaeger.[14] Five of the 15 screened strains showed lipolytic activity on agar plates: DSM 405, 406, 411, 461, and 730. These strains were selected and further cultivated in fluid media with different pHs and nitrogen sources and supplemented with different concentrations of $CaCl_2$ and olive oil as inducers.[12] Cultivation in media containing 1.3% (w/v) nutrient broth (Difco, Detroit, MI) or 3% (w/v) trypticase (Difco), 0.05 or 0.01% (w/v) $CaCl_2$, and 5% (w/v) olive oil at pH 6.5 and 7.5 resulted in low exolipase production (0.03–0.05 U/ml) compared to the higher lipase production observed on the plates. *Bacillus thermocatenulatus* (DSM 730) produced twice as much extracellular lipase as the others on induction with olive oil and therefore seemed to be the most promising strain for further work.

For this reason we turned our attention to optimizing the medium for exolipase production by *B. thermocatenulatus.* Reducing the nutrient broth concentration in the medium from 1.3% (w/v) to 0.325% (w/v) resulted in a twofold increase in lipase production. Maximal lipase production up to 1 U/ml after 16 hr of cultivation of 63° was obtained by adding 1% (w/v) gum arabic to a medium consisting of 0.325% (w/v) nutrient broth, 0.1% (w/v) $CaCl_2$, and 1.5% (w/v) olive oil. An increase in exolipase formation by different polysaccharides and other macromolecules has also been reported for lipases from *Pseudomonoas aeruginosa* and *Bacillus subtilis.*[15–19] Winkler and Stuckmann[20] proposed the detachment theory to explain this phenomenon.

The large-scale production of lipase BTL-1 was carried out by fermentation on a 100-liter scale at 63° and pH 6.5 for 14.5 hr with the above-described medium. Lipase activity in the culture supernatant during cultivation and later during purification and characterization were determined using *p*-nitrophenyl palmitate (pNPP) as a substrate at 63° and pH 7.5.[12]

Purification and Properties of 16-kDa Native Lipase BTL-1

After lipase production by fermentation a crude enzyme solution is obtained by removing the cells by centrifugation and concentrating the culture supernatant with a film evaporator (Typ SA1; Westfalia Separator

[14] G. Kouker and K.-E. Jaeger, *Appl. Environ. Microbiol.* **53,** 211 (1987).
[15] E. Lesuisse, K. Schanck, and C. Colson, *Eur. J. Biochem.* **216,** 155 (1993).
[16] G. Schulte, L. Bohne, and U. Winkler, *Can. J. Microbiol.* **28,** 636 (1982).
[17] K.-E. Jaeger and U. K. Winkler, *FEMS Microbiol. Lett.* **19,** 59 (1983).
[18] K.-E. Jaeger and U. K. Winkler, *FEMS Microbiol. Lett.* **21,** 33 (1984).
[19] J. Wingender and U. K. Winkler, *FEMS Microbiol. Lett.* **21,** 63 (1984).
[20] U. K. Winkler and M. Stuckmann, *J. Bacteriol.* **138,** 663 (1979).

TABLE I

SUMMARY OF PURIFICATION OF LIPASE BTL-1 FROM *Bacillus thermocatenulatus*[a]

Step	Total protein (mg)	Total units[b] (U)	Specific activity (U/mg)	Yield (%)	Purification factor
Culture concentrate	1617	696	0.4	100	1
Calcium soap[c]	427	517	1.2	74	2.8
Hexane extraction (intermediate phase)	105	108	1.0	13	2.3
Methanol precipitation	34	70	2.0	10	4.7
Q-Sepharose (pooled fractions)	2.6	75	29	11	67.4

[a] Reprinted from *Biochim. Biophys. Acta,* **1214,** C. Schmidt-Dannert, H. Sztajer, W. Stöcklein, U. Menge, and R. D. Schmid, pp. 43–53, Copyright 1994 with kind permission of Elsevier Science–NL, Sara Burgerhartstraat 25, 1055 KV Amsterdam, The Netherlands.

[b] Activity was measured by the photometric assay with pNPP as substrate (60° and pH 7.5).

[c] Determination of protein and lipase activity of the soap (newly formed after storage of culture concentrate at 4° and obtained after filtration of the culture concentrate) was not possible; the values were therefore deduced from those of the filtrate compared to those of the concentrate.

AG, Germany) to a final volume of 17 liters and a yield of 71% activity. The resulting concentrate is stored for 2 days at 4°. During this time solid calcium soap is formed by oleic acid and calcium ions. Most of the lipase activity is adsorbed to the calcium soap. Further purification of the adsorbed lipase BTL-1 is achieved by extracting the soap with hexane, which results in the formation of an interphase containing the extracted lipase and located between the organic and aqueous phases. The extracted lipase is then subjected to precipitation with 74% (v/v) methanol. The lipase solution thus obtained is purified to homogeneity by anion-exchange chromatography with Q-Sepharose at pH 6.5, using 30% (v/v) methanol and 1 *M* NaCl in phosphate buffer (pH 6.5) to elute the strongly bound lipase.[12] The pure lipase is obtained with a yield of 11% and a specific activity of 29 U/mg (Table I).

Sodium dodecyl sulfate–polyacrylamide gel electrophoresis (SDS–PAGE) of the pure enzyme resulted in three bands of 24, 48, and 16 kDa with 0.6% (w/v) SDS in the applied sample. However, increasing the SDS concentration up to 2.1% (w/v) in the applied sample converted the higher molecular weight band into the 16-kDa band (Fig. 1). Its molecular weight of 16,000 makes BTL-1 one of the smallest lipases ever published, even among other *Bacillus* lipases, which in general have low molecular weights of about 20,000.[21–24]

[21] A. Sugihara, T. Tani, and Y. Tominaga, *J. Biochem.* **109,** 211 (1991).

[22] W. Kugimiya, Y. Otani, and Y. Hashimoto, Japanese Patent JP 62-233531 (1987).

[23] V. Dartois, A. Baulard, K. Schank, and C. Colson, *Biochim. Biophys. Acta* **1131,** 253 (1992).

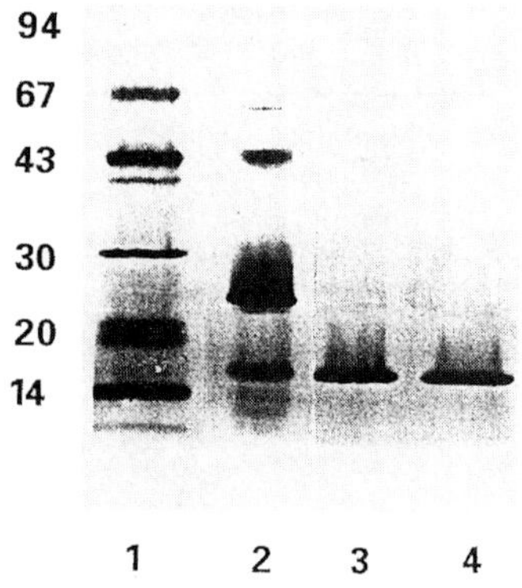

FIG. 1. SDS–polyacrylamide gel electrophoresis: Conversion of different molecular forms of lipase into the monomeric 16-kDa form by increased SDS concentrations in the applied enzyme sample (pooled lipase fractions after Q-Sepharose). Lane 1, molecular weight standard; lane 2, 0.6% (w/v) SDS; lane 3, 1.8% (w/v) SDS; lane 4, 2.1% (w/v) SDS. The gel is stained with silver for protein. (Reprinted from *Biochim. Biophys. Acta,* **1214,** C. Schmidt-Dannert, H. Sztajer, W. Stöcklein, U. Menge, and R. D. Schmid, pp. 43–53, Copyright 1994 with kind permission of Elsevier Science–NL, Sara Burgerhartstraat 25, 1055 KV Amsterdam, The Netherlands.)

The N-terminal amino acid sequence of lipase BTL-1 showed no homology to other sequences of esterases and lipases deposited in the EMBL and Swiss-Prot databases, or to the three other published *Bacillus* lipases.[22–24]

Lipase BTL-1 showed a strong tendency to form large aggregates of more than 47 monomers, as revealed by native PAGE. Strong aggregation and an almost irreversible adsorption of this lipase to hydrophobic surfaces (e.g., matrices that have hydrophobic cross-linkers, including Sepharose, Amberlit, and Serolit) attest to the extreme hydrophobicity of the lipase BTL-1. This makes handling the enzyme complicated. Aggregate formation, for example, makes it difficult to determine the real amount of lipase by activity measurement, as only the lipase molecules at the surface of the aggregate will have access to the emulsified substrate. Thus, aggregation might also be responsible for the apparent low specific activity of this lipase. However, the lipase of *B. subtilis* also showed a low specific activity of 116 U/mg.[15] Lipases of *Pseudomonas* strains[25–27] and the lipase of *B. subtilis*[15] have also been reported to be hydrophobic, with a tendency to aggregate. An analysis of the amino acid composition of lipase BTL-1 revealed a high

[24] B. Moeller, V. Roman, D. Wilke, and B. Foullois, Alkaline lipases from *Bacillus* and the cloning of their genes. Patent DE 41 11 21 A1 (1991).

[25] K.-E. Jaeger, F.-J. Adrian, H. E. Meyer, R. E. W. Hancock, and U. K. Winkler, *Biochim. Biophys. Acta* **1120,** 315 (1992).

[26] A. Dünhaupt, S. Lang, and F. Wagner, *Biotechnol. Lett.* **14,** 953 (1992).

[27] J. Kötting, D. Jürgens, and H. Huser, *J. Chromatogr.* **281,** 253 (1983).

TABLE II
AMINO ACID COMPOSITION OF VARIOUS *Bacillus* STRAINS

Amino acid	BTL-1[a] (mol%)	BTL-2[b] (mol%)	*B. stearothermophilus*[b] (mol%)	*B. species*[a] (mol%)	*B. subtilis*[b] (mol%)
Asx	17.7	10.8	7.7	10.2	14.4
Ala	12.2	8.5	5.7	12.1	6.6
Ile	11.0	7.8	6.5	3.3	5.5
Glx	10.4	8.0	15	7.3	5.0
Gly	9.3	9.8	7.3	11.4	13.3
Leu	8.4	9.3	8.5	10.4	8.3
Thr	7.6	5.9	3.6	10.4	5.5
Ser	6.1	6.4	4.0	6.5	7.2
Val	6.0	7.5	6.5	8.3	9.9
Lys	2.9	3.1	7.7	2.1	6.0
Pro	2.7	4.9	5.3	4.3	2.2
Arg	1.8	6.7	2.8	3.1	2.8
Tyr	1.7	4.9	5.6	4.6	5.0
Phe	1.0	4.1	4.0	1.9	2.2
His	0.8	3.4	3.2	2.6	2.8
Met	0.4	1.5	4.0	0.3	2.2
Cys	0	0.5	0.8	0.4	0
Trp	0	2.3	1.6	0.8	1.1

[a] Data obtained from the experimentally estimated amino acid composition.[12,21]
[b] Data obtained from the nucleotide sequences.[13,22,23]

content (39.4%) of the hydrophobic amino acids isoleucine, valine, leucine, phenylalanine, methionine, and alanine. In other *Bacillus* lipases, the corresponding content is only around 35%.[21–24] Table II shows the amino acid composition of lipases from different *Bacillus* strains.

As expected for a thermophilic enzyme, the lipase BTL-1 showed maximum activity at elevated temperatures of 60–70° with pNPP and olive oil as substrates. Also, BTL-1 showed a good temperature stability (50% residual activity after 30 min at 60°) and a good stability toward different organic solvents and detergents.[12] At a neutral pH of 7.0–8.0 the enzyme was both maximally stable and active. The lipase BTL-1 prefers pNP esters with acyl chain lengths of C_{10} to C_{12} as substrates, whereas pNP esters with acyl chain lengths below C_8 are poor substrates or are not hydrolyzed at all. A similar preference for middle acyl chain-length substrates has been described for other lipases from *Pseudomonas* and *Bacillus*.[15,28] Like many other lipases BTL-1 shows a lower activity toward the position-2 ester bonds of triglycer-

[28] A. Sugihara, M. Ueshima, Y. Shimada, S. Tsunesawa, and Y. Tominaga, *J. Biochem.* **112,** 598 (1992).

TABLE III
COMPARISON OF PROPERTIES OF NATIVE LIPASE BTL-1 AND RECOMBINANT LIPASE BTL-2 FROM *Bacillus thermocatenulatus*[a]

Properties	BTL-1	BTL-2
MW	16,000	43,000
p*I*	ND	7.2
T_{opt} (°C)	60–70	60–70
pH_{opt}	7.0–8.0	8.0–9.0
T_{stab} (°C)	40	50
pH_{stab}	7.0–8.0	9.0–10.0
Specific activity (U/mg)		
pNPP	29	1,314
Tributyrin	ND	54,887
Detergent stability	Moderate	High
Organic solvent stability	Moderate	High

[a] Abbreviations: BTL, *Bacillus thermocatenulatus* lipase; ND, not determined; MW, molecular weight; T_{opt}, temperature optimum of activity; T_{stab}, temperature stability after 30 min of incubation; pH_{opt}, pH optimum of activity; pH_{stab}, pH stability; pNPP, *p*-nitrophenyl palmitate.

ides.[12] A summary of the properties of the lipase BTL-1 in comparison to the lipase BTL-2 is given in Table III.

Lipase BTL-2

Molecular Cloning and Sequence Analysis of 43-kDa Lipase BTL-2

We attempted to clone the native 16-kDa lipase by generating an expression library, a method that has been successful for several lipases.[29–31] For this purpose, genomic DNA of *B. thermocatenulatus* was partially digested with *Sau*3A. The DNA fragments obtained after digestion were separated on 1% (w/v) agarose gel. Those having a size of 2–6 kb were extracted from the gel and ligated into pUC18, which had been previously cut with *Bam*HI. The ligated vector DNA was then introduced into *E. coli* DH5α

[29] G. H. Chung, Y. P. Lee, G. H. Jeohn, O. J. Yoo, and J. S. Rhee, *Agric. Biol. Chem.* **55,** 2359 (1991).

[30] W. Kugimiya, Y. Otani, Y. Hashimoto, and T. Takagi, *Biochem. Biophys. Res. Commun.* **141,** 185 (1986).

[31] X. Li, S. Tetling, U. K. Winkler, K.-E. Jaeger, and M. J. Benedik, *Appl. Environ. Microbiol.* **61,** 2674 (1995).

and the transformed cells were plated on LB (Luria broth) plates containing emulsified tributyrin (1%, w/v) and ampicillin (100 μg/ml). Among the 2000 obtained transformants, one colony formed a clear zone on LB–tributyrin plates, indicating lipase or esterase production. The expression of true lipase activity by this transformant is additionally confirmed on LB plates containing rhodamine B and olive oil.[13]

Restriction enzyme analysis of the recombinant plasmid, designated as pLIP1, resulted in the map shown in Fig. 2A. A 4.5-kb DNA fragment in pLIP1 was found to be responsible for lipase expression. Several deletion clones of pLIP1 were generated by restriction enzyme digestion with *Bam*HI and *Pst*I, with *Bam*HI and *Eco*RI, and with *Sma*I to identify the region responsible for lipase expression. Transformants with a 2.1-kb deletion after *Sma*I digestion in pLIP1 and those having a 2.7-kb deletion after *Bam*HI and *Eco*RI digestion no longer expressed lipase on LB–tributyrin plates. However, plasmid pLIP1 with a *Bam*HI–*Pst*I deletion did exhibit lipase production. Further subcloning using a nested deletion kit (Pharmacia, Piscataway, NJ) localized the lipase gene on a 2.3-kb insert of a partially deleted subclone designated as pLIP2 (Fig. 2B).

The nucleotide sequence of the 2293-bp insert in pLIP2 was determined

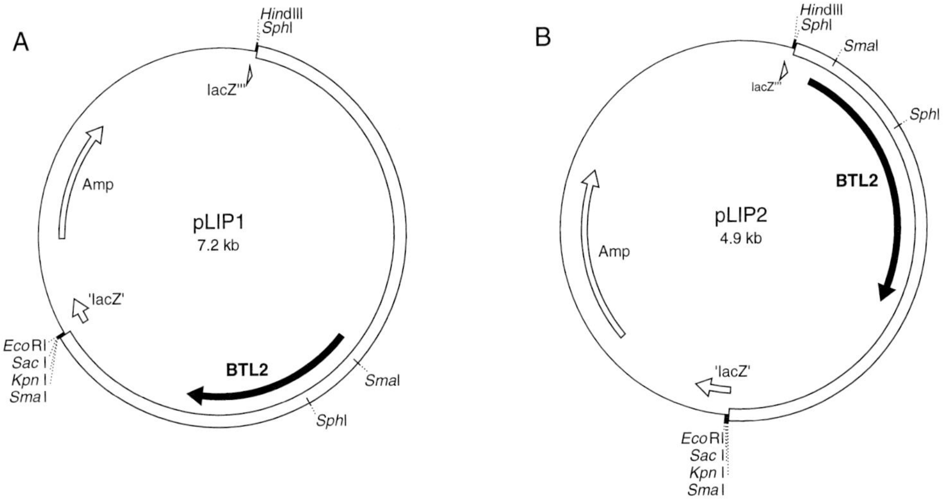

FIG. 2. (A) Physical map of pLIP1 obtained after ligating *Sau*3AI fragments into pUC18. (B) Physical map of pLIP2 obtained after generation of nested deletion within pLIP1. The arrows in both maps indicate the region encoding the lipase gene and the direction of transcription. (Reprinted from *Biochim. Biophys. Acta,* **1301,** C. Schmidt-Dannert, M. L. Rúa, H. Atomi, and R. D. Schmid, pp. 105–114, Copyright 1997 with kind permission of Elsevier Science–NL, Sara Burgerhartstraat 25, 1055 KV Amsterdam, The Netherlands.)

```
                                                -35
TCACAGAAAAACCCGACAATTGCCGGGATTGAATCAGTTGGTTGATATATATAGAATATT 60
    -10                                       RBS        M  M
CAGGTAATTATGAACAAAAAGACTCCGTTTATGTGAGGGGAGGAGAAGGATAGGATGATG 120
 K  G  C  R  V  M  V  V  L  L  G  L  W  F  V  F  G  L  S  V
AAAGGCTGCCGGGTGATGGTTGTGTTGCTTGGATTATGGTTTGTGTTCGGCCTGTCGGTC 180
 P  G  G  R  T  E  A ▼A  S  P  R  A  N  D  A  P  I  V  L  L
CCGGGAGGGCGGACGGAAGCGGCATCCCCACGCGCCAATGATGCACCCATCGTGCTTCTC 240
 H  G  F  T  G  W  G  R  E  E  M  L  G  F  K  Y  W  G  G  V
CATGGGTTTACAGGATGGGGGCGAGAGGAAATGCTTGGATTCAAATATTGGGGCGGCGTG 300
 R  G  D  I  E  Q  W  L  N  D  N  G  Y  R  T  Y  T  L  A  V
CGTGGCGATATCGAACAATGGCTGAACGACAACGGATATCGTACGTATACGCTGGCGGTC 360
 G  P  L  S  S  N  W  D  R  A  C  E  A  Y  A  Q  L  V  G  G
GGACCGCTCTCGAGCAACTGGGACCGGGCGTGTGAAGCGTACGCCCAGCTTGTCGGCGGA 420
 T  V  D  Y  G  A  A  H  A  A  K  H  G  H  A  R  F  G  R  T
ACGGTCGATTATGGCGCGGCCCATGCGGCGAAGCACGGCCATGCGCGGTTTGGCCGCACG 480
 Y  P  G  L  L  P  E  L  K  R  G  G  R  V  H  I  I  A  H  S
TATCCGGGCTTGCTGCCGGAATTGAAAAGAGGCGGCCGCGTCCATATCATCGCTCATAGC 540
 Q  G  G  Q  T  A  R  M  L  V  S  L  L  E  N  G  S  Q  E  E
CAAGGAGGACAGACGGCCCGCATGCTTGTGTCTCTCCTAGAGAACGGAAGCCAAGAAGAG 600
 R  E  Y  A  K  A  H  N  V  S  L  S  P  L  F  E  G  G  H  H
CGGGAGTACGCCAAGGCGCATAACGTGTCGTTGTCGCCGTTGTTTGAAGGCGGACATCAT 660
 F  V  L  S  V  T  T  I  A  T  P  H  D  G  T  T  L  V  N  M
TTTGTGTTGAGTGTGACAACCATTGCCACTCCTCATGACGGGACGACACTTGTCAATATG 720
 V  D  F  T  D  R  F  F  D  L  Q  K  A  V  L  K  A  A  A  V
GTCGATTTCACTGATCGCTTCTTTGACCTGCAAAAAGCGGTGTTGAAAGCGGCGGCTGTC 780
 A  S  N  V  P  Y  T  S  Q  V  Y  D  F  K  L  D  Q  W  G  L
GCCAGCAATGTGCCGTACACGAGTCAAGTATACGATTTTAAGCTCGACCAATGGGGGCTG 840
 R  R  Q  P  G  E  S  F  D  H  Y  F  E  R  L  K  R  S  P  V
CGCCGCCAGCCGGGTGAATCGTTCGACCATTATTTTGAACGGCTCAAACGATCCCCTGTT 900
 W  T  S  T  D  T  A  R  Y  D  L  S  I  P  G  A  E  K  L  N
TGGACGTCGACGGATACTGCCCGCTACGATTTATCCATTCCCGGAGCTGAGAAGTTGAAT 960
 Q  W  V  Q  A  S  P  N  T  Y  Y  L  S  F  S  T  E  R  T  H
CAATGGGTGCAAGCCAGCCCGAATACGTATTATTTGAGCTTTTCCACCGAACGGACGCAC 1020
 R  G  A  L  T  G  N  Y  Y  P  E  L  G  M  N  A  F  S  A  V
CGCGGAGCGCTCACCGGCAACTATTATCCCGAACTCGGAATGAATGCATTCAGCGCGGTC 1080
 V  C  A  P  F  L  G  S  Y  R  N  E  A  L  G  I  D  D  R  W
GTATGCGCCCCGTTTCTCGGCTCGTACCGCAATGAGGCGCTCGGCATTGACGACCGCTGG 1140
 L  E  N  D  G  I  V  N  T  V  S  M  N  G  P  K  R  G  S  S
CTGGAGAACGATGGCATCGTCAACACGGTTTCGATGAACGGTCCAAAGCGTGGATCAAGC 1200
 D  R  I  V  P  Y  D  G  T  L  K  K  G  V  W  N  D  M  G  T
GATCGGATCGTGCCGTATGACGGGACGTTGAAAAAAGGAGTTTGGAATGATATGGGGACG 1260
 Y  N  V  D  H  L  E  V  I  G  V  D  P  N  P  S  F  D  I  R
TACAACGTCGACCATTTGGAAGTGATCGGCGTTGACCCGAATCCGTCATTTGATATCCGC 1320
 A  F  Y  L  R  L  A  E  Q  L  A  S  L  R  P ******
GCCTTTTATTTGCGACTTGCCGAGCAGTTGGCGAGTTTGCGGCCTTAATGAAGGGATGAT 1380
CAGCCAAACGGGCAAAAAGCCATCTCTGTCAGATGGCTCTTTTTTTGTGGAGCAAGCACC 1440
TGAGTTGTATGACGGCCGCGGATGTCCGATGGTATAATAAGGGTAAAGCAAGCGGATGGG 1500
```

by sequencing the subclones obtained after nested deletion reactions, using the M13 forward and M13 reverse sequencing primer, respectively. Both strands of the lipase gene in pLIP2 were sequenced using synthetic oligodeoxynucleotide primer, corresponding to the lipase sequence, with the fluorescence-based dideoxy DNA cycle sequencing method (Applied Biosystems, Foster City, CA).[13]

A single open reading frame of 1251 bp, extending from position 115 to 1371, was found to encode the lipase (Fig. 3). Downstream of the successive termination codons TAA and TGA is a stem–loop structure, while upstream of the likely initiation codon ATG, at position 115, the putative −35 and −10 promoter sequences are located. The fact that the transcriptional direction of the lipase gene in pLIP1 and pLIP2 is opposite to the direction dictated by the *lac* promoter and that the sequence upstream of the lipase gene showed a promoter sequence suggests that this native promoter controls the expression of the lipase gene.

The deduced sequence of the preprotein contains 417 amino acids and corresponds to a molecular weight of 46,230. The rules for signal peptide sequences[32] suggest a 29-amino acid signal sequence and a cleavage site between Ala-29 and Ala-30. The mature lipase contains 388 amino acids, which corresponds to a molecular weight of 43,090. The difference in size of the previously isolated native 16-kDa lipase (BTL-1) and the cloned 43-kDa lipase (BTL-2) demonstrates that lipase BTL-2 is a novel lipase that had not been previously detected in the culture broth of *B. thermocatenulatus* under the cultivation conditions used. Also, no homology of lipase BTL-2 to the 45 N-terminal amino acids of BTL-1 was found, indicating that BTL-1 is not a processed form of BTL-2.

A comparison of the deduced amino acid sequence of lipase BTL-2 with several lipases in the region of the catalytic triad histidine, serine, aspartate (glutamate) and around the oxyanion hole[33] allowed us to determine that Ser-113, Asp-317, and His-351 are the residues forming the cata-

[32] G. Heijne, *Nucleic Acids Res.* **14,** 4683 (1986).
[33] K.-E. Jaeger, S. Ransac, H. B. Koch, F. Ferrato, and B. Dijkstra, *FEBS Lett.* **332,** 143 (1993).

FIG. 3. Nucleotide sequence of the 2293-bp insert in pLIP2 and the deduced amino acid sequence of the encoded lipase BTL-2 (sequence shown from 1 to 1500 bp). Stop codons are marked by asterisks and the putative transcription and translation signals by single underlines (RBS, ribosome-binding site). Arrows under nucleotides show the pairing region of the termination hairpin. The N-terminal amino acids obtained after protein sequencing of the expressed lipase are double underlined. The presumed amino acids of the catalytic triad and the consensus sequence are in boldface. (Reprinted from *Biochim. Biophys. Acta,* **1301,** C. Schmidt-Dannert, M. L. Rúa, H. Atomi, and R. D. Schmid, pp. 105–114, Copyright 1997 with kind permission of Elsevier Science–NL, Sara Burgerhartstraat 25, 1055 KV Amsterdam, The Netherlands.)

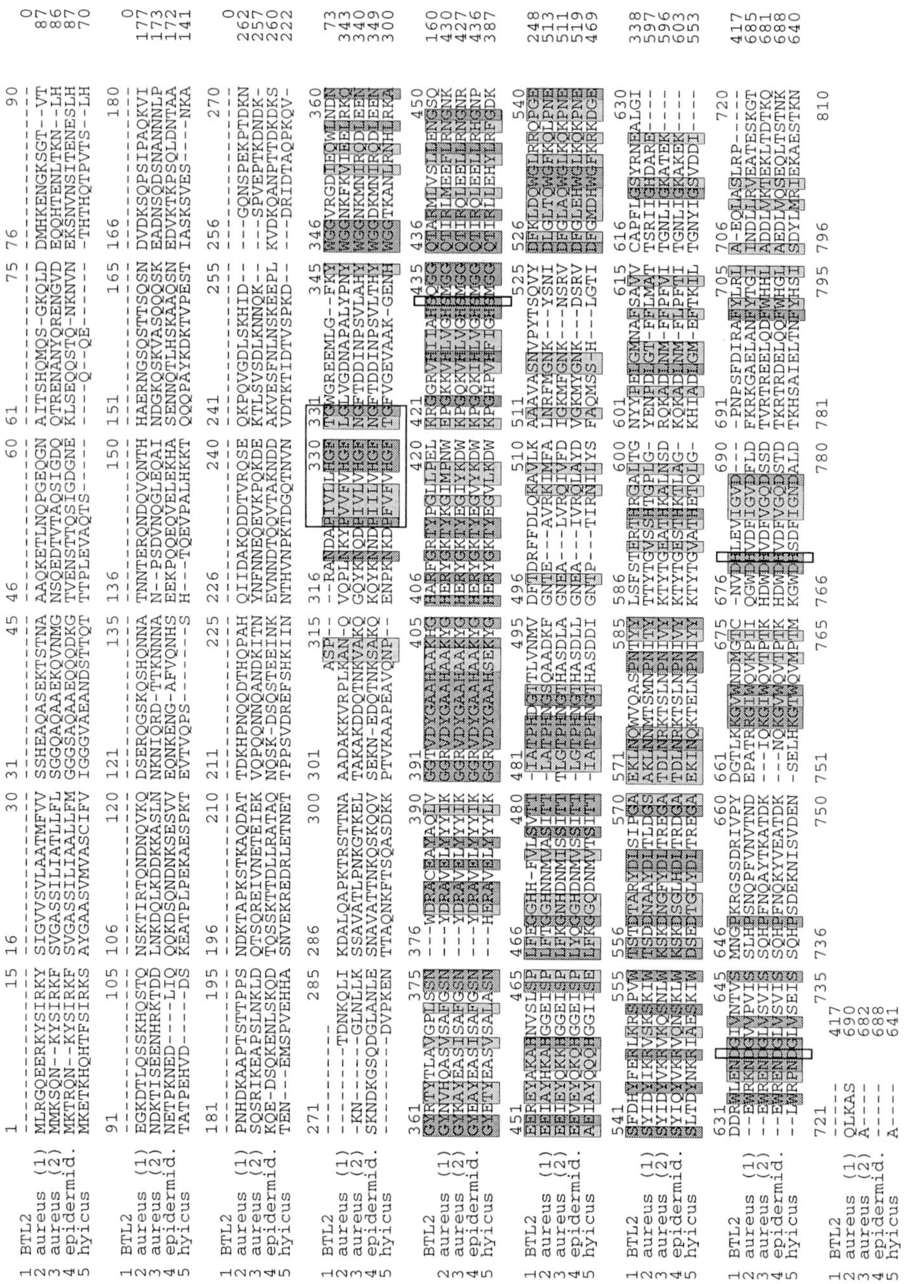

lytic triad and that the oxyanion hole is formed by the residues around His-14. However, no significant homology apart from the region of the nucleophilic serine was found with most other lipases and esterases in the EMBL and Swiss-Prot databases. But lipase BTL-2 shows a significant homology of 30–35% with the mature lipases of *Staphylococcus hyicus, S. aureus,* and *S. epidermidis,*[34–37] which also have a similar molecular weight of around 45,000. An alignment of the mature lipase BTL-2 amino acid sequence with that of the three *Staphylococcus* lipases is given in Fig. 4. Like the lipases from *Bacillus pumilus* and *B. subtilis,*[23,24] an alanine replaces the first glycine in the pentapeptide Gly–*X*–Ser–*X*–Gly, which is conserved among microbial and mammalian lipases. This replacement might be a common feature of *Bacillus* lipases.

The lipase production obtained under the control of the native promoter was rather low (600 U/g cells). Nevertheless, we tried to purify this lipase and investigate some of its properties. The isolation of the lipase started with the disruption of the cells by sonication followed by the incubation of the suspension for 25 min at 65° (heat precipitation). Precipitated protein and cell debris were then removed by centrifugation and the clear supernatant was applied to coupled Q- and S-Sepharose columns that had been previously equilibrated with Tris buffer, pH 7.5. Under these conditions the lipase did not bind, but many other proteins did. Unbound lipase was collected and loaded on a phenyl-Sepharose column that had been equilibrated with 0.5 *M* $(NH_4)_2SO_4$ in Tris buffer, pH 7.5. A final elution of bound lipase with 2% (w/v) octyl glucoside in distilled water resulted in a pure enzyme with a specific activity of 10,225 U/mg, a total yield of 49%, and a purification factor of 329.

SDS–PAGE showed a single protein band corresponding to a molecular weight of 32,000 for the purified lipase, even though the deduced sequence indicates a molecular weight of 43,000 for the mature lipase. But N-terminal sequencing of the purified lipase showed that during expression the signal peptide is correctly cleaved at the position previously assumed and that

[34] C. Y. Lee and J. J. Iandolo, *J. Bacteriol.* **166,** 385 (1986).
[35] K. Nikoleit, R. Rosenstein, H. M. Verheij, and F. Götz, *Eur. J. Biochem.* **228,** 221 (1995).
[36] A. M. Farrell, T. J. Foster, and K. T. Holland, *J. Gen. Microbiol.* **139,** 267 (1993).
[37] F. Götz, F. Popp, E. Korn, and K. H. Schleifer, *Nucleic Acids Res.* **13,** 5895 (1985).

FIG. 4. Alignment of the mature lipase BTL-2 from *B. thermocatenulatus* with unprocessed lipases from *Staphylococcus aureus*[34,35] [(1), strain PS54; (2), strain NCTC8530], *S. epidermidis,*[36] and *S. hyicus.*[37] Residues of BTL-2 homologous to at least one of the *Staphyloccus* lipases are highlighted in light gray and those found in all five lipases are highlighted in dark gray. The region around the oxyanion hole and the amino acids of the catalytic triad are boxed.

the 32,000 band on the SDS gel is the lipase. Abnormal migration during SDS–PAGE might be responsible for this discrepancy in molecular weight, which has also been reported for other proteins.[38,39] A first brief characterization of this purified recombinant lipase BTL-2 showed that the properties of lipase BTL-2 differ significantly from those of the lipase BTL-1 and most other *Bacillus* lipases (Table III). For example, BTL-2 has a molecular weight of 43,000, whereas the molecular weight of BTL-1 and most other *Bacillus* lipases ranges from 16,000 to 22,000.[21–24] In addition, BTL-2 shows maximum activity at alkaline pH, but BTL-1 and most other lipases[21,28,40–42] show maximum activity in a lower pH range of pH 5.6–8.0. The lipase BTL-2 was remarkably stable in the presence of most detergents and organic solvents. Many other lipases would be quickly denatured under these conditions.[42] A more detailed characterization of lipase BTL-2 is given in the following sections.

Overexpression of BTL-2 in Escherichia coli

The low level of BTL-2 expression under the control of its native promoter in pLIP2 made it necessary to develop an efficient expression system that would allow us to produce large amounts of this promising lipase for crystallization and further characterization, including its potential in biotechnological applications. For this purpose, we subcloned the BTL-2 gene in the *E. coli* expression vector pCYTEXP1,[43] which contains the strong λ promoter P_L together with the temperature-sensitive λ repressor *c*I857. Expression of the gene is induced by shifting the cultivation temperature from 30 to 42° (heat shock) for 3–4 hr after reaching an optical density (OD) of 0.8–0.9 at 578 nm. The vector pCYTEXP1(pT1) was generated from the expression vector pT1-ompAROL,[44] which also contains the lipase gene from *Rhizopus oryzae* and the presequence of the outer membrane protein OmpA of *E. coli.*

Two general expression strategies were developed[45]: (1) expression of mature BTL-2 lipase in the cytoplasm of *E. coli* and (2) secretion of BTL-2

[38] R. E. Blankenship, *Antonie van Leeuwenhoek* **65,** 311 (1994).

[39] R. E. Daley and W. Wickner, *J. Biol. Chem.* **260,** 15925 (1985).

[40] T. Iizumi, K. Nakamura, and F. Fukase, *Agric. Biol. Chem.* **54,** 1253 (1990).

[41] H. Sztajer, H. Lünsdorf, H. Erdmann, U. Menge, and R. D. Schmid, *Biochim. Biophys. Acta* **1124,** 253 (1992).

[42] W. Stöcklein, H. Sztajer, U. Menge, and R. D. Schmid, *Biochim. Biophys. Acta* **1168,** 181 (1993).

[43] T. N. Belev, M. Singh, and J. E. G. McCarthy, *Plasmid* **26,** 147 (1991).

[44] H. D. Beer, Ph.D. thesis. University of Hannover, Hannover, Germany, 1994.

[45] M. L. Rúa, H. Atomi, C. Schmidt-Dannert, and R. D. Schmid, submitted (1997).

lipase in the periplasm of *E. coli* by the use of either the original leader sequence of BTL-2 or the OmpA signal sequence. Toward this end three different expression vectors were constructed and introduced into *E. coli* BL321: (1) pT1-BTL2, containing the mature lipase gene, (2) pT1-preBTL2, containing the prelipase gene, and (3) pT1-OmpABTL2, containing the mature lipase gene fused to an OmpA leader sequence. Figure 5 outlines the cloning strategies for the construction of these three expression vectors. Appropriate restriction sites for the ligation of vector and insert were created by polymerase chain reaction (PCR) using oligonucleotide primers containing the desired sites. In all three cases, the initiation codon ATG of the lipase gene (with or without leader sequence) is located at the *Nde*I site of pT1. Transformants were checked for lipase expression on LB–tributyrin plates. Plasmids of transformants exhibiting lipase activity of LB–tributyrin plates were isolated and the correct insertion of the lipase gene was checked by restriction analysis and sequencing of the insert. One plasmid of each construction (pT1-BTL2, pT1-preBTL2, and pT1-OmpABTL2) with the correct insertion was chosen for further experiments.

To determine the amount of active lipase expressed by *E. coli* BL321 containing plasmid pT1-BTL2, pT1-preBTL2, or pT1-OmpABTL2, cells from 200-ml cultures were harvested by centrifugation after 3 hr of heat shock and disrupted by sonication. After the removal of cell debris by centrifugation, the lipase activity in the supernatant was determined using tributyrin as a substrate. After 30 min of heat shock, the production of lipase started and reached a maximum after 3 hr. A comparison of total lipase activity obtained with the three expression vectors after 3 hr of heat shock is shown in Table IV. *Escherichia coli* BL321 cells harboring the plasmid pT1-ompABTL2 produced larger amounts of lipase activity (30,000 U/g cells) than cells harboring pT1-BTL2 (7000 U/g cells) or pT1-preBTL2 (9000 U/g cells). These expression levels represent a 12- to 50-fold improvement compared to pLIP2 (600 U/g cells) containing the BTL-2 gene under the control of its native promoter. The expression vector pT1-OmpABTL2 was also introduced into two more *E. coli* host cells, JM105 and DH5α. The expression level (Table IV) obtained with DH5α was no different than that obtained with BL321, but a nearly 25-fold increase in the production of soluble lipase was reached when JM105 was used as the host cell (660,000 U/g cells).

Total cellular protein obtained before and after the induction of lipase expression by heat shock for 3 hr was analyzed by SDS–PAGE. The proteins were stained with Coomassie Brilliant Blue (Fig. 6A). To determine the activity of the recombinant lipase directly on SDS gels (Fig. 6B), we developed a method to remove the SDS and at the same time renature the lipase

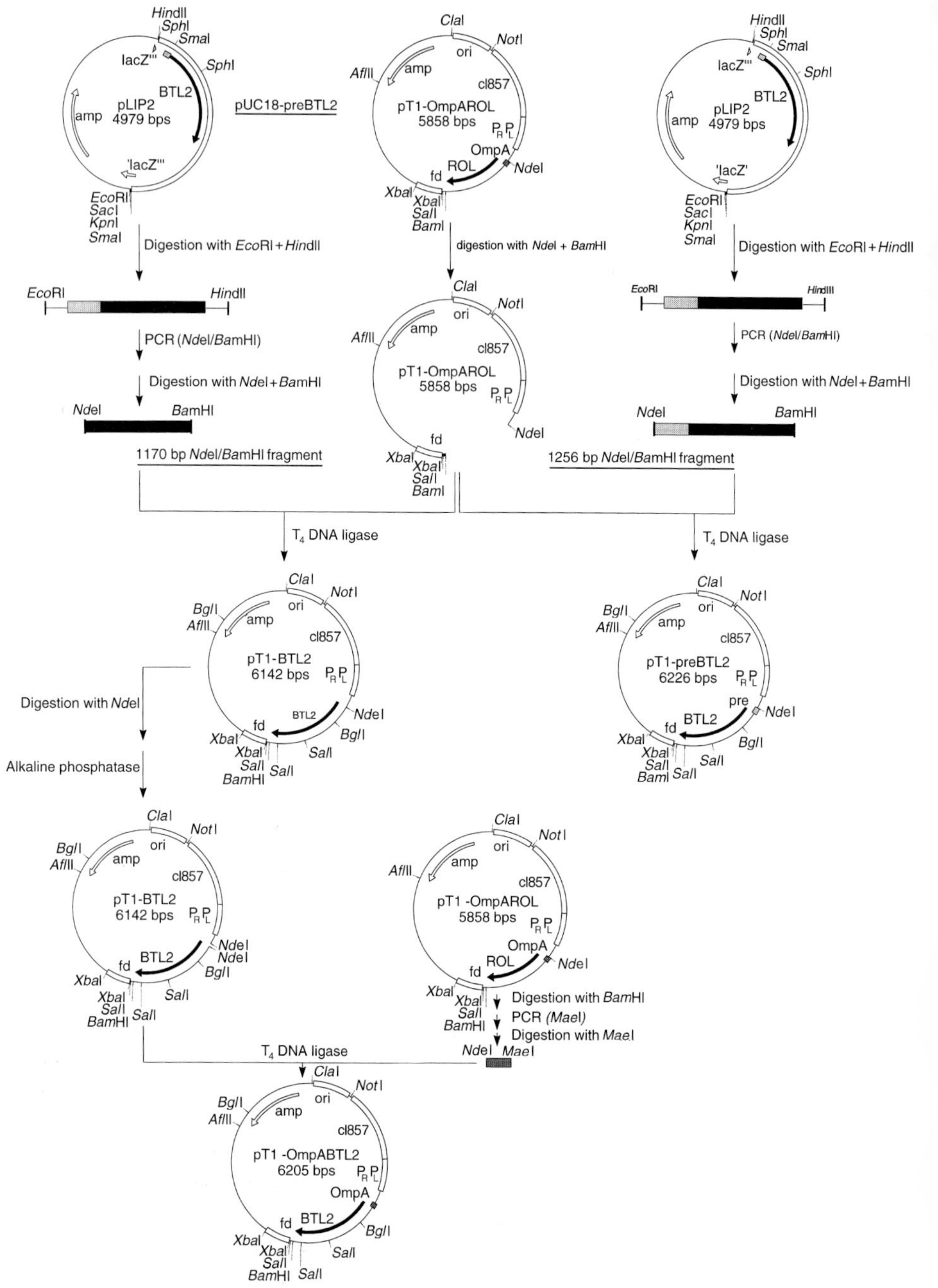

HindII
SphI
SmaI
lacZ'''
SphI
BTL2
pLIP2
4979 bps
amp
'lacZ'''
EcoRI
SacI
KpnI
SmaI
pUC18-preBTL2
ClaI
NotI
ori
AflII
amp
cI857
pT1-OmpAROL
5858 bps
$P_R P_L$
OmpA
ROL
NdeI
fd
XbaI
XbaI
SalI
BamI
lacZ'''
BTL2
pLIP2
4979 bps
'lacZ'
Digestion with EcoRI + HindII
digestion with NdeI + BamHI
Digestion with EcoRI + HindII
EcoRI
HindII
HindIII
PCR (NdeI/BamHI)
Digestion with NdeI + BamHI
NdeI
BamHI
1170 bp NdeI/BamHI fragment
1256 bp NdeI/BamHI fragment
T_4 DNA ligase
T_4 DNA ligase
BglI
pT1-BTL2
6142 bps
pT1-preBTL2
6226 bps
pre
Digestion with NdeI
Alkaline phosphatase
BamHI
Digestion with BamHI
PCR (MaeI)
Digestion with MaeI
NdeI
MaeI
pT1 -OmpAROL
pT1 -OmpABTL2
6205 bps

TABLE IV
EXPRESSION LEVEL OF BTL-2 LIPASE IN VARIOUS *Escherichia coli* EXPRESSION VECTORS AND HOST CELLS[a]

Plasmid	*E. coli* host	Expression level (U/g of cells)[b]
pT1-BTL2	BL321	7,000
pT1-preBTL2	BL321	9,000
pT1-OmpABTL2	BL321	30,000
	DH5α	30,000
	JM105	660,000

[a] From *Biochem. Soc. Trans.* **25,** 178 (1997), with permission.
[b] pH-stat assay with tributyrin as substrate (60° and pH 7.5).

by incubating the gel for 20 min at room temperature in a 20 m*M* Tris buffer, pH 8.5, containing 0.5% (w/v) Triton X-100.[45] After renaturation the gel was stained for lipolytic activity with 1-napthyl acetate as the substrate.[41]

After 3 hr of heat shock *E. coli* BL321 pT1-preBTL2 produced a new protein band with a molecular weight of 43,000 (Fig. 6A, lane 5). However, despite the fact that the soluble lipase activities produced by both plasmids pT1-BTL1 and pT1-preBTL are similar (Table IV), *E. coli* BL321 pT-BTL2 failed to produce a band that was clearly visible after staining for protein. Activity staining indicated that cells harboring pT1-BTL2 produced a band of similar intensity and the same molecular weight 41,000, as those with pT1-preBTL2 (Fig. 6B, lanes 4 and 5, respectively). However, plasmid pT1-preBTL2 expressed an additional band of 43 kDa that was also visible after Coomassie staining (Fig. 6A, lane 5). This protein band was always retained in the pellet of the cell lysate. N-Terminal sequencing of both the 41-kDa lipase produced by plasmids pT1-BTL2 and pT1-preBTL2 and the insoluble 43-kDa lipase band expressed by plasmid pT1-preBTL2 revealed that the insoluble 43-kDa lipase represents the unprocessed lipase BTL2 still containing its hydrophobic signal peptide. The N-terminal sequence of the 41-kDa soluble lipase produced by both plasmids pT1-BTL2 and pT1-preBTL2 was identical with that of the mature lipase BTL-2. Thus, *E. coli* was unable to process the preenzyme fully. Plasmid pT1-OmpABTL2 gen-

FIG. 5. Constructions of plasmids pT1-preBTL2, pT1-BTL2, and pT1-OmpABTL2. Plasmid pLIP2 contains a 1300-bp DNA fragment from *B. thermocatenulatus* in pUC18, which includes the coding sequence for the mature BTL-2 lipase (black arrow) and the N-terminal signal peptide (▨). Plasmid pT1-OmpAROL is a secretion cloning vector, wherein the expression of a foreign gene, cloned in the multiple cloning site, is under the control of the heat-inducible λP_L promoter. (■) Coding region for the OmpA signal peptide.

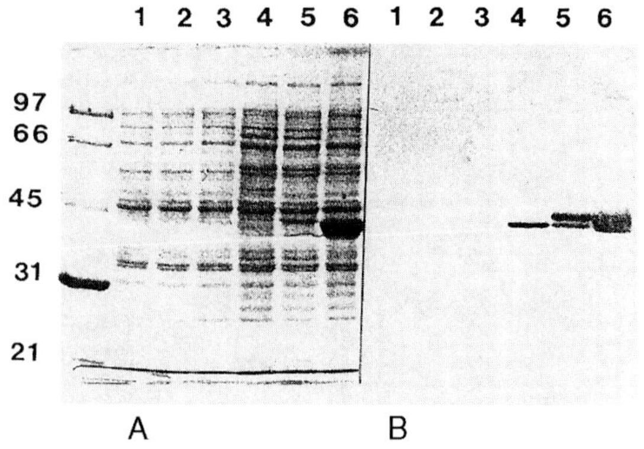

FIG. 6. SDS–polyacrylamide gel electrophoresis of total cellular proteins stained for protein (A) and activity (B): *E. coli* BL321 harboring pT1-BTL2 (lanes 1 and 4), pT1-preBTL2 (lanes 2 and 5), and pT1-OmpA BTL2 (lanes 3 and 6) were grown at 30° in LB medium containing ampicillin (100 μg/ml) to an optical density of 0.8–0.9 at 578 nm. The temperature was then raised to 42° to induce the expression of the lipase and the culture was incubated for another 3 hr. The cells were solubilized as described in text and subjected to SDS–polyacrylamide electrophoresis. The gels were stained with Coomassie Brilliant Blue for protein (A) and with 1-naphthyl acetate for activity (B). The molecular weights of the standards ($\times 10^{-3}$) are indicated on the left.

erated a new and, as expected from the produced lipase activity, broad protein band of 44 kDa on an SDS gel after staining for protein and activity (Fig. 6A and B, lanes 6). N-Terminal sequencing of the produced lipase showed that, similar to pT1-preBTL2, the OmpA signal sequence was not cleaved at all by *E. coli.*

With both pT1-preBTL2 and pT1-OmpABTL2 we tried to increase the amount of mature lipase BTL-2 by decreasing the heat-shock temperature from 42 to 37°, conditions that have been described to reduce the rate at which newly synthesized proteins are released by the protein synthesis machinery into the cell cytoplasm.[46,47] In the case of pT1-preBTL2, the amount of precursor produced was significantly decreased, but the increase in expressed mature lipase was only small, even after 24 hr of heat shock. A decrease in the cultivation temperature for lipase expression had no influence on the processing of OmpA-BTL2, however. In the case of pT1-OmpABTL2 it is likely that the cell machinery responsible for the processing of the protein was simply overloaded by such a high level of lipase

[46] C. H. Schein and M. H. M. Notebom, *Bio/Technology* **6,** 291 (1988).
[47] R. C. Hockney, *Trends Biotechnol.* **12,** 456 (1994).

production, as has also been reported for staphylococcal nuclease A.[48] When expressed in *E. coli* JM105, pT1-OmpABTL2 produced 80 times more soluble lipase than in either *E. coli* BL321 or DH5α (Table IV). But with SDS–PAGE no difference could be detected in the expression levels of the three strains. Because a large amount of OmpA-lipase activity is still associated with the precipitate of the cell lysate from all three strains, it is possible that in *E. coli* JM105 a larger amount of soluble OmpA-lipase is released after cell disruption.

To utilize the large amounts of lipase expressed by *E. coli* JM105 pT1-OmpABTL2, it was necessary to cleave the OmpA leader sequence from the lipase BTL-2. Daley and Wickner[39] reported the successful cleavage of the leader sequence from the OmpA protein using proteinase K. We followed their example and used proteinase K to cleave the OmpA signal peptide. First, the insoluble lipase was extracted from the precipitate of the cell lysate with 1% (w/v) cholate in 50 m*M* Tris buffer (pH 8.0) for 16 hr at 4°. In this way, 2.2 million units of lipase (using tributyrin as substrate) was solubilized from 2 g of *E. coli* JM105 cells in a relatively pure state with a specific activity of 16,208 U/mg. Next, the digestion of the OmpA lipase with proteinase K in a final concentration of 0.7 U/ml for 15 min at 0° (the reaction was stopped by addition of 10 m*M* phenylmethylsulfonyl fluoride [PMSF]) yielded 82% of the lipolytic activity before digestion. Small peptide fragments produced by the proteinase digestion were removed by ultrafiltration through a 100-kDa membrane. A 16-fold concentrated enzyme solution with 1.152 million units and a specific activity of 36,000 U/mg could be obtained with this simple purification procedure. SDS–PAGE of the digested lipase showed a protein band corresponding to a molecular weight about 2,000, smaller than the lipase before digestion (Fig. 7). The OmpA signal sequence consists of 21 residues and has a molecular weight of 2,400, which is consistent with the smaller size of the digested lipase.

Large-Scale Production and Purification of BTL-2

For crystallization it must be ensured that only the mature lipase form is present in the protein solution. Hence, *E. coli* BL321 pT1-BTL2 was chosen for cultivation in a 100-liter fermenter. The medium consisted of LB supplemented with glucose (10 g/liter) and ampicillin (100 mg/liter) and adjusted to pH 7.3. The cells were grown at 30° for 5 hr until the end of the exponential growth phase ($OD_{578\,nm}$ 1.65). Lipase expression was induced by raising the temperature to 42° (heat shock) and reached a

[48] M. Takahara, D. W. Hibler, P. J. Barr, J. A. Gerlt, and M. Inouye, *J. Biol. Chem.* **260,** 2670 (1985).

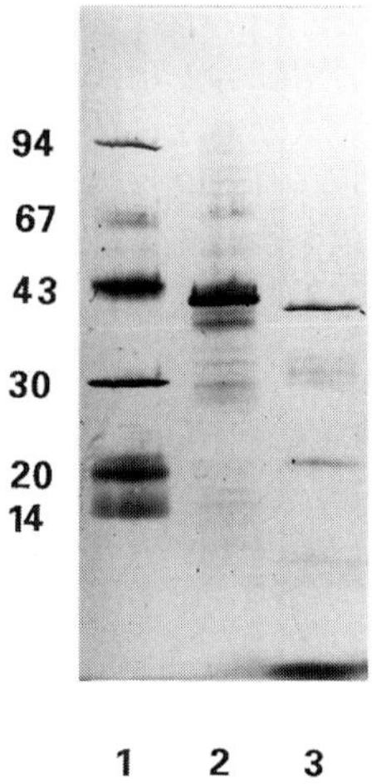

FIG. 7. SDS–polyacrylamide gel electrophoresis: Digestion of OmpA lipase with proteinase K after cholate extraction from insoluble precipitate of the cell lysate. Lane 1, molecular weight standard ($\times$ 10^{-3}); lane 2, sample after cholate extraction of OmpA lipase; lane 3, sample of proteinase K-digested lipase. The gel is stained with silver for protein.

maximum after 4 hr of induction, at which time the cells were harvested. The final cell mass was 4.8 g/liter and the lipase yield was 53,700 U/g cells.[49]

Lipase, which is located in the cytoplasm of *E. coli* cells, was isolated by disrupting the cells in a high-pressure homogenizer followed by centrifugation. The crude enzyme solution was then loaded on a butyl-Sepharose column that had been equilibrated with a Tris buffer, pH 8.0, containing 0.1 *M* NaCl. After washing the column with equilibration buffer, bound lipase was eluted with 1% (w/v) cholate in Tris buffer, pH 8.0. Fractions with activity were pooled and concentrated by ultrafiltration through a 50-kDa membrane. Final purification of the concentrated lipase solution was achieved by gel filtration using a TSK G300 column (Tosohaas, Tokyo, Japan). The purification procedure resulted in a highly active and pure lipase with a yield of 32% and a specific activity of 54,887 U/mg.[49] Table V summarizes the results of the purification. The key purification step was the hydrophobic interaction chromatography with butyl-Sepharose, which resulted in a 28-fold increase in specific activity. On SDS–PAGE (Fig. 8) a single band with a molecular weight of 40,000 could already be detected by silver staining after this purification step. After ultrafiltration and gel filtration, the same major band was visualized together with three minor

[49] M. L. Rúa, C. Schmidt-Dannert, S. Wahl, A. Sprauer, and R. D. Schmid, *J. Biotechnol.*, in press (1997).

TABLE V
SUMMARY OF PURIFICATION OF RECOMBINANT LIPASE BTL-2 PRODUCED WITH *Escherichia coli* BL321 pT-BTL2[a]

Step	Total units (U)	Total protein (mg)	Specific activity (U/mg)[b]	Purification factor	Yield (%)
Cell breakage	942,270	2,151	438	1	100
Butyl-Sepharose	481,815	38	12,354	28	51
Ultrafiltration	403,875	15.8	25,562	58	42
Gel filtration	296,391	5.4	54,887	125	32

[a] From *Biochem. Soc. Trans.* **25,** 178 (1997), with permission.
[b] pH-stat assay with tributyrin as substrate (60° and pH 7.5).

bands by both protein and activity staining (activity staining not shown). The fact that all four bands showed lipase activity means that the minor bands might possibly represent aggregate forms of the lipase.

N-Terminal sequence analysis showed the identity of the purified protein with the mature lipase BTL-2 previously expressed in only small quantities and with the amino acid sequence predicted from the nucleotide sequence.

Physicochemical Properties of BTL-2

The pure and highly active enzyme was characterized and its physicochemical properties were compared to those of the BTL-2 lipase previously expressed only weakly in *E. coli* under the control of its native promoter (Table III).[49] Both lipases have the same N-terminal sequence and a p*I* of 7.2. A difference of 8,000 was found in the molecular weight, as estimated

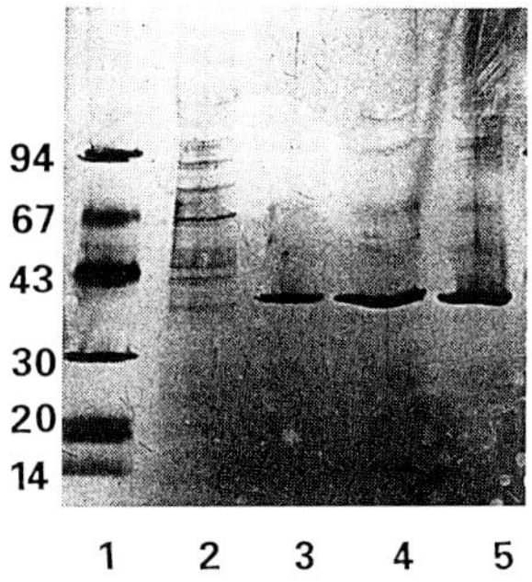

FIG. 8. SDS–polyacrylamide gel elecrophoresis of BTL-2 lipase at different purification steps: Lane 1, molecular weight standards (× 10^{-3}); lane 2, sample after cell breakage; lane 3, sample after butyl-Sepharose; lane 4, after sample ultrafiltration; lane 5, sample after TSK G3000 column. The gel is stained with silver for protein.

by SDS–PAGE, between the weakly expressed lipase (32,000) and that of the overexpressed lipase (40,000). The molecular weight predicted on the basis of the amino acid sequence was 43,000. The reason for this discrepancy might be attributed to an abnormal migration of the weakly expressed lipase during SDS–PAGE, as previously explained. The major difference between the two enzymes was found in their specific activity toward tributyrin. That of the overexpressed enzyme (54,788 U/mg) was five times higher than that of the weakly expressed lipase (10,225 U/mg). The reason for this discrepancy is not clear at present, but might be due to conformational differences. Also, the presence of inactive enzyme in the enzyme preparation of the weakly expressed lipase might contribute to this difference. Similar results have also been reported for native and overexpressed recombinant rhodanase from bovine liver.[50] The specific activity of the overexpressed lipase BTL-2 is 320 to 1675 times higher than that of BTL-1 and other small *Bacillus* lipases.[21–24] In addition, most lipases have specific activities below 3000 U/mg.[40–42]

As expected, the lipase BTL-2, like BTL-1, reaches maximum activity between 55 and 75°. With triolein as substrate maximal activity was found at 75° and with tributyrin at 60°. Published thermostable lipases from mesophiles exhibit only low activities at temperatures above 60°. Also, BTL-2 is thermostable, like most lipases from *Bacillus* and *Pseudomonas*.[9,15,18,21,28,51] The overexpressed lipase BTL-2 was stable up to 50° when incubated for 30 min at pH 9.0. Both lipases, BTL-1 and the weakly expressed lipase BTL-2, were slightly less stable below 40°.

The pH activity of BTL-2 at 60° depends on the substrate. With triolein maximum activity was found at pH 9.0, decreasing rapidly so that no activity could be detected any longer at pH 6.0. With tributyrin, however, a maximum was reached at pH 8.0 and, in contrast to triolein, 70% residual activity was detected at pH 6.0. After 14 hr of incubation at 30°, BTL-2 was stable within the pH range of 9.0–11.0. Thus, unlike BTL-1 and most other lipases,[21,28,40–42,51] BTL-2 is a thermoalkalophilic lipase.

Most detergents enhanced the activity, with the increase ranging from 50% {3-[(3-cholamidopropyl)-dimethylammonio]-1-propanesulfonate (CHAPS) and octyl-β-D-glucopyranoside (OGP)} to 80% (Triton X-100 and cholate) when added to the enzyme solution of overexpressed lipase. After 1.5 hr of incubation at 30° in the presence of these detergents, no significant loss of pNPP-hydrolyzing activity was observed. But the addition of SDS, Lubrol PX, Tween 20, or Tween 80 immediately inactivated the

[50] D. M. Miller, G. P. Kurzban, J. A. Mendoza, J. M. Chirgwin, S. C. Hardies, and P. M. Horowiz, *Biochim. Biophys. Acta* **1121,** 286 (1992).

[51] H.-K. Kim, M.-H. Sung, H.-M. Kim, and T.-K. Oh, *Biosci. Biotech. Biochem.* **58,** 961 (1994).

enzyme. The weakly expressed lipase showed a similar behavior, but Triton X-100 inhibited this lipase, whereas Tween 20 did not.

Lipases from both preparations showed only about a 20% decrease in activity when incubated in the presence of 30% (v/v) methanol, propanol, or acetone for 1 hr at 30°. After prolonged incubation for 4 days at room temperature in the presence of 30% (v/v) propanol no further decrease of activity was detected, indicating that the lipase BTL-2 is exceptionally stable in the presence of organic solvents.

With respect to substrate specificity, BTL-2 exhibited the highest activity with tributyrin, both at pH 7.5 and 8.5 (Fig. 9). The lipase hydrolyzed triacylglycerols up to C_{18} with nearly the same efficiency at both pH values. The activity on triacylglycerols was between 40 (C_6 or C_8) and 20% (C_{12} or C_{16}) that of the activity on tributyrin. Triacetin (C_2) was a poor substrate, especially at pH 8.5, where no activity was detected. However, the correla-

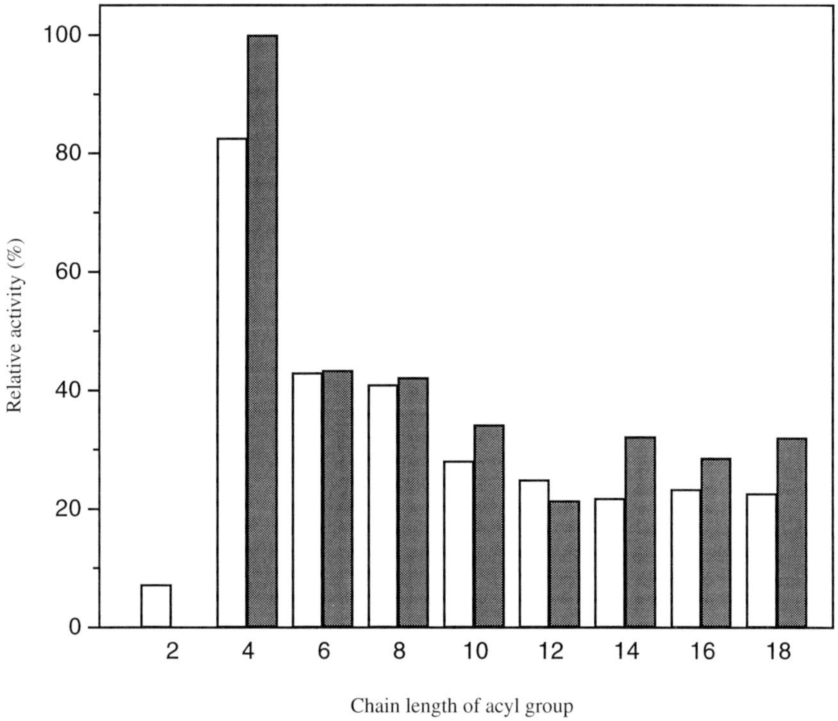

FIG. 9. Activity of the purified lipase on triacylglycerols with various chain lengths of the acyl group was determined by the pH-stat assay at 60° and pH 7.5 (open columns) or pH 8.5 (filled columns).

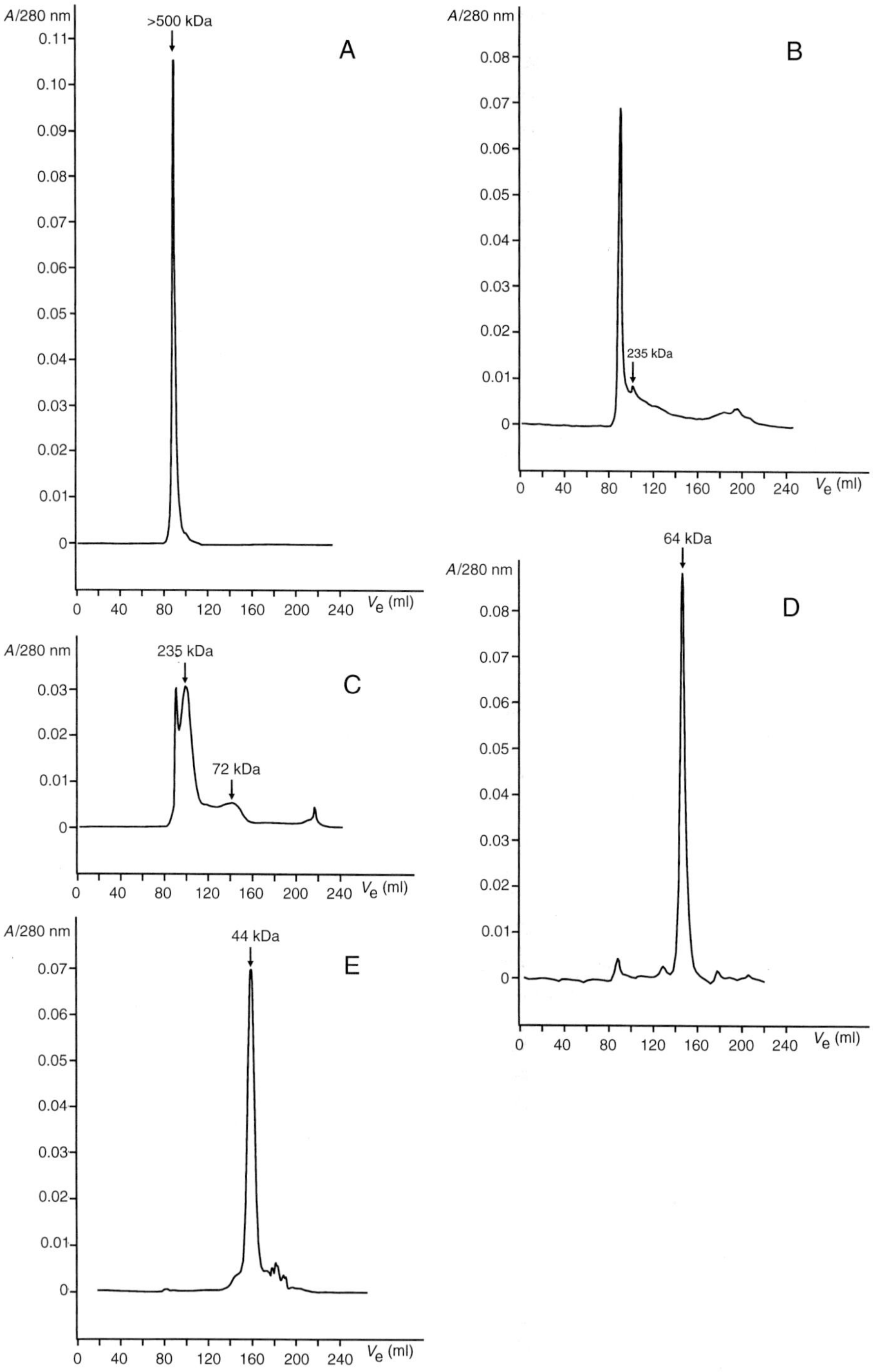
A
A/280 nm
>500 kDa
0.11
0.10
0.09
0.08
0.07
0.06
0.05
0.04
0.03
0.02
0.01
0
0 40 80 120 160 200 240
Ve (ml)
B
A/280 nm
0.08
0.07
0.06
0.05
0.04
0.03
0.02
0.01
0
235 kDa
0 40 80 120 160 200 240
Ve (ml)
C
A/280 nm
235 kDa
72 kDa
0.03
0.02
0.01
0
0 40 80 120 160 200 240
Ve (ml)
D
A/280 nm
64 kDa
0.08
0.07
0.06
0.05
0.04
0.03
0.02
0.01
0
0 40 80 120 160 200 240
Ve (ml)
E
A/280 nm
44 kDa
0.07
0.06
0.05
0.04
0.03
0.02
0.01
0
0 40 80 120 160 200 240
Ve (ml)

tion between acyl chain length and activity of the weakly expressed lipase changed with pH, resulting in an increase in activity on long-chain triacylglycerides at higher pHs.[12] Lipase BTL-2 favors tributyrin as substrate, as do lipases from *S. hyicus*[52] and *S. epidermidis*.[53] The sequence similarities among these three may be responsible for this similar substrate preference.

Some further similarities exist between lipase BTL-2 and a purified lipase from a thermophile *Bacillus* sp. 398 strain.[51] This lipase, like BTL-2, is a large lipase of 50 kDa, with a high specific activity of 25,300 U/mg. It, too, reaches maximum activity at 60–70°.

Studies on Aggregation Behavior of Lipase BTL-2

The tendency to form aggregates is a well-known property of several lipases.[15,26,41,54] Both lipases of *B. thermocatenulatus* form aggregates, as can be observed during native PAGE or gel filtration. The high hydrophobic amino acid content of lipase BTL-1 (39.4%) might be responsible for this aggregation behavior (Table II). Other lipases from *Bacillus* or *Pseudomonas* strains with similar hydrophobic residue contents have also been reported as tending to aggregate.[15,21,25,26] In the case of lipase BTL-2, however, the overall content of hydrophobic residues is significantly lower (33.2%).

Under different conditions, drastic changes in the aggregation tendency of BTL-2 lipase and, hence, its specific activity were observed. Gel chromatography on TSK G3000 (Tosohaas) columns under varying conditions was used to study the aggregation behavior of BTL-2. Aliquots of concentrated lipase obtained after hydrophobic chromatography were loaded onto the column. As illustrated in Fig. 10, lipase BTL-2 changed its aggregation state in response to the running conditions. The highest molecular mass (>500 kDa) is observed when a low ionic strength buffer (10 m*M* Tris) is used. Under the conditions used for purification [50 m*M* Tris buffer (pH 7.5),

[52] M. G. Van Oort, A. M. Th. J. Deever, R. Dijkman, M. L. Tjeenk, H. M. Verheij. G. H. De Haas, E. Wenzig, and F. Götz, *Biochemistry* **28,** 9278 (1989).

[53] G. Pablo, A. Hammon, S. Bradley, and J. E. Fulton, *J. Invest. Dermatol.* **63,** 231 (1974).

[54] M. A. Taipa, M. R. Aires-Barros, and J. M. S. Cabral, *J. Biotechnol.* **26,** 111 (1992).

FIG. 10. Elution behavior of BTL-2 lipase on a TSK G3000 column under different running conditions: (A) 10 m*M* Tris-HCl (pH 7.5), (B) 50 m*M* Tris-HCl (pH 7.5) containing 0.1 *M* NaCl, (C) 50 m*M* Tris-HCl (pH 7.5) containing 0.1% (w/v) cholate, (D) 50 m*M* Tris-HCl (pH 7.5) containing 1% (w/v) cholate, and (E) 50 m*M* Tris-HCl (pH 7.5) containing 30% (v/v) propanol. In all experiments, 1 ml of sample was loaded on the column and eluted at room temperature at 4 ml/min, except in (E), in which the flow rate was 3 ml/min.

0.1 *M* NaCl], an overall lipase aggregate size of 235 kDa was observed. The addition of 0.1% (w/v) cholate to the running buffer led to a partial disaggregation of the lipase (lipase peaks at 235 and 72 kDa), whereas with 1% (v/v) cholate one sharp peak corresponding to a molecular weight of 64,000 was observed. In the presence of 30% (v/v) propanol during gel filtration, a sharp peak corresponding to a molecular weight of 44,000 for the lipase monomer was eluted, which is very close to the molecular weight of 43,000 estimated from the amino acid sequence. The deaggregation of lipases in the presence of propanol was also reported for the lipases of *Penicillium* sp.[41] and *Pseudomonas cepacia.*[26]

These observations clearly show that lipase BTL-2 is aggregated in the absence of detergents or organic solvents. Broad peaks during gel filtration indicate the existence of a heterogeneous mixture of aggregate sizes. In the presence of cholate, where lipase peaks of 64 and 72 kDa eluted that do not correspond to the molecular weight of the monomer, it is likely that the lipase is associated with cholate molecules. Other detergents, such as 1% (w/v) Brij, failed to dissociate completely the molecular aggregates of the lipase. At present it is not clear whether lipase BTL-2 interacts with bile salts in a specific way, although only bile salts (cholate, taurocholate, deoxycholate, or CHAPS) were able to elute the lipase from the butyl-Sepharose column during purification. Other detergents (octyl glucoside, Brij, and Triton X-100), organic solvents (propanol, acetone, or methanol) or guanidine, all compounds that are effective in weakening hydrophobic interactions, were ineffective.

With the lipase from *P. cepacia* a strong negative correlation between aggregation and activity has been described, possibly because the aggregation involves the active sites of the lipase molecule.[26] This was also found in the case of the lipase BTL-2. For this investigation, lipase fractions with different aggregation states obtained after gel filtration were incubated for 5 min at room temperature in the presence or absence of 1% (w/v) cholate, after which the activity was checked at 30° using triolein as the substrate (Table VI). Under those conditions, in which monomers of the lipase existed in solution [30% (v/v) propanol], the activity did not change on incubation with the detergent. For this reason, the ratio between the specific activities measured before and after incubation with detergent was one.

In contrast, the activity of the lipase with the highest molecular weight (>500,000) after gel filtration increased by a factor of 6.8 after incubation with 1% (w/v) cholate. Lipase forms eluted from the gel-filtration column with a molecular weight of 235,000 and 72,000 exhibited intermediate increases in activity after incubation with the bile salt (by a factor of 4.1 and 1.9, respectively). Nevertheless, the effect of aggregation was strongly dependent on the substrate employed for the activity assay (tributyrin or

TABLE VI

COMPARISON OF ESTIMATED MOLECULAR MASS OF BTL-2 LIPASE OBTAINED AFTER GEL-FILTRATION CHROMATOGRAPHY UNDER SEVERAL RUNNING CONDITIONS AND SPECIFIC ACTIVITY RATIOS[a,b]

Running conditions	MW (estimated)	$(U/mg)^+/(U/mg)^-$ [c]
Tris buffer (pH 7.5, 10 m*M*)	>500,000	6.8
Tris buffer (pH 7.5, 50 m*M*), 0.1% (w/v) cholate	235,000	4.1
Tris buffer (pH 7.5, 50 m*M*), 1% (w/v) cholate	72,000	1.9
Tris buffer (pH 7.5, 50 m*M*), 30% (v/v) propanol	44,000	1.0

[a] From *Biochem. Soc. Trans.* **25,** 178 (1997), with permission.
[b] Ratios in units per milligram, measured before (−) and after (+) 5 min of incubation with 1% (w/v) cholate.
[c] pH-stat assay with triolein as substrate (30° and pH 8.5).

triolein) and the temperature during the assay (30 or 60°). Whereas the tributyrin-hydrolyzing activity of aggregated BTL-2 (molecular weight > 500,000) was unchanged at 30 and 60° by incubation with 1% (w/v) cholate, a strong enhancement of the triolein-hydrolyzing activity was produced, especially at 30°. One explanation for this could be that during tributyrin hydrolysis, mono- and dibutyrin are produced. Monoacylglycerides are nonionic surfactants that are widely used as emulsifiers.[55] It might be possible that, because of its water solubility (in contrast to the long-chain monoolein), monobutyrin is itself capable of breaking down the lipase aggregates, so that lipase monomers are always present in the assay. A more specific effect of tributyrin cannot be excluded.

Conclusion

We have found a strain of the thermophile *B. thermocatenulatus* that has at least two different lipases. One of these lipases, BTL-1, exhibited similarities to other *Bacillus* lipases previously published. But the second lipase, BTL-2, which was not detected during the cultivation of *B. thermocatenulatus,* is different from most other lipases of *Bacillus.* Moreover, this lipase BTL-2 shows homology to lipases from *Staphylococcus* strains, which are also gram positive. Until now no structure of such a large prokaryotic lipase has been available. Hence, we developed an efficient expression system and a simple purification procedure that allow the production of the large amounts of pure enzyme needed for crystallization. Crystallization and X-ray diffraction analysis of the lipase BTL-2 are in progress. Lipase

[55] K. Holmberg and E. Osterberg, *J. Am. Oil Chem. Soc.* **65,** 1544 (1988).

BTL-2 proved to be a stable enzyme, with optimal activity at elevated temperatures and at alkaline pH, which are promising properties for biotechnological applications.

The discovery of two different lipases in *B. thermocatenulatus* prompted us to ask why this strain needs two lipolytic enzymes, what their physiological function is, and why only BTL-1 was detected in the culture broth of *B. thermocatenulatus* on induction with olive oil. It would be worthwhile investigating other *Bacillus* strains from which either a small lipase[21–24] or a higher molecular weight lipase[51] is known to determine if a second lipolytic enzyme also exists in these strains.

[12] *Vernonia* Lipase: A Plant Lipase with Strong Fatty Acid Selectivity

By PATRICK ADLERCREUTZ, THOMAS GITLESEN, IGNATIOUS NCUBE, and JOHN S. READ

Introduction

Lipases are useful tools in lipid processing. One important advantage of lipase-catalyzed processing compared to chemical methods is the inherent selectivity of the enzymatic reactions. The 1,3-regioselectivity of several lipases has been used extensively. A typical example is the production of cocoa butter substitutes, which is carried out on an industrial scale. Furthermore, there is great interest in the use of lipases for the preparation of chiral building blocks in the production of pharmaceuticals, agrochemicals, etc. In these applications, the often broad substrate specificity and high stereoselectivity of the lipases are utilized.

The fatty acid selectivity of lipases has great potential for the enrichment of different fatty acids from natural fats and oils. Several promising reports have appeared in the literature. The fatty acid selectivity of *Geotrichum candidum* lipase was observed long ago.[1] The lipase was selective for fatty acids having a *cis*-9 double bond, for example, oleic acid. Although the originally described strain is no longer available, lipases from other *G. candidum* strains express similar selectivity.[2]

[1] A. A. Alford and D. A. Pierce, *J. Food Sci.* **26,** 518 (1961).

[2] M. W. Baillargeon, R. G. Bristline, and P. E. Sonnet, *Appl. Microbiol. Biotechnol.* **30,** 92 (1989).

0076-6879/97 $25.00

Several different lipases have been used to enrich γ-linolenic acid.[3-6] In all cases the lipase was less active on γ-linolenic acid than on most other fatty acids. Either esterification of a mixture of free fatty acids was carried out so that γ-linolenic acid was enriched in the free fatty acid fractions,[3-5] or the triglyceride was partially hydrolyzed so that the γ-linolenic acid was enriched in the glyceride fraction.[5] In a similar way erucic acid has been enriched from rapeseed oil,[7,8] and C_{20} *cis*-5 polyunsaturated fatty acids have been enriched from *Biota orientalis* oil.[9] Furthermore, lipase from rape has been shown to discriminate against *cis*-4 and *cis*-6 unsaturated fatty acids,[10] and *Rhizopus* lipase has been shown useful in the enrichment of docosahexaenoic acid in fish oil fractions as well.[11]

Lipase fatty acid chain length profiles are not as well investigated as the effect of unsaturation on lipase activity. Studies have been performed on hydrolysis of triacylglycerols in aqueous emulsions[12-14] and hydrolysis of fatty acid alkyl esters in organic solvent.[15] Unfortunately, artifacts can be introduced by the physical properties of the substrate or the reaction system. Investigations on hydrolysis of triglycerides in organic solvent by five different lipases showed that the effect of chain length on hydrolytic activity is in general much smaller than the effect of unsaturation.[16]

It has been shown that some plant seeds contain lipases that show selectivity for the most common fatty acids in that particular kind of seed.[17] It can thus be of interest to study lipases from plant seeds that contain unusual fatty acids. In this chapter, we describe a lipase from seed of *Vernonia galamensis.* These seeds contain about 40% oil and the oil contains up to 80% vernolic acid (*cis*-12,13-epoxy-*cis*-9-octadecenoic acid). Much of

[3] M. J. Hills, I. Kiewitt, and K. D. Mukherjee, *Biotechnol. Lett.* **11,** 629 (1989).

[4] T. A. Foglia and P. E. Sonnet, *J. Am. Oil Chem. Soc.* **72,** 417 (1995).

[5] M. S. K. Syed Rahmatullah, V. S. K. Shukla, and K. D. Mukherjee, *J. Am. Oil Chem. Soc.* **71,** 563 (1994).

[6] Deleted in proof.

[7] P. E. Sonnet, T. A. Foglia, and S. H. Feairheller, *J. Am. Oil Chem. Soc.* **70,** 378 (1993).

[8] G. P. McNeill and P. E. Sonnet, *J. Am. Oil Chem. Soc.* **72,** 1301 (1995).

[9] M. S. F. Lie Ken Jie and M. S. K. Syed Rahmatullah, *J. Am. Oil Chem. Soc.* **72,** 245 (1995).

[10] M. J. Hills, I. Kiewitt, and K. D. Mukherjee, *J. Am. Oil Chem. Soc.* **67,** 561 (1990).

[11] V. B. Yadwad and L. C. Noronha, *Biotechnol. Bioeng.* **38,** 956 (1991).

[12] M. Iwai, S. Okumura, and Y. Tsujisaka, *Agric. Biol. Chem.* **39,** 1063 (1975).

[13] M. Sugiara and M. Isobe, *Chem. Pharm. Bull.* **23,** 1226 (1975).

[14] M. Iwai, *in* "Lipases" (B. Borgström and H. L. Brockman, eds.), p. 416. Elsevier Science, New York, 1984.

[15] M. S. Rangheard, G. Langrand, C. Triantaphylides, and J. Baratti, *Biochim. Biophys. Acta* **1004,** 20 (1989).

[16] M. Berger and M. P. Schneider, *Biotechnol. Lett.* **13,** 641 (1991).

[17] Y. H. Lin, C. Yu, and A. H. C. Huang, *Arch. Biochem. Biophys.* **244,** 346 (1986).

the vernolic acid is present in the form of trivernolin. In the following sections, we describe the purification and properties of the lipase as well as purification and properties of trivernolin and vernolic acid, which are used in the studies of the fatty acid selectivity of the lipase.

Purification of *Vernonia galamensis* Seed Lipase

Lipase Assay Method

Principle. The *Vernonia* lipase is assayed by measuring the release of resorufin from 1,2-*O*-dilauryl-rac-glycero-3-glutaric acid-resorufin ester.

Reagents

1,2-*O*-Dilauryl-rac-glycero-3-glutaric acid-resorufin ester
Dioxane (analytical reagent)

Procedure. The assay mixture contains 1.35 ml of enzyme solution and 0.15 ml of 1,2-*O*-dilauryl-rac-glycero-3-glutaric acid-resorufin ester (1 mg/ml) in dioxan. The mixture is incubated at 37°. Samples of 250 μl are withdrawn after 0.5 and 10 min and added to 1 ml of dioxane. The mixture is centrifuged at 3000 rpm for 5 min at room temperature. The amount of resorufin released is determined by measuring absorbance at 572 nm.

Definition of Unit and Specificity. One unit of enzyme is the amount that catalyzes the release of 1 μmol of resorufin per minute at 37°. Specific activity is defined as units per milligram of protein.

Purification Procedure

The extraction and all purification steps are carried out at room temperature and all centrifugation steps are done at 7°.

Step 1: Extraction of Lipase from Vernonia Seeds and Differential Centrifugation. Vernonia galamensis seeds are washed several times in distilled water and soaked in distilled water for about 15 hr. The seeds are then washed several times with distilled water and 20 g of seed is gently homogenized using a pestle and mortar in 80 ml of buffer [150 m*M* Tris-HCl (pH 7.5), 2 m*M* dithiothreitol (DTT), 0.5 m*M* EDTA, and 0.4 *M* sucrose]. The homogenate is filtered through cheesecloth and centrifuged at 1000 *g* for 5 min. The lipid layer is carefully removed. The resulting supernatant is centrifuged at 20,000 *g* for 30 min. About three-quarters of the activity is recovered in the pellet. The pellet is washed three times by resuspending in the homogenizing buffer and centrifuging at 20,000 *g* for 20 min. The lipase in the pellet is solubilized by suspending the pellet in 10 ml of homogenizing buffer containing 1% (w/v) deoxycholic acid but without sucrose. Material that is not solubilized is removed by centrifugation at 20,000 *g* for 15 min.

Step 2: Sephacryl S-300 Gel Filtration. The solubilized lipase (3 ml) is loaded onto a Sephacryl S-300 (60 × 2.5 cm) column and eluted with elution buffer [50 m*M* Tris-HCl (pH 7.5), 100 m*M* NaCl, 2 m*M* DTT, 0.5 m*M* EDTA, and 1% (w/v) deoxycholic acid] at a flow rate of 25 ml/hr. Fractions of 4.2 ml are collected. The lipase activity is eluted within the void volume fractions (Fig 1). The purification is summarized in Table I.

Properties of Lipase

Solubility, Molecular Weight, Stability, pH Optimum. The lipase is highly insoluble in nondetergent buffers but becomes soluble in 1% (w/v) deoxycholate. As a result of this poor solubility the lipase activity is eluted at an apparent aggregate molecular weight of 1×10^6. A molecular weight of 116,000 is obtained for the lipase from sodium dodecyl sulfate–

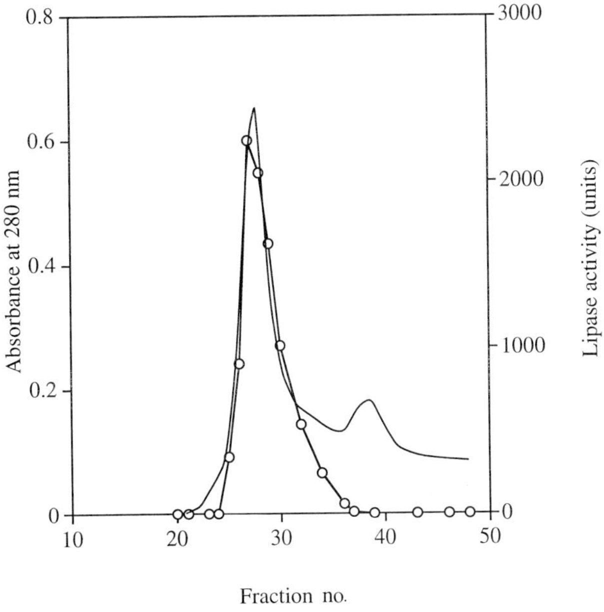

FIG. 1. Elution profile for the elution of *Vernonia galamensis* lipase. Absorbance at 280 nm (—) and lipase activity (○). The solubilized lipase (3 ml) was loaded onto a Sephacryl S-300 column (60 × 2.5 cm) and eluted at a flow rate of 25 ml/hr with fractions of 4.2 ml being collected. [Reprinted from *Biochim. Biophys. Acta* **1257,** I. Ncube, T. Gitlesen, P. Adlercreutz, J. S. Read, and B. Mattiasson, p. 149 (1995). With kind permission of Elsevier Science–NL, Sara Burgerhartstraat 25, 1055 KV Amsterdam, The Netherlands.]

TABLE I
PURIFICATION OF *Vernonia galamensis* LIPASE[a]

Purification state	Purification (fold)	Specific activity (U/mg protein)	Activity yield (%)
1000 *g* supernatant	1	163	100
20,000 *g* pellet	34	5,556	60
Gel-filtration eluate	73	11,905	28

[a] Data from Ncube *et al.* (1995).[22]

polyacrylamide gel electrophoresis (SDS–PAGE) (Fig. 2). The lipase shows maximum activity at pH values between pH 8.0 and 9.0.

Immobilization

Lipase fractions from the Sephacryl S-300 gel filtration are pooled and 40 ml of the pool is added to 1 g of porous polypropene that has been wetted with 3 ml of ethanol. The mixture is shaken overnight on a reciprocal shaker. The polypropene with the adsorbed lipase is collected by filtration and washed with several volumes of distilled water followed by 100 ml of 10 m*M* sodium phosphate buffer, pH 8.0. The preparation is dried overnight under reduced pressure. The lipase is stable for weeks if stored in a desiccator at room temperature.

Extraction of *Vernonia* Oil and Purification of Trivernolin

Oil from *Vernonia* seeds (50 g) is extracted by homogenizing in 200 ml of cold acetone (−20°). Solid material is removed by vacuum filtration and the acetone is evaporated under nitrogen gas at 50°. The oil obtained (15 g) is dissolved in 50 ml of hexane and centrifuged at 3000 rpm for 10 min to remove the precipitated material. Fatty acids are removed from the oil by washing with an equal volume of methanol–bicarbonate buffer (pH 10) (1:1, v/v). Hexane is removed from the oil by evaporation under nitrogen gas. Oil (2 g) is then loaded onto a silica gel 60 (230–400 mesh) column (25 × 3 cm) that has been equilibrated with hexane. The oil is washed into the column with hexane. Nonpolar material is eluted with 200 ml of hexane–ether (95:5, v/v). The triglycerides are eluted with hexane–ether (70:30, v/v). To obtain an elution profile of the triglycerides small fractions of about 1 ml are collected every 5 min for monitoring of total lipid content and fatty acid composition of the fractions. Bulk fractions of about 25 ml are collected and aliquots of 100 μl withdrawn from the fractions for gas chromatography (GC) determination of the fatty acid composition. Frac-

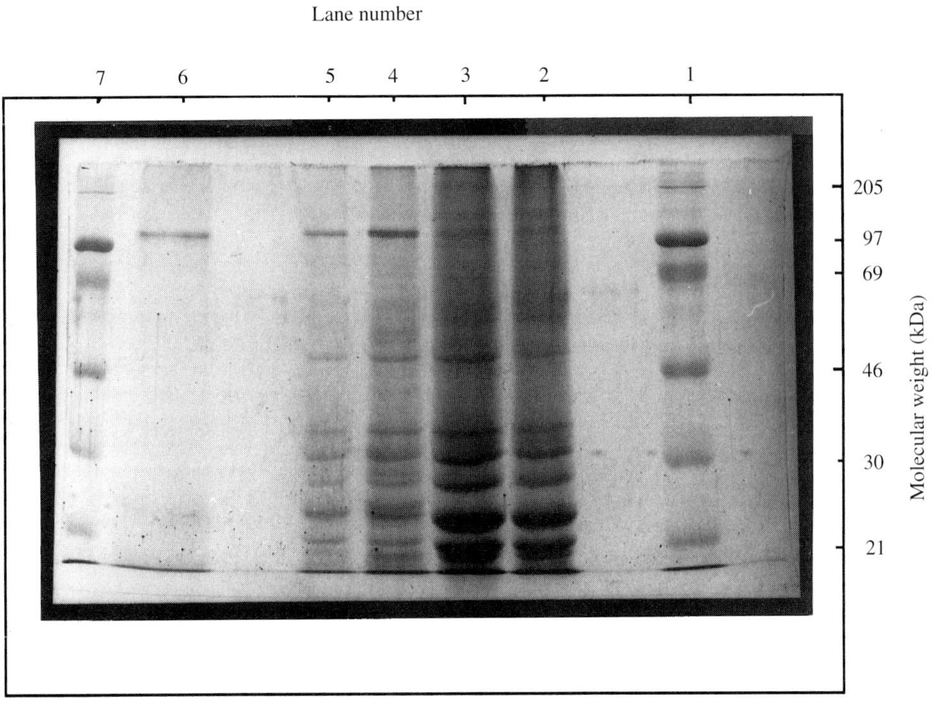

FIG. 2. SDS–polyacrylamide gel electrophoresis of *Vernonia* lipase on 12.5% (w/v) acrylamide gels. Lanes 1 and 7, molecular weight markers; lane 2, crude extract; lane 3, 20,000 *g* supernatant; lane 4, resuspended pellet; lane 5, solubilized pellet; lane 6, gel-filtration lipase active pool. [Reprinted from *Biochim. Biophys. Acta* **1257,** I. Ncube, T. Gitlesen, P. Adlercreutz, J. S. Read, and B. Mattiasson, p. 149 (1995). With kind permission of Elsevier Science–NL, Sara Burgerhartstraat 25, 1055 KV Amsterdam, The Netherlands.]

tions that fall within the pure trivernolin peak are pooled and the solvent evaporated under nitrogen gas to obtain pure trivernolin.

Preparation of Vernolic Acid by Enzymatic Hydrolysis of Trivernolin

Trivernolin (300 mg) is dissolved in 1 ml of water-saturated isooctane and 100 mg of Lipozyme (Novo Nordisk, Bagsvaerd, Denmark) is added. The mixture is incubated in a shaking water bath set at 45° and at a shaking speed of 150 rpm. The extent of hydrolysis is followed by silica gel thin-layer chromatography (TLC), using a mobile phase of hexane–ether–acetic acid (80:20:1, by volume). After complete hydrolysis of trivernolin the solution is removed from the Lipozyme and added to 2 ml of methanol–chloroform (1.4:1.2, v/v). The vernolic acid is extracted into aqueous methanol by addition of 1 ml of 200 mM bicarbonate buffer, pH 10. The aqueous methanol phase is removed and the vernolic acid extracted by adding 1 ml of n-hexane followed by 100 μl of 6 M HCl. The hexane layer is withdrawn and the hexane evaporated in a stream of nitrogen gas to obtain pure vernolic acid.

Properties of Trivernolin and Vernolic Acid

Physical Appearance and Stability

Trivernolin is a clear, light oil that is stable for weeks if stored at $-20°$. Vernolic acid is a white solid at room temperature and remains stable at $-20°$.

General Considerations Concerning Determination of Fatty Acid Selectivity

Lipases are active on many types of substrates. The natural substrates are triglycerides, but partial glycerides and many other esters can also be hydrolyzed by lipases. Furthermore, ester synthesis and transesterification reactions can be carried out, especially if organic reaction media are used. Lipases can express fatty acid selectivity in all these types of reactions and the selectivity is not necessarily identical in all reactions.

It is not straightforward to determine fatty acid selectivity using a triglyceride substrate. If the fatty acids are unevenly distributed in the three positions in the triglyceride molecules, which they indeed are in most natural fats and oils, the apparent fatty acid selectivity can be biased by the regioselectivity of the lipase.[18] It is also difficult to obtain a synthetic triglyceride substrate with completely randomly distributed fatty acids, and thus other substrates must be used. The approach taken here is to use a mixture of synthetic triglycerides containing the same fatty acid in all three positions.

The fatty acid specificity can also be measured using free fatty acids as substrates. Here, this kind of fatty acid specificity was studied in an acidolysis reaction: The fatty acids were incorporated into the triglyceride substrate, which originally consisted of tricaprylin.

Some important parameters in the analyses are discussed in the following sections.

Solvent

The properties of solvents and especially their ability to dissolve water and substrate will affect both reaction rate and equilibrium. Simple triglycer-

[18] P. Villeneuve, M. Pina, D. Montet, and J. Graille, *J. Am. Oil Chem. Soc.* **72,** 753 (1995).

ides containing fatty acids with carbon chain lengths longer than 10 cannot be mixed at room temperature. An organic solvent must be used to obtain homogeneous substrate mixture. Lipases operate *in vivo* at the interface between a polar water phase and an apolar substrate phase. Apolar solvents such as hexane methyl *tert*-butyl ether or diisopropyl ether can be used to mimic the apolar substrate phase. The substrate solubility in these solvents is in the range of 10 m*M*. In the present study isooctane was used as solvent.

Water

Water plays an important role in lipase-catalyzed reactions both as a reactant and an enzyme activator. In the case of hydrolysis there is a net consumption of water during the reaction whereas for acidolysis there is no change in water content of the reaction mixture, but water is needed as an intermediate substrate. Water is also needed by the enzyme as an activator to sustain enzyme activity. To study effects of reaction conditions on enzymatic reaction rates, it is necessary to keep the hydration of the enzyme constant. Otherwise the effects observed are often due only to redistribution of water in the system. Even small changes such as changing the substrate concentration can change the solubility of water in the reaction medium and thereby influence enzyme hydration. To keep enzyme hydration constant, control the thermodynamic water activity in the reactor. This can easily be done by equilibration with saturated salt solutions.[19] The enzyme preparation and the substrate solution can be separately equilibrated to the same water activity before the reaction is started. If the reaction is to be studied over a long period of time, it is advisable to control the water activity continuously.[20] In the present study preequilibration of enzyme preparations and substrate solutions was carried out and initial reaction rates were measured.

Enzyme Preparation

Lipases are not soluble in pure organic solvents and will be present as a separate phase. The lipase preparation can either be used directly as a powder or be immobilized in several ways onto carriers. The latter approach makes the reaction system easier to handle and reduces the mass transfer limitations in the system.

[19] H. L. Goderis, G. Ampe, M. P. Feyten, B. L. Fouwé, W. M. Guffens, S. M. Van Cauwenbergh, and P. P. Tobback, *Biotechnol. Bioeng.* **30,** 258 (1987).

[20] E. Wehtje, I. Svensson, P. Adlercreutz, and B. Mattiasson, *Biotechnol. Tech.* **7,** 873 (1993).

Methods for Determination of Fatty Acid Selectivity

Hydrolysis of Triglyceride Mixtures in Isooctane

Method. Hydrolysis reactions are done at a water activity of 0.97 obtained by preequilibrating both enzyme preparation and substrate solution with a saturated solution of K_2SO_4[19]). The standard substrate solution is a mixture of simple triglycerides of each of the fatty acids C4 : 0, C6 : 0, C8 : 0, C10 : 0, C12 : 0, C14 : 0, C16 : 0, C18 : 1, and trivernolin dissolved in isooctane, each to a concentration of 10 m*M*. The reaction is started by adding 100 mg of lipase preparation to 2 ml of substrate solution. The reactions are carried out at 40° in a shaking water bath that is set at a shaking speed of 150 rpm. Samples of 100 μl are withdrawn for high-performance liquid chromatography (HPLC) analysis of the released fatty acids.

High-Performance Liquid Chromatography Analysis. The HPLC analysis of fatty acids is carried out in an HPLC system equipped with an ultraviolet (UV) detector using a Licrospher RP-C_{18} (5 μm) column (250 × 4 mm; Merck, Darmstadt, Germany). Samples from the enzyme reaction are diluted in 400 μl of acetonitrile. Tridecanoic acid is added as an internal standard. Ten microliters of phenacylation reagent consisting of phenacyl bromide and 18-crown-6 ether (10 mg/ml of molar ratio 10 : 1 in acetonitrile) is added. The reagent is added in slight deficit to avoid artifact peaks from the reagent overlapping the fatty acid peaks of the short fatty acid at the beginning of the chromatograms. About 20 mg of KCO_3 (*s*) is added to each reaction vial to convert the free fatty acids to their reactive potassium salts. The samples are heated to 80° for 1 hr, cooled to room temperature, and filtered through filter paper to remove remaining triglycerides.[21] Elution is done at room temperature using an acetonitrile–water gradient. Elution is started with an initial concentration of acetonitrile of 60%, which is raised to 99% over 10 min and held at 99% for 2 min. Fatty acid phenacyl esters are monitored by determining UV adsorption at 254 nm.

Results. Among the saturated fatty acids, the *Vernonia* lipase showed rather pronounced selectivity for caprylic acid (C8 : 0, Fig. 3). The reaction rate obtained with triolein was quite low, while trivernolin was hydrolyzed rapidly, more than four times faster than C8 : 0. When *Candida rugosa* lipase was investigated in the same way, trivernolin was hydrolyzed more rapidly than the other triglycerides but the differences were much smaller.[22]

[21] A. W. Reed, H. C. Deeth, and D. E. Clegg, *J. Assoc. Anal. Chem.* **67,** 718 (1984).

[22] I. Ncube, T. Gitlesen, P. Adlercreutz, J. S. Read, and B. Mattiasson, *Biochim. Biophys. Acta* **1257,** 149 (1995).

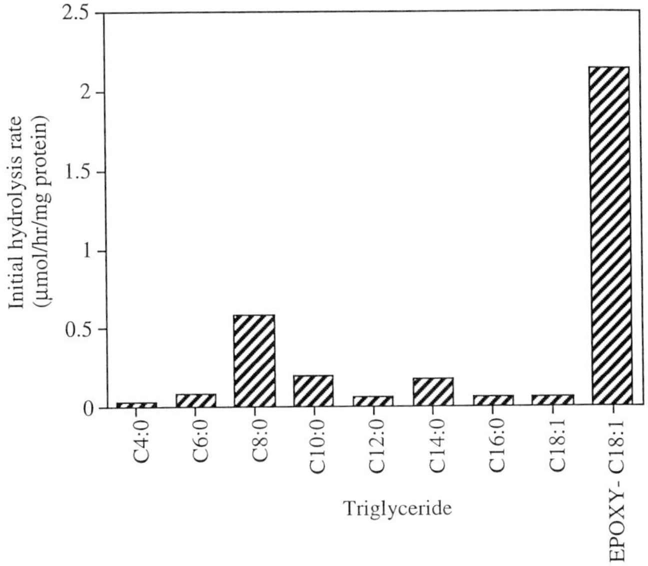

FIG. 3. Fatty acid specificity of *Vernonia galamensis* lipase in triacylglycerol hydrolysis in isooctane. Equimolar mixtures of monoacid-triacylglycerols (10 m*M* each) were hydrolyzed using lipases adsorbed on polypropene. [Data from Ncube *et al.* (1995).[22]]

It is thus clear that the *Vernonia* lipase expresses a strong selectivity for vernolic acid in the form of trivernolin.

Transesterification of Tricaprylin with Fatty Acid Mixtures

Method. The substrate mixture is 100 m*M* tricaprylin and 30 m*M* in each of the fatty acids C7:0, C10:0, C12:0, C14:0, C16:0, C18:1, and vernolic acid dissolved in isooctane. Shorter fatty acids cause inhibition of the enzyme and are therefore not included in the test. The reaction is carried out in 4-ml vials and the reaction mixture consists of 2 ml of substrate solution and 100 mg of lipase preparation. The reactions are carried out at a water activity of 0.43 (attained by preequilibration with a saturated solution of K_2CO_3) on a water bath at a temperature of 40° and shaking speed of 150 rpm. Samples (100 μl) are withdrawn for GC analysis of the esterified fatty acids.

Gas–Liquid Chromatographic Analysis. The fatty acids in glycerides are converted to the corresponding methyl esters using sodium methoxide[23]

[23] I. Svensson, P. Adlercreutz, and B. Mattiason, *Appl. Microbiol. Biotechnol.* **33,** 255 (1990).

with tridecanoin as an internal standard. Analysis of the methyl esters is done on a Shimadzu (Tokyo, Japan) GC14A gas chromatograph equipped with a flame ionization detector (FID) using an SP2380 column (Supelco, Bellefonte, PA) (30 m, 0.32-mm i.d., and 0.2-μm film thickness). In the analysis of the fatty acids of *Vernonia* oil, the initial column temperature is 170° for 2 min and is raised to 190° at 5°/min and held at this temperature for 8 min. In the analysis of fatty acids from the transesterification reaction samples the initial temperature is 40° and is raised to 190° at 15°/min and held at this temperature for 15 min. Helium is used as the carrier gas. In both cases the vernolic acid is eluted well after the other fatty acids.

Results. The *Vernonia* lipase incorporated vernolic acid into the triglyceride fraction more rapidly than any of the other fatty acids (Fig. 4). Among the other fatty acids, heptanoic acid was incorporated faster than the others. The specificity of the *Vernonia* lipase for vernolic acid is thus not limited to the natural substrate, the triglyceride, but is also expressed for the free fatty acids.

Michaelis–Menten Kinetics of Hydrolysis of Triolein and Trivernolin

Method. Candida rugosa lipase was immobilized by a method similar to that used with the *Vernonia* lipase. In the kinetic study, the reaction

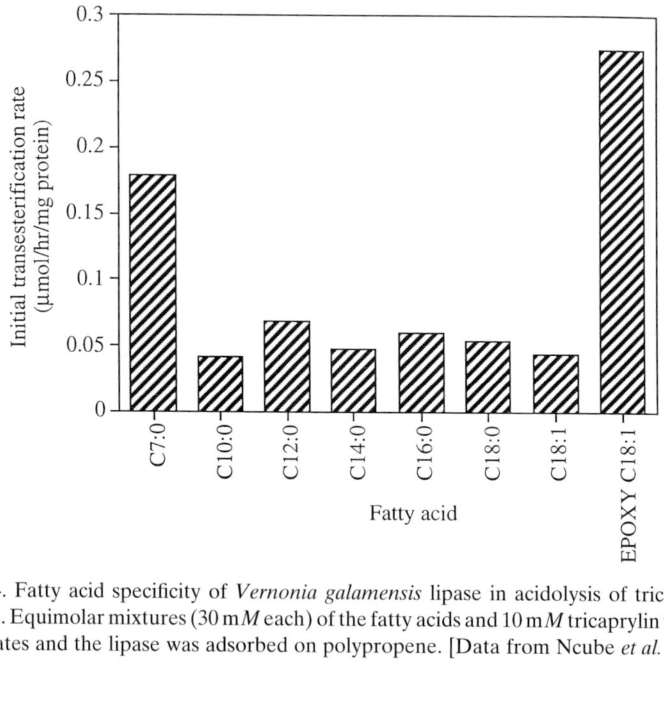

FIG. 4. Fatty acid specificity of *Vernonia galamensis* lipase in acidolysis of tricaprylin in isooctane. Equimolar mixtures (30 m*M* each) of the fatty acids and 10 m*M* tricaprylin were used as substrates and the lipase was adsorbed on polypropene. [Data from Ncube *et al.* (1995).[22]]

TABLE II

KINETIC CONSTANTS FOR HYDROLYSIS OF TRIVERNOLIN AND TRIOLEIN BY *Vernonia* LIPASE AND *Candida* LIPASE[a]

	K_m (mM)		V_{max} (μmol · hr^{-1} · mg $protein^{-1}$)	
Lipase	Trivernolin	Triolein	Trivernolin	Triolein
Candida	39 ± 12	28 ± 3	121± 13	130 ± 4
Vernonia	68 ± 10	58 ± 10	18 ± 1	3.8 ± 3

[a] Data from Ncube *et al.* (1995).[22]

mixtures are 50 mg of lipase preparation in 1.0 ml of substrate solution for *Vernonia* lipase and 25 mg in 500 μl of substrate solution for *Candida* lipase. The substrate solutions consist of either trivernolin or triolein at concentrations ranging from 0 to 200 mM in isooctane. The initial water activity in all solutions is kept the same (0.97) by preequilibration of both enzyme preparation and substrate solution over saturated solutions of K_2SO_4. To determine the initial rates of hydrolysis, samples of 100 μl are withdrawn for HPLC analysis of the released oleic acid or vernolic acid.

Results. The hydrolysis of triglycerides in isooctane catalyzed by the lipases from *V. galamensis* seeds and from *C. rugosa* could be described by the Michaelis–Menten equation. Because vernolic and oleic acid differ only in that the former acid contains an epoxy group, it was of interest to compare the hydrolysis of triolein and trivernolin. It was found for both enzymes that the K_m values were slightly higher for trivernolin compared to triolein (Table II). For the *Candida* lipase, the V_{max} values for the two

TABLE III

FATTY ACID SELECTIVITY OF SOME PLANT SEED LIPASES

Lipase source	Best substrate	Ref.
Corn	Trilinolein, triolein	*a*
Castor bean	Triricinolein	*a*
Rapeseed	Trierucin	*a*
Elm seed	Tricaprin	*a*
Oat seed	Triolein	*b*
Vernonia seed	Trivernolin	This work

[a] Y. H. Lin, C. Yu, and A. H. C. Huang, *Arch. Biochem. Biophys.* **244,** 346 (1986).

[b] G. J. Piazza, A. Bilyk, D. P. Brower, and M. J. Haas, *J. Am. Oil Chem. Soc.* **69,** 978 (1992).

substrates were similar. However, the V_{max} value for trivernolin was about five times higher than that for triolein for the *Vernonia* lipase. It is thus this difference in V_{max} that causes the selectivity for trivernolin. It would be interesting to investigate the mechanism behind this large rate enhancement caused by the epoxy groups in the fatty acids. An important step in the elucidation of this mechanism would be to determine the three-dimensional structure of the *Vernonia* lipase.

Discussion

It is quite natural that the lipases present in plant seeds are effective in hydrolyzing the triglycerides that are predominant in the seeds. This kind of specificity has been observed previously in lipases from a few plant seeds (Table III). However, this work is the first to describe selectivity for epoxide-containing fatty acids. In looking for lipases with novel specificities it can thus be worthwhile to investigate lipases from plant seeds containing triglycerides built up of unusual fatty acids.

[13] Hepatic Lipase: High-Level Expression and Subunit Structure Determination

By John S. Hill, Richard C. Davis, Dawn Yang, Michael C. Schotz, and Howard Wong

Introduction

Through its ability to catalyze the hydrolysis of triglycerides and phospholipids, hepatic lipase (HL) influences the metabolism of chylomicron remnants, intermediate-density lipoproteins (IDLs), and high-density lipoproteins (HDLs).[1] Hepatic lipase is synthesized and secreted by hepatocytes, and binds to the lumenal surface of the endothelium in liver sinusoids through an ionic interaction with heparan sulfate proteoglycans.[2] The detection of HL catalytic activity but not protein synthesis in the ovary and adrenal gland suggested that HL is transported and binds specifically to these tissues as well.[3,4] In addition, there is evidence to suggest that, indepen-

[1] T. Olivecrona and G. Bengtsson-Olivecrona, *Curr. Opin. Lipidol.* **4,** 187 (1993).
[2] T. Kuusi, E. A. Nikkila, I. Virtanen, and P. K. J. Kinnunen, *Biochem. J.* **181,** 245 (1979).
[3] M. H. Doolittle, H. Wong, R. C. Davis, and M. C. Schotz, *J. Lipid Res.* **28,** 1326 (1987).
[4] E. A. Hixenbaugh, T. R. Sullivan, J. F. Strauss III, E. A. Laposata, M. Komaromy, and L. G. Paavola, *J. Biol. Chem.* **266,** 4222 (1989).

0076-6879/97 $25.00

dent of its catalytic activity, HL can act as a ligand to facilitate the uptake of remnant lipoproteins through its interaction with cell surface proteoglycans and the low-density lipoprotein receptor-related protein (LRP).[5–8]

Although HL appears to have a number of roles in lipoprotein metabolism, it is evident in human HL deficiency that remnant metabolism is primarily affected.[9] This observation is consistent with animal studies in which the inhibition of HL activity by administration of HL antibodies decreased the rate of chylomicron remnant uptake by the liver.[10,11] However, there are several *in vitro* studies that demonstrate that HL also participates in the remodeling of HDL.[12–14] This has been clearly demonstrated in gene-targeted and transgenic HL animals in which the metabolism of the predominant plasma lipoprotein species, HDL, is significantly affected.[15–17]

Mature human HL is a 476-amino acid glycoprotein with a calculated polypeptide molecular weight of 53,431, whereas the purified denatured enzyme has an apparent molecular weight of 65,000, reflecting the contribution of N-linked carbohydrate.[18–20] Hepatic lipase is a member of a lipase family that also includes lipoprotein lipase (LPL) and pancreatic lipase

[5] P. Diard, M.-I. Malewiak, D. Lagrange, and S. Griglio, *Biochem. J.* **299,** 889 (1994).

[6] A. Nykjaer, M. Nielson, A. Lookene, N. Meyer, H. Røigaard, M. Etzerodt, U. Beisiegel, G. Olivecrona, and J. Gliemann, *J. Biol. Chem.* **269,** 31747 (1994).

[7] M. Z. Kounnas, D. A. Chappell, H. Wong, W. S. Argraves,and D. K. Strickland, *J. Biol. Chem.* **270,** 9307 (1995).

[8] A. Krapp, S. Ahle, S. Kersting, Y. Hua, K. Kneser, M. Nielsen, J. Gliemann, and U. Beisiegel, *J. Lipid Res.* **37,** 926 (1996).

[9] R. A. Hegele, J. A. Little, C. Vezina, G. F. Maguire, L. Tu, T. S. Wolever, D. J. A. Jenkins, and P. W. Connelly, *Arterioscler. Thromb.* **13,** 720 (1993).

[10] F. Sultan, D. Lagrange, H. Jansen, and S. Griglio, *Biochim. Biophys. Acta* **1042,** 150 (1990).

[11] S. Shafi, S. E. Brady, A. Bensadoun, and R. J. Havel, *J. Lipid Res.* **35,** 709 (1994).

[12] J. R. Patsch, S. Prasad, A. M. Gotto, and W. Patsch, *J. Clin. Invest.* **80,** 341 (1987).

[13] M. A. Clay, H. H. Newnham, and P. J. Barter, *Arterioscler. Thromb.* **11,** 415 (1991).

[14] A. Barrans, X. Collet, R. Barbaras, B. Jaspard, J. Manent, C. Vieu, H. Chap, and B. Perret, *J. Biol. Chem.* **269,** 11572 (1994).

[15] S. J. Busch, R. L. Barnhart, G. A. Martin, M. C. Fitzgerald, M. T. Yates, S. J. T. Mao, C. E. Thomas, and R. L. Jackson, *J. Biol. Chem.* **269,** 16376 (1994).

[16] J. Fan, J. Wang, A. Bensadoun, S. J. Lauer, Q. Dang, R. W. Mahley, and J. M. Taylor, *Proc. Natl. Acad. Sci. U.S.A.* **91,** 8724 (1994).

[17] G. E. Homanics, H. V. de Silva, J. Osada, S. H. Zhang, H. Wong, J. Borensztajn, and N. Maeda, *J. Biol. Chem.* **270,** 2974 (1995).

[18] G. Stahnke, R. Sprengel, J. Augustin, and H. Will, *Differentiation* **35,** 45 (1987).

[19] S. Datta, C.-C. Luo, W.-H. Li, P. Vantuinen, D. H. Ledbetter, M. A. Brown, S.-H. Chen, S.-W. Lui, and L. Chan, *J. Biol. Chem.* **263,** 1107 (1988).

[20] G. A. Martin, S. J. Busch, G. D. Merideth, A. D. Cardin, D. T. Blakenship, S. J. T. Mao, A. E. Rechtin, C. W. Woods, M. M. Racke, M. P. Schafer, M. C. Fitzgerald, D. M. Burke, M. A. Flanagan, and R. L. Jackson, *J. Biol. Chem.* **263,** 10907 (1987).

(PL).[21-23] Amino acid homology, including the conservation of disulfide bridges among all three enzymes, suggests that they share similar three-dimensional structures.[24] However, on comparison of their phylogenetic relationships, it is apparent that HL and LPL are more closely related to each other than either enzyme is to PL.[22]

The low protein yields associated with the purification of HL from human postheparin plasma have not permitted a more detailed physical analysis of this enzyme. For example, the only previous estimations of the functional molecular weight of HL have relied on gel-filtration techniques. The primary disadvantage of conventional size-exclusion chromatography methods is that the elution position depends not only on the molecular weight of the molecule, but also on its shape. In addition, if the protein has a tendency to adhere to the column matrix, the molecular weight may be underestimated.

In this chapter we describe procedures developed for a large-scale production and purification of recombinant HL. The purification of sufficient quantities of this enzyme has enabled the analysis of its subunit structure by three independent means: intensity light scattering, sedimentation equilibrium, and radiation inactivation.

DNA Transfection and Production of Recombinant Human Hepatic Lipase

Full-length human HL cDNA[18] is cloned into the *Hind*III and *Not*I sites of the expression vector pcDNAI/Neo (Invitrogen, San Diego, CA). Chinese hamster ovary (CHO)-Pro5 cells are maintained in Dulbecco's modified Eagle's medium (DMEM) supplemented with sodium pyrufate, glutamine, nonessential amino acids, 10% fetal bovine serum (FBS), and antibiotics. To mediate the transfection of CHO cells, coprecipitates of plasmid DNA and $CaPO_4$ are prepared.[25] The calcium phosphate–DNA mixture is incubated at room temperature for 30 min before it is added to a 50% confluent CHO monolayer. Stably transfected cells are selected by growth in the presence of geneticin (G418 sulfate, 400 μg/ml) and surviving

[21] O. Ben-Zeev, C. M. Ben-Avram, H. Wong, J. Nikazy, J. E. Shively, and M. C. Schotz, *Biochim. Biophys. Acta* **919,** 13 (1987).

[22] T. G. Kirchgessner, J.-C. Chaut, C. Heinzmann, J. Etienne, S. Guilhot, K. Svenson, D. Ameis, C. Pilon, L. D'Auriol, A. Andalibi, M. C. Schotz, F. Galibert, and A. J. Lusis, *Proc. Natl. Acad. Sci. U.S.A.* **86,** 9647 (1989).

[23] W. A. Hide, L. Chan, and W.-H. Li, *J. Lipid Res.* **33,** 167 (1992).

[24] Z. S. Derewenda and C. Cambillau, *J. Biol. Chem.* **266,** 23112 (1991).

[25] C. Chen and H. Okayama, *Mol. Cell. Biol.* **7,** 2745 (1987).

colonies are selected and expanded. Cell clones expression maximal quantities of HL are identified by enzyme activity analysis.

After growth to confluency in sixteen 225-cm^2 flasks, cells are washed with DMEM and fresh DMEM supplemented with 1% (v/v) Nutridoma (serum substitute to minimize contaminating proteins, Boehringer-Mannheim, Indianapolis, IN) and heparin (10 units/ml) are added to each flask. The medium is harvested and replaced every 24 hr for a 10-day period. After centrifugation at 1000 *g* for 10 min at 4° to remove cellular debris, the harvested medium is stored at −80°.

Purification of Recombinant Human Hepatic Lipase

Three liters of thawed medium is mixed with NaCl and benzamidine to a final concentration of 0.5 *M* and 0.5 m*M*, respectively. A four-step procedure is utilized, consisting of octyl–Sepharose, heparin–Sepharose, hydroxylapatite, and dextran sulfate–Sepharose. Hepatic lipase activity is monitored as described previously[26] using radiolabeled triolein substrate. The apparent K_m and V_{max} values for a triolein substrate are determined from a linear regression analysis of a double-reciprocal plot of lipolytic activity (μmol/min/mg) versus triolein concentration (m*M*). Protein concentration is determined by the Bradford method.[27] All purification steps are carried out at 4°.

Step 1: Octyl–Sepharose Chromatography. The medium (3 liters) is loaded at a flow rate of 150 ml/hr onto an octyl–Sepharose column (2.5 × 25 cm) previously equilibrated with 5 m*M* (1.03 g/liter) barbital buffer, pH 7.2, containing 0.5 *M* (29.22 g/liter) NaCl and 20% (v/v) glycerol (buffer A). Following a wash with 800 ml of buffer A, the lipase is eluted with 700 ml of 5 m*M* barbital buffer, pH 7.2, containing 0.35 *M* NaCl (20.45 g/liter), 20% (v/v) glycerol, and 1.2% (w/v) Triton N-101.

Step 2: Heparin–Sepharose Chromatography. The eluate from the octyl-Sepharose column is applied to a heparin–Sepharose column (2.5 × 20 cm), which is then washed with 800 ml of buffer A prior to elution with 500 ml of 1 *M* (58.44 g/liter) NaCl, 20% (v/v) glycerol, 5 m*M* barbital (pH 7.2).

Step 3. Hydroxylapatite Chromatography. The eluted lipase is passed through a 2-ml hydroxylapatite column to remove contaminating proteins and then diluted with an equal volume of 20% (v/v) glycerol, 5 m*M* barbital (pH 7.2) before it is loaded onto a dextran sulfate–Sepharose column (1 × 5 cm).

[26] M. H. Doolittle, H. Wong, R. C. Davis, and M. C. Schotz, *J. Lipid Res.* **28,** 1326 (1987).
[27] M. D. Bradford, *Anal. Biochem.* **72,** 248 (1976).

TABLE I
PURIFICATION OF HUMAN HEPATIC LIPASE SECRETED BY HEPATIC LIPASE-TRANSFECTED CHO CELLS[a,b]

	Volume (ml)	Protein (mg)	Specific activity (μmol/min/mg)	Purification (fold)	Recovery (%)
Culture medium	3300	131	1.5	1	—
Hydroxylapatite flowthrough	500	1.10	109	73	61
Dextran sulfate–Sepharose eluent	50	0.62	152	101	48
HL concentrate	1	0.55	146	97	41

[a] Reproduced with permission from J. S. Hill, R. C. Davis, D. Yang, J. Wen, J. S. Philo, P. H. Poon, M. L. Phillips, E. S. Kempner, and H. Wong, *J. Biol. Chem.* **271,** 22931 (1996).
[b] As described in text, HL was purified using sequential chromatography consisting of four steps: octyl–Sepharose, heparin–Sepharose, hydroxylapatite, and dextran sulfate–Sepharose. The data presented are representative of several experimental purifications. The presence of Triton N-101 did not permit the measurement of activity present in the octyl- and heparin–Sepharose eluant fractions. Specific activity is expressed as nanomoles of free fatty acid released per minute per microgram of protein.

Step 4. Dextran Sulfate–Sepharose Chromatography. The dextran sulfate–Sepharose column is washed with 50 ml of 0.5 *M* NaCl, 5 m*M* barbital (pH 7.2) and then eluted with 50 ml of 1 *M* NaCl, 5 m*M* barbital (pH 7.2). The collected eluant is concentrated in an Amicon (Danvers, MA) filtration unit using a YM30 membrane to a final volume of 1–2 ml. To preserve enzyme activity, freeze–thaw cycles should be avoided and the purified enzyme should be stored in the presence of 20% (v/v) glycerol.

The results of a typical purification from 3300 ml of medium from transfected CHO cells are shown in Table I. Following concentration of the final eluant, approximately 0.5 mg of protein was obtained with an overall yield based on activity of about 40% with a calculated HL specific activity of 146 μmol/min/mg. This specific activity is similar to previous reports of HL purified from human postheparin plasma[28,29] and recombinant HL made by stably transfected rat hepatoma McARH7777 cells.[30] Kinetic analysis of the purified enzyme with a triolein emulsion substrate indicated an apparent V_{max} of 1.4 ($\pm$ 0.3) mmol/min/mg and an apparent K_m of 16

[28] Y. Ikeda, A. Takagi, and A. Yamamoto, *Biochim. Biophys. Acta* **1003,** 254 (1989).
[29] C. F. Cheng, A. Bensadoun, T. Bersot, J. S. T. Hsu, and K. H. Melford, *J. Biol. Chem.* **260,** 10720 (1985).
[30] Z.-S.Ji, S. J. Lauer, S. Fazio, A. Bensadoun, J. M. Taylor, and R. W. Mahley, *J. Biol. Chem.* **269,** 13429 (1994).

($\pm$ 2) mM (n = 3). The purified HL bound to heparin–Sepharose and eluted as a single peak with maximum elution at 0.7 M NaCl (data not shown). This affinity for heparin is consistent with previous reports for human HL.[29,31] Thus, the properties of the recombinant enzyme were the functional equivalent of the native enzyme.

Determination of Subunit Structure of Hepatic Lipase

Intensity Light Scattering

Size-exclusion chromatography is performed with a Superose 12 (Pharmacia, Piscataway, NJ) column (30 × 0.6 cm) equilibrated with 0.1 M sodium phosphate buffer containing 0.5 M NaCl, using a 500-μl sample loop operated at a flow rate of 0.5 ml/min. The eluant is monitored using three online detectors in series: a light-scattering detector (Wyatt Minidawn, Santa Barbara, CA), a refractive index detector (Polymer Laboratories PL-RI, Vienna, VA), and a UV absorbance monitor at 280 nm. The polypeptide molecular weight of a glycoprotein can be calculated using Eq. (1):

$$MW_p = K(UV)(LS)/\varepsilon_p(RI)^2] \qquad (1)$$

where UV, LS, and RI are the signals from the absorbance, 90° light scattering, and differential refractive index detectors, respectively, ε_p is the extinction coefficient (the absorbance of a solution containing 1 mg of polypeptide/ml for a 1-cm pathlength), MW_p is the polypeptide molecular weight, and K is a calibration constant that depends on instrument conditions. The calibration constant, K, is measured by running bovine serum albumin (Sigma, St. Louis, MO) and ovalbumin (Sigma) molecular weight standards.[32,33] The extinction coefficient of human HL is calculated as 1.22 using the Gill and von Hippel method.[34]

The elution characteristics of purified HL are analyzed using a Superose 12 gel-filtration column in the presence of 0.1 M sodium phosphate and various NaCl concentrations. A single protein peak is obtained with the salt concentration ranging from 0.15 to 1 M NaCl (data not shown). Following calibration of the light-scattering/size-exclusion chromatography system using molecular weight standards, a 500-μl sample of purified enzyme (150 μg) is injected to assess the subunit structure of HL. The chromatograms

[31] H. L. Dichek, C. Parrott, R. Ronan, J. D. Brunzell, H. B. Brewer, Jr., and S. Santamarina-Fojo, *J. Lipid Res.* **34,** 1393 (1993).
[32] T. Takagi, *J. Chromatogr.* **280,** 124 (1983).
[33] T. Arakawa, K. E. Langley, K. Kameyama, and T. Takagi, *Anal. Biochem.* **203,** 53 (1992).
[34] S. C. Gill and P. H. von Hippel, *Anal. Biochem.* **182,** 319 (1989).

from all three online detectors and corresponding activity assay of each fraction are shown in Fig. 1. The first peak, eluting at about 6.4 ml, has considerable light scattering but lacks absorbance, indicating that there is a small amount of aggregates (Fig. 1A). The second peak at about 9.6 ml also corresponds to the ultraviolet absorbance and refractive index measurements and has a calculated molecular weight [Eq. (1)] of 107,000, in agreement with the expected value for a dimer of hepatic lipase (the monomer polypeptide formula weight of HL is 53,431). Also, this second light-scattering peak corresponds to fractions determined to be lipolytically

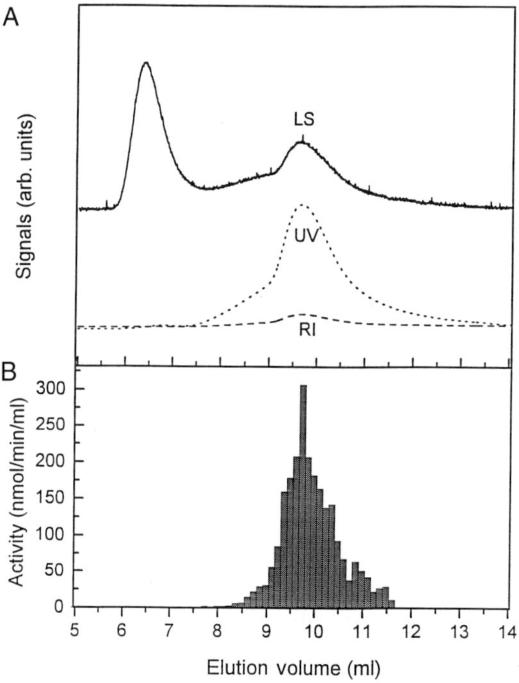

FIG. 1. Size-exclusion chromatogram of hepatic lipase and the lipolytic activity of corresponding fractions. A 500-μl sample of purified enzyme (150 μg) was injected into the light-scattering/size-exclusion chomatography system. The chromatograms from all three online detectors (A) and activity assay of corresponding fractions (B) are shown. LS (—), RI (· · ·), and UV (– – –) are signals from 90° light-scattering, refractive index, and absorbance detectors, respectively. The first peak, eluting at about 6.4 ml, has considerable light scattering but lacked absorbance, indicating that there is a small amount of aggregated protein. The second peak, at about 9.6 ml, has a molecular weight that was calculated to be 107,000 ± 3000 (mean ± SD). [Reproduced with permission from J. S. Hill, R. C. Davis, D. Yang, J. Wen, J. S. Philo, P. H. Poon, M. L. Phillips, E. S. Kempner, and H. Wong, *J. Biol. Chem.* **271,** 22931 (1996).]

active (Fig. 1B). The inability to detect HL monomer suggests that the dimer conformation of the enzyme predominates and is relatively stable in solution.

Sedimentation Equilibrium

Sedimentation equilibrium experiments are performed in a model E analytical ultracentrifuge (Beckman, Palo Alto, CA) equipped with a UV optical system and a photoelectric scanner interfaced to an IBM PC computer. Experiments are performed at 4–7°, at a rotor speed of 5200 rpm over 3 days. During each run, three sample chambers are monitored, two with solutions of HL and one with a solution of PL. For the determination of molecular weights, partial specific volumes are calculated from amino acid composition and carbohydrate content of HL ($\nu = 0.72$).[19,20] Slopes are calculated from the least-squares lines of the log OD (optical density) versus $1000 \times r^2\omega^2/2RT$ plots, where r is the distance from the axis of rotation, ω is the angular velocity, R is the gas constant, and T is the absolute temperature.

Sedimentation equilibrium experiments of HL are analyzed simultaneously with human PL, which is recognized as a monomer[35] and thus acts as an internal control. The slopes of the least-squares lines of the $\log_e$OD versus $1000 \times r^2\omega^2/2RT$ plots obtained at 5200 rpm of two different concentrations of HL are similar, whereas the slope observed for PL is clearly distinct and approximately one-half the magnitude in comparison to HL (Fig. 2A). The molecular weight of PL is confirmed to be 54,000, and indicates that HL is at least twice as large. The observed "tailing" of the HL data at higher OD readings (Fig. 2A) is indicative of sample heterogeneity and the light-scattering findings, which showed the presence of a small amount of aggregates and asymmetry in the main protein peak (Fig. 1A). To improve the quality of the data, only fractions including and immediately adjacent to the protein peak obtained from Superose 12 chromatography are subjected to sedimentation equilibrium analyses (Fig. 2B). Measured at two wavelengths (230 and 235 nm), these data show greater homogeneity and are consistent with a single species with an average molecular weight of 121,000. Unlike the light-scattering analysis, which does not represent the contribution of HL carbohydrates, sedimentation equilibrium results represent the total glycoprotein molecular mass of the dimer [~130 kDa based on sodium dodecyl sulfate-polyacrylamide gel electrophoresis (SDS–PAGE) analysis]. The accuracy of the sedimentation equilibrium result (121 vs 130 kDa) as well as that obtained from light-scattering experiments

[35] M. Rovery, M. Boudouard, and J. Bianchetta, *Biochim. Biophys. Acta* **525,** 373 (1978).

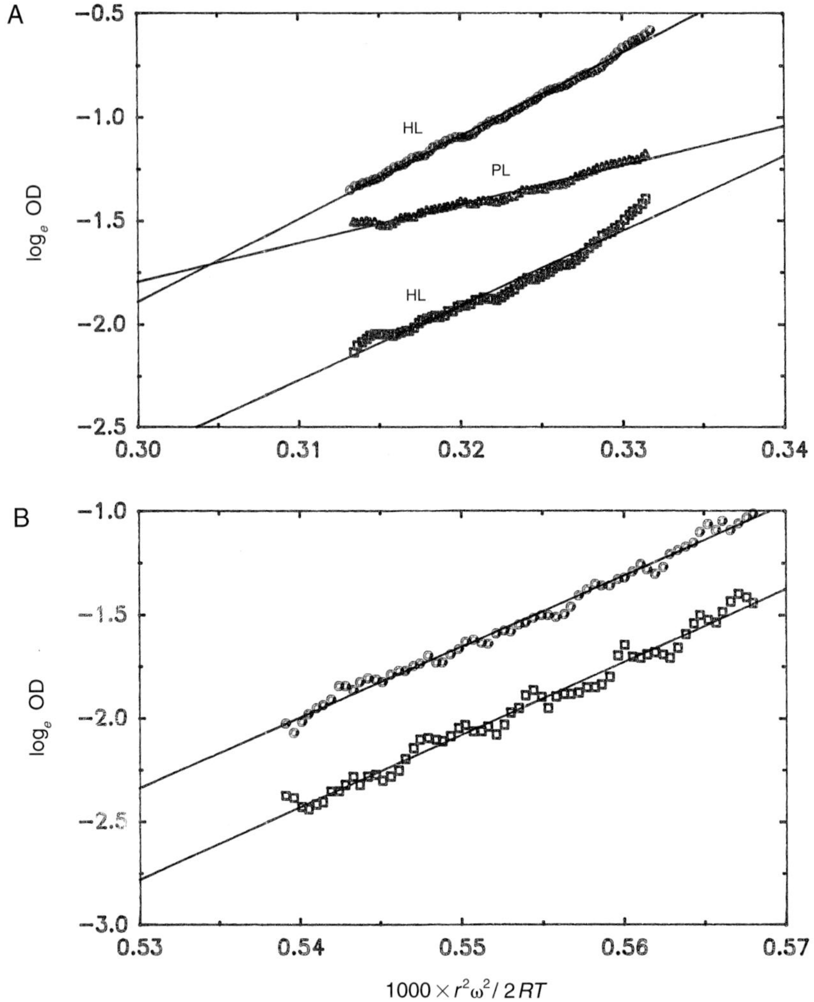

FIG. 2. Sedimentation of HL and PL. (A) Results obtained from the equilibrium distributions of HL and PL at 5200 rpm, 7°. Two solutions of HL of initial OD_{280} 0.5 and 0.25, and one solution of PL of initial OD_{280} of 0.3, were placed in chambers of the same rotor. (B) Results obtained for HL at two wavelengths [230 nm (○) and 235 nm (□)] after aggregates had been removed by gel filtration. The molecular mass of HL was calculated to be 121 (±7) kDa (mean ± SD). [Reproduced with permission from J. S. Hill, R. C. Davis, D. Yang, J. Wen, J. S. Philo, P. H. Poon, M. L. Phillips, E. S. Kempner, and H. Wong, *J. Biol. Chem.* **271,** 22931 (1996).]

(107 vs 106.8 kDa) clearly demonstrates that HL is a dimeric molecule. However, it is important to note that neither light-scattering nor sedimentation equilibrium analyses necessarily indicate the functional molecular weight of HL, that is, the smallest enzyme species capable of activity.

Radiation Inactivation

To determine the smallest functional unit necessary for lipolytic activity, radiation inactivation and target size analysis are performed. Radiation inactivation of a molecule enables the determination of the molecular mass or "target size" of the functional unit. Target theory states that irradiated molecules are physically degraded and rendered totally inactive by a single electron hit, whereas undamaged molecules remain completely active.[36] Thus, by irradiating an enzyme with different doses, enzymatic inactivation curves are produced from which the target size and hence functional molecular weight is calculated. The logarithm of the fraction of enzyme activity remaining after increasing radiation doses is plotted as a linear function of radiation dose (in rads), and the slope of this function, K, is determined by linear regression analysis. The molecular weight of he enzyme unit is calculated from Eq. (2),

$$\text{Molecular weight} = 6.4 \times 10^{11}\, S_t K \qquad (2)$$

where S_t is a temperature factor, determined experimentally with standard proteins. It should be noted that the carbohydrate contribution to the molecular mass of protein will not be reflected by radiation inactivation analysis.

Preparation and Irradiation of Samples. Samples of purified HL (0.3–0.6 mg/ml) are prepared for irradiation by adding 2000 units (2.8 mg) of glucose-6-phosphate dehydrogenase per milliliter (from *Leuconostoc mesenteroides;* Sigma). It should be noted that errant target sizes may be observed if protein concentrations are too low.[37] Therefore, it is recommended that irradiations be performed at no less than 2 mg of total protein per milliliter. Aliquots of purified HL (80 μl) in 0.5-ml microcentrifuge tubes are frozen immediately on dry ice and stored at −80° until shipped on dry ice for irradiation. These samples are irradiated at −135° with 13-MeV electrons for various times as described previously.[36] Following irradiation the samples are returned on dry ice and stored at −80° until thawed for assays. Nonirradiated samples are assayed as controls. Glucose-6-phosphate dehydrogenase activity is assayed by measuring absorption at 340 nm at 25° and

[36] E. S. Kempner, *Adv. Enzymol. Relat. Areas Molec. Biol.* **61,** 107 (1988).

[37] E. S. Kempner and J. H. Miller, *Anal. Biochem.* **216,** 451 (1994).

TABLE II
TARGET SIZES[a,b]

	Average ± SD (*n*)
HL activity	109 ± 24 (5)
HL structural unit	63 ± 11 (4)
G6PDH activity[c]	121 ± 13 (4)

[a] Reproduced with permission from J. S. Hill, R. C. Davis, D. Yang, J. Wen, J. S. Philo, P. H. Poon, M. L. Phillips, E. S. Kempner, and H. Wong, *J. Biol. Chem.* **271,** 22931 (1996).
[b] Molecular masses (in kDa) were determined in individual experiments as described in text.
[c] Control enzyme.

pH 7.8 to monitor the production of NADPH in the presence of glucose 6-phosphate and NADP.[38]

Electrophoresis and Protein Staining. To monitor the radiation-induced degradation of HL monomers, irradiated samples are analyzed by SDS–PAGE. Sample aliquots are mixed with a half-volume of buffer containing 2% (w/v) SDS, 0.1 *M* Tris-HCl (pH 6.8), 50% (v/v) glycerol, 10% (v/v) 2-mercaptoethanol, 0.05% (v/v) bromphenol blue. The mixture is placed in boiling water for 5 min prior to loading onto a 10% (w/v) acrylamide gel run at a constant current of 15 mA/gel. After fixation in 40% (v/v) methanol–10% (v/v) acetic acid, gels are stained overnight with a fluorescent dye (SYPRO orange; Molecular Probes, Eugene, OR) in the presence of 7.5% (v/v) acetic acid. Protein bands are visualized using a standard UV transilluminator. The relative intensity of the signal for individual bands is quantitated using the Ambis optical imaging system and software. Protein concentration standards are produced using the Bradford protein assay and serial dilutions of nonirradiated protein samples.

Prior to irradiation, an internal standard, glucose-6-phosphate dehydrogenase (G6PDH) from *L. mesenteroides,* is added to purified preparations of HL to assess the validity of the experimental protocol. Measurement of G6PDH activity after different doses of radiation allows calculation of a functional molecular mass of 121 (± 13) kDa (Table II), consistent with previously reported values for the molecular mass of the dimeric enzyme.[36] Following irradiation of purified HL with different doses, enzymatic inactivation curves are obtained. The enzyme activity decays as a single exponen-

[38] C. Olive and R. Levy, *Methods Enzymol.* **41,** 196 (1975).

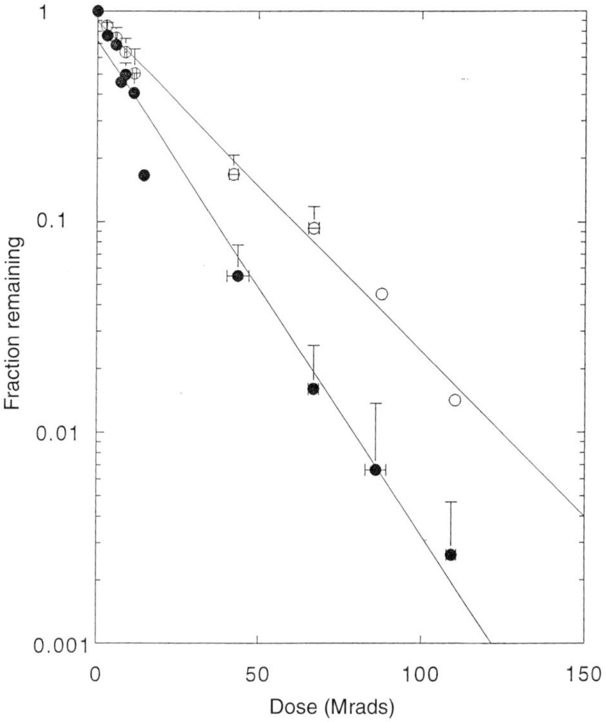

FIG. 3. Radiation inactivation of HL. Recombinant HL was purified from culture medium, prepared, irradiated, and assayed for lipolytic activity as described in text. The fractions of surviving activity at similar doses were combined from all experiments ($n = 5$) and plotted as mean ± SD (●). The fraction of remaining full-size HL protein was determined as described in text. Data were combined from all experiments ($n = 3$) and plotted as a mean ± SD (○). [Reproduced with permission from J. S. Hill, R. C. Davis, D. Yang, J. Wen, J. S. Philo, P. H. Poon, M. L. Phillips, E. S. Kempner, and H. Wong, *J. Biol. Chem.* **271,** 22931 (1996).]

tial function of radiation dose to less than 0.01% remaining activity (Fig. 3). The functional molecular mass for HL, determined from four independent experiments, is calculated to be 109 (± 24) kDa (Table II). Like light scattering, radiation inactivation analyses do not include the contribution of carbohydrate to the molecular mass of the functional unit. Therefore, the 109-kDa value obtained compares favorably to the expected 106.8 kDa of a theoretical dimer. To determine that the affinity of HL for its triolein substrate is not affected by radiation exposure, the apparent K_m of the enzyme is determined in two separate experiments for both nonirradiated samples and exposed samples with a 50% reduction in activity. The values determined are not significantly different (data not shown), suggesting that there are no HL molecules with altered activity as a result of irradiation.

The size of the HL structural unit is determined by measuring the reduction in full-size HL monomers observed by protein staining in SDS–polyacrylamide gels. To ensure that the observed staining is proportional to the quantity of the protein, standard curves are produced using serial dilutions of nonirradiated protein samples. Intact HL monomer decays as a simple exponential function of radiation dose (Fig. 3), revealing a target size of 63 ($\pm$ 11) kDa (Table II). Because the carbohydrate moiety of glycoproteins does not contribute to the target size,[36] these values are in close agreement with the theoretical target size of HL dimer and monomer, 106.8 and 53.4 kDa, respectively. The results of radiation inactivation strongly indicate that two monomers form the active HL dimer.

Radiation inactivation studies of a closely related lipase, bovine LPL, demonstrate that the active dimeric structure of this lipase is observed in the presence or absence of heparin or lipid substrate.[39] In the same manner, we believe that the dimeric target size of purified HL reported here strongly suggests that the same subunit structure exists whether in the presence of lipoprotein substrate or on the surface of the liver endothelium. Thus, it appears that HL and LPL are each functionally active as a homodimer.

The crystal structure of PL has revealed a two-domain structure composed of an NH_2-terminal domain containing the catalytic site joined by a short spanning region to a smaller COOH-terminal domain.[40] On the basis of sequence homology, the conservation of disulfide bridges, and similarity of lipolytic function, LPL and HL are believed to have a similar three-dimensional structure.[24] Multiple functional regions of HL have been identified including the catalytic site, lid domain, heparin affinity, lipid- and LRP-binding properties. Several of these functions have been localized to either of the two domains that compose the enzyme. For example, site-directed mutagenesis studies within the NH_2-terminal domain of HL demonstrate the essential function of serine contained within the Ser-Asp-His catalytic triad and of the 22-amino acid loop ("lid") that restricts access of the substrate to the active site.[41,42] Characterization of functional HL–LPL chimeras suggests that HL and LPL are both active dimers and that the COOH-terminal domain of HL is necessary for catalyzing lipid substrate

[39] T. Olivecrona, G. Bengtsson-Olivecrona, J. C. Osborne, Jr., and E. S. Kempner, *J. Biol. Chem.* **260,** 6888 (1985).

[40] F. K. Winkler, A. D'arcy, and W. Hunziker, *Nature* (*London*) **343,** 771 (1990).

[41] R. C. Davis, G. Stahnke, H. Wong, M. H. Doolittle, D. Ameis, H. Will, and M. C. Schotz, *J. Biol. Chem.* **265,** 6291 (1990).

[42] K. A. Dugi, H. L. Dichek, G. D. Talley, H.B. Brewer, Jr., and S. Santamarina-Fojo, *J. Biol. Chem.* **267,** 25086 (1992).

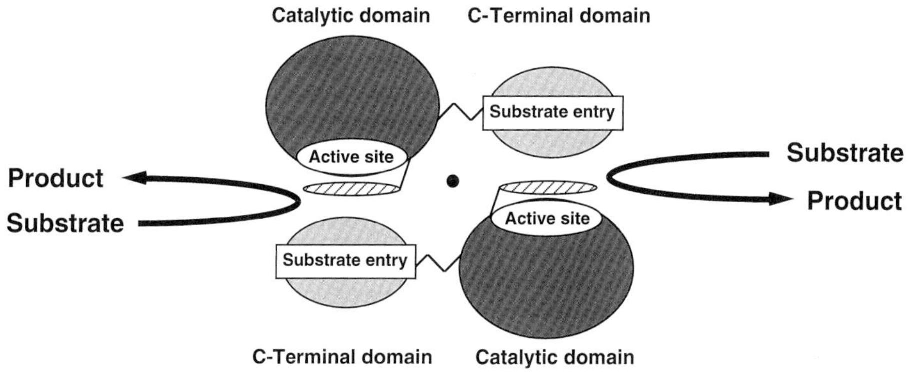

FIG. 4. Model for active, dimeric hepatic lipase. Hepatic lipase monomers are arranged in a head-to-tail fashion, with a twofold rotational axis of symmetry perpendicular to the plane of the page (single dot) that positions the COOH-terminal domain of one subunit in close proximity to the catalytic domain from the other subunit. The catalytic domain contains the active site residues, which in the absence of substrate are covered by a lid domain (hatched oval). As indicated, lipid substrates interact with elements in the COOH-terminal domain for presentation to the catalytic cleft and subsequent hydrolysis. [Reproduced with permission from J. S. Hill, R. C. Davis, D. Yang, J. Wen, J. S. Philo, P. H. Poon, M. L. Phillips, E. S. Kempner, and H. Wong, *J. Biol. Chem.* **271,** 22931 (1996).]

hydrolysis and contained binding sites for heparin.[31,43,44] The COOH-terminal domain of LPL has been shown to bind to the LRP,[45] but such binding remains to be established for HL.

These HL functions must now be considered in the context of a dimer conformation. As we have previously proposed for LPL,[46] a head-to-tail arrangement of two HL monomer subunits is in agreement with both structural and functional studies of the enzyme (Fig. 4). The model predicts that the COOH-terminal domain of one subunit is juxtaposed to the NH_2-terminal or catalytic domain of the opposing subunit. The initial interaction of the enzyme with lipid substrates may be facilitated by requisite contact with the COOH-terminal domain, which enables the catalytic reaction to take place at the active site found within the neighboring catalytic domain originating from the other subunit. The presence of multiple functional

[43] H. Wong, R. C. Davis, J. Nikazy, K. E. Seebart, and M. C. Schotz, *Proc. Natl. Acad. Sci. U.S.A.* **88,** 11290 (1991).

[44] R. C. Davis, H. Wong, J. Nikazy, K. Wang, Q. Han, and M. C. Schotz, *J. Biol. Chem.* **267,** 21499 (1992).

[45] S. E. Williams, I. Inoue, H. Tran, G. L. Fry, M. W. Pladet, P.-H. Iverius, J.-M. Lalouel, D. A. Chappell, and D. K. Strickland, *J. Biol. Chem.* **269,** 8653 (1994).

[46] H. Wong, R. C. Davis, T. Thuren, J. W. Goers, J. Nikazy, M. Waite, and M. C. Schotz, *J. Biol. Chem.* **269,** 10319 (1994).

domains proposed by this model is consistent with the putative role of HL in simultaneous interactions with circulating lipoproteins and cell surface proteoglycans (or to the LRP).[7] In the absence of crystal structure information, this model provides the opportunity to test hypotheses regarding function location and interdomain interaction.

An in-depth analysis of the structure and function of human HL has been limited by the lack of sufficient quantities of purified lipase for analysis. The discovery of the properties of HL subunit structure reported here is essential to probe further the nature of the specific structure–function relationships of this important enzyme of lipoprotein metabolism.

Acknowledgments

Research referred to in this chapter was supported by the Veterans Affairs Merit Review and National Institutes of Health Grant HL28481. J. S. H. was a recipient of a Fellowship from the Medical Research Council of Canada. We thank the following collaborators who contributed to this work: M. H. Doolittle, J. S. Philo, V. N. Schumaker, and E. S. Kempner.

[14] Cloning, Sequencing, and Expression of *Candida rugosa* Lipases

By LILIA ALBERGHINA and MARINA LOTTI

Introduction

Candida rugosa[1] is an imperfect hemiascomycetous fungus known as a good producer of lipases marketed by several manufacturers. These enzymes (CRLs) are monomeric proteins of ca. 60 kDa composed of 534 amino acids that do not require cofactors for activity. Lipases are moderately glycosylated and the carbohydrate moiety is suggested to be of importance for stability and activity.[2] A wide overview of the industrial applications is given in Refs. 3 and 4.

[1] This yeast has been previously named *C. cylindracea* and this classification is still in use in several laboratories. We refer specifically to strain 14830 by the American Type Culture Collection (Rockville, MD).

[2] P. Grochulski, Y. Li, J. D. Schrag, F. Bouthillier, P. Smith, D. Harrison, B. Rubin, and M. Cygler, *J. Biol. Chem.* **268,** 12843 (1993).

[3] P. Eigtved, *in* "Advances in Applied Lipid Research," Vol. 1, p. 1. JAI Press, 1992.

[4] K. Faber and M. C. R. Franssen, *Trends Biotechnol.* **11,** 461 (1993).

0076-6879/97 $25.00

Candida rugosa lipase isoenzymes are the products of different genes[5,6] and, owing to the striking similarity of their overall biochemical properties, are not easy to separate by conventional media, at least on a preparative scale. A reproducible chromatographic procedure, set up to separate crude CRL preparations into two major fractions A and B, is now in use in several laboratories.[7] Fraction B can be further divided into four different proteins by isoelectrofocusing based on the different isoelectric points of the constituent isoforms (see below).[7] This last step, although useful analytically, can hardly be scaled up to obtain substantial amounts of isoenzymes, so that the production of pure CRLs is still far from straightforward. Nevertheless, analysis of the two main lipase fractions has provided interesting hints, in particular because lipases A and B have been shown to be catalytically nonequivalent in that lipase B is a better catalyst for lipidic substrates and lipase A is more specific for soluble esters.[7]

The three-dimensional structure of the major lipase form (CRL1) has been solved in its open and closed conformation as well as in complexes with substrate analogs, providing clues to the structure and action of this enzyme.[2,8,9]

In this chapter we describe work concerning the cloning and expression of CRLs, which are peculiar among fungal lipases for two reasons: (1) the large number of encoding genes and (2) a deviation from the universal genetic code in the source yeast. The importance and effects of this phenomenon for lipase structure and function, as well as for the expression of cloned genes in host systems, are discussed.

Cloning of Lipase Genes

Cloning of lipase-encoding genes has been achieved using different genomic banks through a sequential approach required by the unexpected discovery of a very large gene family.[5,6] The first genomic library, composed of about 70,000 clones, was screened by colony hybridization with the aid of two synthetic oligonucleotides based on the sequence of a cDNA lipase

[5] S. Longhi, F. Fusetti, R. Grandori, M. Lotti, M. Vanoni, and L. Alberghina, *Biochim. Biophys. Acta* **1131,** 227 (1992).

[6] M. Lotti, R. Grandori, F. Fusetti, S. Longhi, S. Brocca, A. Tramontano, and L. Alberghina, *Gene* **124,** 45 (1993).

[7] M. L. Rua, T. Diaz-Maurino, V. M. Fernandez, C. Otero, and A. Ballesteros, *Biochim. Biophys. Acta* **1156,** 181 (1993).

[8] M. Cygler, P. Grochulski, R. J. Kazlauskas, J. D. Schrag, F. Bouthillier, B. Rubin, A. N. Serreqi, and A. K. Gupta, *J. Am. Chem. Soc.* **116,** 3180 (1994).

[9] P. Grochulski, F. Bouthillier, R. J. Kazlauskas, A. N. Serreqi, J. D. Schrag, E. Ziomek, and M. Cygler, *Biochemistry* **33,** 3494 (1994).

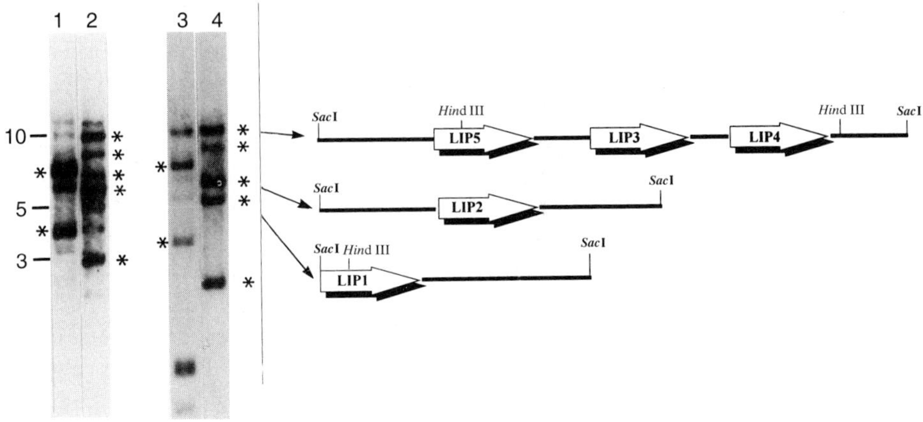

FIG. 1. Analysis of the lipase gene family in *C. rugosa*. *Left:* A Southern blotting analysis of total yeast DNA cut with *Hin*dIII (lanes 1 and 3) or *Sac*I (2 and 4) and hybridized with probes specific either for the 3′ (lanes 1 and 2) or the 5′ (lanes 3 and 4) end of the genes. Asterisks denote bands positive to both probes. *Right:* Genes cloned from selected fragments. Coding sequences are represented by arrows.

clone isolated previously in another laboratory, and specific for the 5′ and 3′ ends of this sequence.[10] This led to the isolation of the first coding sequence termed *LIP1*.

LIP1-derived fragments were then employed to perform a detailed hybridization analysis of the *C. rugosa* DNA to ascertain the presence of other related sequences. Duplicate blots were hybridized with [^{32}P]dCTP-labeled probes of about 200 bp encompassing either the 5′ or the 3′ end of *LIP1*. These experiments were performed with relatively high stringency to avoid interfering signals. As shown in Fig. 1, multiple reactive bands were generated, with the intensity of the signal suggesting a close homology to the DNA probes. We interpreted these results as consistent with the existence of a family of genes very closely related in terms of their sequences.

Hybridization bands positive to both probes were assumed to contain complete lipase genes. Genomic DNA digested with the appropriate nucleases and size-selected by gel electrophoresis before cloning into the plasmid vector was used to construct "minibanks" of 3000 to 5000 clones for the isolation of a definite DNA fragment, selected on the basis of the hybridization experiments.

[10] Y. Kawaguchi, H. Honda, J. Taniguchi-Morimura, and S. Iwasaki, *Nature* (*London*) **341,** 164 (1989).

Altogether, three different clones were isolated, two of them containing a single gene (*LIP1* and *LIP2*) and a third carrying three other coding sequences (*LIP3*, *LIP4*, and *LIP5*) oriented head to tail and separated by spacers of about 1 kbp. Sequencing (see below) confirmed that clones bore functional genes. At the time, this was not obvious because the complexity of the hybridization pattern might have been ascribed to related but different sequences (i.e., esterases) or to pseudogenes.[10] Even the heterogeneity in protein composition reported by several groups did not necessarily imply the existence of multiple genes; on the contrary, other instances of lipase mixtures, such as that of *Rhizopus,* were satisfactorily explained with post-translational modifications of a single gene product.[11]

Later, to isolate additional sequences that possibly escaped cloning, the *C. rugosa* DNA was amplified by the polymerase chain reaction (PCR) technique with oligonucleotide primers annealing to the ends of lipase genes and designed on the basis of the alignment of the *LIP1–LIP5* nucleotides.[12] Whereas the 5′ end is relatively conserved, allowing the identification of a consensus sequence for the forward primer, the 3′ end is more variable. Therefore, degenerate oligonucleotide primers were synthesized. Experimental conditions were set up so as to favor annealing of the primers to unknown but possibly highly related sequences. The PCR amplification was performed on 500 ng of total DNA in a 50-μl mixture containing 2.5 U of *Taq* polymerase for 35 cycles as follows: 94°, 5 min; 94°, 1 min; 65°, 45 sec; 72°, 1 min 50 sec; and a final extension at 72° for 10 min. Amplified DNAs were subcloned and discriminated from the genes already characterized on the basis of their restriction pattern. Sequencing of the first 300 nucleotides of each clone unambiguously ascertained that two further lipase sequences (*LIP6* and *LIP7*) had been isolated.

Moreover, to account for the complex hybridization pattern observed in Southern blots with DNA digested with combinations of restriction endonucleases (not shown) we should acknowledge the existence of another one to three lipase sequences.

The redundancy of the CRL gene family is unusual within fungal lipases, but not exceptional for other *Candida* species, as, for example, *Candida albicans* (at least seven genes encode aspartic proteinases[13]), and *C. maltosa* and *C. tropicalis* (multiple genes encode cytochrome *P*-450[14,15]).

[11] E. Boel, B. Huge-Jensen, M. Christensen, L. Thim, and N. P. Fiil, *Lipids* **23,** 701 (1988).

[12] S. Brocca, R. Grandori, D. Breviario, and M. Lotti, *Curr. Genet.* **28,** 454 (1995).

[13] M. Monod, G. Togni, B. Hube, and D. Sanglard, *Mol. Microbiol.* **13,** 357 (1994).

[14] M. Ohkuma, T. Tanimoto, K. Yano, and M. Takagi, *DNA Cell Biol.* **10,** 271 (1991).

[15] W. Seghezzi, C. Meili, R. Ruffiner, R. Kuenzi, D. Sanglard, and A. Fietcher, *DNA Cell Biol.* **10,** 767 (1992).

Chromosomal Localization

To gain more information about the structure of the gene family, the *Candida* genome was characterized and the genes localized on the yeast chromosomes.[12] The genome size was determined by laser flow cytometry using exponentially growing cells stained with propidium iodide. In this procedure, cells are mildly sonicated, washed, and resuspended in 70% (v/v) ethanol, in which they can be preserved at 4°. Staining is performed on cells washed with phosphate-buffered saline (PBS), treated with RNase (1 mg/ml, 90 min, 37°), washed with PBS, and incubated for 20 min on ice in 50 m*M* Tris-HCl, pH 7.7, containing 15 m*M* $MgCl_2$ and 46 m*M* propidium iodide. Fluorescence intensity is measured with a FACS^{Star} Plus (Becton Dickinson, Mountain View, CA) equipped with a 2-W argon-ion laser (excitation wavelength, 488 nm; laser power, 200 mW). The total DNA content of *C. rugosa* is estimated by comparison with a reference strain to be about 20 Mb (Fig. 2).

Yeast chromosomes were separated by pulsed-field gel electrophoresis, a technique of growing importance for the taxonomy of fungal and plant

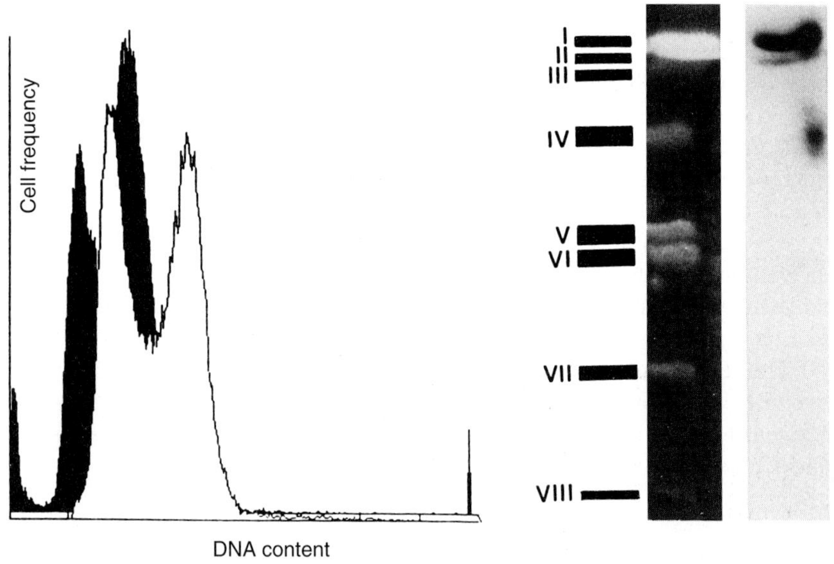

FIG. 2. Structure of the *C. rugosa* genome and localization of lipase genes. *Left:* Profile of *Candida* DNA (white) and of a *Saccharomyces cerevisiae* reference strain (black) as obtained by laser flow cytometry. *Right:* Electrophoretic separation of *C. rugosa* chromosomes and hybridization thereof with a lipase-specific probe.

species. Intact DNA is extracted from *Candida* cells using the following solutions.

Buffer I: 50 m*M* EDTA, 10 m*M* Tris-HCl (pH 7.5)
Buffer II: 500 m*M* EDTA, 10 m*M* Tris-HCl (pH 7.5)
Buffer III: 500 m*M* EDTA, 10 m*M* Tris-HCl (pH 7.5), 1% (w/v) sodium *N*-lauroyl sarcosinate, Proteinase K (1 mg/ml)

Cells are collected from 3-ml aliquots of early stationary-phase cultures, washed three times with buffer I, and incubated overnight at 37° in 150 μl of the same buffer. Lyticase (50–100 U; Sigma, St. Louis, MO) and 250 μl of 1% (w/v) low melting point agarose in 125 m*M* EDTA, pH 7.5, are added under gentle mixing. Following polymerization in a mold chamber, agarose plugs are incubated overnight at 37° in buffer II and then for 18–24 hr at 50° in buffer III. Finally, plugs are applied to 0.9% (w/v) agarose gels and run in Tris–borate buffer at 11° using an electrophoretic system with a hexagonal electrode array. Separation is achieved in 48 hr at 140 V with a 2-min switch.

The *C. rugosa* genome is organized in eight chromosomes, whose sizes were evaluated to be 0.1, 0.72, 0.95, 1.1, 1.3, 1.7, 1.9, and more than 2.0 Mb (Fig. 2). Resolution of the two larger chromosomes is difficult to achieve and a longer electrophoresis, in which the smaller chromosomes are lost from the gel, may be required. This result allowed us to estimate the aploid genome to about 10 Mb, arguing for a diploid status of *C. rugosa* cells. When separated chromosomes were hybridized with a lipase-specific probe, a strong hybridization signal was observed only on the larger chromosome (Fig. 2). This location may suggest a possible origin of the gene family through gene duplication.

Sequence Analysis

All genes isolated or amplified in the previous step were sequenced and analyzed to infer some knowledge about the lipase isoenzymes.[6] The availability of predicted amino acid sequences is regarded as being a valuable tool, because properties of CRLs have been determined only on raw enzyme preparations containing mixtures of lipases or, more recently, on lipases A and B.[7] Figure 3 shows an alignment of the LIP1–LIP5 complete primary sequences and the amino-terminal stretches of LIP6 and LIP7. The two latter sequences, being incomplete, are not considered further.

Mature CRLs are obtained on cleavage of a leader sequence of 14–15 residues and contain 534 amino acids. Residues conserved in all five sequences make up 66% if identity is considered and 84% if amino acids are grouped on the basis of their physicochemical similarity. Pairs of isoenzymes share 77–88% identity. In spite of their close relatedness, each isoenzyme

```
CRL1   APTATLANGDTITGLNAIINEAFLGIPFAEPPVGNLRFKDPVPYSGSLDGQKFTSYGPSCMQQNPEGTIEENLPK
CRL2   ....T.........L.AIVNEK....F.E....T....P.V..SASLN.QQ.TSY..S.M.MN.MGSFEDT.PK 75
CRL3   ....K.........L.AIINEA....F.E....N....D.V..SGSLN.QK.TSY..S.M.QN.EGTFEEN.GK
CRL4   ....T.........L.AIINEA....F.Q....N....P.V..SASLN.QK.TSY..S.M.MN.LGNWDSS.PK
CRL5   ....T.........L.AIINEA....F.E....N....D.V..RGSLN.QS.TAY..S.M.QN.EGTYEEN.PK
CRL6                ..L.AIINEA....F.E....N....D.V..RGSHN.ES.TAY..S.I.QN.EGTYEEN.PN
CRL7                ..I.LLGVDQ....Y.E....S....A.I..KVVHN.QT.KSD..T.Y.MD.IVSWQDN.AS

CRL1   AALDLVMQSKVFEAVSPSSEDCLTINVVRPPGTKAGANLPVMLWIFGGGFEVGGTSTFPPAQMITKSIAMGKPII 150
CRL2   NARHLVL...I.QVVL.ND....TI..IR.P.TRAS.G....L........L.GSSL..GD..VAK.VL....V.
CRL3   TALDLVM...V.QAVL.QS....TI..VR.P.TKAG.N....L........I.SPTI..PA..VTK.VL....I.
CRL4   AAINSLM...L.QAVL.NG....TI..VR.S.TKPG.N....V........V.GSSL..PA..ITA.VL....I.
CRL5   VALDLVM...V.QAVL.NS....TI..VR.P.TKAG.N....L........I.SPTI..PA..VSK.VL....I.
CRL6   VALDLVM...V.QAVL.NS....TI..VR.P.NKA
CRL7   NVVNLVL...I.QVIY.QS....SL..SV.P.TKA
                                                                      ☆
                                                                     209
CRL1   HVSVNYRVSSWGFLAGDEIKAEGSANAGLKDQRLGMQWVADNIAAFGGDPTKVTIFGESAGSMSVMCHILWNDGD 225
CRL2   ..SM...VA.W....PD.QN...G....H....AM.......G.....S....Y........TFVHLV..D..
CRL3   ..AV...VA.W....DD.KA...G....K....GM.......G.....S....F........VLCHLI..D..
CRL4   ..SM...VA.W....PD.KA...G....H....GL.......G.....S....F........VMCQLL..D..
CRL5   ..AV...LA.F....PD.KA...S....K....GM.......G.....S....F........VLCHLL..G..
                     241   247                                           282
CRL1   NTYKGKPLFRAGIMQSGAMVPSDAVDGIYGNEIFDLLASNAGCGSASDKLACLRGVSSDTLEDATNNTPGFLAYS 300
CRL2   ...N.......A.....C.....P...T..TE.YNQVVASA..G...D......GL.QDT.YQ..SD...V.A.P
CRL3   ...K.......G.....A.....P...T..NE.YDLFVSSA..G...D......SA.SDT.LD..NN...F.A.S
CRL4   ...N.......A.....A.....P...P..TQ.YDQVVASA..G...D......SI.NDK.FQ..SD...A.A.P
CRL5   ...K.......G.....A.....P...T..TQ.YDTLVAST..S...N......GL.TQA.LD..ND...F.S.T
                                                ☆
CRL1   SLRLSYLPRPDGVNITDDMYALVREGKYANIPVIIGDQNDEGTFFGTSSLNVTTDAQAREYFKQSFVHASDAEID 375
CRL2   .....Y......TF....YA...D..Y.HV............TL.GL....V..D.Q.RA.FKQ..I..S..E.D
CRL3   .....Y......KN....YK...D..Y.SV............TI.GL....V..N.Q.RA.FKQ..I..S..E.D
CRL4   .....F......TF....FK...D..C.NV............TV.AL....V..D.Q.RQ.FKE..I..S..E.D
CRL5   .....Y......AN....YK...D..Y.SV............FL.GL....T..E.D.EA.LRK..I..T..D.T
                                                                                ☆
CRL1   TLMTAYPGDITQGSPFDTGILNALTPQFKRISAVLGDLGFTLARRYFLNHYTGGTKYSFLSKQLSGLPVLGTFHS 450
CRL2   T.MA..TS.I.........F..I...F...S.L....A.TLA......YYQ..T..............VL..F.G
CRL3   T.MA..PQ.I.........F..I...F...S.V....A.IHA......HFQ..T..............IM..F.A
CRL4   T.MA..PS.I.........F..I...F...A.V....A.TLP......HFQ..T..............VI..H.A
CRL5   A.KA..PS.V.........L..L...L...N.V....T.TLS......HYT..P..............IL..F.A

CRL1   NDIVFQDYLLGSGSLIYNNAFIAFATDLDPNTAGLLVKWPEYTSSSQSGNNLMMINALGLYTGKDNFRTAGYDAL FSNPPSFFV 534
CRL2   ...IW.DY.VGSG.VI.........N.....K...WTN..TY.....S.N..MQ.NG...........PDA.S.. FSN.PS...
CRL3   ...VW.DY.LGSG.VI.........T.....T...LVN..KY.....S.N..MM.NA...........TAG.D.. MTN.SS...
CRL4   ...VW.DF.VSHS.AV.........N.....K...LVN..KY.....S.N..LQ.NA...........TAG.D.. FTN.SS...
CRL5   ...VW.HF.LGSG.VI.........T.....T...SVQ..KS.....A.D..MQ.SA...........TAG.N.. FAD.SH...
```

can be distinguished from the others by the unique amino acids that occur along the polypeptide chain. The results obtained from sequence analysis account for the experimental problems encountered in the purification procedures. In fact, overall biochemical features of CRLs are expected to be similar. Notable exceptions are hydrophobicity and, but unfortunately only in one case (LIP3), the amino-terminal sequence. An interesting difference among CRLs, particularly in view of the suggested role of sugars in activity, is the pattern of sites for N-glycosylation: three in LIP1, one in LIP2, three in LIP3, one in LIP4, and three in LIP5. The only biochemical feature unique to each isoenzyme is the isoelectric point: 4.5, 4.9, 5.1, 5.7, and 5.5, respectively. All of these features are being exploited to develop an experimental protocol for the separation of CRL isoforms.

In the absence of a precise catalytic characterization of each protein, the availability of a group of sequences highly related to each other, as CRLs are, provides a valuable tool with which to approach the study of the structural and functional roles of individual residues and stretches of amino acids within the protein.[16] In the *Candida* lipase family, residues ensuring correct protein folding and catalysis, as identified by Cygler and colleagues[17] by comparing enzymes belonging to the lipase/esterase family, are strictly conserved. On the other hand, a higher variability seems to be allowed in other regions of the protein. In particular, we focused our attention on protein regions that directly contact substrates, as indicated by crystallography.[8] Besides the amino acids composing the lid, which is extremely nonconserved in lipases, other positions of CRLs are occupied by residues clearly nonequivalent in their properties.[16] From this preliminary analysis, with due caution because of the absence of experimental support, one can argue that CRLs might differ in the fine tuning of their catalytic activity. This prediction fits well with the results obtained with partly purified CRLs, which showed a different behavior in hydrolysis reactions.[7]

It is well known that lipases of different origins are extremely diversified in their sequence. In fact, CRLs do not share homology with other lipases, with the exception of proteins secreted by the mold *Geotrichum candidum*

[16] M. Lotti, A. Tramontano, S. Longhi, F. Fusetti, S. Brocca, E. Pizzi, and L. Alberghina, *Protein Eng.* **7**(4), 531 (1994).

[17] M. Cygler, J. D. Schrag, J. L. Sussman, M. Harel, I. Silman, M. K. Gentry, and B. P. Doctor, *Protein Sci.* **2**, 366 (1993).

FIG. 3. Multiple alignment of mature CRLs. Amino acids identical in all isoenzymes are marked by dots and residues of the catalytic triad by stars. N-Glycosylation sites are underlined. CUG serines are in boldface. Positions mentioned in text are indicated by their numbers.

(GCL),[18] an organism relatively distant from *C. rugosa* in the phylogenetic tree. Lipases from other candidas have not been isolated, with the only exception being those from *C. antarctica,* which are nevertheless completely different in size (30 kDa) and sequence.[19] At present, with the sparse information available, it is difficult to decide whether lipase variability is extended even within the same genus or whether, as might be possible considering some confusion still present in the *Candida* taxonomy, the current classification of *C. antarctica* should be somehow revised. However, similarity with GCLs is high both in sequence (40% identity) and in the conservation of secondary structure elements.[6] Homology is shared also with several esterases and acetylcholinesterases grouped, together with CRLs and GCLs, in the so-called lipase/esterase family.[17]

Physiology and Regulation of *Candida rugosa* Lipase Expression

Candida rugosa lipase production has been studied in detail using different growth conditions, with particular regard to the carbon and nitrogen sources.[20] In its general pattern, the synthesis of lipase is stimulated by fatty acids and inhibited by glucose.

However, we also observed a basal synthesis of enzyme in *C. rugosa* cultures growing in glucose-based medium [2% (w/v) glucose, 2% (w/v) peptone, 1% (w/v) yeast extract] in the absence of inducers. This suggests that lipase genes might be subjected to different regulatory mechanisms, some being constitutive and others inducible. Consensus sequences possibly involved in regulation have not been identified upstream of the *LIP* genes. Nevertheless, cross-inhibition experiments showed that the synthesis of constitutive lipases is blocked by fatty acids and, conversely, the synthesis of inducible lipases is inhibited by glucose, thus suggesting a complex regulatory pattern. These preliminary results, obtained in our laboratory, might find application in the production of CRL preparations enriched with definite sets of isoenzymes.

Expression of Cloned Genes in Host Cells

Genetic and molecular features of *C. rugosa* are at present almost unknown. Indeed, this yeast has been used to date just as a factory for the

[18] M. C. Bertolini, L. Laramée, D. Y. Thomas, M. Cygler, J. D. Schrag, and T. Vernet, *Eur. J. Biochem.* **219,** 119 (1994).

[19] J. Uppenberg, M. T. Hansen, S. Patkar, and T. A. Jones, *Structure* **2,** 293 (1994).

[20] N. Obradors, J. L. Montesinos, F. Valero, F. J. Lafuente, and C. Solà, *Biotechnol. Lett.* **15**(4), 357 (1993).

TABLE I
Candida SPECIES USING CUG AS A CODON FOR SERINE

Species	Ref.
Candida parapsilosis	21
Candida zeylanoides	21
Candida albicans	21
Candida cylindracea JMC1613	21
Candida rugosa ATCC 14830	10
Candida melibiosica	21
Candida maltosa	20a
Candida guillermondii	[a]
Candida tropicalis	[a]
Candida viswananthii	22

[a] Sequence of the $tRNA^{Ser}$ CAG reported in database.

production of lipolytic enzymes, without concern for understanding its molecular biology. However, in 1989 *C. rugosa* suddenly gained attention when it was discovered to deviate from the universal genetic code in the use of the leucine triplet CUG for serine.[10] Later, working on the heterologous expression of CRLs, we came to the conclusion that this organism is quite dissimilar from other yeasts, as becomes evident in the following.

Occurrence and Frequency of CUG Serines

CUG is employed as a serine codon by several *Candida* species, listed in Table I,[10,20a,21,22] all belonging to a monophyletic group as shown from a phylogenetic analysis based on the alignment of their 18S RNA sequences.[21,22] Other species not closely related to this group, such as *C. glabrata, C. kefyr,* and *C. krusei,* follow the usual code. Nonuniversal reading of the genetic code has been described in mitochondria and also in nuclear genes of bacteria and protozoans, but usually concerns the use of stop codons. CUG serine is the first reported instance of a change in a codon specifying an amino acid. To explain the genesis of this deviation in an evolutionary context, it has been suggested that in the course of evolution the codon became unassigned and was later reassigned to serine following

[20a] H. Sugiyama, M. Ohkuma, Y. Masuda, S. M. Park, A. Ohta, and M. Takagi, *Yeast* **11,** 43 (1995).

[21] T. Ohama, T. Suzuki, M. Mori, S. Osawa, T. Ueda, K. Watanabe, and T. Nakase, *Nucleic Acids Res.* **21,** 4039 (1993).

[22] G. Pesole, M. Lotti, L. Alberghina, and C. Saccone, *Genetics* **141,** 903 (1995).

TABLE II
FREQUENCY OF SERINE CODONS IN *Candida rugosa* LIPASES

	LIP1	LIP2	LIP3	LIP4	LIP5
CUG	0.42	0.38	0.41	0.36	0.37
AGU	0.02	0.09	0.09	0.04	0.06
AGC	0.25	0.29	0.28	0.29	0.28
UCG	0.17	0.13	0.13	0.16	0.11
UCA	0.02	0.02	0	0.02	0
UCU	0	0.02	0.02	0.02	0.02
UCC	0.11	0.07	0.06	0.08	0.15

the appearance of a $tRNA^{Ser}$ with anticodon CAG.[23] Such fascinating theories are outside the scope of this chapter and are not further discussed. What is relevant in this context are the consequences to the heterologous expression of genes cloned from CUG-Ser candidas. In fact, the translational machinery of the host cells would introduce leucine residues at any CUG codon, where a serine was present in the original protein. Indeed, several genes cloned from *Candida* belonging to the CUG-Ser group have been expressed in universal hosts, without any appreciable effect on expression or activity. The occurrence of unusual codons was noticed only later. In fact, CUG codons are extremely rare in these proteins and, in general, CUG is quite uncommon, accounting for about 2–3% of all serine codons in more than 170 proteins from *Candida* available in databases.[22] Thus, insertion of wrong amino acids in a few positions may be tolerable and not induce dramatic effects on the folding and stability of the recombinant protein.

This is not the case for *C. rugosa,* in which CUG is preferentially used in addition to the other six codons for serine, accounting for more than 40% of all serines (Table II). To our knowledge, *C. rugosa* is unique in its bias for CUG, which is likely correlated with the high G + C content of its genome (63%). A closer analysis of the lipase sequences isolated in our laboratory suggests an important structural role for at least some of the CUG-encoded serine residues (Fig. 3), one of them being the catalytic Ser-209. Worth mentioning also are position 282, where a serine is maintained within the esterase/lipase family, and position 247, which is invariantly a serine also in lipases from *G. candidum.* Apart from these few amino acids, it is interesting to note that several CUG serines are conserved in the family

[23] T. Suzuki, T. Ueda, T. Yokogawa, K. Nishikawa, and K. Watanabe, *Nucleic Acids Res.* **22,** 115 (1994).

of CRL isoenzymes, showing once more that the use of this codon is not disfavored, at least in lipase genes.

Heterologous Expression of Candida rugosa Lipases

To obtain expression of recombinant CRLs, two main issues must be faced: (1) the development of an efficient expression system and (2) the mutagenesis of CUG codons.

The budding yeast *Saccharomyces cerevisiae* is a popular host for the expression of cloned eukaryotic genes. This is due in particular to its ability to perform posttranslational modifications (such as glycosylation) of the recombinant proteins, to the availability of advanced fermentation protocols, and, not least, to the acceptability of yeast as a GRAS organism for expression. For all these reasons, expression of CRL genes in *S. cerevisiae* would be desirable. The expression vector used was a shuttle *Escherichia coli*–yeast plasmid containing the inducible *GAL1-GAL10/CYC1* promoter.[24] Expression of genes cloned under the control of this sequence is repressed by the presence of glucose in the culture medium and induced up to 1000-fold by galactose. The *LIP1* gene was therefore cloned in the expression vectors and cells harboring the constructs were induced by growth in galactose-based medium. As a rule, these experiments are performed in two steps: cells are first grown for 48 hr in glucose medium [0.67% (w/v) yeast nitrogen base supplemented with the appropriate amino acids at 50 mg/liter and 2% (w/v) glucose], collected by centrifugation, washed, and resuspended in minimal medium containing 2% (w/v) galactose. The two-step procedure is intended to obtain biomass before induction, while growing cells in repressing medium until all glucose is consumed avoids the so-called catabolite repression, i.e., a delay in galactose-induced expression due to the repressing action of glucose catabolites. The time course of production is followed by Western blotting of samples taken at different times in culture growth. In these experiments, recombinant lipase was not detected, although high levels of specific mRNA were revealed in Northern blots (not shown).

After considering several possibilities (low translatability of the message, protein instability, etc.), it was decided to replace the first 15-amino acid coding region of the lipase gene, i.e., its leader sequence (LS) for secretion.[25] This signal is extremely efficient in *Candida,* because lipase proteins are quickly transported to the exterior of cells, and it is not significantly different from LSs of several yeast extracellular proteins characterized to date. Nevertheless, it is known that the role of the signal sequence

[24] C. Baldari, J. A. C. Murray, P. Ghiara, G. Cesareni, and V. Galeotti, *EMBO J.* **6,** 229 (1987).
[25] F. Fusetti, S. Brocca, D. Porro, and M. Lotti, *Biotechnol. Lett.* **18**(3), 281 (1996).

A

```
cgatcactataggagctc ATG GAA CTC GCT CTT GCG CTC CTG CTC ATT GCC TCG GTG GCT GCT GCC CCC ACC
                   Met Glu Leu Ala Leu Ala Leu Leu Leu Ile Ala Ser Val Ala Ala Ala Pro Thr

ATG AAT ATA TTT TAC ATA TTT TTG TTT TTG CTG TCA TTC GTT CAA GGT ACC CTG GCT GCT GCC CCC ACC
Met Asn Ile Phe Tyr Ile Phe Leu Phe Leu Leu Ser Phe Val Gln Gly Thr Leu Ala Ala Ala Pro Thr
```

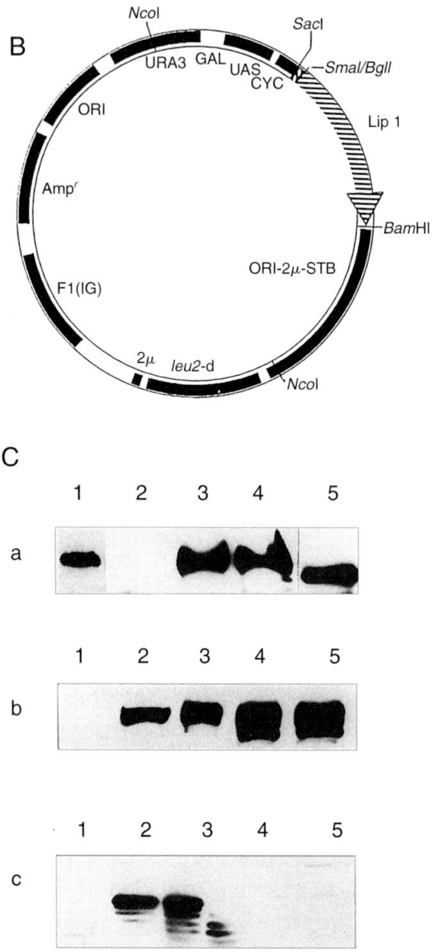

FIG. 4. Expression of recombinant LIP1 in *S. cerevisiae.* (A) Sequence of leader peptides in the wild-type protein (*top*) and after fusion with the *K. lactis* sequence (bottom). (B) Schematic representation of the expression plasmid employed; the inducible promoter is indicated as GAL-UAS CYC. URA3 and *leu2*-d are the selection markers and ORI 2μ-STB is the yeast origin of replication. (C) Western blotting of cell extracts (1×10^7 cells) from cultures harboring the *K. lactis*–lipase fusion. (a) Experiment in batch. Cells grown in glucose (lane 2) or in galactose medium to the exponential and stationary phases of growth (lanes 3

might overcome the simple transport of the protein through the secretory pathway. The extent and meaning of this role are not really understood, but it has been hypothesized that the N-terminal part of the protein may somehow stabilize the peptide within cells. Relying on this hypothesis, the *LIP1* gene was cut so as to remove the original sequence and join it to plasmid pEMBLYexSecI, in frame with a 16-residue-long leader sequence derived from the killer toxin produced by the yeast *Kluyveromyces lactis.* This construct was similar to the previous construct, differing only in the signal preceding the coding sequence for the mature proteins. In this case, the induction of specific mRNA was accompanied by the intracellular accumulation of recombinant protein at a level of about 10–20 mg/liter of culture (Fig. 4). Treatment of the recombinant protein with endoglycosidase F showed that it has been glycosylated (Fig. 4).

These observations about the significance of the leader peptide may be of general interest to people working on heterologous expression in yeast. Moreover, they suggest that *C. rugosa* is somehow dissimilar from other yeasts not only in its genetic code, but also in the use and recognition of transcription/translation signals.

The process was scaled up in a bioreactor using a two-stage computer-controlled fed-batch fermentation previously developed in our laboratory.[25] The control system is based on the determination of ethanol concentration in the bioreactor outflow gases, allowing us to regulate the supply rate of glucose so as to prevent high ethanol production. To induce expression, we used a mixed substrate in the feeding as follows.

1. Start with 50% (w/v) glucose (repression).
2. After 40 hr shift to 25% (w/v) glucose–25% (w/v) galactose (no repression).
3. After 80 hr shift to 50% (w/v) galactose (induction).

The recombinant protein was accumulated intracellularly by the transformed culture over several hours of fermentation, reaching levels as high as 1 g/liter (Fig. 4). However, it was inactive in enzyme assay. This was quite obvious owing to the substitution of key serine residues, such as Ser-209, -247, and -282 with leucine. Less evident is the reason why the protein was accumulated intracellularly, although glycosylation was accomplished

and 4, respectively). Lane 1 contains 1 μg of purified lipase and lane 5 contains the recombinant protein after enzymatic deglycosylation. (b) Fed batch process with 50% glucose in the medium (lane 1), glucose–galactose (25% each, w/v) (lane 2), and 50% (w/v) galactose (lanes 3, 4, and 5; sample taken at different times). (c) Localization of recombinant LIP1 in fractions from yeast cells grown in glucose-based medium (lane 1) or in galactose-based medium (lanes 2–5). Membrane fractions from cells in the stationary (lane 2) and exponential (lane 3) phase of growth. Cytoplasm fractions of cells in stationary (lane 4) and exponential (lane 5) phase.

by the host cells. The cellular localization of the recombinant protein was better clarified on separation of cell fractions followed by Western blotting. Cell components can be easily fractionated by incubating 1×10^8 cells in 1 ml of buffer containing 50 m*M* Tris-HCl (pH 7.5), 1.2 *M* sorbitol, 40 m*M* 2-mercaptoethanol, 1 m*M* phenylmethylsulfonyl fluoride (PMSF), 10 m*M* pepstatin, and 100 U of glucanase (Quantazyme, Cellon, Strassen, Luxembourg) until spheroplasts are formed. Centrifugation at a low rate (2000 rpm, 15 min) separates a supernatant containing the periplasmic content. The pellet is resuspended in 50 m*M* Tris-HCl, pH 8.0, containing PMSF and pepstatin and incubated 20 min on ice. A 30-min centrifugation, as before, allows the separation of the membranes from a supernatant containing the soluble cytoplasmatic components. Fractions are then analyzed by Western blotting for the presence of lipase. By this procedure, it was possible to demonstrate that the recombinant enzyme was entrapped in the membrane fraction (Fig. 4). While in polyacrylamide gel electrophoresis (PAGE) the recombinant protein was indistinguishable from native lipase, it displayed a truly different migration under nondenaturing conditions, suggesting that it had been incorrectly folded.

CUGs were subsequently mutagenized to universal serine codons frequently employed by *S. cerevisiae.* We observed that restoration of the three serines previously mentioned as strictly conserved within lipases and esterases was not sufficient to obtain activity and secretion, neither was it sufficient to mutate another group of six residues selected among the serines that have an obvious structural significance or are conserved in the CRL family. On the basis of these observations, the hypothesis is emerging that CRL folding is subject to complex and general constraints. If this is the case, two approaches are open for the production of a recombinant protein: (1) the substitution of all nonuniversal codons (17 to 19 in CRLs), possibly also by chemical synthesis of the gene, or (2) the development of an expression system in *C. rugosa* or in organisms phylogenetically related. We are exploring the possibility of expressing CRL genes in *Candida* cells using the same genetic code as *C. rugosa.* By this means we have obtained the production and secretion of recombinant lipases. However, the uniqueness of *C. rugosa* in terms of the frequency of CUG, on the one hand, and its apparent use of signals different from those of other organisms (including other candidas), on the other hand, still requires us to optimize the expression systems to be able to produce recombinant CRL on a large scale.

Acknowledgments

This work was supported by the Progetto Finalizzato Biotecnologie e Biostrumentazioni of the Italian National Research Council and by the EC BRIDGE Program "Lipases."

[15] Influence of Various Signal Peptides on Secretion of Mammalian Acidic Lipases in Baculovirus–Insect Cell System

By LILIANE DUPUIS, STEPHANE CANAAN, MIREILLE RIVIÈRE, and CATHERINE WICKER-PLANQUART

Introduction

Mammalian acidic lipases belong to a family of lipases that share the ability to be resistant and active under acidic conditions. This family includes preduodenal lipases and lysosomal lipase and shows no sequence homology with other known lipase families.[1,2] Preduodenal lipases form a group of closely related enzymes originating either from the stomach (gastric lipase[3]) or oral tissues (lingual lipase[4–6] and pharyngeal lipase[7,8]). They share low pH optimum, absence of requirements for a specific cofactor, and resistance to pepsin. Human gastric lipase (glycerol ester hydrolase, EC 3.1.1.3, HGL) is secreted by chief cells of the fundic part of the stomach,[9] where it initiates the digestion of triacylglycerols.[10] This enzyme plays a crucial role in newborns, because pancreatic lipase is not fully developed.[11] Human gastric lipase has an apparent molecular mass of 50 kDa and is a highly glycosylated molecule (10–15% of the protein mass). Despite the close similarities of human lysosomal acid lipase (HLAL) and HGL in their amino acid sequences[12–14] (they share 59% identical amino acids), HGL lacks the cholesteryl ester hydrolase activity.

[1] S. B. Petersen, and F. Drabløs, *in* "Lipases: Their Biochemistry, Structure and Application" (P. Woolley and S. Petersen, eds.), p. 23. Cambridge University Press, Cambridge, 1994.

[2] M. Aoubala, F. Carrière, S. Ransac, R. Verger, and A. de Caro, *Regards Biochim.* **2,** 13 (1995).

[3] C. Tiruppathi and K. A. Balasubramanian, *Biochim. Biophys. Acta* **712,** 692 (1982).

[4] M. Hamosh, D. Ganot, and P. Hamosh, *J. Biol. Chem.* **254,** 12121 (1979).

[5] R. Field and R. O. Scow, *J. Biol. Chem.* **258,** 14563 (1983).

[6] I. M. Roberts, R. K. Montgomery, and M. C. Carey, *Am. J. Physiol.* **247G,** 385 (1984).

[7] S. Bernbäck, O. Hernell, and L. Blackberg, *Eur. J. Biochem.* **148,** 233 (1985).

[8] J. de Caro, F. Ferrato, R. Verger, and A. de Caro, *Biochim. Biophys. Acta* **1252,** 321 (1995).

[9] H. Moreau, R. Laugier, Y. Gargouri, F. Ferrato, and R. Verger, *Gastroenterology* **95,** 1221 (1988).

[10] F. Carrière, J. A. Barrowman, R. Verger, and R. Laugier, *Gastroenterology* **105,** 876 (1993).

[11] M. Hamosh, "Lingual and Gastric Lipases: Their Role in Fat Digestion." CRC Press, Boca Raton, Florida, 1990.

[12] D. Ameis, M. Merkel, C. Eckerskorn, and H. Greten, *Eur. J. Biochem.* **219,** 905 (1994).

[13] R. A. Anderson and G. N. Sando, *J. Biol. Chem.* **266,** 22479 (1991).

[14] M. W. Bodmer, S. Angal, G. T. Yarranton, T. J. R. Harris, A. Lyons, D. J. King, G. Piéroni, C. Rivière, R. Verger, and P. A. Lowe, *Biochim. Biophys. Acta* **909,** 237 (1987).

0076-6879/97 $25.00

Human lysosomal acid lipase (EC 3.1.1.13) hydrolyzes not only triglycerides that are delivered to the lysosomes by low-density lipoprotein receptor-mediated endocytosis, but also cholesteryl esters.[15] Cholesterol liberated by this reaction has an important regulatory role in cellular sterol metabolism. Defective HLAL activity has been associated with two rare autosomal recessive traits: Wolman's disease (WD) and cholesteryl ester storage disease (CESD). In WD,[16] lack of HLAL activity results in a pronounced accumulation of cholesteryl esters and triacylglycerols in lysosomes of most tissues of the body. The patients usually succumb to hepatic and adrenal failure within the first year of life. Cholesteryl ester storage disease, the other clinically recognized phenotypic expression of HLAL deficiency, follows a more benign clinical course,[17] and a residual HLAL activity has been detected. To investigate the regulation of this acid lipase, its role in lipid hydrolysis, and its relationship to lipid storage disorders, the purification of HLAL has been attempted from several sources including human liver,[12,18] placenta,[19] aorta,[20] leukocytes,[21] and fibroblasts.[22] The low amounts of HLAL found within mammalian cells has, however, limited its structure–function characterization and explains the reported discrepancies in the molecular mass of HLAL: either 29 kDa[18] or 41 and 56 kDa[12] for human liver, 30.6 and 102.5 kDa for human placenta,[19] and 47–49 kDa for the secreted acid lipase from human fibroblasts.[22] Molecular cloning of HLAL has been reported[12,13] as well as the entire structure of the HLAL gene,[23] which consists of 10 exons and the DNA sequence of the putative promoter region.

Our aim was to produce sufficient amounts of both enzymes, in a secreted and active form, to allow biochemical characterization of these two lipases. The baculovirus expression system has been chosen, because of its efficiency in producing heterologous proteins on a large scale.[24] This system takes advantage of the fact that the polyhedrin gene, which governs the expression at a high level of a late gene product in infected cells, is nonessen-

[15] J. L. Goldstein, and M. S. Brown, *Annu. Rev. Biochem.* **46,** 897 (1977).

[16] A. D. Patrick and B. D. Lake, *Nature* (*London*) **222,** 1067 (1969).

[17] J. A. Burke and W. K. Schubert, *Science* **176,** 309 (1972).

[18] T. G. Warner, L. M. Dambach, J. J. Shin, and J. S. O'Brien, *J. Biol. Chem.* **256,** 2952 (1981).

[19] B. K. Burton and H. W. Mueller, *Biochim. Biophys. Acta* **618,** 449 (1980).

[20] T. Sakurada, H. Orimo, H. Okabe, A. Noma, and M. Murakami, *Biochim. Biophys. Acta* **424,** 204 (1976).

[21] R. Rindler-Ludwig, W. Patsch, S. Saimler, and H. Braunsteiner, *Biochim. Biophys. Acta* **488,** 294 (1977).

[22] G. N. Sando and L. M. Rosenbaum, *J. Biol. Chem.* **260,** 15186 (1985).

[23] C. Aslanidis, H. Klima, K. J. Lackner, and G. Schmitz, *Genomics* **20,** 329 (1994).

[24] P. Sridhar, A. K. Awasthi, C. A. Azim, S. Burma, S. Habib, A. Jain, B. Mukerjee, A. Ranjan, and S. E. Hasnain, *J. Biosci.* **19,** 603 (1994).

tial for infection or replication of the virus and is controlled by an extremely efficient promoter (polyhedrin can account for up to 50% of the cell protein mass in the terminal stages of infection[25]). Recombinant baculovirus is generated by replacing the polyhedrin gene with a foreign gene through homologous recombination. The reported yields of proteins produced in insect cells, however, vary widely, from 1 to 1220 mg/liter.[26–28] We report here on the influence of the signal peptide on the production of a recombinant protein (acid lipase) using a baculovirus–insect cell expression system.

Production of HGL and HLAL Genes

Human gastric lipase and human lysosomal acid lipase (HLAL) cDNAs are synthesized by reverse transcriptase-polymerase chain reaction (RT-PCR) amplification using human stomach and liver mRNAs (Clontech Laboratories, Palo Alto, CA) and avian myeloblastosis virus reverse transcriptase for 60 min at 37°.[29] The HGL and HLAL coding regions (1.2 kilobase pairs [kbp]) are amplified by polymerase chain reaction using primers synthesized to contain restriction sites (primers 1, 2, 7, and 8; see Table I). *Taq* polymerase (Perkin-Elmer, Norwalk, CT) is used with 30 cycles of denaturation (30 sec, 94°), annealing (1 min, 52°), and extension (2 min, 72°). The coding regions of HGL and HLAL generated by polymerase chain reaction are digested with *Pst*I and *Bam*HI, and with *Bgl*II and *Nco*I, respectively, and gel purified on GeneClean (Bio 101, La Jolla, CA). Ligation of cleaved pUC18 plasmid and the PCR fragment corresponding to the coding region of HGL, and ligation of cleaved pMOS *Blue* T plasmid (Amersham, Arlington Heights, IL) and the fragment corresponding to the coding region of HLAL, are performed and the resulting plasmids used to transform *Escherichia coli* cells (strain JM 83). Recombinant clones are screened and the coding regions are entirely sequenced by the Sanger[30] dideoxy-mediated chain-termination method.

Construction of Recombinant Baculoviruses and Expression in Insect Cell Cultures

The 1.2-kbp synthetic gene encoding HGL is cloned by ligation into the *Pst*I and *Bam*HI sites of transfer vector PVL1392 downstream from

[25] V. A. Luckow and M. D. Summers, *Biotechnology* **6,** 47 (1988).
[26] B. Maiorella, D. Inlow, A. Shauger, and D. Harano, *Biotechnology* **6,** 1406 (1988).
[27] A. W. Caron, L. Archambault, and B. Massie, *Biotechnol. Bioeng.* **36,** 1133 (1990).
[28] S. Sheriff, H. Du, and G. A. Grabowski, *J. Biol. Chem.* **270,** 27766 (1995).
[29] T. Maniatis, E. F. Fritch, and J. Sambrook, "Molecular Cloning: A Laboratory Manual." Cold Spring Harbor Press, Cold Spring Harbor, New York, 1982.
[30] F. Sanger, S. Nicklen, and A. R. Coulson, *Proc. Natl. Acad. Sci. U.S.A.* **74,** 5463 (1977).

TABLE I
SYNTHETIC OLIGONUCLEOTIDES USED AS PRIMERS TO SYNTHESIZE HGL AND HLAL CONSTRUCTIONS WITH VARIOUS SIGNAL PEPTIDES[a]

Oligonucleotide	Sequence (5′ → 3′)	Restriction enzyme
1. HGL forward	GAGGAAACTGCAGGTCC	*Pst*I
2. HGL reverse	AATTCTTGGATCCAGAACTACT	*Bam*HI
3. Melittin signal peptide	TTGTACTGCAGATGAAATTCTTAGTCAACGTT GCCCTTGTTTTATGGTCGTGTACATTTCTT ACATCTATGCGTTGTTTGGAAAATTACATCC	*Pst*I
4. HPL forward signal peptide	GAACTCTGCAGATGCTGCCACTTTGG	*Pst*I
5. HPL reverse signal peptide	GTTCGAGCTTAACTTCTTTTCCTGC	*Alu*I
6. HGL forward	TTGGACAGCTGCATCCTGGAAGCCC	*Pvu*II
7. HLAL forward	AGGACAGATCTCCAGAATGAAAATGCG	*Bgl*II
8. HLAL reverse	CTGATCCATGGTACACAGCTCAAG	*Nco*I
9. HPL forward signal peptide	GAACTAGATCTATGCTGCCACTTTGGACTCTTT CACTGCTGCTGGGAGCAGTAGCAGGAAAA GAAGTTAAACTGACAGCTGTGGAT	*Bgl*II
10. HLAL reverse	GAGGAAATATCAGTGAAAGCGGCCGCTGAGC	*Not*I

[a] The restriction enzyme sites are underlined.

the strong *polh* promoter. The 1.2-kbp synthetic gene encoding HLAL is cloned by ligation into the *Bgl*II and *Nco*I sites of transfer vector pBlue Bac III. The recombinant transfer vectors (HGL-SP/HGL and HLAL-SP/HLAL; SP, signal peptide) are cotransfected with a baculovirus (BaculoGold virus; Pharmingen, San Diego, CA).

Sf21 cells are grown at 27° in TNM-FH insect medium (Sigma, St. Louis, MO) supplemented with 5% (v/v) fetal bovine serum (Sigma). The Sf21 cells are cotransfected with BaculoGold DNA and recombinant transfer vector as indicated by the manufacturer. BaculoGold DNA is a modified virus, deleted of an essential part of its genome in the neighborhood of the polyhedrin locus. Cotransfection of the lethally deleted virus (BaculoGold) and recombinant transfer vector rescues the lethal deletion of this virus DNA and results in a recombination efficiency that is virtually 100%. The recombinant virus stock is amplified by several rounds of culture and a high-titer virus stock solution is harvested. To express the protein on a larger scale, BTI-TN-5B1-4 insect cells are grown in a serum-free medium (Ex-cell 405; JRH Sciences, Lenexa, KS) prior to the infection at a multiplicity of infection (MOI) of 3.

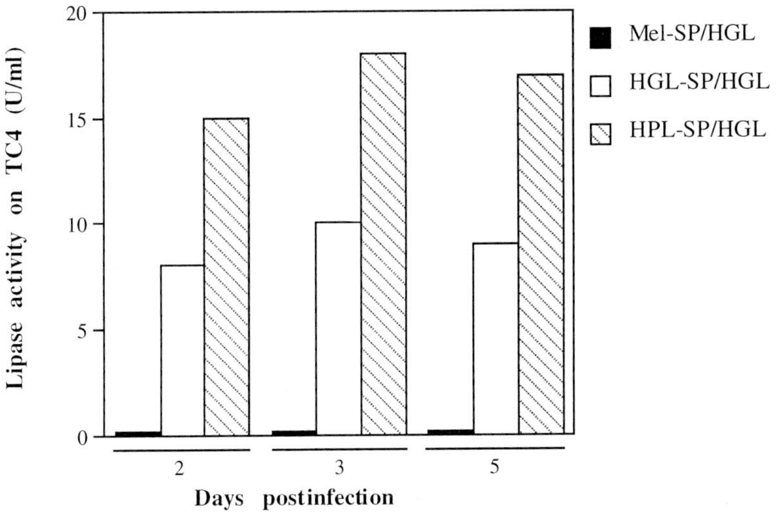

FIG. 1. Time course of lipase activity in the supernatant of insect cells infected with HGL-SP/HGL, HPL (human pancreatic lipase)-SP/HGL, and Mel (melittin)-SP/HGL viruses. BTI-TN-5B1-4 insect cells were grown in 75-ml flasks as a monolayer at 27° in a serum-free (Ex-cell 405) medium. About 10^6 cells in 20 ml of medium were infected with a recombinant baculovirus at an MOI of 3. The cells were washed with the same culture medium 4 hr after infection. Culture medium was collected on various days (days 2, 3, and 5) following infection and assayed for lipase activity with tributyrin as substrate.

From the second day onward after infection with HGL-SP/HGL recombinant baculovirus (Fig. 1), lipase activity is detected in the supernatant of insect cells (7 U/ml), using the standard lipase activity test[31] [mechanically stirred emulsion of 0.5 ml of tributyrin (Fluka Buchs, Switzerland), 14.5 ml of 0.9% (w/v) NaCl, 0.01% (w/v) bovine serum albumin (BSA), 2 m*M* sodium taurodeoxycholate (NaTDC) at 37° with pH maintained (TTT 80 radiometer, Radiometer, Copenhagen, Denmark) at pH 5.7]. The activity reaches a plateau value by day 3 postinfection (around 10 U/ml of cell culture) (Fig. 1). Immunoblotting of insect recombinant HGL is performed using polyclonal anti-HGL antibody as previously described.[32] A single immunoreactive band of approximately 45 kDa is detected (Fig. 2, lane 5).

[31] Y. Gargouri, G. Piéroni, C. Rivière, J.-F. Saunière, P. A. Lowe, L. Sarda, and R. Verger, *Gastroenterology* **91,** 919 (1986).

[32] M. Aoubala, J. Bonicel, C. Bénicourt, R. Verger, and A. de Caro, *Biochim. Biophys. Acta* **1213,** 319 (1994).

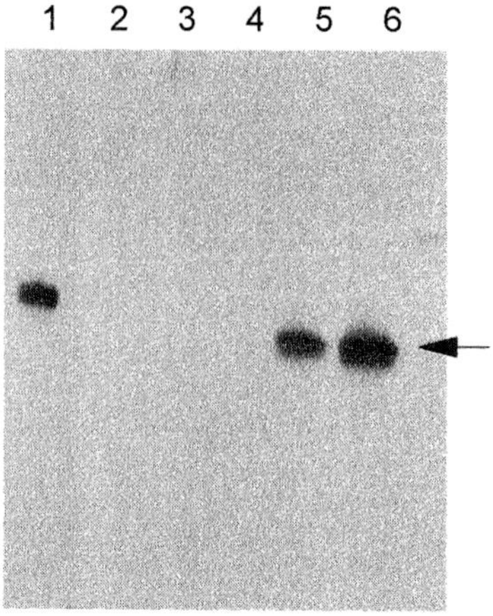

FIG. 2. Western blot analysis of secreted HGL in the culture medium. The cell supernatants were collected 3 days after infection, and subjected to 12% (w/v) polyacrylamide gel electrophoresis. Proteins from SDS gels were electrophoretically transferred to a nitrocellulose membrane, rinsed three times with 10 m*M* phosphate (pH 7.5) containing 150 m*M* NaCl (PBS buffer). The remaining free sites on the nitrocellulose membrane were blocked by a solution of PBS buffer containing 3% (w/v) powdered milk (commercial grade) for 1 hr. Thereafter, 1:5000 (v/v) diluted rabbit antiserum directed against HGL was incubated with the membrane for 1 hr at room temperature. The immunoreactive proteins were detected using a goat anti-rabbit IgG conjugated with horseradish peroxidase by covering the membrane with enhanced chemiluminescence (ECL) reagents (Amersham) and exposing it briefly to blue light-sensitive autoradiography film (Hyperfilm ECL; Amersham). Lane 1, natural HGL; lane 2, medium supernatant: insect cells (control); lane 3, medium supernatant: insect cells infected with wild baculovirus; lane 4, medium supernatant: insect cells infected with Mel-SP/HGL baculovirus; lane 5, medium supernatant: insect cells infected with HGL-SP/HGL baculovirus; lane 6, medium supernatant: insect cells infected with HPL-SP/HGL baculovirus. The arrow indicates the position of the 43-kDa marker protein.

The molecular mass of the recombinant protein is lower than that of the native HGL (M_r 50 kDa; see Fig. 2, lane 1), which probably reflects incomplete glycosylation by the insect cells. Despite N-linked glycosylation in the endoplasmic reticulum of both mammalian and insect cells, the glycoproteins produced in the latter system have relatively simple, unbranched sugar side chains with a high mannose content.[33–35] Insect cells are thus unable to produce the complex processed carbohydrate patterns of mammalian proteins. A lower apparent molecular mass is also reported for the avian leukemia virus envelope glycoprotein, when expressed in insect cells

[33] K. Kuroda, H. Geyer, R. Geyer, W. Doerfler, and H. D. Klenk, *Virology* **174,** 418 (1990).
[34] T. D. Butters and R. C. Hughes, *Biochim. Biophys. Acta* **640,** 655 (1981).
[35] T. D. Butters, R. C. Hughes, and R. C. Visher, *Biochim. Biophys. Acta* **640,** 672 (1981).

(45–65 kDa for the recombinant protein instead of 85 kDa for the native protein).[36]

No standardized procedure has been established yet for the assay of acid lipase/cholesteryl ester hydrolase. We have attempted to measure HLAL activity by using the gastric lipase test conditions on tributyroylglycerol. However, when HLAL is expressed in the insect cell system with its own homologous signal peptide, no activity can be measured in the supernatant of the cell culture, whereas weak activity (0.9 U in a pellet of 10^8 cells) is detectable on tributyroylglycerol as substrate. To determine more accurately this weak lipase activity, we have tested both the pellet and the supernatant of the culture medium with the fluorogenic substrate 4-methylumbelliferyl heptanoate (4-MUH). Three days after infection, the medium and the cells are separated by centrifugation (500 *g*, 10 min, 4°). The cell pellet is washed twice in 50 m*M* Tris-HCl buffer, pH 7, and then treated in 50 m*M* Tris-HCl, pH 7, containing 10 m*M* 2-mercaptoethanol, 0.25% (v/v) Triton X-100, and 1 m*M* EDTA with ultrasonic irradiation at 4° using a microtip (20 sec, threefold, 4°). The sonicates are centrifuged (800 *g*, 15 min, 4°), and the supernatant (20 to 100 μl) is used for hydrolytic activity determination. The enzyme assay (200-μl final volume) is performed in 0.2 *M* succinate, pH 4.5, by using 2 m*M* 4-MUH dissolved in ethanol and the fluorescence intensity is quantified on Fluoroskan II apparatus (Labsystems, Finland).[37] We have detected a slight activity in the supernatant of the insect cells infected by the HLAL-SP/HLAL recombinant virus (Table II), which differed significantly from the value measured in the supernatant of noninfected cells (data not shown).

We have succeeded in expressing two active acidic lipases: HGL and HLAL in the baculovirus insect cell system. However, it is worth noticing that HGL is expressed mainly as a secretory protein, whereas HLAL is produced predominantly in association with the insect cells (see Table II). Small amounts of immunoreactive material also accumulate inside the HGL-SP/HGL-infected cells (data not shown), as a species of ~40-kDa, which could correspond to an unglycosylated form of the protein. This could reflect the need for carbohydrate residues in the correct targeting to the secretory pathway of the cell. To increase the secretion levels of both enzymes, we have replaced the natural signal peptides of both proteins with signal peptides of other highly expressed secreted proteins in the baculovirus–insect cell system. The fusion of melittin signal peptide to HLAL and expression in the insect cell system have been reported,[28] re-

[36] M. H. M. Noteborn, G. F. de Boer, A. Kant, G. Koch, J. L. Bos, A. Zantema, and A. J. van der Eb, *J. Gen. Virol.* **71,** 2641 (1990).

[37] A. E. Negre, R. S. Salvayre, A. Dagan, and S. Gatt, *Clin. Chim. Acta* **149,** 81 (1985).

TABLE II

LIPASE ACTIVITY IN SUPERNATANT OR PELLET OF INSECT CELLS INFECTED WITH HLAL-SP/HLAL, HPL-SP/HLAL, HGL-SP/HGL, AND HPL-SP/HGL VIRUSES[a]

Product released from 4-MUH (nmol)	Time (min)		
	1	10	25
Supernatants			
HLAL-SP/HLAL	2.2	4.5	14.1
HPL-SP/HLAL	29.8	45.3	95.7
Pellets			
HLAL-SP/HLAL	0.8	0.9	8.3
HPL-SP/HLAL	0.6	1.2	4.7
HGL-SP/HGL	2.4	7	18.6
HPL-SP/HGL	2.9	5.3	20.8

[a] BTI-TN-5B1-4 insect cells were grown in a six-well microtiter plate (10^5 cells/well) as a monolayer at 27° in Ex-cell 405 medium. They were infected with a recombinant baculovirus at an MOI of 5. The cells were washed with the same culture medium (3 ml) at 4 hr after infection. Culture medium and cell pellet were collected 60 hr following infection and assayed for lipase activity with 4-MUH as substrate at various times after the addition of the substrate (1, 10, and 25 min). The amounts of product released from 4-MUH in the supernatant of cells infected with HGL-SP/HGL and HPL-SP/HGL are not indicated, as they reached a plateau after 1 min of reaction with the substrate.

sulting in the production of an intracellular HLAL recombinant protein. Only 24% of the recombinant protein is secreted into the medium. However, the efficiency of secretion of the papain precursor is five times increased by using the honeybee melittin signal peptide[38] fused to the papain cDNA.[39] We have therefore fused the melittin signal peptide to HGL. The natural signal peptide of either HGL or HLAL has also been substituted by the

[38] C. Mollay, U. Vilas, and G. Kreil, *Proc. Natl. Acad. Sci. U.S.A.* **79,** 2260 (1982).

[39] D. C. Tessier, D. Y. Thomas, H. E. Khouri, F. Laliberté, and T. Vernet, *Gene* **98,** 177 (1991).

human pancreatic lipase signal peptide[40] (HPL-SP/HGL; HPL-SP/HLAL), because secretion of HPL in insect cells has been reported to reach a level of about 40 mg/liter of cell culture.[41]

Construction of HGL and HLAL Variants with Heterologous Signal Peptides

The oligonucleotides used for mutagenesis are listed in Table I. One construction (SP-Mel/HGL) encoding the melittin signal peptide fused to mature HGL is synthesized by PCR using oligonucleotides 2 and 3 (Table I). The other construction (HPL-SP/HGL), encoding the human pancreatic lipase signal peptide followed by the first three residues of HPL (Lys–Glu–Val), is fused to HGL starting from Lys-4 through Lys-379 (see Table III). For such construction, we use oligonucleotides 4 and 5 for the synthesis of the HPL moiety, and oligonucleotides 2 and 6 for the synthesis of the HGL moiety. The preceding construction has been evaluated according to the von Heijne method[42] for the prediction of human pancreatic signal sequence cleavage sites and we have found that the cleavage probability is more likely when the first three amino acids of human pancreatic lipase (KEV) are used instead of those of HGL (LFG). The HPL moiety corresponding to HPL-SP followed by the three first amino acid of human pancreatic lipase is purified and digested with *Pst*I and *Alu*I and ligated with the *Pvu*II- and *Bam*HI-digested HGL fragment, followed by ligation with the *Pst*I–*Bam*HI-digested pUC18 vector.

The natural putative signal peptide length of HLAL has been reported to be either 21 amino acids[13] or 27 amino acids.[12] When aligning the HLAL sequence with that of HGL, the putative signal peptide cleavage point occurs between amino acids 21 and 22.[13] However, the assignment of the first amino acid of mature LAL protein to residue 28 is supported by purification and N-terminal sequencing of HLAL in human liver.[12] Proteolytic degradation during purification can account for this observation. We have chosen to align both sequences and to begin the sequence of mature HLAL at residue 22. Moreover, the site of cleavage between amino acids 21 and 22 is more probable according to the von Heijne predictive method than a cleavage between amino acids 27 and 28. The HPL-SP/HLAL construction (see Table III) encoding the human pancreatic lipase signal peptide followed by the first three residues of HPL fused to HLAL from

[40] M. E. Lowe, J. L. Rosemblum, and A. W. Strauss, *J. Biol. Chem.* **264,** 20042 (1989).

[41] K. Thirstrup, F. Carrière, S. Hjorth, P. B. Rasmussen, H. Wöldike, P. F. Nielsen, and L. Thim, *FEBS Lett.* **327,** 79 (1993).

[42] G. von Heijne, *Nucleic Acids Res.* **14,** 4683 (1986).

TABLE III

NUCLEOTIDE SEQUENCES AND DEDUCED AMINO ACID SEQUENCES OF HGL AND HLAL CONSTRUCTIONS WITH VARIOUS SIGNAL PEPTIDES[a] FROM HGL,[b] HLAL,[c] MELITTIN,[d] AND HPL[e]

HGL-SP/HGL

```
                         HGL-SP                                HGL
-19                                                     -1  +1
 M  W  L  L  L  T  M  A  S  L  I  S  V  L  G  T  T  H  G  L  F  G  K  L  H
ATGTGGCTGCTTTTAACAATGGCAAGTTTGATATCTGTACTGGGACTACACATGG TT TGTTTGGAAAATTACAT
```

Mel-SP/HGL

```
                      Melittin-SP                                    HGL
-21                                                           -1  +1
 M  K  F  L  V  N  V  A  L  V  F  M  V  V  Y  I  S  Y  I  Y  A  L  F  G  K  L  H
ATGAAATTCTTAGTCAACGTTGCCCTTGTTTTTATGGTCGTGTACATTTCTTACATCTATGC GT TGTTTGGAAAATTACAT
```

HPL-SP/HGL

```
                  HPL-SP                       HPL       HGL
-16                                      -1  +1    +3  +4
 M  L  P  L  W  T  L  S  L  L  L  G  A  V  A  G  K  E  V  K  L  H
ATGCTGCCACTTTGGACTCTTTCACTGCTGCTGGGAGCAGTAGCAGG AA AAGAAGT TAAGTTACAT
```

HLAL-SP/HLAL

```
                      HLAL-SP                                        HLAL
-21                                                           -1 +1
 M  K  M  R  F  L  G  L  V  V  C  L  V  L  W  T  L  H  S  E  G  S  G  G  K  L  T
ATGAAAATGCGGTTCTTGGGGTTGGTGGTCTGTTTGGTTCTCTGGACCCTGCATTCTGAGGG G TCTGGAGGAAAACTGACA
```

HPL-SP/HLAL

```
                  HPL-SP                       HPL       HLAL
-16                                      -1  +1    +3  +4
 M  L  P  L  W  T  L  S  L  L  L  G  A  V  A  G  K  E  V  K  L  T
ATGCTGCCACTTTGGACTCTTTCACTGCTGCTGGGAGCAGTAGCAGG AA AAGAAGT TAAACTGACA
```

[a] SP, Signal peptide. We have labeled the residues that are located at the hinge between either signal peptide and mature protein or between two mature proteins.

[b] See Ref. 14.

[c] See Ref.13.

[d] See Refs. 37 and 38.

[e] See Ref. 39.

Lys-4 through Gln-378 is realized using oligonucleotides 9 and 10 (Table I). The cleavage probability is again more likely between the human pancreatic signal peptide and the first three amino acids of human pancreatic lipase (KEV) than those of human lysosomal acid lipase (SGG). Two restriction sites have been created: *Bgl*II at the 5′ end and *Not*I at the 3′ end. All constructions have been subjected to sequence analysis. The 1.2-kbp synthetic genes encoding HPL-SP/HGL and Mel-SP/HGL are cloned by ligation into the *Pst*I and *Bam*HI sites of transfer vector PVL1392. The HPL-SP/HLAL construction is inserted directly into the *Bgl*II and *Not*I restriction sites of the transfer vector PVL1392. The recombinant transfer vectors are cotransfected with the BaculoGold virus from Pharmingen.

Expression of HPL-SP/HGL, Mel-SP/HGL, and HPL-SP/HLAL Viruses in Insect Cells

We failed to detect any lipase activity on tributyrin when the melittin signal peptide was fused to the HGL gene (Fig. 1). The absence of activity in the supernatant of infected cells correlated with the lack of any immunoreactive material in the cell supernatant (Fig. 2). However, two immunoreactive bands were revealed using polyclonal antibodies against native HGL in the pellet of infected insect cells with the melittin variant (data not shown). The major band had a molecular mass of approximately 30 kDa and probably resulted from an incorrect signal peptide cleavage of HGL. The other band had the expected molecular mass of 50 kDa. A similar negative result was obtained by Tessier *et al.*[39] by substituting papain natural signal peptide by *Drosophila* α-amylase signal peptide and expressing the modified papain sequence in the baculovirus–insect cell system. It has always been reported that the fusion of foreign proteins with a consensus signal peptide does not always result in increased amounts of secreted protein from insect cells.[43] On the other hand, when using the human pancreatic lipase signal peptide to direct HGL secretion, lipase activity in the cell supernatant was twice that measured when infecting the cells with HGL and its homologous natural signal peptide. This increase in lipase activity was probably due to an increased secretion of the enzyme, because the amounts measured by activity determination (pH-stat) correlate well with the intensity of the various bands in Western blot analysis (Fig. 2). The fusion of HLAL variant with the human pancreatic lipase signal peptide resulted in a 7- to 10-fold increase in the esterase activity in the cell supernatant (Table II). After several rounds of virus amplification, the maximum

[43] J. J. Devlin, P. E. Devlin, R. Clark, E. C. O'Rourke, C. Levenson, and D. F. Mark, *BioTechnology* **7,** 286 (1989).

activities obtained with infected cell supernatant were 0.08 U/ml (tributyrin as substrate) and 0.18 U/ml (trioctanoin as substrate) at pH 5.5. A weak activity was detected in the pellets of both HLAL- and HGL-infected cells, which probably represents the proteins about to be secreted (Table II). Despite the strong homology between HLAL and HGL, no cross-reactivity with polyclonal antibodies (against native HGL) was observed with either the pellet or cell supernatant.

It is possible to enhance the secretion of both HGL and HLAL by fusing the human pancreatic lipase signal peptide to the mature proteins. A twofold increase in HGL secretion was found in the supernatant of insect cell culture, reaching 18 U/ml in the tributyrin assay system. The effect of this pancreatic lipase signal peptide on HLAL secretion was even more pronounced because the rise in activity was 7- to 10-fold. However, the absolute amounts of expressed HLAL still remained low, because a maximal activity of 0.18 U/ml in the supernatant of cell culture was measured with trioctanoin as substrate. This probably explains why HLAL was always produced in low amounts in heterologous systems such as Cos-1 cells[13] and insect cells.[28] However, the amounts of recombinant acid lipases secreted by the baculovirus–insect cell system can thus be increased by a change in the signal peptide.

[16] Large-Scale Purification and Kinetic Properties of Recombinant Hormone-Sensitive Lipase from Baculovirus-Insect Cell Systems

By Cecilia Holm, Juan Antonio Contreras, Robert Verger, and Michael C. Schotz

Hormone-sensitive lipase (HSL) catalyzes the rate-limiting step in the breakdown of stored adipocyte lipids, i.e., the hydrolysis of the primary ester bond of triacylglycerol.[1–3] It also catalyzes the hydrolysis of diacylglycerols and monoacylglycerols, although a second enzyme, monoacylglycerol lipase, is required for complete hydrolysis of monoacylglycerols.[4]

[1] M. Vaughan, J. E. Berger, and D. Steinberg, *J. Biol. Chem.* **239,** 401 (1964).

[2] P. Belfrage, G. Fredrikson, P. Strålfors, and H. Tornqvist, *in* "Lipases" (B. Borgström, and H. L. Brockman, eds.), p. 365. Elsevier, Amsterdam, 1984.

[3] P. Strålfors, H. Olsson, and P. Belfrage, *in* "The Enzymes" (P. D. Boyer and E. G. Krebs, eds.), Vol. 18, p. 147. Academic Press, New York, 1987.

[4] G. Fredrikson, H. Tornqvist, and P. Belfrage, *Biochim. Biophys. Acta* **876,** 288 (1986).

0076-6879/97 $25.00

The activity of HSL is under hormonal and neural control through a mechanism involving phosphorylation by cAMP-dependent protein kinase.[5,6] The regulatory aspects of HSL are described in more detail in [35] in this volume.[6a]

The purification of HSL from adipose tissue of several species has been described previously.[7-10] In addition, some of the biochemical properties of this lipase have been reported on the basis of the limited amounts of pure protein available. cDNAs for HSL have been isolated from rat, human, and mouse adipose tissue and the amino acid sequences encoded by these cDNAs show between 83 and 94% identity.[11-13] The human HSL gene contains 9 exons, encoding the adipose tissue HSL protein of 775 amino acids. The amino acid sequence shares homology with lipase 2 of the antarctic bacterium *Moraxella* TA144 and several other bacterial proteins.[12,14-16] The homology region spans exons 5–9, and includes exon 6, which encodes the active site serine.[17] The sequence alignments to the *Moraxella* lipase, together with limited proteolysis experiments, suggest that HSL has at least two domains: an N-terminal domain, which is unique to HSL compared to other lipases and important for interaction with lipid substrates, and a C-terminal domain, harboring the catalytic triad and a regulatory module.[18] A three-dimensional model of the catalytic domain of HSL was generated on the basis of secondary structure predictions and subsequent sequence

[5] P. Strålfors and P. Belfrage, *J. Biol. Chem.* **258,** 15146 (1983).

[6] P. Strålfors, P. Björgell, and P. Belfrage, *Proc. Natl. Acad. Sci. U.S.A.* **81,** 3317 (1984).

[6a] C. Holm, D. Langin, V. Hanganielle, P. Belfrage, and E. Degerman, *Methods Enzymol.* **284,** [35], 1997 (this volume).

[7] G. Fredrikson, P. Strålfors, N. Ö. Nilsson, and P. Belfrage, *J. Biol. Chem.* **256,** 6311 (1981).

[8] S. Nilsson and P. Belfrage, *Anal. Biochem.* **158,** 399 (1986).

[9] S. Nilsson, C. Holm, and P. Belfrage, *Biomed. Chromatogr.* **3,** 82 (1989).

[10] G.-K. Lee, S. Yeaman, G. Fredrikson, P. Strålfors, and P. Belfrage, *Comp. Biochem. Physiol.* **80B,** 609 (1985).

[11] C. Holm, T. G. Kirchgessner, K. L. Svenson, G. Fredrikson, S. Nilsson, C. G. Miller, J. E. Shively, C. Heinzmann, R. S. Sparkes, T. Mohandas, A. J. Lusis, P. Belfrage, and M. C. Schotz, *Science* **241,** 1503 (1988).

[12] D. Langin, H. Laurell, L. Stenson Holst, P. Belfrage, and C. Holm, *Proc. Natl. Acad. Sci. U.S.A.* **90,** 4897 (1993).

[13] Z. Li, M. Sumida, A. Birchbauer, M. C. Schotz, and K. Reue, *Genomics* **24,** 259 (1994).

[14] D. Langin and C. Holm, *Trends Biochem. Sci.* **18,** 466 (1993).

[15] H. Hemilä, T. T. Koivula, and I. Palva, *Biochim. Biophys. Acta* **1210,** 249 (1994).

[16] G. Feller, M. Thiry, and C. Gerday, *DNA Cell Biol.* **10,** 381 (1991).

[17] C. Holm, R. C. Davis, T. Østerlund, M. C. Schotz, and G. Fredrikson, *FEBS Lett.* **344,** 234 (1994).

[18] T. Østerlund, B. Danielsson, E. Degerman, J. A. Contreras, G. Edgren, R. C. Davis, M. C. Schotz, and C. Holm, *Biochem. J.* **319,** 411 (1996).

alignments to members of the cholinesterase family.[19] The model implies that exons 5, 6, and 9 encode the core of the α/β-hydrolase fold, found in all known esterases and lipases, and constituting the catalytic domain of HSL. Furthermore, it has been proposed that parts of exons 7 and 8 encode a regulatory module, containing the phosphorylation sites of HSL, probably inserted into this lipase during the course of evolution.

The available HSL cDNAs have been used to establish transient expression systems in COS cells for the purpose of mapping functional domains of the HSL protein using site-directed mutagenesis.[17,20] The cDNAs have also been used to establish large-scale expression systems for both rat and human HSL utilizing baculovirus–insect cell technology. This chapter describes the production and purification of recombinant HSL on a large scale using a baculovirus–insect cell expression system, and also provides an update on the available activity assays for HSL, including the use of monomolecular films.

Production and Purification of Recombinant Rat Enzyme

Generation of Recombinant Baculovirus

A full-length HSL cDNA, lacking all but 14 of the 5′ untranslated nucleotides,[11,21] is inserted into the plasmid pVL1393 downstream of the polyhedrin promoter. This plasmid is cotransfected with wild-type baculovirus DNA into *Spodoptera frugiperda* (Sf9) cells by calcium phosphate coprecipitation.[22] Recombinant baculovirus-expressing HSL is then isolated by a combination of plaque morphology screening and dot-blot hybridization. The identity of recombinant virus is confirmed by Northern and Western blot analyses, sodium dodecyl sulfate–polyacrylamide gel electrophoresis (SDS–PAGE), and HSL activity assays. Infected Sf9 cells expressed an 84-kDa protein, consistent with the size of rat adipose tissue HSL.[7,8] Hormone-sensitive lipase is the major protein expressed and accounts for approximately 5% of total cellular protein, compared to 0.02% for rat adipocytes. Activity measurements both of homogenates of infected

[19] J. A. Contreras, M. Karlsson, T. Østerlund, H. Laurell, A. Svensson, and C. Holm, *J. Biol. Chem.* **271,** 31426 (1996).

[20] C. Holm, R. C. Davis, G. Fredrikson, P. Belfrage, and M. C. Schotz, *FEBS Lett.* **285,** 139 (1991).

[21] C. Holm, T. G. Kirchgessner, K. L. Svenson, A. J. Lusis, P. Belfrage, and M. C. Schotz, *Nucleic Acids Res.* **16,** 9879 (1988).

[22] M. D. Summers and G. E. Smith, *in* "A Manual of Methods for Baculovirus Vectors and Insect Cell Culture Procedures," Bull. 1555. Texas Agricultural Experiment Station, College Station, Texas, 1987.

cells and culture medium show that HSL is expressed only intracellularly, in an enzymatically active form.

Large-Scale Suspension Cultures

Large amounts of recombinant HSL are produced in suspension cultures in autoclavable polycarbonate Erlenmeyer flasks (Corning, Corning, NY). The medium used for culturing the Sf9 cells is Ex-Cell 401 with glutamine (JRH Biosciences, Lenexa, KS) supplemented with 10% fetal bovine serum (GIBCO, Grand Island, NY). Cells, at a density of 2×10^6/ml (100 ml/500-ml flask), are infected with recombinant virus stock and incubated at 27° with orbital shaking (110–120 rpm). The optimal time of harvest determined by the highest expression of HSL activity, is 60 hr postinfection. Furthermore, the level of expression reaches a plateau at a multiplicity of infection (MOI) of about 2. On the basis of these results, an MOI of 8–10 is used and cells are harvested after 60 hr. Approximately 40 mg of HSL protein is produced per liter of culture medium (2×10^9 cells), i.e., corresponding to the amount found in the fat pads of 10,000 Sprague-Dawley rats.

Preliminary results using cells cultured in the absence of fetal bovine serum indicate that the production of HSL is at least as high as in the presence of serum.[23]

Purification

The purification procedure described here produces an enzyme of more than 95% purity. Typically the purification starts with a homogenate representing 12 liters of cell culture, requires 5 days to complete, and yields 220 mg of pure enzyme. Table I[18,23a,23b] shows a summary of the procedure. Steps 1 and 2 are performed at 4°, steps 3 through 5 at 10°. Enzyme activity is monitored using monoacylmonoalkylglycerol as substrate (see below).

Step 1: Preparation of Insect Cell Homogenate. Cells are harvested by centrifugation at 1200 *g* for 10 min at 4° and then homogenized on ice in 3 vol of 0.25 *M* sucrose containing 1 m*M* EDTA, 1 m*M* dithioerythritol, leupeptin (20 μg/ml), antipain (2 μg/ml) and pepstatin (1 μg/ml), using 20 strokes with a glass pestle in a glass homogenizer.

[23] C. Holm, unpublished data (1996).

[23a] M. M. Bradford, *Anal. Biochem.* **72,** 248 (1976).

[23b] D. Wessel and U. I. Flügge, *Anal. Biochem.* **138,** 141 (1989).

TABLE I

PURIFICATION OF RAT HORMONE-SENSITIVE LIPASE FROM INSECT CELLS INFECTED WITH RECOMBINANT BACULOVIRUS[a]

Purification step	Specific activity (U/mg protein)[b]	Purification (fold)	Recovery (%)
1. Cell homogenate	6.0	1	100
2. Detergent solubilization + 10,000 *g* centrifugation	5.3	0.9	92
3. Q-Sepharose	37	6.2	80
4. Phenyl-Sepharose	217[c]	36	45
5. Q-Sepharose	217[c]	36	40

[a] Results are from one purification, which is representative of four. Reproduced with permission from Ref. 18.

[b] Enzymes activity determined using the monoether diacylglycerol analog and protein determined according to Bradford,[23a] unless otherwise indicated.

[c] Protein determined using ultraviolet absorption at 280 nm and the calculated extinction coefficient[23b] for rat HSL; 50, 490 M^{-1} cm^{-1}.

Step 2: Detergent Solubilization. The homogenate (~500 ml) is solubilized with 1% (v/v) $C_{13}E_{12}$[24] and 10 m*M* NaCl by sonicating with a Branson sonifier 250 (Danbury, CT) at setting 1–2 (3–4 min in 30-sec pulses; cooling with ice). Insoluble material is removed by centrifugation at 10,000 *g* for 10 min at 4°. The clear supernatant is dialyzed overnight at 4° against 20 m*M* Tris–acetate (pH 7.5), 20% (v/v) glycerol, 1 m*M* dithioerythritol, 0.2% (v/v) $C_{13}E_{12}$, leupeptin (2 μg/ml).

Step 3: First Q-Sepharose Chromatography. Anion-exchange chromatography on Q-Sepharose FF (Pharmacia, Piscataway, NJ) is carried out essentially as reported for purification of rat adipose tissue HSL utilizing high-performance liquid chromatography (HPLC) on Mono Q,[8] except that the chromatography is performed in a conventional manner using the Gradi-Frac system from Pharmacia. Q-Sepharose, packed in a 5 × 15-cm column in 1 *M* Tris-HCl, pH 7.5, is equilibrated with buffer A [50 m*M* Tris–acetate (pH 7.5), 1 m*M* dithioerythritol, 20% (v/v) glycerol, 0.2% (v/v) $C_{13}E_{12}$].

[24] $C_{13}E_{12}$ is a polydisperse preparation of alkyl polyoxyethylenes with the indicated average composition; C, alkyl carbons; E, oxyethylene units. Detergents of this type used to be, but are no longer, available from Berolkemi AB (Stenungssund, Sweden). The authors still have a limited supply of $C_{13}E_{12}$ from this company. Other polyoxyethylene ethers, although not exactly $C_{13}E_{12}$, are available from Sigma and other companies. The authors have good experience with all tested variants of polyoxyethylene ethers with regard to their behavior with the purified enzyme, but recommend types with numbers of alkyl carbons and oxyethylene units similar to $C_{13}E_{12}$ for steps 2–4 of the purification scheme, because the exchange of $C_{13}E_{12}$ in these steps has not been worked out.

The dialyzed sample from step 2 is loaded and the column washed with 10 vol of buffer A, at a flow rate of 6 ml/min. A 400-ml gradient from 0 to 0.15 *M* sodium acetate in buffer A, pH 7.0, is applied, followed by a plateau (400 ml) at 0.1 *M* sodium acetate in buffer A, pH 7.0. The HSL is eluted from the column by a second gradient (800 ml) from 0.15 to 0.3 *M* sodium acetate in buffer A, pH 7.0. Hormone-sensitive lipase elutes in a reproducible manner at 0.2 *M* sodium acetate. Protease inhibitors (leupeptin, 20 μg/ml; antipain, 2 μg/ml; and pepstatin, 1 μg/ml) are added to the collected fractions (28 ml). Fractions containing more than 15 U of enzyme activity are pooled. The peak of activity correlates well with a peak on the recording of absorbance at 280 nm, and, alternatively, the pooling is based on this absorbance.

Step 4: Phenyl-Sepharose Chromatography. The pooled material from step 3 is subjected to hydrophobic interaction chromatography on phenyl-Sepharose CL-4B (Pharmacia), fundamentally as described for purification of bovine adipose tissue HSL utilizing HPLC on phenylsilica.[9] As for step 3, the chromatography is performed in a conventional manner using the Pharmacia Gradi-Frac system. Phenyl-Sepharose, in a 3.2 × 12-cm column, is equilibrated successively with 20 vol of 5 m*M* potassium phosphate (pH 7.4), 2 *M* NaCl, 2% (v/v) $C_{13}E_{12}$, 1 m*M* dithioerythritol, 10% (v/v) glycerol, and 10 vol of buffer B [0.1 *M* sodium phosphate (pH 7.4), 0.8 *M* NaCl, 1 m*M* dithioerythritol, 0.005% (v/v) $C_{13}E_{12}$, 10% (v/v) glycerol]. Salt (0.8 *M* NaCl, final concentration) is added to the pooled enzyme from step 3 and it is applied to the column at a flow rate of 6 ml/min. After washing with 10 vol of buffer B, HSL is eluted with a four-component gradient (200 ml) in buffer B: sodium phosphate from 1 *M* to 5 m*M*, NaCl for 1 to 0 *M*, $C_{13}E_{12}$ from 0.005 to 0.2% (v/v), and glycerol from 10 to 50% (v/v), at a flow rate of 3 ml/min. The HSL enzyme usually elutes at the end of the gradient. The point of elution is less reproducible for this column than for the Q-Sepharose column (step 3). For this reason the column is always washed with at least 2 column volumes of 5 m*M* sodium phosphate (pH 7.4), 1 m*M* dithioerythritol, 0.2% (v/v) $C_{13}E_{12}$, and 50% (v/v) glycerol, to ensure optimal recovery. Fractions (15 ml) containing more than 7 U of enzyme activity are pooled. Usually the first few fractions are omitted, because these frequently contain small amounts of contaminating proteins, as judged by SDS–PAGE.

Step 5: Second Q-Sepharose Chromatography (Optional). To concentrate the enzyme from step 4 and/or exchange the detergent, a second Q-Sepharose chromatography is performed on the pooled material from step 4. The conductivity of the pooled material is checked and if it exceeds the conductivity of buffer A (step 3) it is dialyzed as in step 2. Typically, 35 mg of enzyme from step 4 (i.e., 15% of the total material) is applied to a 1 × 3.8-cm column, which has been packed and equilibrated at a flow rate of 1.5 ml/min. The equilibration buffer is 50 m*M* Tris–acetate (pH 7.5), 1 m*M* dithioerythritol, 5% (v/v) glycerol and 0.2% (v/v) $C_{13}E_{12}$ (buffer C). After the column is washed with 5–10 vol of buffer C, the HSL enzyme elutes in a narrow peak and fractions (2 ml) are pooled according to the online recording of absorbance at 280 nm. The final protein concentration is usually between 1 and 4 mg/ml.

Purification of Human Enzyme

More recently, a recombinant baculovirus has also been generated for human HSL. The level of production of human HSL on infection of Sf9 cells with this virus is similar to the level of rat HSL described above. Despite the high degree of amino acid sequence homology (83%) between the rat and human enzymes, the rat HSL purification scheme had to be modified before it could be used for the human enzyme. In the Q-Sepharose chromatography steps, i.e., steps 3 and 5, the human enzyme binds more strongly to Q-Sepharose and does not elute until 0.6 *M* sodium acetate. Furthermore, the capacity of the Q-Sepharose column is three- to fourfold lower for the human compared to the rat enzyme. Therefore, the starting material for the large-scale purification described above is typically 150 ml for the human enzyme. Using the same buffers, the same flow rates, and the same size columns, step 3 is performed in the following manner for the human enzyme: After application of the sample, the column is washed with 10 vol of buffer A. The first gradient is from 0 to 0.2 *M* sodium acetate, followed by a plateau at 0.2 *M*. During the second gradient, from 0.2 to 1 *M* sodium acetate, human HSL elutes in a reproducible manner at 0.6 *M* sodium acetate. The later elution of human HSL affects the purity of the enzyme adversely. Finally, to obtain homogeneous enzyme, it is necessary to pool the eluted fractions over a narrow range, and strictly according to their purity as judged by SDS–PAGE. This also affects the overall recovery of purified human HSL, such that it is usually between 25 and 30%, instead of 40% as observed with the rat enzyme (Table I). The phenyl-Sepharose chromatography (step 4) is performed as described for the rat enzyme, except that buffer B contains 1 *M* NaCl (instead of 0.8 *M*) to ensure complete binding of the human enzyme to the column. Regarding step 5 above, half as much enzyme is loaded and 1 *M* sodium acetate is used in the elution buffer.

Comments on Purification Procedure

Owing to its amphiphilic character, hormone-sensitive lipase is strictly dependent on the presence of detergent throughout the purification scheme, and also during subsequent studies and analyses of the purified protein.[25] Virtually all commercially available detergents have been tested with regard to their effect on enzyme activity. Detergents from the alkyl polyoxyethylene group are especially useful and HSL activity shows a high tolerance to the length of the alkyl carbon and oxyethylene group of this type of detergent. Among the other detergents, in the presence of which HSL activity is retained, 3-[(3-cholamidopropyl)-dimethyl-ammonio]-1-propanesulfonate (CHAPS, a zwitterionic bile-salt-derived detergent) in our hands has shown the most reproducible results. However, the solubility of HSL appears to be lower in CHAPS than in the alkyl polyoxyethylene detergents, a problem that is evident at HSL concentrations of several milligrams per milliliter, especially on freezing–thawing. With regard to purification of HSL, the only detergent used routinely in our laboratories is $C_{13}E_{12}$. On the basis of the successful use of this type of detergent with HSL, it is reasonable to assume that $C_{13}E_{12}$ can be exchanged for other types from the same group, although the authors would recommend types with numbers of alkyl carbons and oxyethylene units similar to $C_{13}E_{12}$ for steps 2–4 of the purification scheme, to minimize the modifications needed in the purification scheme. With regard to step 5 of the purification procedure, the detergent concentration used has been successfully reduced from 0.2 to 0.008%, which is close to the critical micelle concentration (CMC) for $C_{13}E_{12}$. The detergent has also been exchanged for shorter versions of $C_{13}E_{12}$, C_8E_6 (0.4%, v/v) and C_8E_5 (0.35%, v/v), and for CHAPS (9 m*M*), without any loss in the recovery of active enzyme. It should also be mentioned that a procedure for partial purification of CHAPS-solubilized rat adipocyte HSL, utilizing phenyl-Sepharose, has been described by another group.[26]

Properties

Stability

The concentrated enzymes from step 5 retained more than 85% of their activity in 4 months if stored at −80° at neutral pH in 5% (v/v) glycerol with either 0.2% (v/v) $C_{13}E_{12}$ or 9 m*M* CHAPS. Higher glycerol concentrations (up to 50%, v/v) did not improve the stability. It was important to keep the enzyme at −80°, because the stability was considerably reduced on storage at −20°.

[25] C. Holm, G. Fredrikson, and P. Belfrage, *J. Biol. Chem.* **261,** 15659 (1986).

[26] J. J. Egan, A. S. Greenberg, M.-K. Chang, S. A. Wek, M. C. Moos, Jr., and C. Londos, *Proc. Natl. Acad. Sci. U.S.A.* **89,** 8537 (1992).

Other Properties

As described in detail elsewhere, the recombinant rat HSL was compared with the previously characterized rat adipose tissue HSL with regard to specific activity, substrate specificity, and phosphorylation and activation by cAMP-dependent protein kinase.[18] The recombinant enzyme was virtually identical to rat adipose tissue HSL. The recombinant human HSL was compared to recombinant rat HSL with regard to the same properties as described above, and with regard to pH optimum. Preliminary analyses indicate that, except for the lower specific activity observed for the human HSL, there are no major differences between the properties of the rat and human enzymes.

Assay Methods

Hormone-sensitive lipase has broad substrate specificity, hydrolyzing triglycerides, diglycerides, monoglycerides, cholesteryl esters, as well as more water-soluble substrates such as *p*-nitrophenylbutyrate (PNPB) and tributyrin.[7,18] The substrate used routinely in our laboratory, a monoether analog of dioleoylglycerol, has been described in detail previously.[27,28] This substrate has several advantages: (1) It optimizes the sensitivity of the assay method, because diglycerides are, at least under *in vitro* conditions, the preferred lipid substrate for HSL[29]; (2) the substrate for monoacylglycerol lipase is not formed from monoacylmonoalkylglycerol; (3) under the conditions used in the assay, i.e., pH 7.0 and in the absence of apoCII, virtually no lipoprotein lipase activity is measured; and (4) detergent inhibition is less with diacylglycerol (and its analogs) than with triacylglycerol (and its analogs). Points 2 and 3 are critical when measuring HSL activity in crude tissue or cell extracts.

Lipid Substrates

The synthesis of ^{3}H-labeled and unlabeled 1(3)oleoyl-2-*O*-oleylglycerol, as well as the preparation and use of phospholipid-stabilized emulsions of this compound, have been described in detail.[18,27,28] Only the modifications introduced in this assay are discussed in this chapter. The phospholipids used to stabilize the emulsion are optimized by using a mixture of phosphatidylcholine and phosphatidylinositol (weight ratios, 3 : 1). These are from egg yolk and soybean, respectively, and obtained as methanol fractions

[27] H. Tornqvist, P. Björgell, L. Krabisch, and P. Belfrage, *J. Lipid Res.* **19,** 654 (1978).
[28] G. Fredrikson, P. Strålfors, N. Ö. Nilsson, and P. Belfrage, *Methods Enzymol.* **71,** 636 (1981).
[29] G. Fredrikson and P. Belfrage, *J. Biol. Chem.* **258,** 14253 (1983).

from silicic acid chromatography of a chloroform–methanol extract. The buffer used currently is 0.1 *M* potassium dihydrogen phosphate, pH 7.0, instead of the Tris-HCl buffer originally proposed, and the buffer used for diluting the enzyme sample up to 100 μl is 20 m*M* potassium dihydrogen phosphate, 1 m*M* EDTA, 1 m*M* dithioerythritol, and 0.02% (w/v) defatted bovine serum albumin, pH 7.0, instead of Tris-HCl with 0.02% (w/v) albumin. Another modification is that oleic acid, included as carrier in the extraction mixture (methanol–chloroform–heptane),[30] is not used routinely, but only when analyzing series of samples known to have low activity. The liquid scintillator used now is Ready-Safe[31] instead of Instagel-toluene. Last, regarding the sonication of the substrate, the 20% (w/v) defatted bovine serum albumin (in 0.1 *M* potassium dihydrogen phosphate, pH 7.0) is not added until the sonication procedure is completed.

Other Lipid Substrates

A dioleoylglycerol emulsion can be prepared and used exactly as described above.[27] Trioleoylglycerol and cholesteryl oleate are prepared in a similar manner.[18,32] With the important exception of assays utilizing the monolayer technique (see below), trioleoylglycerol is the only substrate that can be used to monitor, *in vitro,* the differences in enzyme activity between unphosphorylated HSL and HSL phosphorylated by cAMP-dependent protein kinase.[5] Emulsified monooleoylglycerol can be used for measuring HSL activity in preparations not containing monoacylglycerol lipase.[33]

Water-Soluble Substrates

For monitoring esterase, rather than lipase activity, *p*-nitrophenyl butyrate (PNPB) and tributyrin can be used under conditions in which they are water soluble. The PNPB assay is performed as described for lipoprotein lipase,[34] except that 1 m*M* dithioerythritol is included in the incubation buffer and a substrate concentration of 2 m*M* is used to ensure V_{max} conditions.[18] In the assay with tributyrin,[35] the same incubation buffer as in the PNPB assay[18] is employed. Despite the fact that three-dimensional modeling work indicates striking similarities between the structures of the catalytic

[30] P. Belfrage and M. Vaughan, *J. Lipid Res.* **10,** 341 (1969).
[31] Scintillator solution is commercially available from Beckman Instruments, Inc.
[32] H. Tornqvist, L. Krabisch, and P. Belfrage, *J. Lipid Res.* **13,** 424 (1972).
[33] H. Tornqvist, P. Nilsson-Ehle, and P. Belfrage, *Biochim. Biophys. Acta* **530,** 474 (1978).
[34] K. Shirai and R. L. Jackson, *J. Biol. Chem.* **257,** 1253 (1982).
[35] K. Shirai, N. Katsuoka, S. Saito, and S. Yosida, *Biochim. Biophys. Acta* **795,** 1 (1984).

domain of HSL and members of the cholinesterase family, including acetylcholine esterase, HSL exhibits low acetylcholine esterase activity.[23]

Monomolecular Films

The availability of recombinant HSL in large quantities theoretically enables the measurement of HSL catalytic activity in assay systems previously unused because of the amount of enzyme required. An example of this is the monolayer system, in which lipase activity is monitored by a decrease in surface pressure of the lipid monolayer, owing to its hydrolysis. This fall in surface pressure is automatically compensated by a continuous movement of the barrier, using a surface barostat device.[36] A detailed description of the method is given in [12] of this volume.[36a] However, the special characteristics of HSL impose practical restrictions not easily overcome. The two major limiting factors for the use of HSL with the surface barostat method are the relatively low specific activity of this enzyme, compared with other lipases, and the requirement for detergent by HSL. These problems are interconnected, because increasing the amount of enzyme injected under the monolayer inevitably also increases the amount of detergent in the system. Despite this, we have succeeded in measuring HSL activity with monolayers of pure 1,2-didecanoyl glycerol (dicaprin) and mixed monolayers of dicaprin and 1,2-diacyl-*sn*-glycero-3-phosphocholine (phosphatidylcholine), as described below. However, it is not possible to measure HSL activity on triglyceride monolayers. The low collapse pressure of triglyceride monolayers limits the working surface pressure to values below 20 mN/m. When HSL is injected in the system, an increase in surface pressure is observed until the collapse pressure of the monolayer was reached. The reason for this resides in the high tensioactive character of the HSL protein itself, because this phenomenon is observed even with samples in which the amount of detergent has been previously reduced to levels not affected the triglyceride monolayer.

Trough. A zero-order trough consisting of a circular reaction compartment with a surface area of 39.2 cm^2 (volume, 33 ml) connected, by a narrow surface channel, to a reservoir compartment 25 $\times$ 1.5 cm in size, is used for all measurements. The special dimensions of this trough increase the sensitivity of the system, because a large displacement of the Teflon barrier is required to compensate a small decrease in the surface pressure.

Buffer. The assay buffer contains 10 m*M* Tris-HCl (pH 7.0), 100 m*M* NaCl, 21 m*M* $CaCl_2$, 1 m*M* EDTA, 1 m*M* dithioerythritol. The buffer in

[36] S. Ransac, H. Moreau, C. Rivière, and R. Verger, *Methods Enzymol.* **197,** 46 (1991).

[36a] S. Ransac and R. Verger, *Methods Enzymol.* **284,** [12], 1997 (this volume).

the reaction chamber is supplemented with β-cyclodextrin (0.5 mg/ml; Sigma, St. Louis, MO) to trap the excess detergent bound to the enzyme. In the presence of β-cyclodextrin, the nonionic detergent $C_{13}E_{12}$ increases only slightly the monolayer surface pressure, which stabilizes rapidly after the injection. The successful use of β-cyclodextrin has been previously described in the case of monolayers of long-chain lipids as substrates.[37,38]

Enzyme. Recombinant rat HSL is used directly from the eluates of step 4 or step 5 [0.01–0.2% (v/v) $C_{13}E_{12}$] of the purification protocol, at a final concentration of about 1 mg of protein/ml. Owing to the small volumes injected under the monolayer, the effect of the nontensioactive components of the buffer (salts, dithioerythritol, etc.) is found to be negligible. In the presence of β-cyclodextrin, the nonionic detergent $C_{13}E_{12}$ increases the surface tension only slightly, and the monolayer stabilizes rapidly after the injection. Hormone-sensitive lipase is phosphorylated by the catalytic subunit of cAMP-dependent protein kinase as described.[6a,18]

Assay Conditions. After spreading and stabilization of the monolayer, 5 to 25 μg of pure recombinant rat HSL is injected into the reaction compartment. Lipolysis is monitored at a constant surface pressure. The best results are obtained at higher surface pressure (30 mN/m). Assays are performed at room temperature, with continuous agitation of the subphase in the reaction compartment by means of a magnetic stirrer set to 250 rpm. When mixed monolayers containing phosphatidylcholine are used, the trough is placed in an atmosphere of argon, to prevent the oxidation of the phospholipids.

Under these conditions, HSL activity can be monitored with pure dicaprin monolayers. The lag time between sample injection and detection of lipase activity varies between 5 and 20 min (depending on the amount of HSL used and the amount of detergent in the HSL sample). This lag time is probably due to two factors: (1) a slow interaction of HSL with the monolayer and (2) an initial increase in the surface pressure caused by the incorporation of detergent into the monolayer. When the system reaches an equilibrium state, linear HSL activity is observed for periods longer than 1 hr. In this system, phosphorylated HSL shows twofold higher activity than nonphosphorylated HSL, similar to what is observed in bulk assays with a radiolabeled triolein.[5,18] The relevance of this result resides in the fact that it is the first time that activation of HSL has been detected using a diglyceride, rather than a triglyceride as substrate.

[37] S. Laurent, M. Ivanova, D. Pioch, J. Graille, and R. Verger, *Chem. Phys. Lipids* **70,** 35 (1994).

[38] M. G. Ivanova, T. Ivanova, R. Verger, and I. Panaiotov, *Colloids Surfaces B: Biointerfaces* **6,** 9 (1996).

The activity of HSL on dicaprin is greatly affected by the presence of phospholipids in the monolayer. Thus, when mixed monolayers of dicaprin and phosphatidylcholine are used, HSL activity increases gradually as the proportion of phospholipid increases, reaching a maximum at a dicaprin–phosphatidylcholine ratio of 1:1 (w/w), and decreasing when the amount of phospholipid exceeds that of dicaprin. This effect of phospholipids is most likely due to an increased binding of HSL to the monolayer, because we have observed a strong binding of HSL to phosphatidylcholine vesicles *in vitro*.[39] No hydrolysis is detected with pure phosphatidylcholine monolayers, i.e., HSL has no detectable phospholipase activity.

Under optimal conditions [i.e., dicaprin–phosphatidylcholine, 1:1 (w/w)], the lag time, after HSL injection into the subphase below the monolayer, is drastically reduced, and the HSL activities obtained are 5 to 10 times higher than with pure dicaprin monolayers. Interestingly, under these conditions, the effect of phosphorylation of HSL vanishes, and both phosphorylated and nonphosphorylated HSL show identical catalytic activities. In conclusion, these results indicate that the ability of HSL to interact with lipid surfaces is greatly affected by the "quality of the interface," as observed for other lipases. Phosphorylation of HSL may affect the ability of the enzyme to interact with interfaces having unfavorable physicochemical properties, whereas the effect of phosphorylation is negligible if the interfacial parameters are already favorable for the binding of HSL.

Acknowledgments

The authors are indebted to Birgitta Danielsson for excellent technical assistance in production and purification of the recombinant enzymes, to Torben Østerlund and Camilla Johansson for their contributions to the characterization of the recombinant enzymes, and to Richard Davis for his contribution in establishing the expression system for the rat enzyme.

This work was supported by grants from the Medical Research Council (No. 112 84 to C. H.), the Medical Faculty of Lund University, the A. Påhlsson Foundation, the E. and W. Cornell Foundation, the Crafoord Foundation, the P. Håkansson Foundation, the C. Tesdorpf Foundation, the Veterans Administration, the National Institutes of Health (HL-28481), and E.U. Project BIOTECH 2-CT94-3041.

[39] M. Karlsson and C. Holm, unpublished data, 1995.

[17] New Pancreatic Lipases: Gene Expression, Protein Secretion, and the Newborn

By MARK E. LOWE

The similarities among the amino acid sequences of lipoprotein lipase, hepatic lipase, and pancreatic triglyceride lipase led to the hypothesis that the genes encoding these proteins evolved from a common ancestor gene.[1] The concept of a lipase gene family was strengthened by the description of two novel cDNAs isolated from a human pancreas library.[2] The protein sequences predicted from these cDNAs were 65 and 68% identical to the pancreatic lipase amino acid sequence, which resulted in these proteins being called pancreatic lipase-related proteins 1 and 2 (PLRP1 and -2). Subsequently, clones with high homology to human PLRP2 were isolated from mouse CD8 lymphocytes (cytotoxic T lymphocyte lipase) and from rat pancreas (the zymogen granule membrane protein GP). A clone with homology to human PLRP1 was also isolated from rat pancreas.[3–6]

Comparison of the predicted amino acid sequences of PLRP1 and PLRP2 with that of pancreatic lipase showed marked conservation of important functional and structural determinants. The Ser–His–Asp catalytic triad residues, the number and position of the cysteines, and the amino acid sequences of important functional domains (such as the lid) were conserved. The presence of the PLRPs raised important questions about their physiological functions, gene regulation, and expression pattern. In this chapter, methods for investigation of the expression pattern of these genes, for identifying the individual proteins in pancreatic secretions, and for expressing recombinant proteins are described.

[1] T. G. Kirchgessner, J.-C. Chuat, C. Heinzman, J. Etienne, S. Guilhot, K. Svenson, D. Ameis, C. Pilon, L. D'Auriol, A. Andalibi, M. C. Schotz, F. Galibert, and A. J. Lusis, *Nuc. Acid Res.* **86,** 9647–9651 (1989).

[2] T. Giller, P. Buchwald, D. Blum-Kaelin, and W. Hunziker, *J. Biol. Chem.* **267,** 16509 (1992).

[3] M. J. Grusby, N. Nabavi, H. Wang, R. F. Dick, J. A. Bluestone, M. C. Shotz, and L. H. Glimcher, *Cell* **60,** 451 (1990).

[4] M. J. Wishart, P. C. Andrews, R. Nichols, G. T. Blevins, C. D. Logsdon, and J. A. Williams, *J. Biol. Chem.* **268,** 10303 (1993).

[5] R. M. Payne, H. F. Sims, M. L. Jennens, and M. E. Lowe, *Am. J. Physiol.* **266,** G914 (1994).

[6] C. Wicker-Planquart and A. Puigserver, *FEBS Lett.* **296,** 61 (1992).

0076-6879/97 $25.00

Gene Expression

Gene expression can be regulated at several levels. Expression can be limited to certain cell types (cell-specific expression). For instance, pancreatic acinar cells express a variety of gene products, including pancreatic lipase, exclusively or in much greater abundance than other cell types, suggesting that these genes are regulated by cell-specific elements. Expression can vary with stages of development (temporal expression). This form of regulation is found in many acinar cell-specific genes, which are not expressed until birth or later. Expression can be increased or decreased in response to various modulators. In the pancreas, gastrointestinal hormones such as secretin or cholecystokinin may increase expression of acinar genes. Each of these regulatory elements helps determine the physiological function of the gene product by determining the amount of protein expressed at a particular time.

Analysis of RNA

RNA Isolation. The first examination of gene expression is generally to determine the presence of mRNA encoding the gene product in various cells, that is, cell-specific expression. This step requires the isolation of RNA from various tissues and an analysis of the RNA to determine if the mRNA of interest is expressed. Total RNA or mRNA may be isolated by various methods.[7] Pancreatic RNA was commonly isolated by the guanidine thiocyanate method, but more recently, pancreatic RNA has been purified with a single-step, acid guanidinium thiocyanate–phenol–chloroform extraction.[8–10] The first method and general precautions for handling RNA were previously described and only the second method and precautions for isolating pancreatic RNA are outlined.[11]

The isolation of intact RNA from pancreas can be difficult because of the high ribonuclease content of this tissue and success depends on several factors. First, the pancreas should be the first tissue dissected when multiple tissues are to be harvested. Delay in removing the pancreas after sacrificing the animal ensures degraded RNA. Once removed, the tissue should be immediately processed or snap frozen in liquid nitrogen for later processing. If frozen, the tissue should not be thawed prior to homogenization with the extraction solution.

[7] S. L. Berger and A. R. Kimmel, eds., *Methods Enzymol.* **152** (1987).
[8] P. Chomczynski and N. Sacchi, *Anal. Biochem.* **162,** 156 (1987).
[9] C. Puissant and L. Houdebine, *BioTechniques* **8,** 148 (1991).
[10] J. M. Chirgwin, A. E. Przybylz, R. J. McDonald, and W. J. Rutter, *Biochemistry* **18,** 5294 (1979).
[11] D. D. Blumberg, *Methods Enzymol.* **152,** 20 (1987).

Acid Guanidinium Thiocyanate–Phenol–Chloroform Extraction

Reagents

Acid guanidinium thiocyanate–phenol solution: Can be prepared or obtained premixed (RNAzol B; Tel-Test, Friendswood, TX)[8,9]

Chloroform, 2-propanol, 75% (v/v) ethanol, and diethylpyrocarbonate (DEPC)-treated water

RNase-free plasticware: Tubes capable of holding 10 ml and withstanding 12,000 *g* centrifugal forces (e.g., polypropylene tubes, 17 × 100 mm; Fisher, Pittsburgh, PA); and 1.7-ml microcentrifuge tubes

Tissue homogenizer: A glass–Teflon homogenizer or motor-driven Polytron (Brinkmann, Westbury, NY) can be used. For frozen tissues, a motor-driven homogenizer or Polytron is preferable

Method

1. Homogenize the tissue in ice-cold extraction solution until well dispersed. Five or six strokes of the homogenizer or 30 sec with a Polytron should be adequate. Two milliliters per 100 mg of tissue is suggested, but 5 ml/100 mg of pancreas gives more consistent results.

2. A one-tenth volume of chloroform is added and the samples are shaken vigorously by hand.

3. The mixture is incubated on ice for 5 min followed by centrifugation at 12,000 *g* for 15 min at 4°.

4. The aqueous phase (top layer) is transferred to a fresh tube and an equal volume of 2-propanol is added. The sample is mixed by inversion and incubated on ice for 15 min.

5. The RNA is pelleted by centrifugation at 12,000 *g* for 15 min at 4°.

6. Remove the supernatant, leaving the pellet behind. Wash the pellet with 75% (v/v) ethanol by vortexing and centrifuging at 7500 *g* for 10 min at 4°.

7. Remove the 70% (v/v) ethanol as completely as possible. Resuspend the RNA in 500 μl or less of DEPC-treated water. Transfer to a 1.7-ml microcentrifuge tube.

8. Add a one-tenth vol of 2 *M* NaCl and 2 vol of ethanol. Incubate at −20° for 1 hr. Pellet the RNA in a microcentrifuge at full speed for 15 min at 4°. Wash with 0.5 ml of 70% (v/v) ethanol by vortexing. Centrifuge at full speed for 5 min at 4°.

9. Remove the 70% (v/v) ethanol and centrifuge at full speed briefly. Remove the remaining 70% (v/v) ethanol with a pipette, being careful not to disturb the pellet. It is not necessary to vacuum dry the pellet at this point. The RNA is resuspended in DEPC-treated water and stored at −70°. The concentration can be determined by the optical density at 260 nm.

Other methods of isolating RNA may be suitable for pancreatic RNA. There are multiple kits available from various sources that isolate RNA from cells or tissues without using phenol or guanidinium thiocyanate, giving them the advantage of avoiding these hazardous chemicals.

Detecting Specific mRNAs

RNA Blots. The first step in preparing a blot is to separate the RNA in any of the denaturing agarose gel systems that are well described in other sources.[12–14] Fifteen to 20 μg of total RNA is suitable for detecting the mRNAs for most exocrine proteins, including pancreatic lipase and both related proteins. If less abundant species are to be identified, then 1–10 μg of poly(A)$^+$ mRNA can be loaded onto the gel.

The second step is to transfer the RNA to a membrane.[15] There are now a variety of nylon-based membranes that are designed for structural stability and use with nucleic acids. In general, transfers are done in 10× SSC [1× SSC is 30 m*M* sodium citrate (pH 7.5), 0.03 *M* NaCl] or 10× SSPE [1× SSPE is 0.18 *M* NaCl, 0.01 *M* sodium phosphate (pH 7.7), 1 m*M* EDTA], but the conditions may vary depending on the choice of membranes and the manufacturer recommendations should be followed. Transfers are complete in 4 hr to overnight. The blot is rinsed with 2× SSC to remove agarose particles and the RNA is then fixed to the membrane by drying in a vacuum oven or by ultraviolet (UV) cross-linking. As with blotting, the manufacturer recommendations for fixing RNA should be followed.

Dot Blots

Dot blotting is a rapid technique for screening multiple samples and the general technique has been previously outlined.[16] This method can give relative amounts of a specific mRNA in various samples or can be adapted to quantitate the amount of mRNA in each sample. Because quantitation with dot blots and RNA blots can be difficult, other methods including solution hybridization, RNase protection, and polymerase chain reaction (PCR) have been utilized.[7,12] If the dot blot is used to compare levels of

[12] J. Sambrook, E. F. Fritsch, and T. Maniatis, "Molecular Cloning: A Laboratory Manual." Cold Spring Harbor Laboratory Press, Cold Spring Harbor, New York, 1989.
[13] L. G. Davis, M. D. Dibner, and J. F. Battey, "Basic Methods in Molecular Biology." Elsevier, New York, 1986.
[14] R. C. Ogden and D. A. Adams, *Methods Enzymol.* **152,** 61 (1987).
[15] G. M. Wahl, J. L. Meinkoth, and A. R. Kimmel, *Methods Enzymol.* **152,** 572 (1987).
[16] C. Costanzi and D. Gillespie, *Methods Enzymol.* **152,** 582 (1987).

mRNA between samples appropriate controls and dilutions are required. Because the RNA is not separated in the dot blot, controls for nonspecific binding are critical. Commonly, samples of RNA isolated from tissues that do not express the test gene or yeast tRNA are included to control for nonspecific binding. Both loading and immobilization efficiency can vary and should be determined. Applying several dilutions of the RNA sample to demonstrate linearity and hybridizing with probes that detect ubiquitous, unregulated mRNA or RNA species present in all tissues, such as mRNA encoding actin or cyclophylin or 18S RNA, help to detect differences in loading or immobilization. Signal strength can be determined by autoradiography with densitometry, radioisotopic scanning, or by cutting out the dot and counting by liquid scintillation.

Probes. When determining the expression pattern of related genes, the specificity of the probes is crucial. For studies examining expression of pancreatic lipase and the related proteins, oligonucleotide probes were selected by visual inspection of the cDNA sequence.[5] This process was aided by computer alignment of the sequences. Regions of below-average homology containing gaps were selected. The specificity of the probes for each species was demonstrated by slot blots containing *in vitro*-transcribed mRNA derived from the respective cDNAs.[5] In this example, the probes were end labeled with [^{32}P]dATP with T4 polynucleotide kinase. Other applications may require cDNA or RNA probes. Labeling methods for these probes are described in several sources.[11–13]

Prehybridization and Hybridization. Multiple conditions for prehybridization and hybridization have been described. The choice of conditions depends on the probe and on the membrane.[11–13] These sources and the membrane manufacturer suggestions should be referred to for specific conditions.

Expression of mRNA Encoding Pancreatic Lipase and Pancreatic Lipase-Related Proteins 1 and 2

Tissue Expression

The isolation of cDNAs encoding pancreatic lipase and both related proteins from pancreatic cDNA libraries provided the first evidence that the genes encoding these proteins were expressed in pancreas.[2] This location has been confirmed and other tissues have been examined by RNA blot analysis. This analysis of pancreatic lipase and PLRP2 mRNA in a variety of human, rat, and mouse tissues showed that expression was limited to the pancreas. The tissue distribution of PLRP1 has not been reported for any species.

Temporal Expression

Many genes are regulated during development, including genes encoding pancreatic acinar cells. The temporal expression of rat pancreatic lipase, PLRP1, and PLRP2 was determined by RNA dot blot (Fig. 1). mRNA

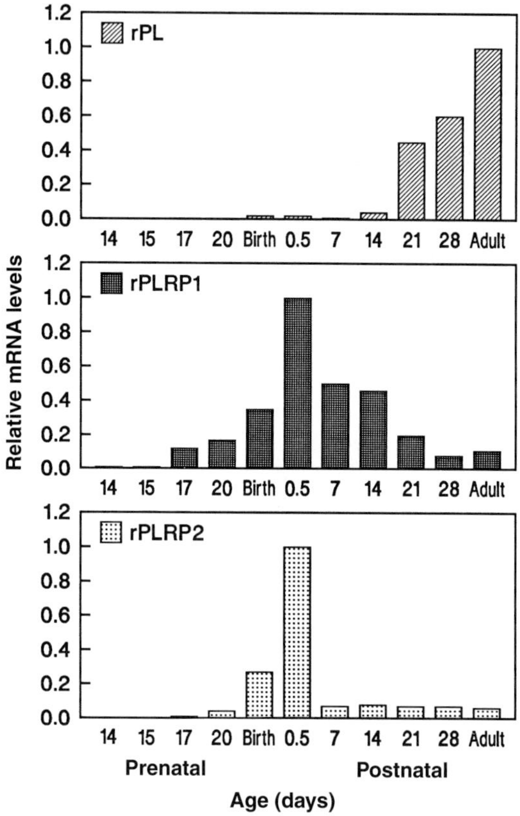

FIG. 1. Temporal expression of the mRNAs encoding pancreatic lipase, PLRP1, and PLRP2. Total RNA was isolated from rat pancreas at various ages. The relative levels of mRNA were determined by dot-blot analysis of 3 μg of total RNA using oligonucleotide probes specific for each mRNA species. Nonspecific binding was determined by hybridizing the probes to dots blots of yeast tRNA and total liver RNA. The loading and blotting efficiency was determined by probing each sample with an oligonucleotide complementary to 18S RNA. The amount of probe hybridizing to each sample was measured by radioanalytic scanning. Only the relative values are givne. Because the probes are of the same length and similar dGdC content and labeled to the same specific activity, the relative amounts of each species can be estimated. The absolute values for the maximum signal are in a ratio of 1 : 1 : 0.5 for pancreatic lipase : PLRP1 : PLRP2. (Adapted from Payne *et al.*[5])

encoding pancreatic lipase was poorly expressed at birth and in the suckling period.[5] Expression increased to adult levels during the suckling–weanling transition. In contrast, mRNA encoding PLRP1 and PLRP2 showed maximal expression after the initial postnatal feedings. The levels fell during the following weeks and remained low relative to the mRNA for pancreatic lipase in adults.

Protein Analysis

The presence of mRNA for a particular protein does not ensure that the protein is expressed or that synthesis is proportionate to the amount mRNA in that tissue. The protein needs to be identified separately by one of several methods. The protein can be purified from tissue or secretions or it can be identified in extracts of tissue or in secretions by specific antibodies against that protein. In general, antibody methods are employed because they offer the advantages of speed, of easily adapting established protocols to a wide variety of cells, and of the ability to screening multiple samples. If closely related proteins, such as pancreatic lipase, PLRP1, and PLRP2, are being screened, then the specificity of the antibodies must be carefully documented. Polyclonal antibodies raised against the entire protein antigen may not have sufficient specificity to distinguish between related proteins and other approaches are then required. The choices usually include making monoclonal antibodies against each protein of interest or making polyclonal antibodies against peptides that are divergent between the proteins of interest. The advantage of producing polyclonal antibodies is that they will generally be useful in a variety of applications including immunoblot, enzyme-linked immunosorbent assay (ELISA), immunoprecipitation, and immunohistochemistry whereas monoclonal antibodies may not work for all applications.

Anti-Peptide Antibodies

Anti-peptide antibodies can be made by injecting animals, rabbits in most cases, with specific peptides linked to a carrier protein such as albumin or keyhold limpet hemocyanin (KLH). Because pancreatic lipase has a lower molecular weight than PLRP1 and PLRP2, it can be separated from the related proteins on sodium dodecyl sulfate–polyacrylamide gel electrophoresis (SDS–PAGE) and identified with a polyclonal antibody against pancreatic lipase. The two related proteins are not well separated on SDS–PAGE and cannot be identified by this method. Thus, anti-peptide antibodies to PLRP1 and PLRP2 were made in rabbits using KLH as the carrier.[17] The peptides were selected by visual inspection of the predicted amino acids sequences of PLRP1 and PLRP2 and by computer alignment of the selected sequences with the complete amino acid sequences of each protein. Synthetic peptides with a cysteine at either the carboxyl- or amino-terminal end to facilitate coupling were linked to maleimide-activated keyhole limpet hemocyanin (Pierce, Rockford, IL). Two milligrams of peptide was coupled to the carrier protein as described by the manufacturer. Rabbits were immunized with 100 μg of carrier protein containing coupled peptide emulsified in Freund's complete adjuvant. The animals were boosted 14 days later with 100 μg in Freund's incomplete adjuvant and again at 1 month and monthly intervals with 10–20 μg in phosphate-buffered saline. The animals were test bled before each boost and the serum tested for antibody by immunoblot against purified, recombinant pancreatic lipase, PLRP1, and PLRP2,[5] Finally, the specificity of the antibodies was demonstrated by immunoblot of purified recombinant pancreatic lipase, PLRP1, and PLRP2.

Analysis of Secretion

Secretion of pancreatic lipase and the related proteins can be examined in tissues, pancreatic secretions from the intact gland or isolated acinar cells, or in tissue culture cell lines. The presence of pancreatic lipase in pancreatic juice was described long ago and documented many times.[18] In contrast, little has been done to establish the presence of PLRP1 and PLRP2 in pancreatic secretions. Rat PLRP2 was originally identified as a zymogen granule membrane protein, GP3. Later, secretion into pancreatic juice was demonstrated by cannulating the pancreatic duct and collecting secretions.[19] The presence of PLRP1 has not been demonstrated in pancreatic juice.

Tissue Culture

All three proteins were found to be secreted by a pancreatic acinar cell line, AR42J.[17] These cells were derived from a tumor of the protodifferentiated pancreatic epithelium because they possess both exocrine and endocrine properties.[20] Despite these dual properties, AR42J cells have been a

[17] J. Kullman, C. Gisi, and M. E. Lowe, *Am. J. Physiol.* **270,** G746 (1996).

[18] R. Verger, *in* "Lipase" (B. Borgstrom and H. L. Brockman, eds.), 1st Ed., pp. 84–150. Elsevier, Amsterdam, 1984.

[19] A. C. Wagner, M. J. Wishart, S. M. Mulders, P. M. Blevins, P. C. Andrews, A. W. Lowe, and J. A. Williams, *J. Biol. Chem.* **269,** 9099 (1994).

[20] J. Christophe, *Am. J. Physiol.* **266,** G963 (1994).

model system for investigating the differentiation, exocrine secretion, and receptors of pancreatic acinar cells.[21–23]

To demonstrate the AR42J cells secreted all three proteins, the cells were incubated for 48 hr in serum-free F12K medium.[19] The cells remained viable for at least 2 days and changing to serum-free medium avoided high albumin concentrations that would alter the mobility of the related proteins. The medium was concentrated about 10-fold over an Amicon (Danvers, MA) macrosolute concentrator. An aliquot of the medium was separated by SDS–PAGE and the proteins transferred to polyvinylidene fluoride membrane. The membrane was incubated with phosphate-buffered saline, 0.1% (v/v) Tween 20, and 1% (w/v) bovine serum albumin for 1 hr to block the nonspecific protein-binding sites. After blocking, the membrane was incubated with one of the anti-peptide antibodies or with the polyclonal anti-pancreatic lipase antibody. Dilutions of 1:2000 in phosphate-buffered saline containing 0.1% (v/v) Tween 20 gave a good signal, but the appropriate dilution for a new antibody needs to be determined by testing various dilutions. The blot was washed three times (5–10 min each wash) with phosphate-buffered saline and 0.1% (v/v) Tween 20 and then incubated for 1 hr with a secondary antibody, alkaline phosphatase-conjugated anti-rabbit IgG, in the same buffer. After this incubation, the blot was washed as described above and incubated with nitroblue tetrazolium and 5-bromo-4-chloro-3-indolylphosphate in 100 m*M* Tris-HCl (pH 9.5), 100 m*M* NaCl, and 5 m*M* $MgCl_2$. The reaction was allowed to proceed until the desired bands were apparent and nonspecific bands were minimal. At that point, the incubation solution was removed and the blot was washed extensively with distilled water. The blot was allowed to dry and was stored in the dark to prevent band fading. If kept dark, the signal is stable for several years. The blots should be photographed if a long-term record is desired. Pancreatic lipase, PLRP1, and PLRP2 were detected in the medium of AR42J cells by immunoblot with this method, suggesting that these cells would be a suitable model system for evaluating the effects of various stimuli on gene expression, protein synthesis, and secretion.

Primary Acinar Cells

Even though a tissue culture cell line secreted PLRP1, it was necessary to demonstrate that this expression was not an artifact of tissue culture or of the selection of this cell line. One approach that allows the analysis of acinar cell secretions and the ability to explore the effect of secretogogues

[21] C. D. Logsdon, *J. Biol. Chem.* **261,** 2096 (1986).
[22] C. D. Logsdon, J. Moessner, J. A. Williams, and I. D. Goldfine, *J. Cell. Biol.* **100,** 1200 (1985).
[23] B. Swarovsky, W. Steinhilber, G. A. Scheele, and H. F. Kern, *J. Cell. Biol.* **47,** 101 (1988).

on the secretion of the exocrine proteins is to study isolated acinar cells in primary culture. Several methods to prepare pancreatic acinar cells have been reported and adapted by others.[24–26] The following method has worked well in our laboratory.[26]

Reagents

Buffer A: 130 m*M* NaCl, 5 m*M* KCl, 12.5 m*M* *N*-2-hydroxyethylpiperazine-*N,N'*-ethanesulfonic acid (HEPES), 10 m*M* glucose; adjust to pH 7.4, then add 0.1% (w/v) bovine serum albumin (BSA), 0.01% (w/v) soybean trypsin inhibitor, 2 m*M* $CaCl_2$, 1.2 m*M* $MgCl_2$

Buffer B: 139 m*M* potassium glutamate, 20 m*M* piperazine-*N,N'*-bis(2-ethanesulfonic acid) (PIPES, pH 6.6), 0.01% (w/v) soybean trypsin inhibitor

Collagenase: Worthington (Freehold, NJ) CLSPA type resuspended in buffer A to 4000 U/ml DNase I: Type II (Sigma, St. Louis, MO) resuspended in distilled water to 20,000 U/ml

Procedure. A 75- to 125-g rat is sacrificed by CO_2 narcosis and the pancreas is removed through an abdominal incision. The pancreas can be located by lifting the spleen. The tail of the gland will be attached to the spleen. The organ is cleaned of fat and connective tissue and the gland is diced into small pieces with a razor blade. If additional organs are harvested the pancreas is placed in 5–10 ml of buffer A and all are diced at the same time. The pieces are placed into a 50-ml Erlenmeyer flask containing 5 ml of buffer A and 400 U of collagenase. The flask is incubated for 10 min at 37° in a shaking incubator at 200 rpm. DNase I (5000 U) is added for the last 3 min of the incubation. DNase I is not necessary when preparing acini from adult animals, but it improves the yield and quality of acini obtained from suckling animals. The entire mixture is placed into a 15-ml siliconized Corex tube, covered with Parafilm, and shaken vigorously by hand until the pancreas is dissociated, about 1 min. The cells should have a fine sand appearance at this point. Additional shaking will damage the cells and basal secretion will be high in subsequent experiments. The cell suspension is filtered through surgical gauze and, next, through a 200-micron nylon mesh (SpectraMesh; Spectrum, Los Angeles, CA) into a beaker. The cell suspension is transferred to a fresh siliconized 15-ml Corex tube and the cells are allowed to settle for 6 min. The supernatant is carefully removed, 10 ml of buffer B is added, and the cells allowed to settle for 3 min. The supernatant should be clear. If it is cloudy, there are too many leaky cells and the

[24] R. Bruzzone, P. A. Halban, A. Gjinovci, and E. R. Trimble, *Biochem. J.* **226,** 621 (1985).
[25] P. M. Brannon, B. M. Orrison, and N. Kretchmer, *In Vitro Cell. Devel. Biol.* **21,** 6 (1985).
[26] P. J. Padfield and N. Panesar, *Am. J. Physiol.* **269,** G647 (1995).

preparation should be discarded. The supernatant is aspirated and replaced with 10 ml of buffer B. The cells are pelleted at 200 *g* for 3 min. This step is repeated two times and the pellet is resuspended in 2–4 ml of buffer A.

Secretion of Pancreatic Lipase-Related Proteins 1 and 2 from Acinar Cells. The presence of pancreatic lipase, PLRP1, and PLRP2 in the medium of isolated acinar cells is determined by immunoblot with the antibodies described above. The isolated acinar cells are incubated in 1 ml of buffer B with shaking, 200 rpm, at 37° for 1–2 hr. Secretogogues (cholecystokinin, secretin, or carbachol) can be included in the incubation if desired. The medium is harvested by pelleting the cells at 1000 *g* for 5 min at 4° and removing the medium. Detection of the proteins is facilitated by concentrating the medium 10-fold over an Amicon B15 macrosolute concentrator. This analysis demonstrates that acinar cells can secrete all three homologs into the medium (Fig. 2). Furthermore, the secretion pattern mirrors the expression pattern of the mRNAs encoding these proteins. PLRP1 and PLRP2 are secreted by acinar cells isolated from suckling animals whereas pancreatic lipase is not detected in the medium of the suckling animals but is secreted by acinar cells isolated from older animals.

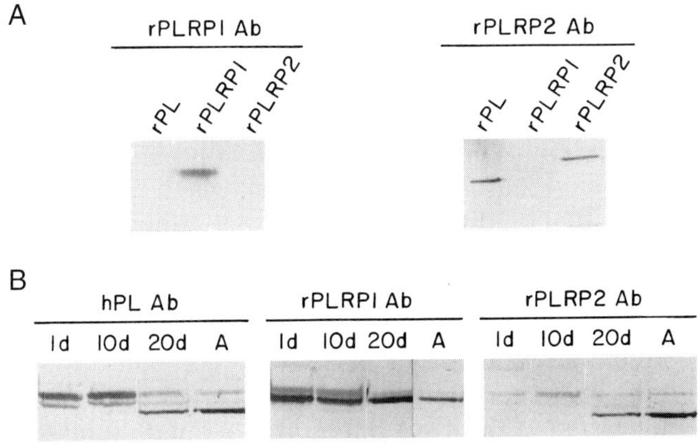

FIG. 2. The secretion of pancreatic lipase, PLRP1, and PLRP2 from pancreatic acinar cells isolated from rats of various ages. (A) The specificity of anti-peptide antibodies against purified recombinant rat pancreatic lipase, PLRP1, and PLRP2. In the left-hand blot, 500 ng of pancreatic lipase (rPL), 500 ng of PLRP2, and 50 ng of PLRP1 were loaded. In the right-hand blot, 500 ng of rPL, 500 ng of PLRP1, and 50 ng of PLRP2 were loaded. The antibody used to stain the blot is given above each blot. (B) The results of probing secretions from acinar cells isolated from rats of various ages. The antibody used is given above each blot and the age of the animal is given above each lane. The PLRP1 antibody detects two bands in the younger animals. An explanation for this observation has not been found. (From M. E. Lowe, unpublished observations, 1996).

Expression and Purification of Recombinant Pancreatic Lipase-Related Proteins 1 and 2

Even though the characterization of gene expression and secretion patterns is an important facet in describing the physiological function of a protein, a complete understanding requires that the biochemical properties of that protein be defined. The enzymatic properties of enzymes are best characterized after they have been purified, a qualification that presents particular problems for purifying the PLRPs from a tissue rich in pancreatic lipase. The similar physical properties and preponderance of pancreatic lipase in adult pancreas make the separation of these proteins more difficult and obtaining a preparation of the PLRPs that is free of pancreatic lipase has been difficult.[2,27] Starting with tissue from younger animals enriches the proportion of PLRPs, but the pancreata of many animals are needed. One approach to eliminate the problem of contamination with pancreatic lipase and of low abundance is to express recombinant proteins.

Expression of Pancreatic Lipase-Related Proteins 1 and 2

The expression of pancreatic lipase, PLRP1, and PLRP2 in baculovirus has been described (see Refs. 5 and 28, and [8] in this volume[29]). Purification from Sf9 cells is accomplished from medium harvested between 72 and 96 hr after infection. The cells are pelleted by centrifugation at 5000 *g* for 15 min at 4° and the medium removed for dialysis against 10 m*M* Tris-HCl, pH 8.0, using hollow fiber dialysis. After dialysis, the medium is centrifuged at 5000 *g* for 15 min at 4° to remove any precipitate and applied to a DEAE-Blue Sepharose column (75-ml bed volume for 1 liter of medium) equilibrated in 10 m*M* Tris-HCl, pH 8.0. Both PLRP1 and PLRP2 bind to the column and can be eluted with a linear gradient from 0.0 to 0.3 *M* NaCl. PLRP1 is located by immunoblot of column fractions and PLRP2 is located by activity against tributyrin. The fractions containing the related protein are pooled and dialyzed against 30 m*M* Tris–succinate, pH 6.2, and the dialysate applied to a CM-Sepharose column (40-ml bed volume) equilibrated in the same buffer. The related proteins are eluted with a linear gradient from 0.0 to 0.3 *M* NaCl. At this point, PLRP1 and PLRP2 are generally homogeneous by SDS–PAGE. If minor contaminants are present, which occurs if the infected cells are harvested late and significant numbers are lysed, they can be removed by gel chromatography on

[27] M. E. Lowe, unpublished observations (1996).

[28] M. L. Jennens and M. E. Lowe, *J. Lipid Res.* **36,** 2374 (1995).

[29] M. E. Lowe, *Methods Enzymol.* **284,** [8], 1997 (this volume).

an S-200 column equilibrated in 10 m*M* Tris-HCl (pH 8.0) and 0.15 *M* NaCl. Yields vary from 3 to 10 mg/liter.

Enzymatic Properties of Pancreatic Lipase-Related Proteins 1 and 2

The purified recombinant PLRP1 and PLRP2 provided suitable preparations for determination of the enzymatic properties of both proteins. An activity for PLRP1 has not been identified despite testing multiple substrates under various conditions. No activity was found against short-, medium-, or long-chain triglycerides, phospholipids, cholesterol esters, or galactolipids.[2,27,30] The triglyceride substrates have been tested in the presence and absence of taurodeoxycholate and colipase and at pH values ranging from pH 6.0 to 8.0. The inability to demonstrate activity for PLRP1 raised concerns about its physiological significance. Although strong arguments can be made that PLRP1 is unlikely to be an inactive homolog of pancreatic lipase, its role in pancreatic function must remain suspect until an activity is identified.

In contrast, activity of PLRP2 against triglycerides, phospholipids, and galactolipids has been demonstrated.[28,30] These results demonstrated that the substrate preference of PLRP2 is broader than that of pancraetic lipase. Of particular interest was the finding that PLRP2 does not show interfacial activation, which implied that the presence of a lid domain alone does not determine if a lipase will be activated by interfaces. The lipolytic activity of PLRP2 and its secretion into pancreatic juice suggest that it plays a role in fat digestion. The temporal pattern of mRNA expression for pancreatic lipase and PLRP2 may be important for fat digestion in the suckling period. Still to be explained is the close association of PLRP2 with the zymogen granule membrane.[4]

Conclusion

The physiological function of the pancreatic lipase homologs, which are expressed in acinar cells, has not been defined. This chapter describes some of the methods that have been applied and can be applied to determine the function of PLRP1 and PLRP2. Because of the marked homology among the nucleotide sequence and among the amino acid sequences of these proteins, it is imperative that further investigations employ reagents that are specific for each species. As more is learned about the biological and enzymatic properties of these pancreatic lipase homologs, their physiological function will be delineated.

[30] L. Andersson, F. Carriere, M. E. Lowe, A. Nilsson, and R. Verger, *Biochim. Biophys. Acta* **1302,** 236–240 (1996).

[18] Structure and Function of Engineered *Pseudomonas mendocina* Lipase

By MATTHEW BOSTON, CAROL REQUADT, STEVE DANKO, ALISHA JARNAGIN, EUNICE ASHIZAWA, SHAN WU, A. J. POULOSE, and RICHARD BOTT

Introduction

Recombinant DNA technology provides the means to introduce site-specific substitutions in the amino acid sequence of a particular gene. The amino acid sequence in turn determines the three-dimensional structure that the protein will adopt and thereby its function. This technology offers the potential to engineer proteins for particular applications and conditions. To do this, it is necessary to develop an understanding of what changes in the structure result in the desired change in function to further improve the enzyme. This approach has been successfully employed for a number of enzymes including subtilisin and lysozyme, into which numerous substitutions have been introduced to alter enzymatic activity and stability.[1,2] To analyze the consequences of the site-specific substitutions, these studies have relied on an extensive database of crystal structures derived from nearly identical crystal forms. The situation is more complex if a single variant enzyme crystallizes in a different form than the native. Different crystal lattice contacts can introduce structural changes and must be differentiated from those that arise from site-specific substitutions.[3] In this chapter, we discuss an engineered lipase that has an altered substrate-binding surface. The engineered enzyme crystallized in a different form than the native enzyme. This necessitated differentiating structural change arising as a consequence of different crystal forms from the changes, arising from the site-specific substitutions, that give improved lipase performance.

Lipases are a diverse class of enzymes that hydrolyze a range of complex ester-linked triglycerides. *Pseudomonas mendocina* (originally reported as *putida*) produces an enzyme (originally classified as a cutinase) that hydrolyzes cutin, a waxy polyester found on the leaf surface.[4] This enzyme is also capable of hydrolyzing ester linkages in a broad variety of compounds

[1] J. A. Wells and D. A. Estell, *Trends Biochem. Sci.* **13,** 291 (1988).

[2] B. W. Matthews, *Annu. Rev. Biochem.* **62,** 139 (1993).

[3] J. Sebastian, A. K. Chandra, and P. E. Kolattukudy, *J. Bacteriol.* **169,** 131 (1987).

[4] R. Bott, J. Dauberman, R. Caldwell, C. Mitchinson, L. Wilson, B. Schmidt, C. Simpson, S. Power, P. Lad, I. H. Sagar, T. Graycar, and D. Estell, *Ann. N.Y. Acad. Sci.* **672,** 10 (1992).

0076-6879/97 $25.00

including mono- and triglycerides. It should be noted that classic lipases display the phenomenon of micellar activation.[5] This phenomenon manifests itself as an increase in enzymatic activity at substrate concentrations approaching the CMC (critical micellar concentration). The *Pseudomonas* enzyme does not display any micellar activation. However, by virtue of the ability of the enzyme to hydrolyze triglycerides, it is referred to here as a lipase and was selected as a parent enzyme to begin engineering efforts to produce an engineered lipase to function as an improved oil-hydrolyzing enzyme for detergent applications.

The presumed physiological function of a cutinase is to hydrolyze cutin encountered by the organism under natural conditions (near neutral pH and ambient temperature). It would be expected that in the course of evolution the *Pseudomonas* enzyme would have become optimized for this function. In contrast, the laundry environment (highly alkaline pH, higher temperatures, and detergency) represents an entirely different environment and one for which the cutinase has not evolved to function efficiently. Similarly, the substrates found in laundry derived from a wide range of synthetic, plant, and animal triglycerides along with wax esters that are different from cutin. The *Pseudomonas* enzyme was selected for engineering as a detergent lipase on the basis of its alkaline performance on triglyceride hydrolysis. The structural gene encoding this enzyme has been cloned and expressed in large quantities from *Escherichia coli*. It is inhibited by diethyl *p*-nitrophenyl phosphate, suggesting that is a serine esterase. The amino acid sequence of this enzyme is not homologous with those of any other reported *Pseudomonas* lipases.[6]

The three-dimensional structure of the *Pseudomonas* lipase shares a tertiary fold that is common to the "α/β hydrolases."[7] This tertiary folding motif consists of a core domain of an eight-stranded β sheet (strands 1–8) and six helices (A–F; Fig. 1). All lipases share a common three-dimensional structure consisting of a central β sheet that is predominantly parallel connected by α helix on both strands of this central sheet. Lipases can be further subdivided into two classes based on the topology of the connection of the β strands in the N-terminal half of their three-dimensional structures. The predominant class consists of those having a "back cross-over" connection such that the third β strand is found between the fourth and fifth strand, on the basis of linear sequence (the α/β hydrolases topology),

[5] L. Sarda and P. Desnuelle, *Biochim. Biophys. Acta* **30,** 513 (1958).

[6] R. Bott, unpublished observation (1996).

[7] D. L. Ollis, E. Cheah, M. Cygler, B. Dijsktra, F. Frolow, S. M. Franken, M. Harel, S. J. Remington, I. Silman, J. Schrag, J. L. Sussman, K. H. G. Verscheuren, and A. Goldman, *Protein Eng.* **5,** 197 (1992).

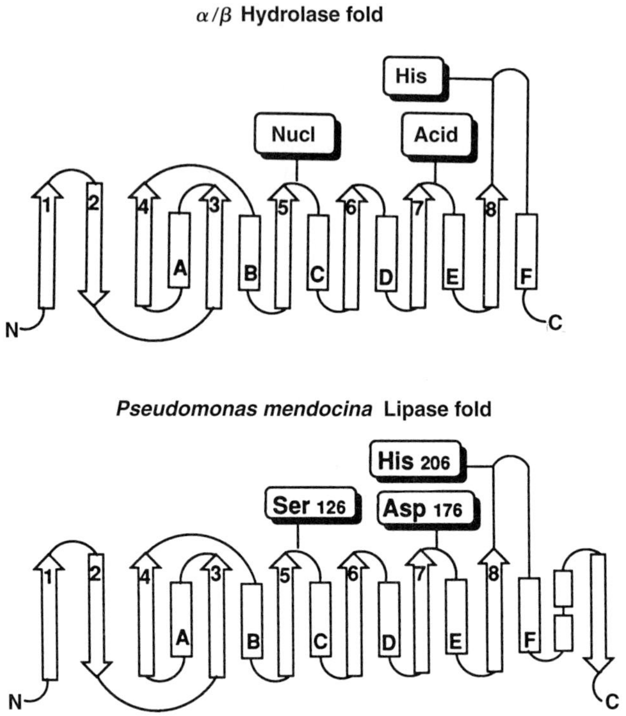

FIG. 1. Schematic diagram of α/β hydrolase fold and tertiary folding topology of *Pseudomonas* lipase.

including the lipases from human pancreas,[8] *Geotrichum candidum,*[9] *Candida rugosa,*[10] *C. antarctica,*[11] *P. glumae*[12] and *Fusarium solani pisi.*[13] The second class includes the structures of *Rhizomucor,*[14] *Humicola lanuginosa,*[15] *Rhizopus delemar*[15] and *Penicillium camembertii.*[15]

[8] F. K. Winkler, A. D'Arcy, and W. Hunziker, *Nature* (*London*) **343,** 767 (1990).

[9] J. D. Schrag, Y. Li, S. Wu, and M. Cygler, *Nature* (*London*) **351,** 761 (1991).

[10] P. Grochulski, Y. Li, J. D. Schrag, F. Bouthillier, P. Smith, D. Harrison, B. Rubin, and M. Cygler, *J. Biol. Chem.* **268,** 12843 (1993).

[11] J. Uppenberg and T. A. Jones, *Structure* **2,** 293 (1994).

[12] M. E. M. Noble, A. Cleasby, L. N. Johnson, M. R. Egmond, and L. G. J. Frenken, *FEBS Lett.* **331,** 123 (1993).

[13] C. Martiniez, P. DeGues, M. Lauwereys, G. Matthyssens, and C. Cambillau, *Nature* (*London*) **356,** 615 (1992).

[14] L. Brady, A. M. Brzozowsky, Z. S. Derewenda, E. Dodson, G. Dodoson, S. Tolley, J. P. Turkenburg, L. Christiansen, B. Huge-Jensen, L. Norskov, L. Thim, and U. Menge, *Nature* (*London*) **343,** 767 (1990).

[15] U. Derewenda, L. Swenson, R. Green, Y. Wei, S. Yamaguchi, R. Joerger, M. J. Haas, and Z. S. Derewenda, *Protein Eng.* **7,** 551 (1994).

The tertiary fold of *Pseudomonas* lipase also contains an additional strand, strand 9 of the core β sheet, preceded by an interrupted helix (G); these features are novel to this enzyme. The *Pseudomonas* enzyme is also an unusual α/β hydrolase in that the β strands and α helices are linked by short loops in contrast to the elaborate loops seen in most other α/β hydrolases and lipases. The majority of lipases have insertions at the loop junctions of β strands and helices that are sometimes longer than the strands of sheet and helical segments they connect. These insertions can form rather elaborate surface features that occlude the active site. The active site of *P. mendocina* lipase, however, is quite solvent accessible; the catalytic triad consisting of residues Ser-126, Asp-176, and His-206 is situated on the surface of the molecule (Fig. 2).

The *P. mendocina* enzyme is one of two such low molecular weight triglyceride hydrolases, the second also characterized as a cutinase from *F. solani pisi.* The three-dimensional structure of the *Fusarium* cutinase consists of a reduced α/β hydrolase fold consisting of five β strands and five α helices (strands 3–7; helices A, B, C, D, and F). The *fusarium* cutainse has also provided some insight into the understanding of micellar activation. As mentioned earlier, most of the lipolytic enzymes have highly elaborated loops blocking access to the active site of the enzyme. Several studies, one of an inhibitor bound to *Rhizomucor* lipase[16] and another analyzing the human lipase–colipase complex,[17] have supported the hypothesis that micelles induce a conformational change in the three-dimensional structure of lipases. In these complexes, loops that occlude the active site shift to an "open" conformation, creating a more hydrophobic surface and an oxyanion hole. This feature, also found in serine proteases, serves to polarize the peptide bond, facilitating nucleophilic attack by the catalytic serine hydroxyl.[18] The *Fusarium* enzyme has a preformed oxyanion hole and by virtue of shorter connecting loops between core β strands and connecting α helices has a solvent-accessible active site mimicking the open conformation of the other lipases.

Using the homologous α/β hydrolase fold, it is possible to align the *Pseudomonas* enzyme and the *Fusarium* cutinase. In the *Pseudomonas* lipase, there is a preformed oxyanion hole where the $O_{\gamma 1}$ atom of Thr-58 occupies the same position found for the O_{γ} atom of Ser-42, which forms the preformed oxyanion hole with the amide nitrogen of the catalytic Ser-120 in the *Fusarium* cutinase. It has been proposed that the *Fusarium*

[16] A. M. Brzozowski, U. Derewenda, Z. S. Derewenda, G. G. Dodson, D. M. Lawson, J. P. Turkenburg, F. Bjorking, B. Huge-Jensen, S. A. Patkar, and L. Thim, *Nature (London)* **351,** 491 (1991).

[17] H. van Tilbeurgh, M.-P. Egloff, C. Martinez, N. Rugani, R. Verger, and C. Cambillau, *Nature (London)* **362,** 814 (1993).

[18] J. Kraut, *Annu. Rev. Biochem.* **46,** 331 (1977).

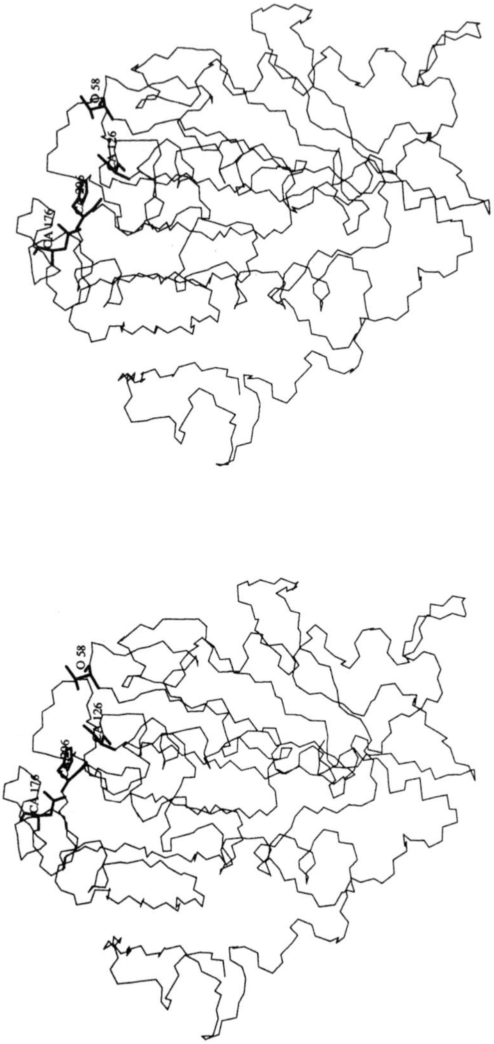

FIG. 2. Stereodiagram showing the location of the active site that is solvent accessible in *Pseudomonas mendocina* lipase. The side chains of Ser-126, Asp-176, and His-206, which form the catalytic triad, along with Thr58 are shown to delineate the active site.

cutinase represents a link between true esterase and true lipases, those triglyceride hydrolases displaying micellar activation.[19] Likewise, the *Pseudomonas* enzyme does not require micellar activation and therefore is an attractive candidate enzyme for laundry applications, because of its ability to hydrolyze lipid soils equally well above or below the CMC.

Construction and Selection of an Engineered Lipase

The engineered lipase is derived by regiospecific mutagenesis around the catalytic histidine, His-206, followed by screening of the library of variants of *P. mendocina* lipase. The lipase gene is cloned into a pUC18 plasmid and expressed in *E. coli* via the *tac* promoter. Unique *Bst*xI and *Bam*HI restriction sites are engineered upstream and downstream of the His-206 active site, using M13 mutagenesis according to the technique described by Sambrook *et al.*[20] After restriction digestion by *Bst*xI and *Bam*HI, a vector is prepared by removing a 41-bp segment around position 206 of the lipase structural gene. The segment encodes residues 197 to 211.

Synthetic primers are designed to mutagenize this region randomly by substituting a mixture of nucleotides for the correct deoxynucleotide every tenth synthetic cycle. This technique is known as *poisoning,* which in this example represents 10% poisoning, that is, every tenth oligonucleotide is "random."[21] The synthetic positive oligonucleotide strands are annealed to complementary negative strands; the *Bst*xI and *Bam*HI sites at the ends of the fragment permit ligation into the lipase gene around the region containing position 206.

A mutagenic library of 14,000 mutants is made, transformed into *E. coli,* and plated on Luria–agar plaets containing triolein emulsified with Nile Blue dye. Potential positives are selected on the basis of the clearing zone or halo, at least 10% larger in diameter than native, that forms around the growing colonies. Variants selected are then screened via a secondary performance screen.

A secondary screening assay is designed to select for improved wash performance on triglyceride substrates. The underlying assumption is that wash performance is associated with hydrolysis of triglycerides that are immobilized on a surface, in this case cloth. Triglyceride hydrolysis is measured by the release of radioactivity from cloth-immobilized [^{3}H]triolein into the wash supernatant.

[19] C. Martinez, A. Nicolas, H. van Tilbeurgh, M.-P. Egloff, C. Cudry, R. Verger, and C. Cambillau, *Biochemistry* **33,** 83 (1994).

[20] J. Sambrook, E. F. Fritsch, and T. Maniatis, *in* "Molecular Cloning: A Laboratory Manual," pp. 15.51–15.80. Cold Spring Harbor Laboratory Press, Cold Spring Harbor, New York, 1989.

[21] J. D. Hermes, S. M. Parekh, S. C. Blacklow, H. Koster, and J. R. Knowles, *Gene* **84,** 143 (1989).

[^{3}H]triolein stock solution is prepared by mixing 1.0 mCi of [^{3}H]triolein [0.5 mCi/ml; 17 μg of triolein per milliliter commercially available from New England Nuclear Products, (Du Pont, Boston, MA)] with 421 mg of unlabeled triolein and diluting to a final volume of 18 ml with 2-propanol. The stock solution is divided into 1-ml aliquots and stored at 4° for future use. Radioactive swatches are prepared by spotting 3 μl of [^{3}H]triolein stock solution (180,000–200,000 cpm) onto the center of 0.5 × 0.5 inch polyester swatches, and allowing them to dry for 1 hr at 22°.

Each swatch is then immersed in 2.8 ml of a 100 m*M* glycine buffer, pH 10, containing 1% (w/v) bovine serum albumin and preequilibrated to 30°. *Escherichia coli*-expressed enzymes representing random variants of *P. mendocina* lipase, obtained in a sucrose wash, are normalized by their rate of hydrolysis of *p*-nitrophenyl butyrate (pNB). Aliquots of variant enzymes with equal pNB activity are then added to the swatches containing tritium-labeled immobilized triolein and incubated in a shaker for 15 min at 32°. After 15 min, a 0.5-ml aliquot from the reaction mixture is removed, quenched with 5 ml of scintillation fluid, and ^{3}H released from the cloth is determined by scintillation. Tritium counts released into solution represent a measure of triolein hydrolysis and therefore a measure of lipase-mediated stain removal. Two enzyme levels are used in each experiment to verify the linearity of the rates measured. The initial rates are taken as a measure of the relative k_{cat}/K_m of the native enzyme and its variants.

The engineered lipase variant Y202L/R203L presents the largest increase in relative activity in terms of surface-immobilized triolein hydrolysis, among more than 200 variants selected from the primary screen. The data for the engineered enzyme vs the native form are summarized in Table I. The engineered variant displays a 220–270% hydrolysis rate of immobilized triolein relative to the native enzyme, at a concentration at which the enzymatic rate of hydrolysis of *p*-nitrophenyl butyrate is held constant.

At position 203, 10 single site-specific variants were constructed. The relative activity of these variants as measured by immobilized triolein hydro-

TABLE I
DATA FOR ENGINEERED ENZYME VERSUS NATIVE FORM

Lipase species	Concentration (ppm)	Lipase activity[a]
Native	0.25	212
L202/L203	0.25	568
Native	0.5	401
L202/L203	0.5	921

[a] Counts per minute of ^{3}H above background per 560 μl of supernatant.

lysis is compared with that of the native enzyme, which has tyrosine at position 203 (see Fig. 3). Substitution of medium to large hydrophobic side chains (R203L and R203V) resulted in the greatest activity. On the basis of the improvements in performance seen for single substitutions at position 203, we conclude that the improvement shown by variant Y202L/R203L is largely due to the R203L.

The activity on emusified substrates that were not immobilized was tested by pH-stat titration. These pH-stat assays were performed on the native and variant enzymes using two emulsified substrates: (1) a tricaprylin (C_8)–sodium dodecyl sulfate (SDS) emulsion, and (2) an olive oil (C_{18}) emulsion. The activity data are normalized to pNB activity and are summarized in Table II. These data demonstrate that the increased activity shown on emulsified triglyceride substrates correlates with activity on immobilized triglyceride on cloth swatches. Thus the Y202L/R203L variant shows increased lipase activity on representative lipid substrates. We have analyzed

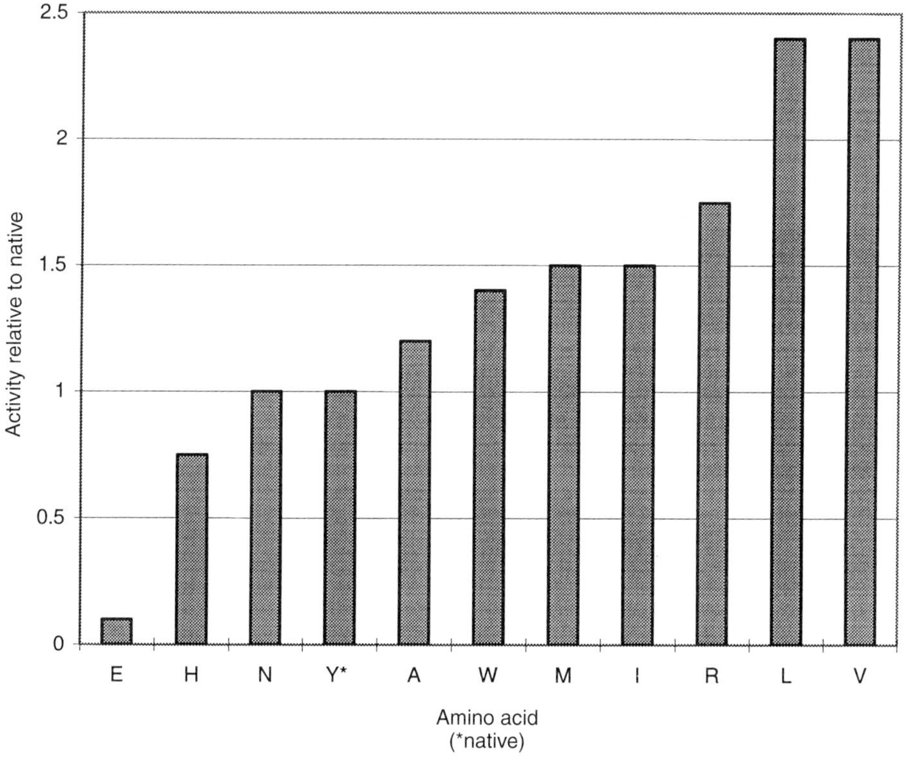

FIG. 3. Comparison of the rates of hydrolysis of cloth-immobilized triolein by variants of *Pseudomonas* lipase with site-specific substitutions at position 203.

TABLE II
ACTIVITY DATA NORMALIZED TO *p*-NITROPHENYL BUTYRATE ACTIVITY[a]

Lipase species	Tricaprylin hydrolysis	Olive oil hydrolysis
Native	0.293 μmol of NaOH/min	0.140 μmol of NaOH/min
L202/L203	0.609 μmol of NaOH/min	0.409 μmol of NaOH/min

[a] Micromoles of NaOH per minute is the amount of NaOH required to neutralize 1 μmol of fatty acid per minute.

the crystallographic structure for some understanding of the structural changes that have occurred to effect this increased lipolytic activity.

X-Ray Crystallography

Crystals of the native enzyme were obtained using hanging drop diffusions, whereby a 5-μl drop containing 4–10 mg of enzyme per milliliter in 50 m*M* sodium acetate buffer, pH 4.5, was mixed with 5 μl of a reservoir solution of 18% (w/v) saturated potassium tartrate. This drop was inverted and sealed over a well in a 4 × 6 Linbro tissue culture plate and allowed to reach equilibrium. The native crystals are of space group $P4_32_12$, with unit cell dimensions summarized in Table III. Crystals of the engineered enzyme were grown by batch technique, whereby a 4-mg/ml solution of

TABLE III
CRYSTALLOGRAPHIC AND REFINEMENT STATISTICS OF NATIVE AND ENGINEERED *Pseudomonas mendocina* LIPASE

Statistic	Enzyme: Native	Enzyme: Engineered lipase
Space group	$P4_32_12$	$C222_1$
Unit cell dimensions		
a (Å)	58.50	54.75
b (Å)	58.50	131.35
c (Å)	145.00	79.65
Resolution (Å)	10–1.8	10–1.8
No. of reflections	19,100	18,715
RMS deviation from ideality		
Bond length (Å)	0.013	0.015
Bond angle (°)	3.1	2.6
R factor	0.178	0.186
No hydrogen atoms	2,061	2,019
Atoms with variable occupancy	114	79

enzyme and 100 mM sodium chloride buffered with 10 mM N-tris(hydroxymethyl)methyl-2-aminoethanesulfonic acid (TES, pH 7.6) was left to stand, resulting in crystals of space group $C222_1$ (Table III). Attempts to crystallize the engineered enzyme in the native crystal form by seeding with small native crystals proved unsuccessful. Data were collected using a Rigaku Raxis-II area detector (monochromatic CuK_α radiation) and a Rigaku RU200 rotating anode generator (0.3-mm fine focus cathode running at 50 mA and 100 kV). Data from two crystals each of the native and variant enzymes in roughly perpendicular orientation were collected to maximize completeness of the data collected in a fixed ϕ, ω scan mode. Data were reduced, merged, and scaled using the manufacturer-supplied software. Both crystal forms diffracted to comparably high resolution (1.8 Å) with 19,100 and 18,715 reflections measured from the crystals of the native and engineered enzyme, respectively.

The structure of the engineered enzyme was determined by molecular replacement, using the coordinates of the native enzyme. The method determines the rotation and translation necessary to orient and position a known molecule, the native lipase, to serve as an initial model for the unknown structure, the engineered lipase. The orientation was determined by rotation function,[22] in which an initial, 5°-interval search identified a single 15σ maximum that subsequently led to an 18σ maximum in a fine, 1° search. Once the orientation is determined, the molecule must then be correctly positioned with respect to the origin in the unit cell of the unknown crystal lattice. A translation function, using a brute force R-factor minimization search program,[23] identified in a translation to position the native enzyme that resulted in an R factor of 0.34 for all data of 10- to 5-Å resolution. These parameters were refined by rigid-body least-squares refinement using the program ROTLSQ, which gave an R factor of 0.299 for all data of 10- to 2.5-Å resolution.

The coordinates of the native enzyme after rotation and translation were then used to calculate $F_o - F_c$ and $2F_o - F_c$ electron-density maps, where F_o and F_c are the observed and calculated structure factors for a given reflection (hkl). The entire structure of the enzyme could be seen in the $2F_o - F_c$ electron-density map. It was also straightforward to position the side-chain atoms of the two substitutions Y202L and R203L. The side chains of these residues adopt quite similar conformations for the common atoms between the old and substituted side chains (Fig. 4). Errors detected in the $F_o - F_c$ map were corrected and solvent added on the basis of peaks that were confirmed to have reasonable geometry and corresponding

[22] M. G. Rossmann and D. M. Blow, *Acta Crystallogr.* **15,** 24 (1962).
[23] R. Bott and R. Sarma, *J. Mol. Biol.* **106,** 1037 (1976).

FIG. 4. Stereodiagram of the electron-density map in the region of the site-specific substitutions at positions 202 and 203 for the engineered *Pseudomonas mendocina* lipase. Leucine side chains have been substituted for tyrosine and arginine at positions 202 and 203, respectively.

density in the $2F_o - F_c$ electron-density map. The resulting model was refined with the program PROLSQ.[24] The refinement statistics of the native and engineered enzyme, after two rounds of adjustment and refinement, are presented in Table III. A representative example of the electron-density map in the vicinity of the active site serine is shown in Fig. 5. The electron density is of sufficient quantity to distinguish easily the characteristic conformation of the side chain of the catalytic serine, Ser-126, as seen in all other lipases.[16] The side-chain density for the catalytic His-206 and Thr-58 can also be seen in Fig. 5. The high quality of the electron-density map suggests that the residues forming the active site are well ordered.

Differentiating Structural Consequences of Crystallization and Site-Specific Mutagenesis

The engineered enzyme has been crystallized under different conditions of pH, buffer, and precipitant, which resulted in a different crystal form and space group compared to the native enzyme. The crystal forms of the native and engineered enzymes have different symmetry elements: a fourfold screw axis in a primitive lattice versus a twofold screw axis in a face-centered C-lattice. Considerable care must be exercised to differentiate the potentially causative factors arising from the site-specific substitutions from the structural changes that are artifacts of different crystal lattice interactions.

Table IV lists residues that are involved in lattice interactions between enzyme molecules in either of the two crystal forms. Differences in the native and engineered enzyme at or near these residues could easily be attributed to altered lattice interactions. The crystal lattice interactions in Table V include a common subset of residues (Ala-9, Thr-58, Gly-59, Pro-107, and Tyr-108) that are involved in lattice interactions in both crystal forms. It is interesting that these residues interact with a different constellation of residues through different symmetry operations in the two crystal forms. For example, Tyr-108 links two different, symmetry-related molecules in both crystal lattices. In the native enzyme crystal Tyr-108 links residues Asp-18, Ser-19, Ser-20, and His-71 from one molecule and Arg-162 from a second molecule; in the engineered enzyme crystal, Tyr-108 links two molecules via different symmetry operations and by linking different segments (residues Gly-43, Gly-45, and Gly-46 of one molecule and Gly-257 from a second molecule). Similar results, in which residues form different lattice interactions in different crystal forms, were also found in different

[24] W. H. Hendrickson and J. H. Konnert, *in* "Biomolecular Structure Conformation Function and Evolution" (R. Srinivasan, ed.), Vol. 1, p. 43. Pergamon Press, Oxford, 1981.

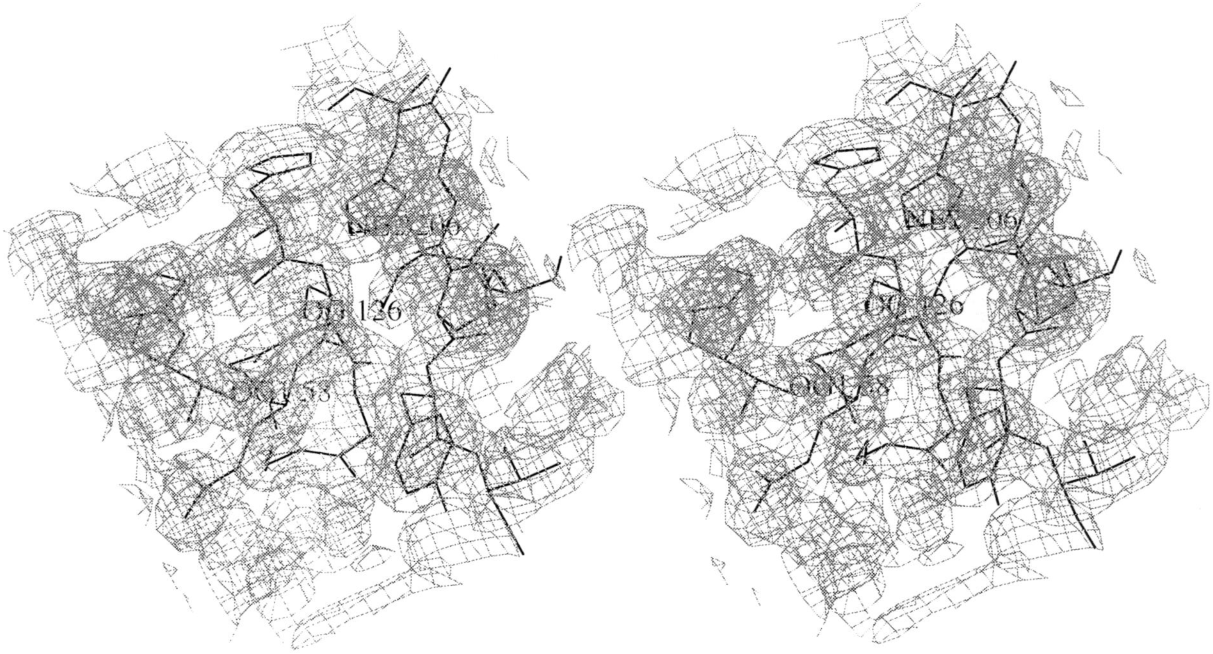

FIG. 5. Stereodiagram of the electron-density map at the active site of the engineered *Pseudomonas* lipase. The electron density for Ser-126 and His-206 from the catalytic triad and Thr-58, which forms part of the oxyanion hole, may be seen. The quality of the electron-density map suggests that the active site is well ordered.

TABLE IV
LATTICE CONTACT RESIDUES IN NATIVE AND ENGINEERED *Pseudomonas* LIPASE[a]

Native lipase			Engineered lipase		
Residue(s)	Symmetry	Residue(s)	Residue(s)	Symmetry	Residue(s)
A1	$y, x, -z$	S29	D5, P7	$\frac{1}{2}+x, \frac{1}{2}-y, -z$	R162
P2	$y, x, -z$	S33			
A9	$\frac{1}{2}+y, \frac{1}{2}-x, \frac{1}{4}+z$	Y203	P7, G8, A9	$\frac{1}{2}+x, \frac{1}{2}-y, -z$	N193
			P10	$\frac{1}{2}+x, \frac{1}{2}-y, -z$	Q164, Q165
D18, S19, S20	$\frac{1}{2}-y, \frac{1}{2}+x, \frac{3}{4}+z$	Y108			
S20	$\frac{1}{2}+x, \frac{1}{2}-y, \frac{1}{4}-z$	R162			
S21, P22	$\frac{1}{2}+x, \frac{1}{2}-y, \frac{1}{4}-z$	R190			
T24	$\frac{1}{2}+x, \frac{1}{2}-y, \frac{1}{4}-z$	Q186			
R40	$\frac{1}{2}+x, \frac{1}{2}-y, \frac{1}{4}-z$	Q186, E254			
Q44	$\frac{1}{2}+x, \frac{1}{2}-y, \frac{1}{4}-z$	Y182			
G45	$\frac{1}{2}+y, \frac{1}{2}-x, \frac{1}{4}+z$	L250	G43, G45, G46	$-x, y, \frac{1}{2}-z$	Y108
R48	$\frac{1}{2}+y, \frac{1}{2}-x, \frac{1}{4}+z$	R201			
T58, G59	$\frac{1}{2}+x, \frac{1}{2}-y, \frac{1}{4}-z$	T58	T58, G59, A60, T64	$x, -y, -z$	F207
H71	$\frac{1}{2}-y, \frac{1}{2}+x, \frac{3}{4}+z$	Y108			
N86	$\frac{1}{2}+x, \frac{1}{2}-y, \frac{1}{4}-z$	I178	G59	$x, -y, -z$	Y177
D105, P107	$\frac{1}{2}+y, \frac{1}{2}-x, \frac{1}{4}+z$	S212	P107, Y108	$\frac{1}{2}-x, \frac{1}{2}-y, \frac{1}{2}+z$	G257
P107	$\frac{1}{2}+y, \frac{1}{2}-x, \frac{1}{4}+z$	R217	P107	$x, y, \frac{1}{2}-z$	L258
L154	$\frac{1}{2}+x, \frac{1}{2}-y, \frac{1}{4}-z$	I178	R190	$-x, y, \frac{1}{2}-z$	V253, E254

[a] All residues that are within van der Waals contact distance are listed.

TABLE V
RESIDUES IN ENGINEERED LIPASE VARIANT DISPLAYING SIGNIFICANT DIFFERENCES COMPARED WITH NATIVE ENZYME

Residue	*Z* score	Probable cause
Arg-35	3.3	Side-chain disorder
Thr-58	3.1	Lattice interaction
Gly-59	3.5	Lattice interaction
Ser-63	3.3	Lattice interaction
Thr-64	3.0	Lattice interaction
Arg-91	4.6	Side-chain disorder
Tyr-108	6.4	Lattice interaction
Gly-109	3.2	Lattice interaction
Thr-110	3.2	Lattice interaction
Tyr-111	3.4	Lattice interaction
Ser-112	3.7	Lattice interaction
Thr-139	3.4	Side-chain disorder
Arg-190	6.6	Lattice interaction
Phe-207	3.3	Lattice interaction

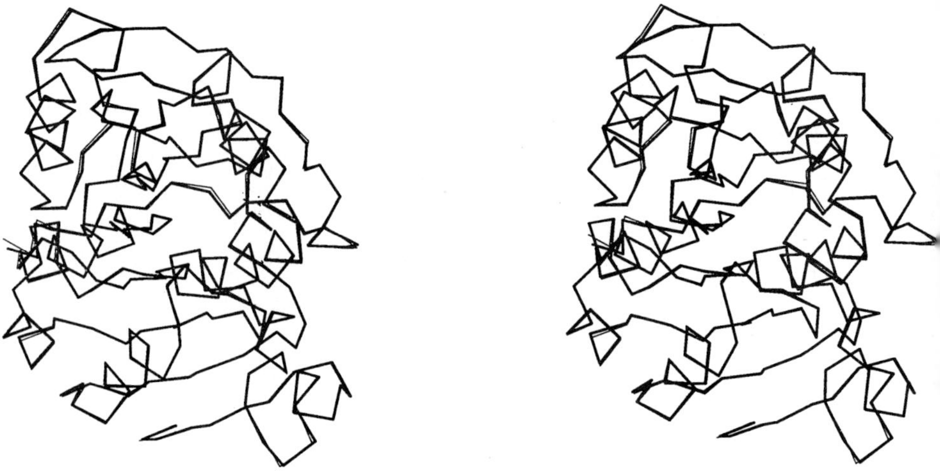

FIG. 6. Stereodiagram comparing the C_α backbone folding of the native (thin lines) and engineered (thick lines) *Pseudomonas mendocina* lipase.

crystal forms of a variant of *Bacillus lentus subtilisin.*[25] It is likely that these lattice contact residues are potentially sticky surfaces and prominent on the surface of the molecules, and the presence of these will require favorable lattice interactions to form for crystallization to occur. The remaining residues form lattice contacs in only one of the two crystal lattices. Approximately similar numbers of overall contacts were found to hold the individual enzyme molecules in place in the two crystal forms.

It is possible to align the native and engineered enzymes to facilitate comparison of their structure. A comparison of the backbone folding of the native and variant enzymes is presented in Fig. 6. The differences between the structures are found in the conformation of side chains or in the subtle alteration in main-chain conformation. It is therefore possible to relate the equivalent atoms in the two structures. This in turns allows for an analysis of variance of the differences in atomic coordinates of equivalent atoms in the native and engineered lipase structures. This analysis is facilitated by a linear relationship between the logarithm of the difference in atomic coordinates and the crystallographic temperature factor.[26]

[25] J. Dauberman, G. Ganshaw, C. Simpson, T. P. Graycar, S. McGinnis, and R. Bott, *Acta Crystallogr.* **D50,** 650 (1994).

[26] R. Bott and J. Frane, *Protein Eng.* **3,** 649 (1990).

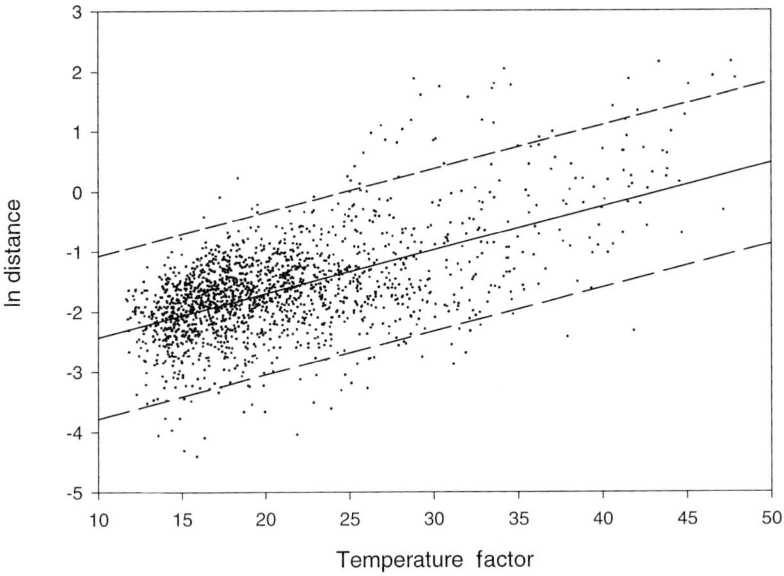

FIG. 7. Plot of the distribution of the natural logarithm of the distance between equivalent atoms in the native and engineered *Pseudomonas mendocina* lipase as a function of the crystallographic temperature factor.

The linear relationship (Fig. 7) provides a means to determine the mean error and, more important, the variance about the mean. The temperature factor is a measure of the relative mobility of different segments within the protein structure. Coordinates of segments with higher temperature factors tend to vary more than segments with low-temperature factors.

Having both an estimate of the mean error as well as the variance about the mean error provides the necessary information to compute statistical *Z* scores for differences between individual residues in the native and variant structures. A residue *Z* score above 3.0 is taken here to indicate that the residue is significantly different, on the basis of the average temperature factor of the same residue in native and engineered lipase. Table V lists 14 positions in engineered lipase giving residue *Z* scores above 3.0 when compared to the native enzyme. Nine of these residues are found in two segments, residues 58–64 and 108–112; these residues are involved in lattice contacts in either or both crystal forms that are potentially influenced by these lattice interactions. The significant differences seen for these residues are likely to be due to the different lattice interactions rather than being a consequence of the substitutions at positions 202 and 203. Three more residues, Arg-35, Arg-91, and Thr-139, are all far removed from the site of substitution and are consistent with the potential side-chain disorder that is seen for lysine and arginine side chains, which are known to adopt multiple conformations and found to have high residue *Z* scores in other comparisons of related protein structures.[26] These may be attributed to disorder, which in different crystal lattices favors different conformers even in the absence of immediate lattice contacts. The side-chain flexibility or disorder of these residues is the probable cause of their high residue *Z* score.

The differences detected for residues 190 and 207, which are involved in lattice contacts in one or both forms, are more problematic. It is likely that the differences in positions 190 and 207 are due to the different lattice interactions. This is indicated in the Probable cause column of Table V. However, it may also be possible, because these residues bracket the substitutions at positions 202 and 203, that one or both of these differences may be a consequence of the substitutions. The latter conclusion, however, is less probable because we do not observe changes in the positions that are closer to the substituted residues.

Relating Structure to Function

Thus there are no structural differences that may confidently be interpreted as being probable consequences of the amino acid substitutions at positions 202 and 203. The improved performance is associated with hydrolysis of triglyceride substrates. Experimental results point to a retention of a common binding site for short-chain substrates for which no appreciable change is seen between native and engineered enzyme. The modification at positions 202 and 203 must selectively alter the performance on triglyceride substrates in a more subtle manner. The finding that hydrolysis of triglyceride substrates is selectively improved led us to consider the need for triglyceride substrates to bind a number of aliphatic hydrocarbon chains simultaneously. From this we postulate that the R202L/Y203L substitution most likely modifies the surface of the enzyme, selectively modifying a site exploited for triglyceride substrates that is not utilized for *p*-nitrophenyl butyrate.

This would ideally be deduced from an analysis of the structure of enzyme–substrate interactions that are not presently viable for lipolytic enzymes. However, in the absence of any three-dimensional structure of enzyme–substrate complexes for lipases, crystallographers often make extensive use of substrate analogs to deduce some models for substrate binding to lipases.[19,27,28] The most extensive work has been conducted

[27] U. Derewenda, A. M. Brzozowski, D. M. Lawson, and Z. S. Derewenda, *Biochemistry* **31,** 1532 (1992).

[28] D. M. Lawson, A. M. Brzozowski, S. Rety, C. Verma, and G. G. Dodson, *Protein Eng.* **7,** 543 (1994).

with *Humicola* lipase.[28] In this work two inhibitors, diethyl *p*-nitrophenyl phosphate and dodecylethylphosphate ethyl ester, were used as a basis for modeling enzyme–substrate complexes that led to the identification of five possible binding sites for a triglyceride substrate having three acyl chains.

Remembering that *Rhizomucor* lipase belongs to the second class of lipases, which are less homologous to *Pseudomonas* lipase, it is beneficial that the three-dimensional structure of a complex having the same inhibitor, diethyl *p*-nitrophenyl phosphate, bound to the *Fusarium* cutinase was determined.[19] Despite differences in specific details, the *Pseudomonas* enzyme three-dimensional structure can be related to the *Fusarium* cutinase structure on the basis of the core sheet segments of the two enzymes and helices C and F. The root-mean-square (RMS) deviation for the α carbons of the catalytic triad after alignment is 0.67 Å. After alignment, the OG1 of Thr-58 can be related to OG of Ser-42, which forms the oxyanion hole in the *Fusarium* enzyme. From this homology, it is possible to identify the location of the oxyanion hole in *Pseudomonas* lipase, which is also preformed in both engineered and native enzymes. Using the alignment of *Pseudomonas* lipase with the *Fusarium* cutinase complex with the inhibitor E600, it is possible to obtain a model of E600 bound to the *Pseudomonas* enzyme. This model represents a plausible binding of the inhibitor to *Pseudomonas* lipase (Fig. 8). Figure 8 also shows the location of the substituted residues at positions 202 and 203, which are at least 16 Å away from the inhibitor-binding site, which is presumed to indicate the acyl chain-binding site for mono- and triglycerides.

The results suggest that both classes of enzymes bind monoglyceride substrates as evidenced by the mode of binding of monoglyceride-like inhibitors at a homologous site formed in the homologous C-terminal half of their respective structures. This allows us to incorporate more confidently the results from *Humicola* and *Rhizomucor.* Positions 202 and 203 are situated in the vicinity of one of four additional binding sites for triglyceride substrates, as determined by modeling studies based on the inhibitor-binding studies with *Humicola* lipase.[28] This site is situated in the C-terminal half of the *Rhizomucor* lipase and *Pseudomonas* enzyme, which are the most homologous of the enzymes. Although these positions are situated far from the active site, they are within the interaction distance of C_{18} hydrocarbon chains found in triolein that are more rapidly hydrolyzed by the engineered enzyme. The replacement of two hydrophilic residues, arginine and tyrosine, with more hydrophobic leucine side chains would increase the hydrophobicity of this binding site. We conclude that creation of a better binding site for one of the extra acyl side chains present in triglyceride substrates is the likely explanation for the increased performance of this engineered lipase in triglyceride hydrolysis.

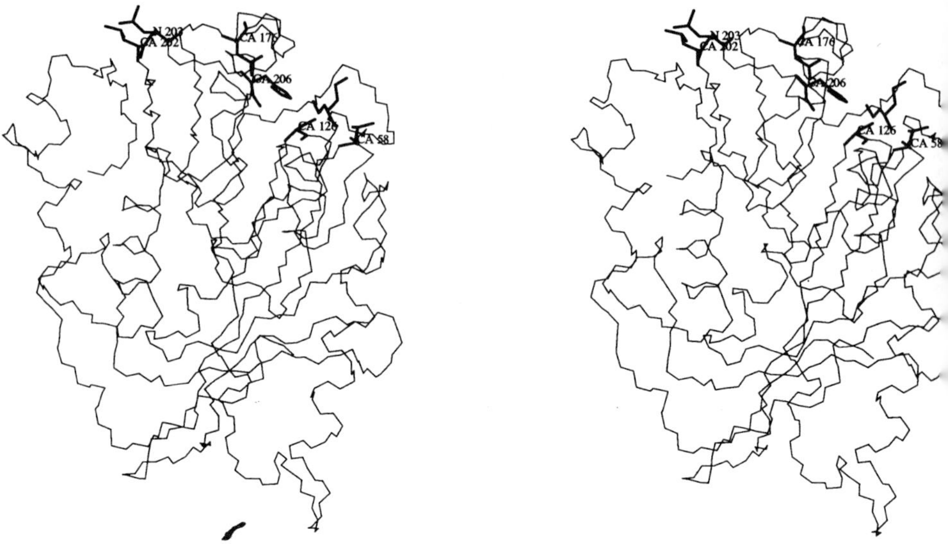

FIG. 8. Model of diethyl phosphonate complexed to the engineered *Pseudomonas mendocina* lipase. The relative locations of the site-specific substitutions at positions 202 and 203 are indicated by side chains along with residues of the catalytic triad Ser-126, Asp-176, and His-206, and Thr-58, which forms the oxyanion hole. It may be seen that positions 202 and 203 are now interactive with the inhibitor, which mimicks small monoglyceride substrates. Instead, positions 202 and 203 are proposed to form part of an ancillary binding site to bind one of the additional acyl side chains present in long-chain triglyceride substrates such as triolein.

Speculation Concerning the Crystallization of Engineered Lipase

It is not clear why the engineered enzyme crystallizes in a different crystal form, given that neither substitution is involved in a lattice contact in either crystal form. The analysis by the necessity of differentiating structural differences arising as "artifacts" of the different subtle lattice deformations in one or the other structure crystal lattice. We also cannot entirely rule out the possibility that differences seen for residues at positions 190 and 207, which bracket the substitutions, may be a consequence of these substitutions, which in fact predispose the engineered enzyme to crystallize in a different crystal form.

Conclusion

Our analysis was complicated by different crystal forms. All of the major structural differences are due to the distortion occurring when molecules crystallize in different lattices. We were able to exploit what appears to be a conserved mode of substrate binding in lipases to identify the probable effect leading to increased lipolytic activity in the engineered enzyme.

Because the activity increase is seen for triglyceride substrates, particularly those with long (C_{18}) acyl side chains, as opposed to *p*-nitrophenyl butyrate, we have introduced increased lipase activity into the *P. mendocina* enzyme. We conclude that the primary benefit of the Y202L/R203L substitutions is to increase the acyl chain binding of a triglyceride substrate not utilized when a monoglyceride substrate is hydrolyzed. These positions are juxtaposed far from the catalytic triad and support the assumption, based on work with other lipases, that triglyceride binding will involve a large binding site having multiple acyl chain-binding sites.

[19] Protein Engineering of Microbial Lipases of Industrial Interest

By Allan Svendsen, Ib Groth Clausen, Shamkant Anant Patkar, Kim Borch, and Marianne Thellersen

Introduction

Protein engineering of industrial enzymes to obtain better enzymes for specific applications has long been a subject of interest. Mainly lipases and proteases, but also amylases, glucoisomerases, cellulases, and other carbohydrate-degrading enzymes, have been targets of protein engineering. The most significant application in the engineering of new variant enzymes has been for detergent use; but the relatively efficient protein-engineering methods make it feasible to work with many other enzymes and applications.

Protein engineering involves recombinant DNA technologies, which make it possible to alter sequences site specifically or randomly.

Lipases have a number of potential industrial applications such as production of esters and specialty fats, removal of resins from pulp, cleaning of hard surfaces, and use in detergents. In future, chiral synthesis of high-value compounds could become a new and important field. Enzymes showing lipolytic activity can in some cases also act as esterases, phospholipases, cholesterolesterases, thioesterases, and cutinases. The specificities of lipases are broad, but each enzyme has a preference. The selectivity for a specific activity of the lipases can be improved by protein engineer-

0076-6879/97 $25.00

ing.[1] Today most of the industrially relevant efforts in protein engineering of lipases have been to improve the hydrolytic efficiency, peracid generation,[2] and detergent and protease stability,[3–5] all aiming at applications for detergents. Other protein-engineering efforts have been published on the analysis of active site residues[6] and on the specificity and activity of lid variants.[7,8]

Structure and Function of Lipases: Lipid Contact Zone

The first descriptions of microbial lipases were of the *Humicola lanuginosa* (now renamed *Thermomyces lanuginosus*) lipase[9] and the *Rhizomucor miehei* lipase.[10]

One of the major differences within the catalytic domains of lipases concerns the loops forming the lipid contact zone of the enzyme. The lipase structure can be described as an enzyme having a scaffold consisting of a β sheet with a Rossman-like fold around the center of the sheet.[11] The α/β fold around the β turn having the active site serine residue is referred to

[1] J. S. Okkels, A. Svendsen, S. A. Patkar, and K. Borch, *in* "Engineering of/with Lipases" (F. X. Malcata, ed.), p. 203. NATO ANSI Series. Kluwer Academic Publishers, Dordrecht, 1996.

[2] R. Bott, J. W. Shield, and A. J. Poulose, *in* "Lipases: Their Structure, Biochemistry and Application" (P. Woolley and S. B. Petersen, eds.), p. 337. Cambridge University Press, Cambridge, 1994.

[3] M. R. Egmond, J. de Vlieg, H. M. Verhey, and G. H. de Haas, *in* "Engineering off/with Lipases" (F. X. Malcata, ed.), p. 193. NATO ANSI Series. Kluwer Academic Publishers, Dordrecht, 1996.

[4] L. G. J. Frenken, M. R. Egmond, A. M. Batenburg, and C. T. Verrips, *Protein Eng.* **6,** 637 (1993).

[5] A. Svendsen, *Inform* **5,** 619 (1994).

[6] J. D. Schrag, T. Vernet, L. Laramee, D. Y. Thomaas, A. Recktenwald, M. Okoniewska, E. Ziomek, and M. Cygler, *Protein Eng.* **8,** 835 (1994).

[7] M. Holmquist, I. G. Clausen, S. Patkar, A. Svendsen, and K. Hult, *J. Protein Chem.* **14,** 217 (1995).

[8] M. Holmquist, M. Martinelle, I. G. Clausen, S.A. Patkar, A. Svendsen, and K. Hult, *Lipids* **29,** (1995).

[9] D. M. Lawson, A. M. Brzozowski, G. G. Dodson, R. E. Hubbard, B. Huge-Jensen, E. Boel, and Z. S. Derewenda, *in* "Lipases: Their Structure, Biochemistry and Application" (P. Woolley and S. B. Petersen, eds.), p. 77. Cambridge University Press, Cambridge, 1994.

[10] L. Brady, A. M. Brzozowski, Z. S. Derewenda, E. Dodson, G. G. Dodson, S. Tolley, J. P. Turkenburg, L. Christiansen, B. Huge-Jensen, L. Nørskov, L. Thim, and U. Menge, *Nature (London)* **343,** 767 (1990).

[11] A. Svendsen, *in* "Lipases: Their Structure, Biochemistry and Application" (P. Woolley and S. B. Petersen, eds.), p. 1. Cambridge University Press, Cambridge, 1994.

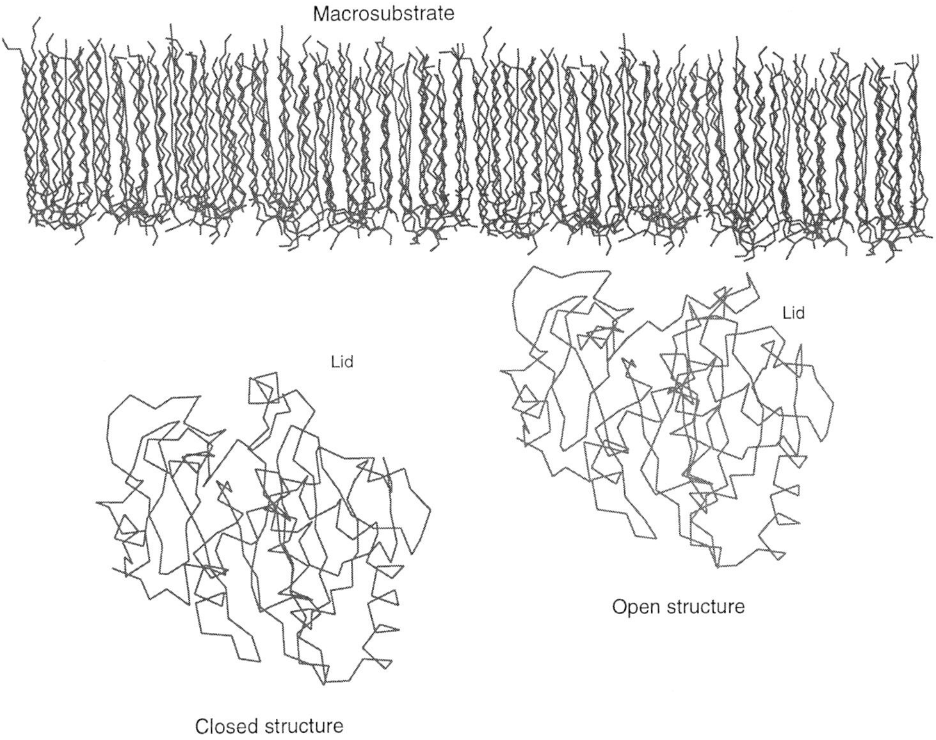

FIG. 1. The lipid contact zone of *Humicola lanuginosa* lipase: side views of the closed and open structures of *Humicola lanuginosa* lipase. The lipid contact zone points toward the lipid, called the macrosubstrate. The open structure (activated lipase) is shown closer to the macrosubstrate than is the closed structure (inactivated lipase), indicating that the proposed activation of the lipase takes place in near contact with the macrosubstrate.

as the α/β elbow.[12] All known structures of lipases have this α/β fold in common. In the vicinity of the active site serine side chain the scaffold consists of several loops and looplike structures that form a region in contact with the macroscopic lipid surface; it also contains the structures referred to as lids or flaps. We call this region the *lipid contact zone* (Fig. 1). The lid(s) changes conformation on binding to the lipid surface. To act on the lipid substrate, the lipase must bind to the macroscopic lipid surface. At least four steps are assumed to be necessary for the lipase to hydrolyze the ester bond between the glycerol and the fatty acid[5]: (1) binding to the

[12] D. L. Ollis, E. Cheah, M. Cygler, B. Dijkstra, F. Frolow, S. M. Franken, M. Harel, S. J. Remington, I. Silman, J. Schrag, J. L. Sussman, K. H. G. Vershueren, and A. Goldman, *Protein Eng.* **5,** 197 (1992).

macroscopic lipid surface, (2) penetration into the lipid or the interphase, with correct orientation of the enzyme, (3) activation of the lipase [i.e., opening of the lid(s)], and (4) hydrolytic activity on the ester bond after a part of the monosubstrate (i.e., the triglyceride) has been bound in the elongated hydrophobic binding cleft and the monosubstrate has been oriented correctly in the active site. These hypothetical processes have not been fully elucidated. Whether the activation happens before or after penetration, and whether the penetration process consists of the enzyme entering the lipid or the substrate entering the lipase is debatable.

We have attempted to obtain a better detergent lipase on the basis of the preceding assumptions.

Engineering Surface Charges in *Humicola lanuginosa* Lipase

To improve lipase activity in detergents for laundry use, the lipase should be stable and active under the actual washing conditions used. The conditions include high pH, the presence of anionic and nonionic surfactants, the presence of builders to remove calcium ions, and frequently also oxidative compounds.

Improved Binding to Negatively Charged Surfaces

Removal or Decrease in Negative Charges

Removal of Negative Charges from Lipase Surface. The negative charges on the surface of the lipase decrease binding to the lipid surface by repulsion of negatively charged fatty acids or anionic surfactant molecules in the lipid layer. The negative charges are dominant at the high pH used in detergents. Therefore, a generally higher p*I* of the enzyme could reflect improved binding to the negatively charged surfaces owing to changes in the overall electrostatic surface of the enzyme.

Removal of Negative Charges from Lipid Contact Zone. The negatively charged amino acids of the lipid contact zone were replaced by predominantly positive or hydrophobic residues, but also by hydrophilic residues. The reason for this kind of substitution was to lower the negative potential on the lipase surface, especially in the lipid contact zone. The negatively charged residues in the lipid contact zone are, e.g., D96, E210, E87, D254, and E56 in *H. lanuginosa* lipase. These residues were initially changed to R, K, L, and I, and later to all other possible residues. The D96L variant in *H. lanuginosa* lipase is clearly improved in wash, as shown below.

Introducing Hydrophobic Residues in Lipid Contact Zone

Lipases have a strikingly hydrophobic character. The increase in hydrophobic surface area of *R. miehei* lipase on activation is approximately 500 Å^2.[13] Increasing the hydrophobicity thus could facilitate the interaction and binding of enzyme to the lipid surface. As is shown below, the surface activity increases in variants of *H. lanuginosa* lipase having an increased surface hydrophobicity of the lipid contact zone.

Improved Opening of Lid: Activation of Lipases

The lid is in the closed conformation when the lipase is in a water solution (see Fig. 1). When bound to the lipid surface the lid is displaced by a special mechanism not fully understood at present. Theoretical studies on lid opening[14,15] indicate that the open conformation of the lid is more favorable at low dielectric constant, as found in lipids or organic solvents, and unfavorable at high dielectric constants, as found in water solution. The contact between the lid and the rest of the lipase in the closed form is mainly hydrophobic in nature, as seen in most globular proteins. When the lid opens it exposes this hydrophobic surface, which probably is in contact with the hydrophobic part of the substrate. On opening, the hydrophilic part of the amphipathic lid helix creates a partly buried hydrophilic interaction within the lipase. This hydrophilic zone, together with an inherent hydrophilic part of the core and the many buried waters,[16] probably stabilizes the lipase in the lipid interphase. To improve opening of the lid, the contact between the lid and the rest of the lipase in open and closed conformation has been subject to alterations. The interaction zone between the lid and the rest of the lipase has also been made more hydrophilic. Other approaches to achieve an activated lipase have been to bind the lid covalently in the open conformation by adding a disulfide bridge or simply by removing the lid.

Considerations for Mutation Suggestions

When improving an enzyme for one purpose, many other parameters may be changed in a negative direction. Therefore, stability and activity

[13] A. M. Brzozowski, U. Derewenda, Z. S. Derewenda, G. G. Dodson, D. M. Lawson, J. P. Turkenberg, F. Bjørkling, B. Huge-Jensen, S. A. Patkar, and L. Thim, *Nature (London)* **351,** 491 (1991).

[14] M. Norin, O. Olsen, A. Svendsen, O. Edholm, and K. Hult, *Protein Eng.* **6,** 855 (1993).

[15] G. H. Peters, T. Frimurer, S. Toxvaerd, O. H. Olsen, and A. Svendsen, *IEEE Comput. Sci. Eng.* **2,** 43 (1995).

[16] U. Derewenda, L. Swenson, R. Green, Y. Wei, G. G. Dodson, S. Yamaguchi, M. I. Haas, and Z. S. Derewenda, *Struct. Biol.* **1,** 36 (1994).

tests, or tests of other properties, must accompany the evaluation of variants. Other properties of interest can include solubility, dependence on the presence of metal ions, dependence of surfactants, etc.

Stability. Stability is of great concern if the enzyme must work close to its natural melting temperature. If, however, the enzyme in question has a melting temperature higher than the working temperature, then instability introduced by mutations may not be a problem.

Activity. When changing the amino acids of an enzyme, e.g., the electrostatic environment of the enzyme, by mutating the charged amino acids, or residues closely involved in vital mechanisms of the enzyme, the hydrolytic efficiency is often changed. The following sections show examples of dramatic altering of specific activity, and of the dependence on the substrate used.

Site-Directed Mutagenesis Methods

Cloning and Expression

The gene from *H. lanuginosa* encoding the triacylglycerol lipase has been cloned, sequenced, and expressed and the lipase has been purified from the medium.[17] Expression of the gene was achieved in *Aspergillus oryzae* by integrative protoplast cotransformation with an expression plasmid and a marker plasmid carrying the acetamidase gene (*amdS*) from *Aspergillus nidulans.* For driving the transcription, the homologous TAKA amylase promoter from *A. oryzae* was used.

To help explain the two methods of site-directed mutagenesis described here, it is noted that the expression plasmid carries the lipase gene on a *Bam*HI–*Nhe*I restriction fragment and that the plasmid has a unique *Sph*I restriction site outside the region responsible for transcription and termination of the lipase gene and outside the lipase gene itself.

Site-directed mutagenesis has been an invaluable tool for the study of molecular biology and protein structure and function. Since one of the most cited protocols for site-directed mutagenesis was published by Zoller and Smith in the early 1980s,[18] many other similar methods have been reported. However, almost any method described using single-stranded DNA must circumvent the most common problem in site-directed *in vitro* mutagenesis, namely, (1) the competition between the oligonucleotide primer carrying the mutation and (2) selection of the newly synthesized mutagenic strand over the unwanted and unmutagenized parental strand.

[17] E. Boel and B. Huge-Jensen, *Lipids* **23,** 701 (1988).
[18] M. J. Zoller and M. Smith, *DNA* **3,** 479 (1984).

Site-Directed Mutagenesis

We have primarily been using two methods for generating the mutations W89X and D96L in the lipase expression plasmid discussed in this chapter.

Polymerase chain reaction (PCR)-based methods for site-directed mutagenesis have been developed. These methods, however, have suffered from the error-prone *Taq* polymerases, giving rise to unwanted mutations besides the one(s) defined in the oligonucleotide primer. Now that sequencing techniques have been improved, it has become easier to sequence potential mutants, but it is still a rather costly way of screening. We have, despite this drawback, been using a method described in Nelson and Long[19] quite successfully. It involves the three-step generation of a PCR fragment containing the desired mutation introduced by using a chemically synthesized DNA strand as one of the primers in the PCRs. From the PCR-generated fragment, a DNA fragment carrying the mutation can be isolated by cleavage with restriction enzymes and reinserted into the expression plasmid.

Three-Step Polymerase Chain Reaction Mutagenesis. The circular lipase expression plasmid is initially linearized with the restriction enzyme *Sph*I. As shown in Fig. 2, three-step mutagenization involves the use of four primers: (1) one mutagenization primer carrying the desired mutation in the lipase-encoding part of the plasmid, (2) a helper 1 primer complementary to a region upstream of the lipase gene and carrying a handle sequence, (3) a helper 2 primer complementary to a region downstream of the coding region, and (4) a PCR handle complementary to about half of the helper 1 primer.

Helper 1 and helper 2 are complementary to sequences outside the coding region, and can thus be used in combination with any mutagenization primer in the construction of a mutant sequence.

All three steps are carried out in buffer containing 10 m*M* Tris-HCl (pH 8.3), 50 m*M* KCl, 1.5 m*M* $MgCl_2$, 0.001% (w/v) gelatin, 0.2 m*M* dATP, 0.2 m*M* dCTP, 0.2 m*M* dGTP, 0.2 m*M* TTP, and 2.5 units or *Taq* polymerase.

In Step 1, 100 pmol of primer A, 100 pmol of primer B, and 1 fmol of linearized plasmid are combined in a 100-μl reaction mixture and 15 cycles consisting of 2 min at 95°, 2 min at 37°, and 3 min at 72° are carried out.

The concentration of the PCR product is estimated on an agarose gel. Step 2 is then carried out: Step 1 product (0.6 pmol) and 1 fmol of linearized plasmid are combined in a total of 100 μl of the previously mentioned buffer and one cycle consisting of 5 min at 95°, 2 min at 37°, and 10 min at 72° is carried out.

[19] R. M. Nelson and G. C. Long, *Anal. Biochem.* **180,** 147 (1989).

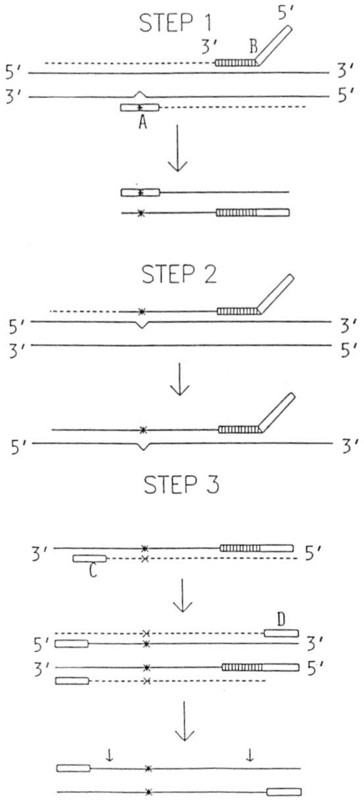

FIG. 2. Outline of three-step PCR mutagenesis.

To the step 2 reaction mixture, 100 pmol of primer C and 100 pmol primer D are added (1 μl of each) and 20 cycles consisting of 2 min at 95°, 2 min at 37°, and 3 min at 72° are carried out. This manipulation composes step 3 in the mutagenization procedure.

Isolation of Mutated Restriction Fragment. The product from step 3 is isolated from an agarose gel and digested with the restriction enzymes *Bam*HI and *Nhe*I. The fragment is isolated from an agarose gel and cloned back into the original expression plasmid.

The advantage of this method is that it is quick and has a high frequency of mutant generation. The major disadvantage as mentioned previously is that the *Taq* polymerase often incorporates errors in the PCR fragment; a problem that is enhanced owing to multiple rounds of PCR amplification. This means that one must minimize the size of the fragment cloned back into the expression plasmid, a process that is limited by the availability of restriction sites.

The advantages and limitations of many other published methods have been discussed.[20,21] We have found that the most severe shortcoming is in fact the unavailability of restriction sites to be used for subcloning of mutagenized fragments. This is a particular problem when mutagenizing the same plasmid repeatedly and at many different sites in a particular gene, because it is necessary to introduce new restriction sites into the gene or even construct a new synthetic gene with the risk of creating a less stable mRNA in the cell, which reduces the yields of the product of interest.

One method solving this problem is the unique site elimination (USE) technique[22] (Fig. 3), in which site-specific mutations are introduced into virtually any plasmid containing a unique, nonessential restriction site. To improve the mutation frequency further, we have modified the USE kit marketed by Pharmacia[23] (Piscataway, NJ) with an earlier method developed by Kunkel *et al.*),[24] in which a uracil-containing template is selected against by uracil *N*-glycosylase, generating apyrimidinic sites in an *Escherichia coli ung*$^+$ *dut*$^+$ host strain. The technique uses two mutagenic primers: One carries the desired mutation, the other carries a mutation in the unique site of the target plasmid. Selection for elimination of the unique site is then used to recover nonselected mutations with high efficiency.[22] This method has been published[25] and gives increased frequencies of mutations, generally up to 75% or more. This means that only two or three colonies need to be screened for the desired mutagenized plasmid. Because the mutant strand is generated by DNA polymerase I, only a minor span around the mutation needs to be sequenced to rule out the possibility of formation of panhandle structures in the mutagenized plasmid due to displacement of the 3′ end of the newly synthesized strand by the DNA polymerase. In essence, the protocol follows the outline provided by Pharmacia with their USE kit; our modification involves preparation of uracil-containing plasmid from *E. coli ung*$^-$ *dut*$^-$ strain CJ236 (Boehringer GmbH, Mannheim, Germany) grown in 2xYT medium[26] supplemented with uridine (6 μg/ml; Sigma Chemical, St. Louis, MO) and ampicillin (100 μg/ml). Although

[20] S. Inouye and M. Inouye, "Oligonucleotide-Directed Site-Specific Mutagenesis Using Double-Stranded Plasmid DNA: Synthesis and Applications of DNA and RNA," p. 181. Academic Press, New York, 1987.

[21] D. H. Jones and B. H. Howard, *Biotechniques* **8,** 178 (1990).

[22] W. P. Deng and J. A. Nickoloff, *Anal. Biochem.* **200,** 81 (1992).

[23] Pharmacia, U.S.E. mutagenesis kit. Pharmacia, Piscataway, New Jersey.

[24] T. A. Kunkel, J. D. Roberts, and R. A. Zakour, *Methods Enzymol.* **154,** 367 (1987).

[25] P. Markvardsen, S. F. Lassen, T. V. Borchert, and I. G. Clausen, *Biotechniques* **18,** 370 (1995).

[26] J. Sambrook, E. F. Fritsch, and T. Maniatis, "Molecular Cloning: A Laboratory Manual." Cold Spring Harbor Laboratory Press, Cold Spring Harbor, New York, 1989.

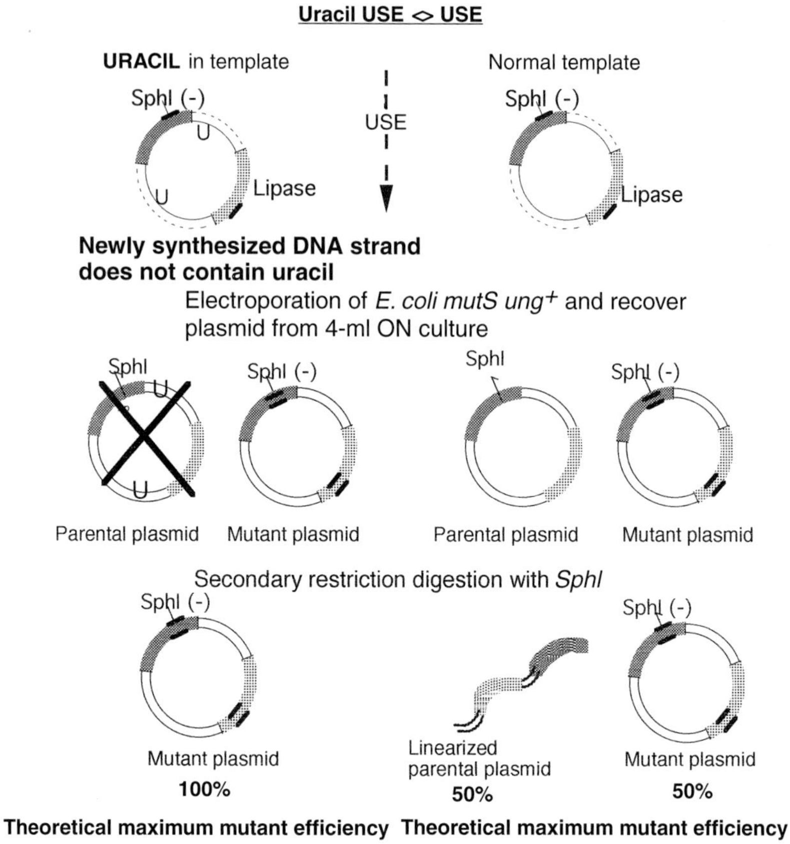

FIG. 3. A comparison of mutation frequencies using the USE method and our improved method.

the preparation of this template needs some optimization, the method is extremely effective and fast.

Purification of *Humicola lanuginosa* Lipase Variants Expressed in *Aspergillus oryzae*

Lipolytic enzymes have been defined as long-chain fatty acid ester hydrolases or as any esterase capable of hydrolyzing esters of oleic acid.[27] Several microbial and mammalian lipases have been purified. A review on

[27] H. Brockersdorf and R. G. Jensen, "Lipolytic Enzymes," p. 1. Academic Press, New York, 1974.

purification of different lipases has been published.[28] The starting materials for lipase purification vary and can originate from tissue, such as liver or pancreas, or may be present in milk or gastric juice. Microbial lipases can exist intracellularly or can be membrane bound, but are generally secreted in extracellular media. In the case of plants, lipases can be found in seeds, latex, fruits, or leaves.

Purification Methods and Strategy

A purification strategy depends on the type of impurity present in the starting material. Traditional purification can be carried out using various methods such as salt precipitation, or aqueous two-phase systems consisting of polyethylene glycol and dextran or polethylene glycol and salt. Only rarely are lipases of native origin purified by simple salting out or two-phase separation. Ion-exchange chromatography and hydrophobic chromatography are always needed to obtain a pure lipase. Use of reverse micelles for extraction of proteins is well described by Hatton.[29] This method involves the ability of reversed micelles to solubilize proteins from an aqueous phase into the aqueous pool of the surfactant aggregates. In a second step the solubilized proteins are back extracted into an aqueous phase by changing the interaction between the proteins and reversed micellar system. The drawback of the method is that some of the lipases can unfold in the presence of surfactants and cannot be renatured. Purification of a lipase from *Chromobacterium viscosum* using reverse micelles was reported by Aires-Barros and Cabral.[30] More sophisticated methods involving immunoaffinity techniques with monoclonal antibodies can be used to isolate and purify lipases, but they are more time consuming because of the need to select and produce the antibodies.

The main parameters to be considered in purification of lipases are isoelectric point, molecular weight, and the stability of the lipase at various pH values. Oxidation stability must also be taken into account. Some of the lipases from *Pseudomonas* species contain a bound calcium ion, and use of citrate or EDTA unfolds the lipases irreversibly.[31] Proteases in the

[28] M. R. Aires-Barros and J. M. S. Cabral, *in* "Lipases: Their Structure, Biochemistry and Application" (P. Woolley and S. B. Petersen, eds.), p. 243. Cambridge University Press, Cambridge, 1994.

[29] T. A. Hatton, *in* "Surfactant Based Separations" (J. F. Scamhorn and J. H. Harwell, eds.), p. 55. Marcel Dekker, New York, 1989.

[30] M. R. Aires-Barros and J. M. S. Cabral, *in* "Lipases: Structure, Mechanism and Genetic Engineering" (L. Alberghina, R. Schmid, and R. Verger, eds.), p. 407. GBF monograph. VCH, Weinheim, Germany, 1991.

[31] A. Svendsen, K. Borch, M. Barfoed, T. B. Nielsen, E. Gormsen, and S. A. Patkar, *Biochim. Biophys. Acta* **1259,** 9 (1995).

media are a common cause of degradation and a protease inhibitor can be employed, if the lipase activity is not affected by the inhibitor. The most common difficulties in purification are always related to the type of impurity present, for example, if the organism produces proteins with similar isoelectric points and molecular weights. Another problem commonly arises owing to the presence of lipase substrate or inhibitor in the medium (such as oil) used for induction of microbial lipases, or fatty acid released from the degradation of the substrate. Furthermore, some organisms produce polysaccharides or endotoxins as shown for the lipase from *Pseudomonas aeruginosa.*[32] These polysaccharides bind the lipase and give rise to many lipase bands on isoelectric focusing. Purification of such complexed lipases is difficult, using traditional purification methods. Only high-voltage isoelectric focusing has been shown to dissociate the polysaccharides from lipase, but the yields are poor.

In the case of lipases from plant origin, there can sometimes be seen associated oil bodies or lipids. It has been shown, in the case of lipase from latex of *Euphorbia characia,* that the lipase was associated with lipid, and dissociation of the lipid from the lipase inactivated the lipase.[33] Reactivation of the lipase took place when the lipid was added back to inactive lipase. This kind of problem might be common in purification but has not been investigated yet owing to loss of activity.

If a specific reversible inhibitor could be used in purification of the lipases, the number of steps in purification could be decreased. A review on lipase inhibitors[34] presents different kinds of reversible and irreversible inhibitors of lipases. It is possible that β-lactone-containing compounds, such as ebalactone or tetrahydrolipstatin, can act as reversible inhibitors that in turn can be employed as an affinity matrix for isolation of lipases. But no such matrix has been reported for successful purifications. As the catalytic triad of lipases is similar to that of proteases, a reversible inhibitor of proteases such as phenylboronic acid might be employed in purification; but in practice it has been shown not to be useful. For some of the mammalian lipases heparin has been shown to be useful in purification owing to its affinity or binding.[35]

As most of the lipases are hydrophobic in nature, they can be extracted in interphase using nonpolar organic solvent and aqueous media. One of the problems in the use of organic solvents is that some of the lipases unfold and lose activity in contact with organic solvents. Most commonly,

[32] K. E. Jaeger, F. J. Adrian, H. E. Meyer, R. E. W. Hancock, and U. K. Winkler, *Biophys. Biochem. Acta* **1120,** 315 (1992).

[33] M. Teissere and C. Bernard, *Proc. Natl. Acad. Sci. U.S.A.* **91,** 11328 (1994).

[34] S. A. Patkar and F. Bjørkling, *in* "Lipases: Their Structure, Biochemistry and Application" (P. Woolley and S. B. Petersen, eds.), p. 207. Cambridge University Press, Cambridge, 1994.

[35] R. Zechner, *Biophys. Biochim. Acta* **1044,** 20 (1990).

hydrophobic matrices such as phenyl, butyl, octyl, or decyl chains cross-linked with different matrices are used in purification. It was reported that esters of polyvinyl alcohol and fatty acids with different chain lengths can be used to purify lipases in one step.[36] *Candida rugosa* lipase isoforms were shown to be separated using a dodecyl fatty acid ester matrix.[37]

Purification of cloned lipases is based on a different strategy than is required for noncloned lipases. Common hosts for expression of cloned lipases are *E. coli,* yeast, *Bacillus,* or a fungal host such as *A. oryzae.* In some cases mammalian lipases are expressed in a baculovirus system.[38] If the lipases are expressed in *E. coli* as inclusion bodies, then breaking of the cells by sonication or by treatment with lysozyme is needed to open the cells. Lipases expressed in other hosts that secrete the lipases in extracellular media are normally more easily purified. One of the major impurities consists of other proteins from the medium, most commonly proteases secreted by the hosts. *Aspergillus oryzae* is known to produce a major protein, an α-amylase with a molecular weight of 55,000 and p*I* of 3. *Aspergillus oryzae* also produces different kinds of exo- and endoproteases with molecular weights between 20,000 and 70,000 and isoelectric points between 3 and 8. The fungus produces a dark-colored pigment that normally binds to anion exchangers. To minimize the degradation by the proteases, it is essential to select conditions such that proteases are inactive. We have previously reported purification of two lipases, from *Candida antarctica* and expressed in *A. oryzae,* that do not follow the rule of binding to ion exchange depending on the p*I* of the lipases.[39] Hence impurities are separated by negative adsorption of the lipases to ion exchangers. Most of the proteins produced by *A. oryzae* have p*I* values between 3 and 5, which makes them difficult to separate from a lipase with p*I* ~4. Final separation of the lipase requires an anion exchanger such as high-performance Q-Sepharose or MonoQ.

Experimental Procedure

Materials

All commonly used chemicals are obtained from Sigma. For filtration of genetically produced fermentation supernatants, pyrogen-reducing

[36] E. Cernia, G. Ortaggi, S. Soro, and M. Castagnola, *Tetrahedron Lett.* **35,** 9051 (1994).

[37] E. Cernia, G. Ortaggi, L. Batteneli, M. T. Bersani, S. Soro, M. Castagnula, and R. Rabino, Poster No. 7 presented at the International Workshop on Microbial Lipases in Biocatalysis, April 11–13, Rome, 1996.

[38] K. Thirstrup, F. Carrier, S. Hjorth, P. B. Rasmussen, H. Wøldike, P. F. Nielsen, and L. Thim, *FEBS Lett.* **12732,** 79 (1993).

[39] S. A. Patkar, F. Bjørkling, M. Zundel, A. Svendsen, H. P. Heldt-Hansen, and E. Gormsen, *Indian J. Chem.* **32B,** 76 (1993).

filters of varying pore sizes are obtained from Seitz-Filter-Werke GmbH. Hydrophobic matrix (Toyopearl-Butyl 6000 C) is purchased from Toso Haas (Pennsylvania, USA). Anion exchangers, Fast Flow DEAE-Sepharose, and high-performance Q-Sepharose are purchased from Pharmacia. Glycerol tributyrate as small-chain substrate and gum arabic as emulsifier are purchased from E. Merck AG (Damstadt, Germany). Olive oil emulsified as long-chain substrate is obtained from Sigma. For sodium dodecyl sulfate–polyacrylamide gel electrophoresis (SDS–PAGE) analysis, precast polyacrylamide gels are purchased from Novex (San Diego, CA).

Standard Assay for Lipase Activity Using Tributyrine as Substrate: pH-Stat Titration Method at pH 7. Equipment required is a VIT90 video titrator manufactured by Radiometer (Denmark). A detailed description of the method follows.

1. Stock gum arabic emulsion is prepared by dissolving 6 g of gum arabic in 1 liter of solution containing 50% (v/v) glycerol, 300 mM NaCl, and 3 mM KH_2PO_4 in ion-exchanged water. After dissolving the emulsification reagent can be stored at 4° for several weeks.
2. Substrate emulsion mixture is prepared fresh daily as follows. Glycerol tributyrate (15 ml) is mixed with 50 ml of the stock gum arabic emulsification reagent and 235 ml of deionized water is added to make a final volume of 300 ml. Final substrate emulsion is prepared by blending the substrate for 30 sec at high speed.
3. Substrate solution (15 ml) is heated to 30° and 0.5 ml of the lipase solution at different dilutions in water is added to the substrate emulsion under constant stirring. Note that no buffering solution should be added to the incubation mixture. The pH of the reaction mixture is adjusted to pH 7 or 9, and butyric acid released by the hydrolysis of the substrate is monitored. Titration is carried out with 0.025 M sodium hydroxide for 5 min using the VIT90 video titrator.
4. Calculation of lipase units using glycerol tributyrate as substrate is carried out by calculating micromoles of sodium hydroxide used per minute. One lipase unit (LU) is defined as the amount of enzyme that releases 1 μmol of fatty acid per minute.

Assay for Lipase Using Olive Oil as Substrate: pH-State Titration at pH 9. The Sigma substrate emulsion (2.5 ml) is mixed with 12.5 ml of the stock solution, pH 9, containing 2 mM Tris, 50 mM NaCl, and 5 mM $CaCl_2$. The reaction is started by adding 0.5 ml of various dilutions of the lipases. Oleic acid released is titrated using 0.025 M NaOH. Blanks are run in exactly the same way, without adding lipase during incubation. Sigma lipase unit

(SLU) is defined as the amount of lipase that releases 1 μmol of fatty acid per minut at pH 9.

SDS–PAGE for characterization of purified lipase variants is carried out using 5 to 20% (w/v) precast gels (Novex) and electrophoresis is carried out as described by the manufacturer.

Purification of Humicola lanuginosa Lipase Variant D96L. Fermentation supernatant is filtered using a coarse filter, then a fine filter, and finally a sterilizing filter under pressure.

Step 1. Precipitation: To 280 ml of supernatant 440 ml of 96% (v/v) ice-cold ethanol is added slowly with constant stirring. When precipitate forms the sample is filtered through a 0.45-μm (pore size) filter. Activity in the supernatant is assayed using tributyrine substrate as described above.

Step 2. Ion exchange: The solution is then applied on a 200-ml DEAE Fast Flow Sepharose column, preequilibrated with 50 m*M* Tris–acetate buffer, pH 7. The column is then washed with the same buffer until the A_{280} of the effluent is less than 0.05. Bound lipase is eluted with a linear salt gradient, using the same buffer containing 0–1 *M* NaCl. Fractions containing lipase activity are pooled. Solid ammonium acetate is added to the pool to a final concentration of 0.8 *M*.

Step 3. Hydrophobic chromatography on butyl-Toyopearl column: The pool is applied to a 100-ml butyl-Toyopearl column, preequilibrated with 0.8 *M* ammonium acetate. The column is then washed with 0.8 *M* ammonium acetate until all of the unbound material is washed out, indicated when ultraviolet (UV) absorption of the effluent at 280 nm is below 0.05. Bound activity is eluted with distilled water. Fractions containing lipase activity are then pooled and diluted with distilled water to adjust conductance to 2 mS and the pH is adjusted to pH 7.

Step 4. High-performance Q-Sepharose chromatography: The lipase-containing pool is then applied on a 50-ml high-performance Q-Sepharose column, which is equilibrated with 50 m*M* Tris–acetate buffer, pH 7. The column is washed with the same buffer until the UV absorption of the effluent is less than 0.05 at 280 nm. Activity is eluted by using a linear salt gradient from 0 to 1 *M*.

Fractions containing lipase activity and with a ratio of A_{280}/A_{260} higher than 1.7 are pooled and assayed for final specific activity of each variant. For a summary of yields of the purified variant, see Table I.

The calculated extinction coefficient for *H. lanuginosa* wild-type lipase is 1.2 and its specific activity on the basis of amino acid analysis, using tributyrine as substrate, is 4500 LU/mg at pH 7. Similarly, the specific activity of the lipase, using olive oil, is calculated at pH 9 to be 4200 Sigma lipase units (SLU).

TABLE I

RECOVERY OF WILD-TYPE LIPASE FROM *Humicola lanuginosa* EXPRESSED IN *Aspergillus oryzae,* AT DIFFERENT PURIFICATION STEPS[a]

	Volume (ml)	Concentration (LU/ml)	Total amount (LU)	Specific activity (LU/OD_{280})	Yield (%)
1	280	600	168,000	nd	100
2	720	230	165,600	nd	99
3	130	1,300	169,200	224	101
4	80	1,790	143,200	2,183	85
5	10	11,651	116,510	3,880	69

[a] 1, crude sample; 2, supernatant from ethanol precipitation; 3, pool from ion-exchange chromatography on DEAE-Sepharose; 4, pool from hydrophobic chromatography on butyl-Sepharose; 5, pool from ion-exchange chromatography on high-performance Q-Sepharose. nd, Not determined.

Recovery of the site-specific variant D96L as 69%. The specific activity of the D96L variant, calculated on the basis of activity on glycerol tributyrate as substrate and UV absorption of the varient lipase at 280 nm, was the same as for the wild-type lipase.

Purification of the site-specific variants of tryptophan residue 89 was carried out in exactly the same way as for the D96L variant. Activities of the W89*X* variants using olive oil and tributyrine are shown in Table II,

TABLE II

ACTIVITIES OF *Humicola lanuginosa* WILD-TYPE LIPASE AND *Humicola lanuginosa* VARIANT LIPASES AT POSITION W89 USING OLIVE OIL AND TRIBUTYRINE AS SUBSTRATES AT pH 9[a]

H. lanuginosa lipase	Olive oil substrate (SLU[b]/mg)	Glycerol tributyrate substrate (LU/mg)
Wild type	4200	4500
W89F	1717	678
W89Y	2361	596
W89L	845	246
W89G	964	213
W89E	84	0

[a] Corrected for tryptophan content.

[b] SLU, Sigma lipase unit.

TABLE III
ACTIVITIES OF *Humicola lanuginosa* WILD-TYPE LIPASE AND *Humicola lanuginosa* VARIANT LIPASES AT POSITION W89 USING GLYCEROL TRIBUTYRATE AS SUBSTRATE AT pH 7[a]

H. lanuginosa lipase	Specific activity (LU/mg)
Wild type	4500
W89F	452
W89Y	455
W89L	171
W89G	137
W89E	75

[a] Corrected for tryptophan content.

and using glycerol tributyrate as substrate at pH 7 (Table III). Purity of the lipase variants by the method used is more than 90% as judged by SDS–PAGE (Fig. 4).

The tryptophan variants have dramatically reduced activity on both the short- and long-chain substrates. It is also noteworthy to see that at pH 9 there is significantly more activity against long-chain substrate than against short-chain substrate. These results suggest that the tryptophan residue at position 89 is involved in binding of the substrate in the hydrophobic pocket.

Influence of Surfactants on Activity of *Humicola lanuginosa* Lipase: A Mixed Monolayer Study

Lipase and Surfactants

It is well known that surfactants in general, under several experimental conditions, influence the activity of lipases (e.g., Gargouri *et al.*[40]). For many lipases reduced stability in the presence of anionic surfactants such as LAS (linear alkylbenzene sulfonate) or SDS (sodium dodecyl sulfate) is a major problem causing decreased activity.[31,41] Instability in the presence of surfactants is, however, not a problem for the *H. lanuginosa* lipase under detergent-relevant conditions.

Another important explanation of the influence of surfactants on the activity of many lipases is that surfactants obviously will have a strong

[40] Y. Gargouri, R. Julien, A. G. Bois, R. Verger, and L. Sarda, *J. Lipid Res.* **24,** 1336 (1983).
[41] M. R. Egmond, J. de Vlieg, and H. M. Verhey, *in* "Engineering of/with Lipases" (F. X. Malcata, ed.), p. 183. NATO ANSI Series. Kluwer Academic Publishers, Dordrecht, 1996.

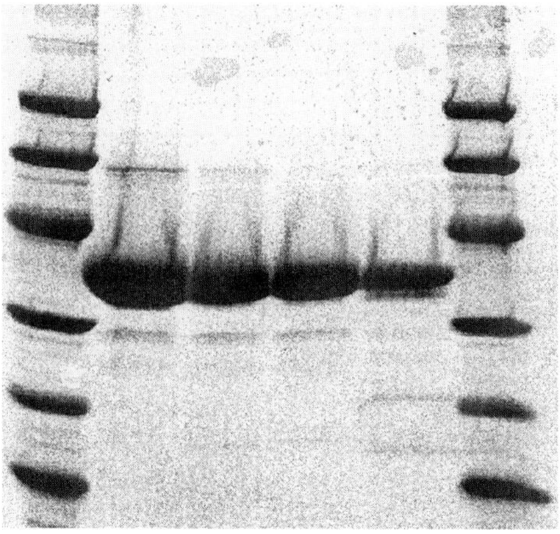

FIG. 4. All samples were adjusted to a concentration of 1 to 3 mg/ml. Samples (each 20 μl) were mixed with 20 μl of sample buffer containing 25 mM Tris buffer (pH 7.2), 2 mM 2-mercaptoethanol, 0.4% (w/v) SDS, and 16% (v/v) glycerol. The samples were then incubated for 5 min at 95°. Twenty microliters of each sample was applied on 5 to 20% SDS–polyacrylamide precast gels and electrophoresis was carried out as described by the manufacturer. Lanes 1 and 6, marker proteins purchased from Pharmacia (92, 67, 45, 30, 20, and 14 kDa). Lane 2, wild type; lane 3, variant D96L; lane 4, variant W89F; lane 5, variant W89E. All the variants were overloaded. The major protein band has a molecular weight of 35,000. In lanes 2 and 3, minor bands with molecular weights of 67,000 and 30,000 are seen as impurities. In lane 4, a small impurity of 30,000 is seen. In lane 5, low molecular weight protein bands 21,000 and 15,000 are seen. These bands might arise from proteolytic digestion of the variant.

tendency to accumulate on the lipase substrate, being very hydrophobic in nature. Thus, in wash liquor, nonionic and anionic surfactants are likely to influence lipase activity by at least one or more of the following mechanisms: (1) influence on substrate–water interfacial tension; (2) reduction of the substrate concentration at the substrate–water interphase (substrate dilution); (3) electrostatic repulsion of the lipase from anionic surfactants or fatty acid products; and (4) physical shielding (adsorption of surfactants and lipase are competing processes).

Influence of Alcohol Ethoxylates Studied in Mixed Monolayers

Having identified that, e.g., alcohol ethoxylates are harmful to the activity of *H. lanuginosa* lipase during washing, it was decided to develop an assay focusing specifically on the effect of mixing certain surfactive detergent components into a substrate phase on the activity of protein-engineered variants of *H. lanuginosa* lipase.

The use of lipid monolayers[42,43] is a well-suited choice for such studies, giving the advantage of excluding the potential effect of bulk surfactants (i.e., as monomers and micelles) on lipase stability, while at the same time giving the opportunity to control surface pressure (interfacial tension) and overall surface composition (apparent area fraction of substrate).

Furthermore, the monolayer technique implies a constant (but diluted) area of the substrate phase on addition of surfactants, in contrast to the use of emulsified substrates where size of emulsion droplets (i.e., substrate area) will change significantly on addition of surfactants.

Method

Using monolayer equipment from KSV Instruments (KSV 5000; Helsinki, Finland) a mixed monolayer in a well-defined overall composition, made of a diglyceride substrate and a monocomponent alcohol ethoxylate, is spread on an aqueous subphase. After addition of lipase, lipolytic action is manifested by the speed with which a mobile barrier compresses the monolayer to maintain constant surface pressure as insoluble substrate molecules are hydrolyzed into more water-soluble reaction products. The procedure is summarized below.

1. A filtered buffer solution is poured into a thoroughly purified and thermostatted Teflon trough.
2. A chloroform solution of the alcohol ethoxylate additive is spread on the carefully purified surface of the buffer solution. Using the computer-controlled mobile barrier the surface pressure is adjusted to the desired value. Stability of the monolayer is verified and the area occupied by the additive is noted (A_1).
3. The monolayer is expanded and a chloroform solution of the lipase substrate is spread into the additive monolayer. The surface pressure is adjusted and maintained at the selected value for 10–15 min, allowing components to mix. Stability of the mixed monolayer is verified and the area of the mixed monolayer prior to addition of lipase is noted (A_2).
4. Lipase is added through the monolayer and lipase activity is recorded until barrier movement (area reduction) stops, reflecting stop of lipolytic activity. The area of mixed monolayer as lipase activity stops is noted (A_3).

Standard Conditions

Buffer: 10 m*M* glycine, 0.1 m*M* EDTA (pH 10.0)
Temperature: 25°

[42] R. Verger and G. H. Haas, *Chem. Phys. Lipids* **10,** 127 (1973).
[43] H. L. Brockman, W. E. Thomsen, and T. Tsujita, *JAOCS* **218,** 4965 (1973).

Substrate: Dicaprin (Nu Chek Prep; Minnesota, USA)
Additive: Heptaethylene monooctadecyl ether (Fluka, Buchs, Switzerland)
Surface pressure: 20 mN/m
Amount of lipase: 10 LU

Calculation and Definitions. Many lipases are unable to hydrolyze all substrate present under these experimental conditions (i.e., $A_3 > A_1$). Lipases are discriminated by the final area fraction of substrate left unhydrolyzed by the lipase, here called β.

By definition: $\beta = (A_3 - A_1)/A_3$ ($\times$ 100%)
Area of additive: A_1
Area of additive and substrate: A_2 (prior to hydrolysis)
Apparent area of substrate: A_4 ($= A_i - A_1$; $i = 2, 3$)
Maximal monolayer area: $A_{max} = 5040\ mm^2$

For *H. lanuginosa* lipase β equals 27% ($\pm$2%) under these condition, indicating that the lipase almost stops hydrolyzing despite the fact that 27% of the monolayer is still covered by substrate.

Using the monolayer assay as described above it has been demonstrated that the D96L variant of *H. lanuginosa* lipase (Lipolase Ultra; Novo Nordisk A/S, Denmark) provides an increased performance in mixed monolayers as reflected by a decrease in β value to approximately 18%.

It has been demonstrated that β is independent of lipase concentration, as indicated by the fact that a 10-fold increase in lipase concentration will not lead to increased hydrolysis as soon as the area fraction of heptaethylene monoocytadecyl ether exceeds approximately 73% (100–27%) for Lipolase (data not shown).

Table IV clearly indicates that increasing the hydrophobicity at position 96 of *H. lanuginosa* lipase significantly increases the ability of this enzyme to hydrolyze at a high concentration of heptaethylene glycol monooctadecyl ether.

Correspondingly, substituting the hydrophobic W89 in *H. lanuginosa* lipase with negatively charged glutamic acid leads to a dramatically higher β value than for the wild type (data not shown).

Evaluation of Wash Effect

Normally lipases are available only in limited amounts; therefore, the washing effect of such lipases must be tested in model systems. Several commercial systems are available, e.g., *terg-o-tometers* (which resemble the top-loading machine with constant temperature used in, e.g., the United States) or *launder-o-meters* (which resemble the drum-type machine with warm-up cycle used in Europe). In this chapter, a method using a homemade

TABLE IV
DEPENDENCE ON NONIONIC SURFACTANT OF *Humicola lanuginosa* WILD-TYPE LIPASE AND *Humicola lanuginosa* VARIANT LIPASES AT POSITION D96[a]

H. lanuginosa lipase	β
Wild type	27
D96N	21
D96L[b]	18
D96F	12
D96W	9

[a] Measured using the mixed monolayer technique, at pH 10, 25°, and constant surface pressure of 20 mN/m; amount of lipase added is 10 LU. The β value is the area fraction of substrate as hydrolysis stops.
[b] Lipolase Ultra.

model wash system, called a *miniwash,* which is downscaled even further, is described.

In principle the choice of equipment mainly means that the type of mechanical action is given, whereas the magnitude of mechanical action, washing temperature, washing time, detergent, water hardness (the content of calcium and magnesium ions), fabric, and soil are factors that should be adjusted to that of interest for the present experiment.

Method: Miniwash

Equipment. The equipment is homemade and consists of 150-ml beakers placed in a thermostatted water bath. Agitation is achieved by placing magnetic stirrers beneath the water bath and triangular magnetic rods inside the beakers. The setup is referred to as the miniwash equipment.

Fabric. Any fabric can be used, but to attain continuity it is best to buy fabric from test fabric companies (see, e.g., Coons *et al.*[44]). The fabric load, that is, "gram of fabric vs gram of washing liquor," differs depending on location (see, e.g., Coons *et al.*[45]).

[44] D. Coons, M. Dankowski, M. Deihl, G. Jakobi, P. Kuzel, E. Sung, and U. Trabitzsch, *in* "Surfactants in Consumer Products" (J. Falbe, ed.), p. 214. Springer-Verlag, Heidelberg, Germany, 1987.
[45] D. Coons, M. Dankowski, M. Deihl, G. Jakobi, P. Kuzel, E. Sung, and U. Trabitzsch, *in* "Surfactants in Consumer Products" (J. Falbe, ed.), p. 247. Springer-Verlag, Heidelberg, Germany, 1987.

Soil. Any soil can be used. Often pure triglycerides such as oils and fats are used, as the lipase effect usually is most enhanced on such soiling. To ease the evaluation after washing the soil can be dyed before application to the textile. One possible dye is Sudan red; e.g., 0.75 mg of dye per gram of lard. Note that presoiled test fabrics can be bought as well (see, e.g., Coons *et al.*[44]).

Conditions. Washington conditions differ greatly among locations (see, e.g., Coons *et al.*[45]). For imitation of European conditions, it is suitable to wash at 30° for 20 min.

Detergents. Either homemade or commercial detergents can be used. When commercial detergents are used the quantity required is usually given on the box. Note that many commercial detergents often contain lipases.

Procedure. A wash cycle consists of the following: The detergent is solubilized in water; 100 ml is added to each beaker of the miniwash equipment and stirring is started. Lipase is added to the first beaker at time −15 sec. The presoiled and numbered fabric is added to the same beaker at time 0 sec. Lipase is added to the next beaker at time 15 sec and the fabric at time 30 sec, and so on. The final wash time, for example, is 20 min. The first beaker is taken out of the miniwash bath and the fabric is rinsed in running tap water; 30 sec later the next beaker is taken off, and so on. After a rinsing period of, e.g., 15 min the swatches are gently squeezed and are allowed to dry under room conditions until the next day. The following day the reflectance of the dyed soil can be measured using a reflectometer.

Often it is necessary to do repeated washes to obtain a reliable result. Therefore the procedure is repeated three times with overnight drying between each wash cycle; this is called a three-cycle miniwash.

Calculation of Result

Dose–response curves are often drawn when comparing lipases. Usually the measured dose–response curves fit the following equation:

$$\Delta R = \Delta R_{\max} \frac{C^{0.5}}{K + C^{0.5}}$$

where ΔR is the effect expressed in reflectance units; C is the enzyme concentration; $\Delta R_{\max}$ is a constant expressing the maximum effect; and K is a constant (K^2 expresses the enzyme concentration at which half of the maximum effect is obtained).

The performance of the lipases in the given test system is characterized by the calculated constants $\Delta R_{\max}$ and K. On the basis of these constants two lipases, e.g., a wild-type lipase and a lipase variant, can be compared. The comparison can be done in several ways.

Comparison I: Comparison of which effect is obtained at a given lipase dose, e.g., $\Delta R_{\text{variant}}/\Delta R_{\text{wild type}}$

Comparison II: Comparison of amount of lipase variant protein needed to obtain the same effect as that obtained with a given amount of the wild-type lipase, e.g., $C_{\text{wild type}}/C_{\text{variant}} \cdot C_{\text{variant}}$ can be calculated from

$$C_{\text{variant}} = \left(K_{\text{variant}} \frac{\Delta R_{\text{wild type}}}{\Delta R_{\text{max variant}}\, \Delta R_{\text{wild type}}} \right)^2$$

Washing Performance of Humicola lanuginosa Lipase and Variants

The washing performance of *H. lanuginosa* lipase (Lipolase) versus *H. lanuginosa* lipase variant D96L (Lipolase Ultra) in the miniwash system under European conditions in a model detergent is given as follows.

Conditions

Volume: 100 ml/beaker

Swatches: Six swatches (3.5 × 3.5 cm) per beaker

Fabric: 100% cotton, Test Fabrics style #400 (Test Fabric, Inc., USA)

Soil: Lard dyed with Sudan red (0.75 mg of dye per gram of lard); 6 μl of lard–Sudan red heated to 70° is applied to the center of each swatch. After application of the stain, the swatches are heated in an

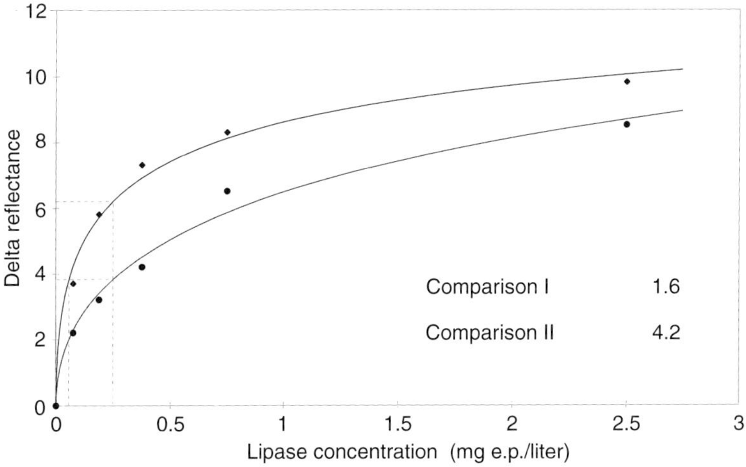

FIG. 5. Evaluation of wash performance of a variant of *Humicola lanuginosa* D96L (Lipolase Ultra) compared with the wild-type lipase (Lipolase). Conditions: Miniwash, three cycles, 30°, 20 min, 18° dH, lard–Sudan red on cotton, European model HDP. (●) Lipolase; (◆) Lipolase Ultra.

oven at 75° for 30 min. The swatches are then stored overnight under room conditions prior to the first wash

Detergent (homemade):

LAS (Nansa 1 1 69/P, 30% a.m.), 1.17 g/liter (Albright & Wilson)
AEO (Alcoholethoylate) (Dobanol 25-7), 0.15 g/liter (Shell)
Sodium triphosphate, 1.25 g/liter
Sodium sulfate, 1.00 g/liter
Sodium carbonate, 0.45 g/liter
Sodium silicate, 0.15 g/liter

pH; 10.2
Lipase concentration: 0.075, 0.188, 0.375, 0.75, and 2.5 mg of lipase protein per liter
Time: 20 min
Temperature: 30°
Water hardness: 18° dH
Rinse: 15 min in running tap water
Drying: Overnight under room conditions
Evaluation: After the third wash, the reflectance at 460 nm is measured

The wash experiment shows that the D96L variant (Lipolase Ultra) is superior to the wild-type *H. lanuginosa* (Lipolase) (Fig. 5).

Conclusion

It has been shown that protein engineering can be successfully used to produce new, commercially interesting products. An understanding of lipase function in general and under specific application conditions is mandatory to make the correct decisions in a protein-engineering strategy.

[20] Glycosylation of Bile Salt-Dependent Lipase (Cholesterol Esterase)

By ERIC MAS, MARIE-ODILE SADOULET, ASSOU EL BATTARI, and DOMINIQUE LOMBARDO

Introduction

Cholesterol esterase produced by pancreatic acinar cells, also called carboxyl ester lipase or bile salt-dependent lipase (BSDL), catalyzes the hydrolysis of ester substrates such as triglycerides, phospholipids, and lyso-

0076-6879/97 $25.00

phospholipids as well.[1,2] In addition, BSDL is the unique enzyme in the pancreatic juice capable of hydrolyzing fat-soluble vitamins and cholesteryl esters. This latter function strongly suggests that BSDL is important for catalyzing the lymphatic absorption of dietary cholesterol and fat-soluble vitamins. Primary bile salts are a prerequisite for the hydrolysis of hydrophobic substrates. Nevertheless, the major physiological role of BSDL in the intestinal lumen is not clear, because many of the lipid substrates of BSDL can also be hydrolyzed by other lipolytic enzymes present in the gastrointestinal tract of adults. These enzymes are gastric or pregastric lipase, pancreatic colipase-dependent lipase, and phospholipase A_2. A bile salt-stimulated lipase identical to pancreatic BSDL[3] is also found in the milk of humans,[4] higher primates,[5] and carnivores such as the dog, cat, and ferret,[6,7] where it represents a significant proportion of total milk proteins and is, at first approximation, the most abundant of the enzymes in milk. Because lipid digestion by pancreatic lipases is low in the infant during the first months of life, it has been suggested that BSDL may serve a compensatory role in fat digestion in the infant. Moreover, milk lipids are present in the form of globules that are resistant to the pancreatic colipase-dependent lipase. The combined actions of intragastric hydrolysis by gastric (or pregastric) lipase, followed by BSDL, appear essential in the digestion of milk lipids by newborns. The importance of BSDL in newborn growth was demonstrated with kittens fed formula with supplemented BSDL; the growth rate of these kittens was twice that of kittens fed with formula alone.[8] In adults, BSDL in combination with the lipolytic enzymes mentioned above could drive lipid hydrolysis to near completeness.[9]

The enzyme is present in human milk at all stages of lactation; nevertheless, in populations living in areas where malnutrition is widespread the amount of secreted BSDL decreases during the first 6 months of lactation.[10] During this time prevalence of diarrheal diseases due to *Giardia lamblia* (a flagellate that colonizes the upper small intestine of newborns) is low as the result of the lethal effect of trophozoides mediated by toxic lipolytic

[1] D. Lombardo, J. Fauvel, and O. Guy, *Biochim. Biophys. Acta* **611,** 136 (1980).

[2] D. Lombardo and O. Guy, *Biochim. Biophys. Acta* **611,** 147 (1980).

[3] L. Bläckberg, D. Lombardo, O. Hernell, O. Guy, and T. Olivecrona, *FEBS Lett.* **136,** 284 (1981).

[4] O. Hernell and T. Olivecrona, *Biochim. Biophys. Acta* **369,** 234 (1974).

[5] E. Freudenberg, *Experientia* **22,** 317 (1966).

[6] L. M. Freed, C. M. York, M. Hamosh, J. A. Sturman, and P. Hamosh, *Biochim. Biophys. Acta* **878,** 209 (1986).

[7] L. A. Ellis and M. Hamosh, *Lipids* **27,** 917 (1992).

[8] C.-S. Wang, M. E. Martindale, M. M. King, and J. Tang, *Am. J. Clin. Nutr.* **49,** 457 (1989).

[9] S. Bernbäch, L. Bläckberg, and O. Hernell, *J. Clin. Invest.* **85,** 1221 (1990).

[10] P. Dupuy, J. F. Saunière, H. L. Vis, M. Leclaire, and D. Lombardo, *Lipids* **26,** 134 (1991).

products.[11] These toxic lipids are likely generated by the lipolysis of milk lipids by BSDL once activated by primary bile salts.[12,13]

In view of these observations and to make a long story short, BSDL may be an important determinant in regulating cholesterol absorption and consequently plasma cholesterol level.[14] Bile salt-dependent lipase is essential for lipid digestion in the newborn but it may also have the capability of reducing the atherogenicity of oxidized low-density lipoproteins (LDLs).[15] Bile salt-dependent lipase may serve other than digestive functions in part because the enzyme is also found in rat liver,[16] in human hepatoma,[17] and in human eosinophils,[18] while a specific cDNA transcript has been detected in all fetal cells examined so far.[19] It appears that the production of sufficient quantities of BSDL by protein engineering may be useful for many reasons. The recombinant protein could be used, first, as adjuvant in formula for newborns and, second, as a material source for the crystallization of the protein. The addition of recombinant BSDL in formula will limit the risk of HIV contamination, still possible when banked human milk is used for nutritional purposes. Crystallization studies will help to design inhibitors that can be used as therapeutic agents for lowering the plasma cholesterol level and the consequent risk of cardiovascular diseases. Because the protein seems to be specifically transferred through enterocytes from the intestinal lumen toward the lamina propria, from which BSDL may reach the general circulation via the thoracic duct,[20] cardiovascular accidents can hypothetically be decreased by modulating the serum level of BSDL. This can be accomplished by means of a diet supplemented with recombinant BSDL.

[11] A. Islam, B. J. Stoll, I. Ljungström, J. Biswas, H. Nazrul, and G. Huldt, *J. Pediatr.* **103,** 996 (1983).

[12] D. S. Reiner, S. C. Wang, and F. D. Gillin, *J. Infect. Dis.* **154,** 825 (1986).

[13] O. Hernell, H. Ward, L. Bläckberg, and M. E. Pereira, *J. Infect. Dis.* **155,** 715 (1986).

[14] R. Shamir, W. J. Johnson, R. Zolfaghari, H. S. Lee, and E. A. Fisher, *Biochemistry* **34,** 6351 (1995).

[15] R. Shamir, W. J. Johnson, K. Morlock-Fitzpatrick, R. Zolfaghari, E. Mas, D. Lombardo, D. W. Morel, and E. A. Fisher, *J. Clin. Invest.* **97,** 1696 (1996).

[16] E. D. Camulli, M. J. Linke, H. L. Brockman, and D. Y. Hui, *Biochim. Biophys. Acta* **1005,** 177 (1989).

[17] K. E. Winkler, E. H. Harrision, J. B. Marsh, J. M. Glick, and A. C. Ross, *Biochim. Biophys. Acta* **1126,** 151 (1992).

[18] F. W. Holtsberg, L. E. Ozgur, D. E. Garsetti, J. Myers, R. W. Egan, and M. A. Clark, *Biochem. J.* **309,** 141 (1995).

[19] S. Roudani, F. Miralles, A. Margotat, M. J. Escribano, and D. Lombardo, *Biochim. Biophys. Acta* **1264,** 141 (1995).

[20] P. Lechêne de la Porte, N. Abouakil, H. Lafont, and D. Lombardo, *Biochim. Biophys. Acta* **920,** 237 (1987).

Molecular Cloning of Bile Salt-Dependent Lipase

The BSDL cDNA sequence has been cloned from the pancreas of several species: human,[21] rat,[22] mouse,[23] cow,[24] rabbit,[25] and salmon.[26] The human[27] and mouse[28] BSDL cDNAs were also cloned from respective mammary glands. The size of the transcript and consequently of the protein varies by species as a consequence of the number of C-terminal tandem repeated sequences. The human enzyme, the largest at 100 kDa, contains 16 repeats of 11 amino acid residues, nine of which are of identical sequence (PVPPTGDSGAP), while others present minor substitutions. The rat, mouse, and cow have four, three and two repeats, respectively, which are absent from the smaller salmon enzyme.[26] The rat, mouse, and human genes have been cloned.[23,29,30] The human genome also contains a truncated pseudogene lacking exons 2 to 7.[31] A transcript of this gene has been detected in all cells.[32] As found by Nilsson *et al.,* the longest open reading frame of the BSDL pseudogene could encode a peptide of 59 amino acids, presumably without any function.[32] Possibly, a third gene with a premature termination codon encoding an active enzyme of a shorter length than the native 100-kDa human BSDL may be present in the human genome. This shorter form of BSDL[30] could be expressed in cancer cells.[33]

An investigation pointing out the presence in human eosinophils of an active 74-kDa form of BSDL,[18] although not yet sequenced, confirmed that more than one gene translated into functional proteins may be present at least in the human genome. It is interesting to note that consensus sequences

[21] K. Reue, J. Zambaux, H. Wong, G. Lee, T. H. Leete, M. Ronk, J. E. Shively, B. Sternby, B. Borgström, D. Ameis, and M. G. Schotz, *J. Lipid Res.* **32,** 267 (1991).

[22] J. H. Han, C. Stratowa, and W. J. Rutter, *Biochemistry* **26,** 1617 (1987).

[23] A. S. Lidmer, M. Kannius, L. Lundberg, G. Bjursell, and J. Nilsson, *Genomics* **29,** 115 (1995).

[24] E. M. Kyger, R. C. Wiegand, and L. G. Lange, *Biochem. Biophys. Res. Commun.* **164,** 1302 (1989).

[25] N. S. Colwell, J. A. Aleman-Gomez, and B. V. Kumar, *Biochim. Biophys. Acta* **1172,** 175 (1993).

[26] D. R. Gjellesvik, J. B. Lorens, and R. Male, *Eur. J. Biochem.* **226,** 603 (1994).

[27] J. Nilsson, L. Bläckberg, P. Carlsson, S. Enerbäck, O. Hernell, and G. Bjursell, *Eur. J. Biochem.* **192,** 543 (1990).

[28] K. Mackay and R. M. Lawn, *Gene* **165,** 255 (1995).

[29] R. N. Fontaine, C. P. Carter, and D. Y. Hui, *Biochemistry* **30,** 7008 (1991).

[30] B. V. Kumar, J. A. Aleman-Gomez, N. Colwell, A. Lopez-Candales, M. S. Bosner, C. A. Spilburg, M. Lowe, and L. G. Lange, *Biochemistry* **31,** 6077 (1992).

[31] U. Lidberg, J. Nilsson, K. Strömberg, G. Stenman, P. Sahlin, S. Enerbäch, and G. Bjursell, *Genomics* **13,** 630 (1992).

[32] J. Nilsson, M. Hellquist, and G. Bjursell, *Genomics* **17,** 416 (1993).

[33] S. Roudani, E. Pasqualini, A. Margotat, M. Gastaldi, V. Sbarra, C. Malezet-Desmoulin, and D. Lombardo, *Eur. J. Cell Biol.* **65,** 132 (1994).

of the glucocorticoid receptor-binding site and the estrogen receptor-binding site were detected within the 5′ flanking sequence of the BSDL gene. Binding sites for nuclear proteins of lactating mammary gland, present in a number of milk protein genes such as the α-lactalbumin and the rat α-casein genes, were also present on the BSDL gene 5′ flanking region along with consensus sequences for a tissue-specific mammary gland factor and for pancreas-specific enhancer element. Whether all these transcriptional elements or binding sites are functional and how BSDL genes are regulated are open to question. The gene for BSDL in human, rat, and mouse consists of 11 exons and 10 introns and spans 9.2, 8.0, and 7.2 kb, respectively.[23,29,30] When examining the organization of the genes in different species, it is evident that they are organized in a similar manner with regard to the position of exons and introns. The sizes of the exons are similar with the exception of exon 11, which encodes the tandemly repeated sequences of variable number. Exon 11 differs also by species with respect to the size of the 3′ untranslated sequence. All of the functional sites are conserved in all species examined so far. Chemical modification[34,35] and site-directed mutagenesis experiments[36–38] have demonstrated that Ser-194, Asp-320, and His-435 form the catalytic triad; these amino acids are encoded by exons 5, 8, and 10, respectively. Exon 3 also contains the sequence encoding the putative heparin-binding site. The cystine residues at position 64 and 80 encoded by the latter exon and forming a disulfide bridge are also conserved, as are cystines forming the second disulfide bridge at positions 246 and 257[39] (encoded by exon 7).

Glycosylation of Bile Salt-Dependent Lipase

Early studies on human pancreatic BSDL have shown that it is a highly glycosylated protein.[40] Carbohydrate composition of the protein is compatible with the presence of N- and O-linked carbohydrate structures.[40] A site for N-linked glycosylation was located on asparagine at position 187, in all known sequences of BSDL.[21–28] The O-linked structures are located on C-terminal tandemly repeated sequences.[41] Consequently, the amount of

[34] D. Lombardo, *Biochim. Biophys. Acta* **700,** 67 (1982).

[35] N. Abouakil, E. Rogalska, and D. Lombardo, *Biochim. Biophys. Acta* **1002,** 225 (1989).

[36] L. P. Di Persio, R. N. Fontaine, and D. Y. Hui, *J. Biol. Chem.* **265,** 16801 (1990).

[37] L. P. Di Persio and D. Y. Hui, *J. Biol. Chem.* **268,** 300 (1993).

[38] L. P. Di Persio, R. N. Fontaine, and D. Y. Hui, *J. Biol. Chem.* **266,** 4033 (1991).

[39] T. Baba, D. Downs, K. W. Jackson, J. Tang, and C.-S. Wang, *Biochemistry* **30,** 500 (1991).

[40] O. Guy, D. Lombardo, and J. G. Brahms, *Eur. J. Biochem.* **117,** 447 (1981).

[41] C.-S. Wang, A. Dashti, K. W. Jackson, J.-C. Yeh, R. D. Cummings, and J. Tang, *Biochemistry* **34,** 10639 (1995).

sugar varies with the species and is correlated in first approximation with the number of repeats.[42]

The structure of N-linked oligosaccharide chains of BSDL was determined on the protein isolated from a normal donor (i.e., without pancreatic pathology).[43] After hydrazinolysis and reduction with $NaB[^3H]_4$, neutral (N) sugar chains were separated from acidic (AN) ones by paper electrophoresis. N-Linked oligosaccharide fractions N and AN were fractionated using serial column chromatographies on *Aleuria aurantia* lectin (AAL)–Sepharose and concanavalin A (ConA)–Sepharose. On AAL–Sepharose column chromatography, the N fraction (72% of N-linked sugar) was separated into two fractions: an unbound [N(AAL$^-$), 54%] fraction and a bound fraction eluted with 1 m*M* fucose [N(AAL$^+$), 18%]. The AN fraction (28% of N-linked sugar) also was divided into an AN(AAL$^-$) fraction (19%) and an AN(AAL$^+$) fraction (9%). All but N(AAL$^+$) were separated into three fractions on ConA–Sepharose: an unbound fraction (ConA$^-$), a weakly bound fraction eluted with 5 m*M* methyl α-glycoside (ConA$^+$), and a strongly bound fraction eluted with 100 m*M* methyl α-mannoside (ConA^{++}). For the N(AAL$^-$) fraction they represented 10, 31, and 13%, respectively. With the AN(AAL$^-$) and AN(AAL$^+$) fractions, only ConA$^-$ and ConA$^+$ material was isolated. The ConA$^-$ material represented 7 and 4%, respectively, of AN(AAL$^-$) and AN(AAL$^+$) fractions, while ConA$^+$ material represented 12 and 5% of AN(AAL$^-$) and AN(AAL$^+$) fractions. The structures of oligosaccharides in each fraction were determined by sequential exoglycosidase digestions. The reaction mixtures at each step were analyzed by Bio-Gel P-4 column chromatography. Eight different structures have been detected[43]; seven of them, accounting for about 60% of N-linked structures, are of the *N*-acetyllactosamine type. The basic structure representative of these glycans is given in Fig. 1.

All oligosaccharides are derived from the basic oligosaccharide (depicted in boldface in Fig. 1) by fucosylation or sialylation on the terminal galactose (Gal) residue (37% of N-linked structures). Poly-*N*-acetyllactosamine represented only 4% of these structures, and 5% have a fucose (Fuc) residue branched on the innermost *N*-acetylglucosamine (GlcNAc) residue. About 13% of the analyzed structures are intermediate, ranging from unprocessed oligomannoside structures to fully trimmed *N*-acetyllactosamine structures. The last structure detected by Sugo *et al.* is hybrid in type and represents only 2% of N-linked structures.[43] These structures are compatible

[42] N. Abouakil, E. Rogalska, J. Bonicel, and D. Lombardo, *Biochim. Biophys. Acta* **961,** 299 (1988).

[43] T. Sugo, E. Mas, N. Abouakil, T. Endo, M.-J. Escribano, A. Kobata, and D. Lombardo, *Eur. J. Biochem.* **216,** 799 (1993).

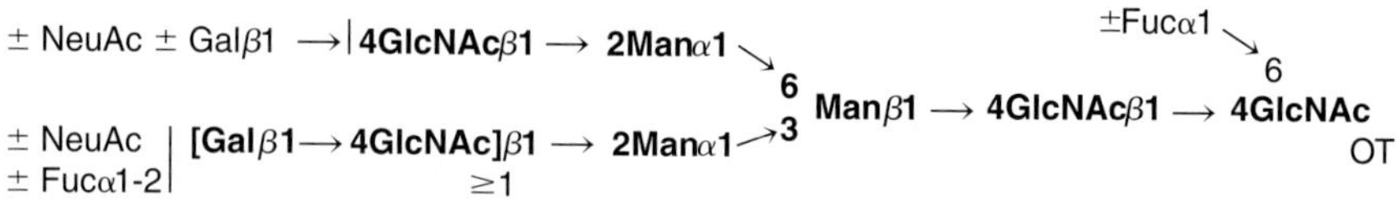

FIG. 1. Basic structure of N-linked oligosaccharides of BSDL. The basic structure of N-linked oligosaccharides of BSDL was deduced from the work of T. Sugo, E. Mas, N. Abouakil, T. Endo, M. J. Escribano, A. Kobota, and D. Lombardo, *Eur. J. Biochem.* **216,** 799 (1993).

with the general trimming of glycoproteins from the endoplasmic reticulum to the *trans*-Golgi vesicles.

Few data are available concerning the structure of O-linked glycan(s). Bile salt-dependent lipase reactivity with lectin such as peanut (*Arachis hypogaea*) agglutinin suggested the presence of galactose linked to *N*-acetylgalactosamine (Galβ1 → 3GalNAc structure). Because prior treatment with neuraminidase is a prerequisite to the binding of the peanut agglutinin to pure BSDL and because lectin reactivity increased after fucosidase treatment one may suspect that the Galβ1 → 3GalNAc structures are sialylated and/or fucosylated. The amount of Gal and GalNAc residues in the normal variant (i.e., isolated from a normal donor) of BSDL is compatible with the presence of 12–14 O-linked structures. These O-linked glycans are bound to threonine residues of the 14 tandem repeat units of identical sequence (i.e., PTTGDS) present in the C-terminal tail of the human BSDL[44] as mentioned above. Nine of these threonine residues have been located.[41] The overestimations of the apparent molecular weight determined by molecular sieving of the human pancreatic and milk BSDL[45,46] are attributable to the mucin-like extended conformation of the C-terminal tail of the protein, which is conferred by both high proline content and O-linked glycosylation.

A subpopulation of BSDL, referred to as the feto-acinar pancreatic protein or FAP, presents an oncofetal carbohydrate-dependent epitope recognized by the J28 monoclonal antibody (MAb J28). Bile salt-dependent lipase and FAP in part differ in their sugar composition both for sugars involved in N- and O-linked oligosaccharides.[47] Because FAP is expressed during the oncogenic process and in cases of inflammatory pathologies of

[44] E. Mas, N. Abouakil, S. Roudani, J. L. Franc, J. Montreuil, and D. Lombardo, *Eur. J. Biochem.* **216,** 807 (1993).

[45] D. Lombardo, O. Guy, and C. Figarella, *Biochim. Biophys. Acta* **527,** 142 (1978).

[46] L. Bläckberg and O. Hernell, *Eur. J. Biochem.* **116,** 221 (1981).

[47] E. Mas, N. Abouakil, S. Roudani, F. Miralles, O. Guy-Crotte, C. Figarella, M. J. Escribano, and D. Lombardo, *Biochem. J.* **289,** 609 (1993).

the human pancreas, the glycosylation of BSDL should differ depending on the physiopathological state of the pancreas. This fact was suggested by the carbohydrate analyses performed on four variants of BSDL isolated from a normal donor and from patients with pancreatic pathologies[44] (Table I).

From these carbohydrate compositions, two main points appear when comparing the normal variant of BSDL (isolated from the pancreatic juice of a normal donor) to pathological variants (isolated from pancreatic juices of two patients with chronic pancreatitis) and to FAP. First, the amount of mannose residues increases and correlates with the prevalence of unprocessed or hybrid N-linked structures in FAP and in pathological variants. Second, the amount of *N*-acetylgalactosamine (GalNAc) and of galactose, two sugars involved in O-linked structures, largely decreases in FAP. The amount of fucose and sialic acid (NeuAc) varies between glycovariants. Interestingly, whatever the origin of BSDL, the ratio of Gal + GalNAc to Fuc + NeuAc is always close to one (0.9 ± 0.09). This suggests or better confirms data obtained with lectin that the minimal O-linked Galβ1 → 3GalNAc structure of the pancreatic BSDL could be sialylated and fucosylated. It is worth noting that fucose prevails over sialic acid in FAP, which correlates with data suggesting that the epitope recognized by MAb J28 is an

TABLE I
CARBOHYDRATE COMPOSITION OF NORMAL AND PATHOLOGICAL VARIANTS OF BILE SALT-DEPENDENT LIPASE[a]

	Normal variant J28$^-$ (mol/mol protein)	Pathological variant		
		PCC J28$^-$ (mol/mol protein)	PCC J28$^-$ (mol/mol protein)	FAP J28$^+$ (mol/mol protein)
Man	5.0	7.2	6.6	8.0
GlcNAc	3.5	2.2	4.4	2.6
Gal	12.7	13.2	11.0	6.7
GalNAc	11.7	11.8	11.7	7.9
Fuc	10.4	11.5	12.8	11.8
NeuAc	10.6	10.3	6.2	3.1

[a] Bile salt-dependent lipase was purified from pancreatic juices from either a normal donor (normal variant) or patients with chronic pancreatitis (pathological variants, PCC), according to the method described by D. Lombardo, O. Guy, and C. Figarella, *Biochim. Biophys. Acta* **527,** 142 (1978). All these preparations were devoid of reactivity with MAb J28 (J28$^-$). The oncofetal variant of BSDL (FAP) reactive with MAb J28 (J28$^+$) was isolated according to E. Mas, N. Abouakil, S. Roudani, F. Miralles, O. Guy-Crotte, C. Figarella, M. J. Escribano, and D. Lombardo, *Biochem. J.* **289,** 609 (1993). Data were compiled from this last reference and from E. Mas, N. Abouakil, S. Roudani, J. L. Franc, J. Montreuil, and D. Lombardo, *Eur. J. Biochem.* **216,** 807 (1993).

O-linked structure involving fucose residues (our unpublished observation). Although differences between BSDL and FAP do not seem to be restricted to sugars,[47] modifications of the glycosylation of BSDL can affect its activity, as can be suspected from the low specific activity of the oncofetal glycovariant measured either on nonspecific substrates (4-nitrophenyl acetate and hexanoate) or cholesteryl esters.[47] This fact is of importance when considering the production of recombinant BSDL in either eukaryotic or prokaryotic cells. The fundamental difference between the DNA, RNA, and protein synthesis, on the one hand, and the glycan synthesis, on the other, is that the first depend on the reproduction of a template, whereas the second is dependent on the control and specificity of a series of glycosidases and glycosyltransferases executing concerted or ordered reactions without the need for a template. Therefore, the glycosylation of any recombinant protein is dependent on the expression system. In this respect, information concerning the role of glycans linked to a protein and in particular to BSDL may be valuable.

Role of N-Glycosylation of Bile Salt-Dependent Lipase

Four experimental approaches are accessible for assessing the role of protein glycosylation. Enzymes that cut glycans to the innermost residue of either N-linked structures (Endo H, PNGase F) or O-linked structures (*O*-glycanase) are provided by some manufacturers and can be used on purified glycoproteins. Some of these enzymes need detergent to catalyze the lysis of glycan structures; under these conditions the denaturation of the protein by the detergent can preclude effects induced by deglycosylation. Chemical methods using, for example, trifluoromethane sulfonic anhydride are sometimes too drastic to be useful. Drugs that affect either the transfer "en bloc"of the oligomannoside structure to the nascent polypeptide or the maturation of N-linked structures during the protein transport can be successfully used when performing studies on cell models expressing the protein constitutively. Drugs that affect O-linked glycosylation (aryl-*N*-acetyl-α-galactosaminides[48]) may have secondary effects because they possibly affect the expression of such structures on plasma membranes and consequently cell-to-cell recognition (our unpublished observation). Mutagenesis directed to the sites of glycosylation and expression of the unglycosylated protein in any kind of cell are also ways to gain information on the role of glycosylation. For BSDL all of these experimental approaches were taken. The two first methods were of little value with BSDL (our unpublished observation), whereas the two last methods were very informative.

[48] S. F. Kuan, J. C. Byrd, C. Bashaum, and Y. S. Kim, *J. Biol. Chem.* **264,** 19271 (1989).

The rat pancreatoma AR4-2J cell line is a unique line capable of synthesis and secretion of BSDL.[49] Tunicamycin, a structural analog of UDP-GlcNAc and the most widely used inhibitor of N-linked glycosylation,[50] blocks the transfer of GlcNAc-1-phosphate to dolichol phosphate and thus the formation of $Glc_3Man_9GlcNAc_2$-pyrophosphoryl dolichol. This drug had no effect on the synthesis of BSDL by AR4-2J cells, but the enzyme was isolated with a lower molecular weight and did not incorporate radiolabeled mannose compatible with the expression of a BSDL lacking its N-linked structure.[51] The nonglycosylated BSDL, although still active, was less efficient than the native enzyme; the former protein was also less stable to temperature changes. It was also shown that the rate of secretion of BSDL by AR4-2J cells exposed to tunicamycin was about two times slower than that of cells grown in the absence of the drug. Nevertheless, drugs that affect the processing of the oligomannose structure, such as castanospermine, 1-deoxymannojirimycin, or swainsonine, had no effect either on the enzyme activity or on its secretion.[51] This result corroborates the observation that the glycosylation of human BSDL may vary with pancreatic pathology without affecting its secretion.[44] Therefore, the transfer of the oligosaccharide precursor to the nascent BSDL is essential for the folding of the fully active BSDL and is required for enzyme secretion. The same effect of tunicamycin on both rate of secretion and activity of BSDL was observed with Chinese hamster ovary (CHO) cells transfected with the complete cDNA encoding the rat BSDL.[52] Nevertheless, controversial studies report on the lack of importance of N-glycosylation for functional properties of BSDL expressed in C127 cells transfected with a cDNA variant with altered N-glycosylation site (Asn-187 to Gln). This variant was fully active, secreted, and displayed the same stability to pH, heat, and proteolysis by trypsin.[53] Whether these differences came from intrinsic properties of the transfected material (e.g., human versus rat) or from cell lines used for the BSDL expression is not yet known.

Role of O-Glycosylation of Bile Salt-Dependent Lipase

On the basis of the sequence of cloned BSDL studied to date, two questions need to be answered: The first concerns the role of the tandem

[49] Y. Huang and D. Y. Hui, *J. Biol. Chem.* **266,** 6720 (1991).

[50] J. A. Pizzey, F. A. Bennett, and G. E. Jones, *Nature* (*London*) **305,** 315 (1983).

[51] N. Abouakil, E. Mas, N. Bruneau, A. Benajiba, and D. Lombardo, *J. Biol. Chem.* **268,** 25755 (1993).

[52] K. R. Morlock-Fitzpatrick and E. A. Fisher, *Proc. Soc. Exp. Biol. Med.* **208,** 186 (1995).

[53] L. Hansson, L. Bläckberg, M. Edlund, L. Lundberg, M. Strömqvist, and O. Hernell, *J. Biol. Chem.* **268,** 26692 (1993).

repeated sequences encoded by exon 11, the number of which varies by species, and the second concerns the importance of the O-linked glycosylation present on these sequences. Each repeated sequence is rich in proline (P), glutamic acid (E), serine (S), and threonine (T). Proteins that are rapidly degraded within eukaryotic cells frequently contain such sequences, referred to as the PEST (for Pro, Glu, Ser, and Thr) region.[54] On BSDL, PEST regions coexist with the C-terminal cluster of O-linked oligosaccharides[21] and are absence in sequences of other known pancreatic enzymes.

The function of this domain of repeated identical sequences remains unknown, but it may function to dictate intracellular processing and transport of BSDL.[55] When this domain was deleted by recombinant DNA techniques, the truncated BSDL was normally expressed and displayed full activity and functional properties such as activation by bile salts and heparin binding.[53,55–58] In contrast to the results of Bläckberg *et al.*,[57] Loomes[56] and Di Persio *et al.*,[55] suggested that the C-terminal repeats may regulate proteolytic degradation of the protein and substrate delivery to the enzyme. Although truncated BSDL was expressed in various cell models, e.g., prokaryotic cells such as *Escherichia coli*[58] or eukaryotic cells such as monkey kidney fibroblast type 1 (COS-1) cells,[55] murine C127 cells,[53,57] or baby hamster kidney (BHK) cells,[56] this cannot be argued to explain the discrepancy in results. However, the size of the recombinant BSDL appeared to be of importance. All truncated forms of BSDL that displayed full activity and functional properties included residues Ala-1 to residue His-536.[55,57,58] Those that were neither catalytically active nor insensitive to proteolysis were shorter and included residues Ala-1 to Tyr-489[55] or residues Ala-1 to Leu-519.[56] Interestingly, BSDL from which the sequence from residue Pro-490 to the end was deleted was not secreted owing to its retention within a particulate fraction of transfected cells.[55] Therefore, it seems likely that the C-terminal tail of BSDL and particularly the sequence between residues Pro-490 and Glu-534 is required for normal intracellular processing and secretion of the enzyme.[55] This domain, encoded by exon 11, has a highly variable sequence between species and precedes the tandemly repeated sequences. Therefore, the intracellular processing of BSDL can be different between species and could be related to the glycosylation of this domain. Interestingly, the rat and human enzymes, which are both O-glycosylated, are associated with intracellular membranes by means of a 94-kDa protein

[54] S. Rogers, R. Wells, and M. Rechsteiner, *Science* **234,** 364 (1986).
[55] L. P. Di Persio, C. P. Carter, and D. Y. Hui, *Biochemistry* **33,** 3442 (1994).
[56] K. M. Loomes, *Eur. J. Biochem.* **230,** 607 (1995).
[57] L. Bläckberg, M. Strömqvist, M. Edlund, K. Juneblad, L. Lundberg, L. Hansson, and O. Hernell, *Eur. J. Biochem.* **228,** 817 (1995).
[58] D. Downs, Y. Y. Xu, J. Tang, and C. S. Wang, *Biochemistry* **33,** 7979 (1994).

immunologically related to the glucose-regulated protein of 94 kDa (Grp-94).[59,60] This association takes place in the endoplasmic reticulum, and BSDL is released from the Grp-94-related 94-kDa protein once fully glycosylated in or after leaving the *trans*-Golgi compartment.[59,60] With this background, it seems difficult to understand the role of the O-linked glycosylation in either the intracellular processing or the enzyme properties using truncated recombinants of BSDL. The use of cells with defects in O-glycosylation, such as the CHO 1d1D cell line, appears more appropriate.[61] This cell line transfected with the full-length cDNA of the rat BSDL has been used in one study.[52] Results indicated that, contrary to the N-linked glycosylation, the O-linked glycosylation did not affect enzyme secretion or the stimulation of its activity by bile salts. We have extended this study to the intracellular processing of BSDL and were unable to detect any degradation of the enzyme in the absence of O-linked glycosylation (our unpublished observation). Consequently, the PEST sequences do not seem to divert non-O-glycosylated BSDL to a degradation cell compartment. A way to reconcile the presence of PEST sequences and the secretion of BSDL instead of its degradation is to propose that the association of BSDL with membranes from the endoplasmic reticulum to the *trans*-Golgi compartment is sufficient to protect the enzyme from degradation independent of the O-glycosylation of PEST sequences.

Conclusion

Despite enormous advances due to recombinant DNA techniques, the understanding of the complex function of protein glycosylation remains primitive. The role of glycans on glycoproteins may be threefold. First, the glycan may participate in the folding of the protein and influence its stability. Second, the glycan may not be directly associated with the function of the glycoprotein but may modulate its half-life by, for example, favoring or inhibiting hepatic clearance. Finally, the glycan may have biological function through its interaction with target molecules such as the homing of lymphocytes via selectins.

As discussed here, the N-linked glycosylation of BSDL seems important for the folding and secretion of BSDL, but further processing of the glycan structure did not affect the secretion and activity of the enzyme. Consequently the expression of recombinant BSDL in baculovirus-infected Sf9 insect cells[62] appeared appropriate for high-level production of BSDL for

[59] N. Bruneau and D. Lombardo, *J. Biol. Chem.* **270,** 13524 (1995).
[60] N. Bruneau, P. Lechêne de la Porte, and D. Lombardo, *Eur. J. Biochem.* **233,** 209 (1995).
[61] D. M. Kingsley, K. F. Kozarsky, L. Hobbie, and M. Krieger, *Cell* **44,** 749 (1986).
[62] L. P. Di Persio, J. A. Kissel, and D. Y. Hui, *Protein Expr. Purif.* **3,** 114 (1992).

further crystallization of the enzyme. With regard to the C-terminal mucin-like tail, we should be more cautious. Transfer of the enzyme from the intestinal lumen to the bloodstream may necessitate the O-linked glycosylation of BSDL, which under those conditions may not only be species specific but may also depend on the cell line used as the model for captation. If through the production of recombinant BSDL in cells that glycosylate BSDL differently, behavior different from the native protein[63] is observed, then a new hypothesis will be required. Overall, results obtained until now indicate, as hypothesized by us 10 years ago, that "the enzyme could be called to assume a function behind the mucosa" (cited from Ref. 20). This function still remains to be characterized, but in the light of more recent data [the ability of circulating BSDL to modify normal human LDL and HDL composition and structure and to reduce the atherogenicity of oxidized LDL, and the presence of sialyl Lewis^x and sialyl Lewis^a antigen on either milk[41] and pancreatic BSDL (our unpublished results)], one wonders whether the C-terminal O-glycosylated tail of the protein may be involved in the interaction of BSDL with apolipoprotein and/or contribute to an adhesive activity of the protein. Although it has been shown that apoB-100 peptides bind to proteoglycans,[64] how BSDL interacts with LDL and what the putative role of BSDL is in adhesive processes must be further analyzed. There is certainly a reason why human eosinophils express a 74-kDa form of BSDL, which can be either the unglycosylated protein or a variant possessing fewer tandem repeats than the 100-kDa human enzyme. This variant must be further characterized before drawing any conclusion. Nevertheless, the role of the eosinophil enzyme, which is localized in granules, is probably to destroy parasites,[18] and likely not to interfere with lipoprotein metabolism. Beside the role of its glycosylation, the intracellular role of BSDL or of its oncofetal glycovariant FAP needs to be further analyzed. Thus it has been suggested that the lipolytic activity of BSDL would be essential for the metabolism of lipids during growth and differentiation of fetal cells[19] and for exocytosis in acinar cells.[65]

Acknowledgments

The research in this laboratory is made possible by institutional funding from the Institut National de la Santé et de la Recherche Médicale (Paris, France) and by grants awarded from the Association pour la Recherche contre le Cancer (Villejuif, France), the Conseil Général des Bouches-du-Rhône (Marseille, France), and the Conseil Régional of Provence-Alpes-

[63] M. Strömqvist, K. Lindgren, L. Hansson, and K. Juneblad, *J. Chromatogr.* **A718,** 53 (1995).
[64] L. Chan, *J. Biol. Chem.* **267,** 25621 (1992).
[65] M. Withiam-Leitch, R. P. Rubin, S. E. Koshlukova, and J. M. Aletta, *J. Biol. Chem.* **270,** 3780 (1995).

Côte-d'Azur (Marseille, France). Eric Mas was a recipient of a fellowship from the Ligue Nationale contre le Cancer (Paris, France). Authors address their special thanks to Mrs. V. Sbarra, Mr. D. Lecestre, and C. Crotte for their generous help. This work is dedicated to the students who were working or are still working in our group.

[21] Stereoselectivity of Lipase from *Rhizopus oryzae* toward Triacylglycerols and Analogs: Computer-Aided Modeling and Experimental Validation

By LUTZ HAALCK, FRITZ PALTAUF, JÜRGEN PLEISS, ROLF D. SCHMID, FRITZ SPENER, and PETER STADLER

Introduction

The genus *Rhizopus* has long been known to produce extracellular lipases. In fact, many *Rhizopus* species such as *R. delemar, R. microsporus, R. niveus,* and *R. arrhizus* were found to be efficient producers of various lipases that could be purified and characterized.[1] All *Rhizopus* lipases studied so far have a molecular mass between 29 and 39 kDa, hydrolyze triacylglycerols preferentially at the *sn*-1 and *sn*-3 positions, and show little selectivity for hydrolysis of triacylglycerols with different fatty acid moieties. As this behavior is similar to the selectivities of human pancreatic lipase, *Rhizopus* lipases have been proposed as substitution products for patients suffering from pancreatic insufficiency.[2]

Cloning of the genes encoding the lipases from *R. delemar*[3,4] and *Rhizopus oryzae*[5] has revealed that their nucleotide sequences are nearly identical, comprising a signal sequence (26 amino acids), a prosequence (97 amino acids), and a region encoding the mature lipase (269 amino acids). The proenzyme (39 kDa), which is fully active, is cleaved by proteolysis *in vivo* to provide the mature form of the lipase (29 kDa). Evidence has been obtained that the prosequence is involved in the correct folding and thus may have a chaperone-like function[5] in this lipase, which contains three disulfide bridges and one free cysteine in the prosequence. On expression

[1] M. Iwai and Y. Tsujisaka, *in* "Lipases" (B. Borgström and H. L. Brockman, eds.), p. 443. Elsevier, Amsterdam, 1984.

[2] A. Noma and B. Borgström, *Scand. J. Gastroenterol.* **6,** 217 (1971).

[3] M. J. Haas, J. Allen, and T. R. Berka, *Gene* **109,** 107 (1991).

[4] R. D. Joerger and M. J. Haas, *Lipids* **28,** 81 (1993).

[5] D. Beer, G. Wohlfahrt, R. D. Schmid, and J. E. G. McCarthy, *Biochem. J.* **319,** 351 (1997).

0076-6879/97 $25.00

in *Escherichia coli,* inclusion bodies are formed that can be reconstituted to fully active lipase.

As *Rhizopus* lipase shows about 60% sequence homology to the lipase from *Rhizomucor miehei,* whose structure has been solved at a resolution of 1.9 Å,[6] a structural model of *R. oryzae* lipase based on the X-ray structure of *R. miehei* lipase can be built and has been corroborated by site-directed mutagenesis.[7] The structure of *R. delemar* lipase was determined by X-ray crystallography at a resolution of 2.6 Å.[8]

Genetic experiments on the substrate specificity of *R. oryzae* lipase showed that substitution of amino acids in the predicted acyl chain-binding groove, in particular Phe95Asp[9] and Val209Trp,[9,10] resulted in enhanced preference for short- and medium-chain triacylglycerols.

The stereopreference of pure *R. arrhizus* lipase (reclassified to *R. oryzae*[11]) has been thoroughly studied using monomolecular films of pure dicaprin.[12] A moderate preference for hydrolysis of the *sn*-1 acyl chain is observed in these experiments. If triacylglycerols and analogs were used in emulsion, stereopreference shifts from *sn*-1 to *s*n-3 depending on the steric requirements of the *sn*-2 substituent of the substrate.[13]

In this chapter, we describe the preparation of *R. oryzae* lipase from recombinant *E. coli,* the analysis of stereoselectivity of this lipase toward triacylglycerols and analogs, and computer modeling to gain insight into the molecular basis of its stereopreference.

Production of *Rhizopus oryzae* Lipase

Cloning and Expression of Rhizopus oryzae Lipase in Escherichia coli

DNA Isolation. The fungus *Rhizopus oryzae* DSM 853 is grown on Czapek-Dox medium[14] supplemented with 2% (w/v) malt extract. Total

[6] A. M. Brzozowski, U. Derewenda, Z. S. Derewenda, G. G. Dodson, D. M. Lawson, J. P. Turkenburg, F. Bjorkling, B. Huge-Jensen, S. A. Patkar, and L. Thim, *Nature (London)* **351,** 491 (1991).

[7] H. D. Beer, G. Wohlfahrt, J. E. G. McCarthy, D. Schomburg, and R. D. Schmid, *Protein Eng.* **9,** 507 (1997).

[8] U. Derewenda, L. Swenson, Y. Wei, R. Green, P. M. Kobos, R. Joerger, M. J. Haas, and Z. S. Derewenda, *J. Lipid Res.* **35,** 524 (1994).

[9] R. D. Joerger and M. J. Haas, *Lipids* **29,** 377 (1994).

[10] H. Atomi, U. Bornscheuer, M. M. Soumanou, H. D. Beer, G. Wohlfahrt, and R. D. Schmid, P. J. Barnes & Associates, High Wycombe, Proc. 21st World Cong. Int. Soc. Fat Res., Vol. 1, 49, 1996.

[11] M. J. Haas and R. D. Joerger, *in* "Food Biotechnology Microorganisms" (Y. H. Hui and G. G. Khachatourians, eds.), p. 564. VCH, New York, 1995.

[12] E. Rogalska, S. Nury, I. Douchet, and R. Verger, *Chirality* **7,** 505 (1995).

[13] P. Stadler, A. Kovac, L. Haalck, F. Spener, and F. Paltauf, *Eur. J. Biochem.* **227,** 335 (1995).

[14] K. Tsuchiya, S. Tada, K. Gomi, K. Kitamoto, C. Kumagai, Y. Jigami, and G. Tamura, *Appl. Microbiol. Biotechnol.* **38,** 109 (1992).

DNA is isolated according to a modified procedure of Raeder and Broda.[15] Southern and colony hybridization are carried out according to standard procedures[16] using two oligonucleotides corresponding to conserved N- and C-terminal regions of the homologous lipases from *Rhizomucor miehei,*[17] *R. niveus,*[18] and *R. delemar*[4] as described by Beer *et al.*[5]

Transformation and Expression Experiments. The cDNA encoding the mature *R. oryzae* lipase is excised from plasmid pCYTEXP1[5,19] as an *Nde*I–*Bam*HI fragment and ligated to the pCRII vector (Invihogen, Leek) for amplification and sequencing of the lipase gene. For plasmid propagation *E. coli* strain InvαF′ is used. The *Nde*I–*Bam*HI fragment is subcloned into the expression vector pET3a, giving pET3-ROL. Transformation with plasmid DNA is accomplished using $CaCl_2$-treated *E. coli* cells as described by Sambrook *et al.*[16]

Escherichia coli strain BL21(DE3) harboring the pET3-ROL recombinant plasmid is grown on solid Luria-Bertani (LB) agar medium containing ampicillin (100 μg/ml). Cells from a single colony are inoculated into LB broth containing the same concentration of ampicillin and grown overnight at 37°. Five milliliters are taken to inoculate 2 liters of LB broth (five 400-ml portions). The cultures are shaken vigorously at 37° until the optical density at 600 nm reaches 0.6 to 1. Isopropyl-β-thiogalactopyranoside (IPTG) is added to a final concentration of 1 m*M*. The cells are harvested 3 hr after induction.

Refolding of Recombinant Rhizopus oryzae Lipase. The recombinant mature *R. oryzae* lipase (ROL) is produced as insoluble inclusion bodies in the cytoplasm. The renaturation procedure is adopted from Joerger and Haas.[4] The cells obtained from a 2-liter culture (about 12 g) are resuspended in 100 ml of 50 m*M* Tris-HCl (pH 8.0), 5 m*M* ethylendiaminetetraacetic acid (EDTA), 10% (w/v) sucrose. Lysozyme is added to a final concentration of 0.8 mg/ml and the mixture is incubated for 30 min at 30°. On addition of 400 ml of buffer [10 m*M* Tris-HCl (pH 8.0), 1 m*M* EDTA, 0.5% (v/v) Triton X-100] the lysate is suspended, sonicated three times for 20 sec with a tip-type sonicator (15 W), and centrifuged at 15,000 *g* for 30 min. The pellet is resuspended in 10 m*M* Tris-HCl (pH 8.0), 1 m*M* EDTA and incubated for 1 hr at 37° with benzonase (90 U/ml) for DNA degradation. After centrifugation at 15,000 *g* for 30 min, the inclusion bodies are dissolved in 200–800 ml of solubilization buffer [8 *M* urea, 25 m*M* phosphate buffer (pH 7.0), 1 m*M* EDTA, and 5 m*M* dithiothreitol (DTT)] at room tempera-

[15] U. Raeder and P. Broda, *Lett. Microbiol.* **1,** 17 (1985).

[16] J. Sambrook, E. F. Fritsch, and T. Maniatis, "Molecular Cloning," 2nd Ed., p. A.1. Cold Spring Harbor Laboratory Press, Cold Spring Harbor, New York, 1989.

[17] E. Boel, B. Huge-Jensen, M. Christensen, L. Thim, and N. Fiil, *Lipids* **23,** 701 (1988).

[18] W. Kugimiya, Y. Otani, and M. Hashimoto, Japanese Patent 64-80290 (1989).

[19] T. N. Belev, M. Singh, and J. E. G. McCarthy, *Plasmid* **26,** 147 (1991).

ture for 1 hr. The protein content is estimated by densitometry after sodium dodecyl sulfate–polyacrylamide gel electrophoresis (SDS–PAGE). The solution is centrifuged at 15,000 *g* for 30 min and cystine is added to the supernatant to a final concentration of 15 m*M*. The pH is adjusted for 10 min between pH 8.5 and 9.0. The solution is added dropwise to 10 to 20 vol of cold (4°) 50 m*M* phosphate buffer (pH 8.0), 1 m*M* EDTA, and 5 m*M* cysteine. The final volume depends on the protein content of inclusion bodies. Best results are achieved at a final protein concentration of 10 μg/ml and 48 hr of renaturation. As judged from densitometry after SDS–PAGE the yield of active mature lipase is 60%.

Purification of Rhizopus oryzae Lipase

Native Rhizopus oryzae Lipase. Rhizopus oryzae lipase (Lot No. 80,000 ST) formerly denoted as *R. arrhizus* lipase,[11] was purchased from Gist Brocades (Delft, The Netherlands). Twenty grams of the crude fermentation product is dissolved in 200 ml of starting buffer (10 m*M* Tris-HCL, pH 6.8). Using fast protein liquid chromatography (FPLC; Pharmacia, Uppsala, Sweden) cation-exchange chromatography is performed on a Fractogel EMD TMAE column (5 × 20 cm; E. Merck AG, Darmstadt, Germany) equilibrated with the same buffer at a flow rate of 4 ml/min. After sample injection the column is washed with 400 ml of starting buffer. Under these conditions, most of the lipase does not bind to the column and elutes directly with the void volume. For regeneration the column is washed with 200 ml of 0.7 *M* NaCl in starting buffer, giving rise to a fraction containing protein contaminants, colorants, and about 20% of the lipase, which is discarded. The protein profile is monitored spectrophotometrically at 280 nm and lipase activity is determined using *p*-nitrophenyl palmitate as the substrate as described by Winkler and Stuckmann.[20]

The pH of the void volume containing the lipase is adjusted to pH 7.2, NaCl is added to a final concentration of 2 *M*, and the sample is subjected to hydrophobic interaction chromatography on a phenyl-Sepharose Fast Flow column (5 × 20 cm; Pharmacia). This low substitution grade material ensures weak binding of the hydrophobic protein and allows for elution without detergents. The lipase retained is eluted with concentration gradients of decreasing NaCl (2–0 *M*) and concomitantly increasing ethylene glycol (0–80%, v/v) at a flow rate of 2 ml/min. The lipase-containing fractions obtained at 65% ethylene glycol are pooled, and salt and ethylene glycol are removed by passing the protein at a flow rate of 50 ml/min over a Sephadex G-25 column (10 × 20 cm; Pharmacia) equilibrated with 5 m*M*

[20] U. K. Winkler and M. Stuckmann, *J. Bacteriol.* **138,** 663 (1979).

calcium acetate, pH 4.7. Subsequently, the solution is applied at a flow rate of 2 ml/min to cation-exchange chromatography on a SP-Sepharose HP column (XK 16/20; Pharmacia). Elution with a linear concentration gradient increasing from 0 to 0.7 *M* NaCl in 5 m*M* calcium acetate, pH 4.7, gives rise to two lipolytically active fractions at 0.47 and 0.54 *M* NaCl, corresponding to a 32-kDa prolipase and the mature 29-kDa lipase, respectively. The yields are 10 and 43 mg for the prolipase and for the mature lipase, respectively. It is well established that *Rhizopus* as well as *Rhizomucor* lipases are expressed as preproenzymes, which are secreted into the periplasm as proenzymes and further processed to the mature lipase.[4] N-terminal sequencing revealed that the N terminus of the purified 32-kDa prolipase corresponds to the N terminus reported for the *R. niveus* lipase.[18] This observation supports the idea that the full-length prolipase (39 kDa) is sequentially cleaved by at least two proteases.

Recombinant Mature Rhizopus oryzae Lipase. The solution containing renatured lipase (10–20 liters) is adjusted to pH 5.5 and applied to cation-exchange chromatography on Fractogel EMD-SO_3^- 650(M) column (5 × 20 cm; E. Merck AG) at a flow rate of 30 ml/min, equilibrated with 50 m*M* sodium phosphate buffer, pH 5.5. The lipase is eluted using a linear concentration gradient increasing from 0 to 1.5 *M* NaCl in 50 m*M* sodium phosphate buffer, pH 5.5. The lipase-containing fractions eluting at 0.25 *M* NaCl are pooled, concentrated, and applied in aliquots of 15 ml at a flow rate of 3 ml/min to size-exclusion chromatography on a Fractogel BioSEC-650(S) Superformance column (2.6 × 60 cm; E. Merck AG) equilibrated with 10 m*M* sodium phosphate buffer, pH 5.5, and 100 m*M* NaCl for final polishing (R_t 68.7 min). Starting from the wet cell paste this procedure yields 150 mg of lipase (estimated yield, 40%).

Enzymatic activity of the purified lipases is determined titrimetrically using a pH-stat apparatus with olive oil[21] or tributyrin as substrate.[13] One unit is defined as the lipase activity that liberates 1 μmol of fatty acids per minute at 37°. The native mature ROL shows a specific activity of 6700 and 2200 U/mg using triolein and tributyrin, respectively, as substrates. These specific activities found for the recombinant mature ROL are essentially the same as those for the native mature lipase as tested for both substrates.

Preparation of Substrates

Synthesis of substrates is described starting from commercially available material (Fig. 1). The purity of substrates and intermediates is routinely

[21] R. Ruyssen and A. Lauwers, "Pharmaceutial Enzymes: Properties and Assay Methods." E. Story-Scientia, Gent, Belgium, 1978.

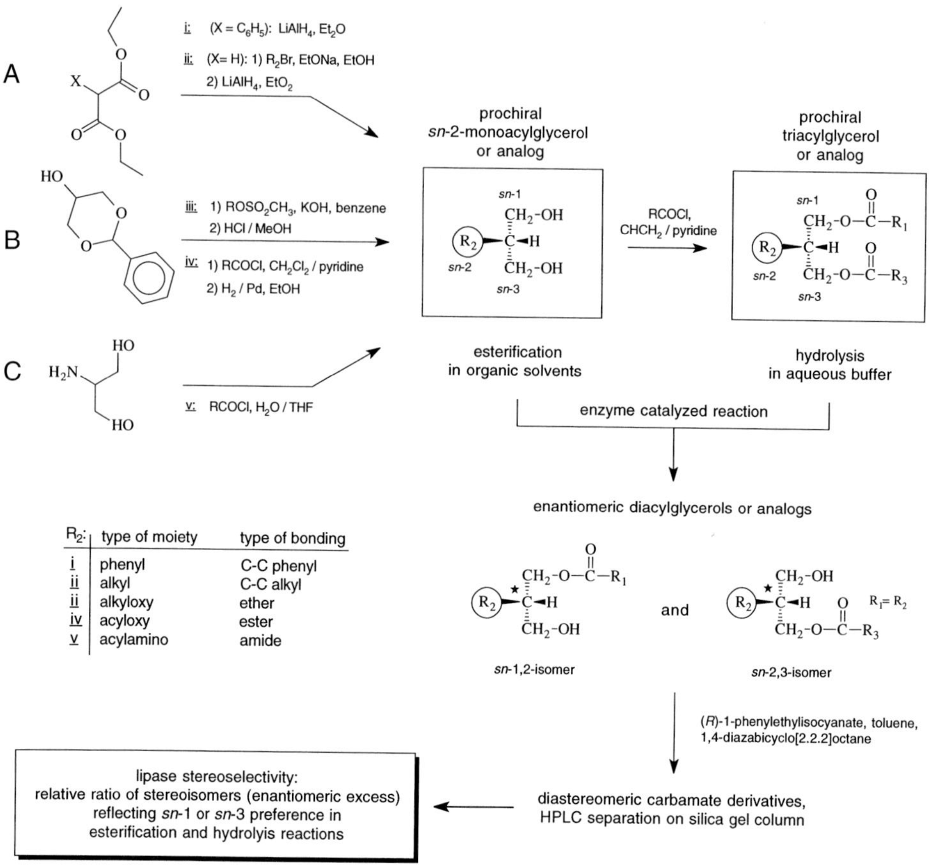

FIG. 1. Scheme for synthesis of substrates and procedure for determination of lipase stereoselectivity in hydrolysis and esterification experiments.

checked by thin-layer chromatography (TLC) on silica gel plates (0.2 mm; E. Merck AG). Spots are visualized by charring at 120° after spraying with 50% (v/v) sulfuric acid. Final products are characterized by their nuclear magnetic resonance (NMR) spectra obtained on a Varian (Palo Alto, CA) Gemini 200 using tetramethylsilane as an internal standard.

1,3-O-Benzylideneglycerol

An equimolar mixture of water-free glycerol (92 g, 1 mol) and freshly distilled benzaldehyde (106 g, 1 mol) in 600 ml of toluene is stirred under reflux in the presence of 1.8 g of *p*-toluenesulfonic acid (2% of glycerol, w/w).[22] The water that is produced during the formation of the benzylidene acetal is removed azeotropically using a water separation funnel. When water evolution has stopped, the cooled reaction mixture is seeded with 1,3-*O*-benzylideneglycerol crystals and stored at 0° overnight. The crystalline product is isolated by filtration and suspended in cold toluene containing sufficient sodium hydroxide (0.5 *N* in methanol) to neutralize the acid catalyst. The product is dissolved by warming and the toluene solution is washed with 1% aqueous sodium dihydrogenphosphate, dried over anhydrous sodium sulfate, and crystallized at 0°. Recrystallization from benzene–hexane (1 : 1) results in 41 g of 1,3-*O*-benzylideneglycerol, mp 58–60° (literature[22] mp 63.5–64.5° for the *cis*-isomer).

Hexadecyl Methane Sulfonate

The alkylmethane sulfonate is obtained in 90% yield by reacting hexadecanol with methane sulfochloride in methylene chloride in the presence of triethylamine as described by Paltauf and Hermetter.[23] The product has a mp of 52–53° (literature[24] mp 54–55°).

2-O-Hexadecyl-sn-glycerol

The 2-*O*-hexadecyl-*sn*-glycerol is prepared in analogy to 1-*O*-alkylglycerols as described by Baumann and Mangold.[24] To a solution of 4.3 g of 1,3-*O*-benzylideneglycerol (23.9 mmol) in 100 ml of benzene 3.6 g of powdered potassium hydroxide (64.3 mmol) is added. The suspension is refluxed for 1 hr using a water separation funnel to remove water by azeotropic distillation. Hexadecyl methanesulfonate (5 g, 15.6 mmol) dissolved in 20 ml of water-free benzene is then added dropwise, and refluxing is continued for 4 hr. After cooling, 100 ml of water is added and the product is extracted with diethyl ether (twice, 100 ml each). The combined organic phases are washed twice with 100 ml of water until neutral and dried over anhydrous sodium sulfate. The solvent is removed under reduced pressure to yield 1,3-*O*-benzylidene-2-*O*-hexadecylglycerol as a light yellow, waxlike product. On TLC this substance shows an R_f of 0.76 with chloroform–acetone–acetic acid (96 : 4 : 1, by volume) as the developing solvent. The crude 1,3-*O*-benzylidene-2-*O*-hexadecylglycerol (7.9 g) is hydrolyzed by stirring for 2 hr at room temperature in a solution consisting of 8 ml of concentrated hydrochloric acid, 33 ml of water, and 100 ml of methanol. After removal of the organic solvent under reduced pressure,

[22] F. H. Mattson and R. A. Volpenhein, *J. Lipid Res.* **3,** 281 (1962).
[23] F. Paltauf and A. Hermetter, *Methods Enzymol.* **197,** 135 (1991).
[24] W. J. Baumann and H. K. Mangold, *J. Org. Chem.* **29,** 3055 (1964).

the residue is extracted with 100 ml of diethyl ether. The ether extract is washed with water until neutral (twice, 100 ml each) and dried over anhydrous sodium sulfate. The sovlent is evaporated and 4.4 g (89% yield based on hexadecylmethane sulfonate) of 2-*O*-hexadecylglycerol is isolated after recrystallization from 100 ml of *n*-heptane. The product shows a single spot on TLC (R_f 0.10) with chloroform–acetone–acetic acid (96 : 4 : 1, by volume) and has a mp of 64–65°. Thin-layer chromatography analysis on boric acid-impregnated plates using the solvent system chloroform–methanol (98 : 2, v/v) confirms the isomeric purity of the 2-*O*-hexadecylglycerol (R_f 0.27); no 1-*O*-hexadecylglycerol is detected (R_f 0.13). ^{1}H NMR ($CDCl_3$): δ, 0.90 (t, 3H, CH_3); 1.20–1.74 (m, 28H, alkyl); 2.03 (m, 3H, β-CH_2); 3.40–3.50 (m, 1H, CH); 3.53–3.59 (t, α-CH_2); 3.62–3.82 (m, 4H, CH_2O).

2-Octadecanoyl-sn-glycerol

To a solution of 500 mg (2.78 mmol) of 1,3-benzylideneglycerol in 5 ml of methylene chloride–pyridine (3 : 1, v/v) 842 mg (2.78 mmol) of octadecanoyl chloride in 3 ml of methylene chloride is added dropwise at 0°. The reaction mixture is then stirred for 3 hr at room temperature. After the addition of 20 ml of methylene chloride the mixture is extracted with 20 ml of ice-cold 2% (v/v) hydrochloric acid. The organic phase is washed subsequently with two 20-ml portions of water and dried over anhydrous sodium sulfate. The solvent is removed under reduced pressure to yield crude 1,3-benzylidene-2-octadecanoylglycerol. The solid material shows an R_f of 0.59 with chloroform–acetone–acetic acid (96 : 4 : 1, by volume) and is used without further purification. The crude intermediate (1.2 g) in 20 ml of ethanol is hydrogenated at room temperature for 2 hr in the presence of 250 mg of 10% palladium–carbon (Pd–C). The catalyst is removed by centrifugation and the solvent is evaporated under reduced pressure. Recrystallization from 20 ml of light petroleum gives 643 mg (65% yield based on 1,3-benzylideneglycerol) of pure *sn*-2-octadecanoylglycerol, mp 70–71°. A TLC analysis of the product shows a single spot with an R_f of 0.09 with the solvent system chloroform–acetone–acetic acid (96 : 4 : 1, by volume). Using boric acid-impregnated plates and the solvent system chloroform–methanol (98 : 2, v/v) the appearance of a single spot (R_f 0.26) indicates the absence of detectable amounts of the corresponding 1-acylglycerol isomer (R_f 0.11). ^{1}H NMR ($CDCl_3$): δ, 0.90 (t, 3H, CH_3); 1.20–1.80 (m, 30 H, alkyl); 2.39 (t, 3H, α-CH_2); 3.85 (d, 4H, CH_2O); 4.87–5.00 (m, 1H, CH).

2-Deoxy-2-hexadecanoylamino-sn-glycerol

Serinol (2 g, 22.0 mmol) dissolved in 20 ml of water and 6.1 g of triethylamine (60.6 mmol) is monoacylated by addition of 5 g (18 mmol) of hexadecanoyl chloride dissolved in 20 ml of tetrahydrofuran in analogy to the procedure described by Dijkman *et al.*[25] The reaction mixture is stirred for 24 hr at 40°. After evaporation of the solvents under reduced pressure the residue is dissolved in 40 ml of chloroform. On washing with 40 ml of 5% hydrochloric acid the crude product precipitates and is isolated by filtration. The pure product (3.25 g, 55% yield) is obtained by recrystallization from hot chloroform. The product (mp 124–126°) has an R_f of 0.29 using the solvent system chloroform–methanol (9:1, v/v). ^{1}H NMR ($CDCl_3$-CD_3OD, 1:1): δ, 0.90 (t, 3H, CH_3); 1.10–1.70 (m, 26 H, alkyl); 2.17 (t, 2H, α-CH_2); 3.52–3.69 (m, 4H, CH_2O); 3.78–3.88 (m, 1H CH).

2-Deoxy-2-hexadecyl-sn-glycerol

To a stirred sodium ethanolate solution [3.2 g of sodium (139 milliatom) in 80 ml of absolute ethanol] 31.6 g (197 mmol) of diethylmalonate is added dropwise. Then 42.6 g (139 mmol) of hexadecyl bromide is added and the reaction mixture is refluxed for 4 hr. After 40 ml of the solvent is removed by distillation the mixture is cooled to room temperature and the precipitated sodium bromide is disoslved by addition of 300 ml of ice-cold water. The organic phase is separated and the aqueous layer is extracted with two 300-ml portions of diethyl ether. The combined organic phases are dried over anhydrous sodium sulfate. The solvent is removed under reduced pressure to give 63 g of the crude alkylated diethylmalonate. Thin-layer chromatography analysis using light petroleum–ethyl acetate (5:1, v/v) shows the alkylation product (R_f 0.74) and unreacted diethylmalonate (R_f 0.42). The crude material is dissolved in 100 ml of diethyl ether and is added dropwise to 8.3 g of lithium alanate (219 mmol) suspended in 220 ml of diethyl ether. The reaction mixture is stirred under reflux for 2.5 hr. After cooling the excess lithium alanate is destroyed by careful dropwise addition of ice-cold water. Precipitated aluminum hydroxides are dissolved with 300 ml of 10% sulfuric acid and the product is extracted with 500 ml of diethyl ether. The organic phase is washed with saturated aqueous sodium chloride and dried over anhydrous sodium sulfate. The solvent is removed under reduced pressure and the crude product (approximatley 30 g) is purified by column chromatography on silica gel. Impurities are eluted using a stepwise gradient of light petroleum–ethyl acetate (80:20 to 50:50, v/v); pure 2-deoxy-2-hexadecylglycerol (22 g, 53% yield based on hexadecyl bromide) is eluted with ethyl acetate as a solvent. The product (mp 79–83°; literature[26] 82–84°) shows a single spot on TLC (R_f 0.15) using heptane–

[25] R. Dijkman, N. Dekker, and G. H. de Haas, *Biochim. Biophys. Acta* **1043,** 67 (1990).
[26] P. L. Anelli, F. Montanari, S. Quici, G. Ciani, and A. Sironi, *J. Org. Chem.* **53,** 5292 (1988).

ethyl acetate (7:3, v/v). ^{1}H NMR ($CDCl_3$-CD_3OD, 1:1): δ, 0.90 (t, 3H, CH_3); 1.10–1.70 (m, 31 H, alkyl); 3.47–3.69 (m, 4H, CH_2O).

2-Deoxy-2-phenyl-sn-glycerol

Reduction of commercially available 2-phenyldiethylmalonate (Fluka, Buchs, Switzerland) with lithium alanate is accomplished in analogy to the procedure described above for the 2-alkyl analog. The pure product is obtained with 78% yield by vacuum distillation (bp. 120°, 10^{-4} bar) and shows a single spot on TLC (R_f 0.09) using chloroform–acetone–acetic acid (96:4:1, by volume).

Acylation of Monoacylglycerol Analogs

Acylation of monoacylglycerol analogs is performed by a procedure exemplified below for the preparation of 1,3-dioleoyl-2-*O*-hexadecylglycerol. To a solution of 1.07 g (3.38 mmol) of 2-*O*-hexadecylglycerol in 20 ml of methylene chloride–pyridine (3:1, v/v) 2.15 g of oleoyl chloride (7.14 mmol) in 3 ml of methylene chloride is added at 0°. The reaction mixture is then stirred for 2 hr at room temperature. The reaction mixture is washed with 20 ml of ice-cold 2% (v/v) hydrochloric acid and then with water (twice, 20 ml each time) until neutral. The organic phase is dried over anhydrous sodium sulfate and the solvent is removed under reduced pressure to give 2.82 g of the crude diacyl derivative. Purification by medium-pressure liquid chromatography (MPLC) on silica gel with light petroleum–diethyl ether (95:5, v/v) gives 2.65 g (92% yield) of the pure product. The oily product shows an R_f of 0.80 using the solvent system chloroform–acetone–acetic acid (96:4:1, by volume).

Assay of Lipase Stereoselectivity

Principle

The procedure described below for the determination of lipase stereoselectivity is based on the method described by Rogalska *et al.*[27] (Fig. 1). The enantiomeric diacylglycerols or analogs (*sn*-1,2 and *sn*-2,3 isomers) resulting from lipase-catalyzed hydrolysis of prochiral triacylglycerols or analogs, or from lipase-catalyzed esterification of prochiral 2-acylglycerols or analogs, are derivatized with (*R*)-phenylethylisocyanate (PEIC) to form the corresponding diastereomeric carbamates, which are separated by high-performance liquid chromatography (HPLC) on a silica gel column.[13] The relative ratio of *sn*-1,2 and *sn*-2,3 stereoisomers reflects the preference of

[27] E. Rogalska, S. Ransac, and R. Verger, *J. Biol. Chem.* **265,** 20271 (1990).

the lipase in hydrolysis or esterification reactions. For example, formation of excess *sn*-2,3 isomer reflects preference for the *sn*-1 position during hydrolysis of a triacylglycerol or analog, but preference for the *sn*-3 position in esterification of a 2-monoacylglycerol or analog. Care must be taken to avoid further hydrolysis of diacylglycerols or analogs during action of the lipase on triacylglycerols or analogs. Similarly, further acylation of diacylglycerols or analogs during esterification of 2-acylglycerols or analogs must be avoided by choosing the appropriate reaction conditions (amount of lipase and reaction time to yield <15% of product). Reaction of prochiral triacylglycerol or 2-acylglycerol (or the respective analogs) yields a mixture of chiral products in a certain ratio from which the enantiomeric excess (ee) can be calculated. This value reflects the stereoselectivity of the first reaction step and remains constant with time as long as subsequent reactions are negligible.

Enzymatic Hydrolysis of Triacylglycerols and Analogs

Reagents

Reaction buffer: 50 m*M* Tris-HCl (pH 7), 0.9% (w/v) NaCl
Substrates: Stock solution, 20 mg/ml in chloroform
Enzyme: Stock solution in 2 m*M* Tris-HCl (pH 7.0) and 0.9% (w/v) NaCl
TLC solvent system A: Chloroform–acetone–acetic acid (96:4:1, by volume)
TLC reagent[28]: 0.8 g of $MnCl_2 \cdot 4H_2O$, 120 ml of H_2O, 120 ml of methanol, 8 ml of concentrated sulfuric acid

Procedure. Into a Teflon-stoppered glass tube 2 mg of the respective substrate is added from a stock solution (20 mg/ml) in chloroform. The solvent is removed under a gentle stream of nitrogen and the lipid is dried *in vacuo.* The substrate is emulsified by vigorous vortexing after addition of 1.5 ml of the reaction buffer. The stirred substrate emulsion is thermostatted in a water bath at 37° and incubated with various amounts (0.01–0.1 U; tributyrin[13]) of lipase. The enzymatic reaction is quenched after certain incubation times (0–10 min) by the addition of 2 ml of chloroform and lipids are extracted under vortexing. The mixture is centrifuged (3 min, 2500 rpm) and the aqueous layer is removed with a Pasteur pipette. The organic phase is dried over anhydrous sodium sulfate and the solvent is removed under a stream of nitrogen. The lipid mixture is dissolved in 200 μl of chloroform for TLC analysis. Aliquots (5 μl) of the lipid mixture are applied to silica gel plates (10 × 10 cm) using a sample applicator (Linomat

[28] H. Jork, W. Funk, W. Fischer, and H. Wimmer, "Thin-Layer Chromatography: Reagents and Detection Methods." VCH, Weinheim, Germany, 1989.

IV; Camag; Muttenz, Switzerland) and the plates are developed using solvent system A. For visualization of spots the dried plates are dipped into the reagent solution for 5 sec using a chromatogram immersion device (Camag), briefly dried, and heated at 120° for 15 min. A semiquantitative measure of hydrolysis products (for R_f values see Table I) is achieved by direct densitometry at 400 nm with a Shimadzu (Kyoto, Japan) CS930 thin-layer chromatography scanner and comparison with corresponding standards.

Esterification of Monoacylglycerols and Analogs with Oleic Acid in Organic Solvent

Reagents

Solvent: *n*-Hexane
Substrates: Oleic acid, monoacylglycerol and analogs, stock solutions in chloroform (10 mg/ml)
Enzyme: Lyophilized lipase

Procedure. 2-Monoacylglycerol or analogs (0.015 mmol) and equimolar amounts of oleic acid from the stock solutions are transferred into a screw-capped Teflon-stoppered glass vial. The solvent is removed under a gentle stream of nitrogen and the residue is dissolved in 2.5 ml of *n*-hexane to give a 5 m*M* substrate solution. The lyophilized lipase (1000 U; olive oil[21]) is suspended in 0.5 ml of *n*-hexane and dispersed in a bath-type sonicator for 10 sec at 300 W and added to the vial. The reaction is carried out under stirring (400 rpm) at 37°. At certain times 200-μl samples are withdrawn, the enzyme is removed by centrifugation (4 min, 14,000 *g*), and the reaction progress is monitored semiquantitatively by TLC on silica gel 60 plates as described in the previous section. To avoid complications due to further acylation of the diacylglycerols or analogs, the samples must be taken at times when the formation of triacylglycerols or analogs has not yet occurred (up to 15% utilization of substrates).

Analysis of Enantiomeric Diacylglycerols or Analogs

Reagents

Toluene (dried over 4-Å molecular sieve)
R-(+)-Phenylethylisocyanate (PEIC, ee > 99%; Fluka)
1,4-Diazabicyclo[2.2.2]octane (DABCO)
TLC solvent A: Chloroform–acetone–acetic acid (96 : 4 : 1, by volume)
TLC solvent B: *n*-Heptane–ethyl acetate (70 : 30, v/v)
HPLC: Eurospher 100 5 μm (250 × 4 mm; Knauer, Berlin, Germany)
Mobile phase A: *n*-Heptane–2-propanol (99.6 : 0.4, v/v)

Mobile phase B: *n*-Heptane–methanol–2-propanol (98.7 : 0.7 : 0.6, by volume)

Procedure. The crude lipid mixture (approximately 1 mg) obtained after hydrolysis of triacylglycerols and analogs or after esterification of 2-monoacylglycerols or analogs is dissolved in 100 μl of dry toluene and reacted with 3 μl of (*R*)-phenylethylisocyanate (PEIC) in the presence of catalytic amounts (approximately 0.1 mg) of 1,4-diazabicyclo[2.2.2]octane. The derivatization is carried out under an argon atmosphere in a sealed Wheaton vial. To ensure complete carbamoylation the mixture is stirred at room temperature for 12 hr. The solvent is removed under a gentle stream of nitrogen and excess isocyanate is evaporated under reduced pressure (1 mtorr, 6 hr) at room temperature. Prior to HPLC analysis the crude carbamate derivatives are purified by preparative TLC on silica gel plates with a fluorescence indicator (0.2 mm, F_{254}; E. Merck AG) using solvent system A or B (Table I). The corresponding bands are visualized under ultraviolet light. The plates are sprayed with water and the respective zone is scraped into a 10-ml glass tube. The product is eluted with two 2-ml portions of *n*-heptane–2-propanol (3 : 2, v/v) under vortexing. After centrifugation (2500 rpm, 3 min) the organic solvent is withdrawn and evaporated under a stream of nitrogen. The residue is dissolved in 200 μl of *n*-heptane for HPLC analysis.

TABLE I

THIN-LAYER CHROMATOGRAPHIC SEPARATION OF PRODUCTS OBTAINED AFTER LIPASE-CATALYZED HYDROLYSIS OF 1,3-DIOLEOYL-2-*X*-GLYCEROL OR LIPASE-CATALYZED ESTERIFICATION OF 2-*X*-GLYCEROL WITH OLEIC ACID AND CORRESPONDING DIASTEREOMERIC 1(3)-OLEOYL-2-*X*-GLYCEROL (*R*)-PHENYLETHYLISOCYANATE DERIVATIVES[a]

	R_f values			
sn-2 Glycerol moiety (*X*)	tri[b]	di[b]	mono[b]	di-PEIC
2-Acyl (e.g., oleoyl)	0.79	0.36 0.47[c]	0.09	0.42[d]
2-*O*-Hexadecyl	0.80	0.39	0.10	0.47[d]
2-Deoxy-2-hexadecyl	0.82	0.45	0.12	0.49[d]
2-Deoxy-2-phenyl	0.78	0.38	0.10	0.39[d]
2-Deoxy-2-aminopalmitoyl	0.64	0.16	0.02	0.41[b]

[a] tri, Substrate triacylglycerol or analog; di, diacylglycerol or analog; mono, monoacylglycerol or analog; di-PEIC, (R)-phenylethylisocyanate derivative of diacylglycerol or analog.

[b] Solvent system A: Chloroform–acetone–acetic acid (96 : 4 : 1, by volume), oleic acid R_f 0.22.

[c] 1,3-Dioleoylglycerol.

[d] Solvent system B: *n*-Heptane–ethyl acetate (7 : 3, v/v).

High-Performance Liquid Chromatography Analysis of Diastereomeric R-(+)-Phenylethylisocyanate Derivatives of Diacylglycerols and Analogs. The resolution of diastereomeric PEIC derivatives of *sn*-1,2- and *sn*-2,3-diacylglycerols and analogs (Fig. 2) is performed on a Hewlett-Packard (Palo Alto, CA) 1050 HPLC system equipped with a silica gel column. Peaks are monitored with a variable ultraviolet (UV) detector set at 210 nm. The mobile phase is *n*-heptane–2-propanol (99.6:0.4, v/v) at a flow rate of 0.8 ml/min for the separation of all derivatives except the 1(3)-oleoyl-2-deoxy-2-palmitoylaminoglycerol carbamates. The latter compounds are separated using *n*-heptane–methanol–2-propanol (98.7:0.7:0.6, by volume) as eluent at a flow rate of 0.8 ml/min. The distribution of *sn*-1,2 and *sn*-2,3 stereoisomers is calculated after automatic integration of the corresponding HPLC peak areas.

Determination of Absolute Configuration. The elution order of the diasteromeric diacylglycerol carbamate derivatives is determined with 1,2-dioleoyl-*sn*-glycerol (Sigma, Munich, Germany) as a standard. The corresponding 1-oleoyl-2-*O*-hexadecyl-*sn*-glycerol is synthesized as described by Stadler *et al.*[13] 1-Oleoyl-2-deoxy-2-palmitoylamino-*sn*-glycerol is prepared from L-serine methyl ester in analogy to Dijkman *et al.*[25] Each of the former optically pure *sn*-1,2 isomers is converted into the respective carbamate derivative and analyzed by HPLC. Retention times are compared to those of the respective derivatives obtained from *rac*-1,2 isomers. The *sn*-1,2 isomers eluted as second peak (R_t *sn*-2,3 < R_t *sn*-1,2). The same retention behavior is observed for 1(3)-acyl-2-deoxy-2-alkylglycerols. The steric configuration of 1-acyl-2-deoxy-2-alkyl-*sn*-glycerols is confirmed by determination of the optical rotation in comparison to literature data as described.[13] The absolute configuration of 1(3)-acyl-2-deoxy-2-phenylglycerols is determined by conversion of the diacylglycerol analogs to tropic acid and by

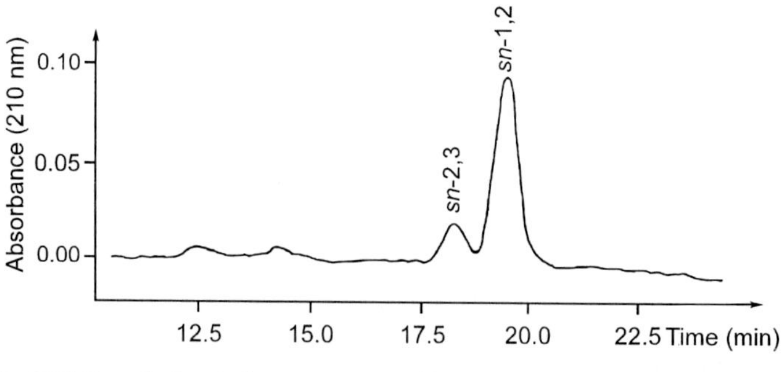

FIG. 2. HPLC analysis of diastereomeric 1(3)-oleoyl-2-*O*-hexadecyl-*sn*-glycerol carbamates. Column, Eurosphere 100, 5 μm, 250 × 4 mm i.d. (Knauer); mobile phase A, a *n*-heptane–2-propanol (99.6:0.4, v/v); flow, 0.8 ml/min; detection, UV absorbance at 210 nm.

comparison of the optical rotation to literature data.[13] It is worth noting that the elution order of 1(3)-acyl-2-deoxy-2-phenylglycerol carbamates is opposite (R_t *sn*-1,2 < R_t *sn*-2,3) to that observed with all other analogs.

Data obtained in hydrolysis and esterification reactions are summarized in Tables II and III, respectively.

The stereoselectivity of the *R. oryzae* lipase is strongly influenced by the substituent in position *sn*-2 of the glycerol- or 1,3-propanediol backbone. The stereochemical preference in hydrolysis switches from position *sn*-1 to *sn*-3 by replacing the 2-acyl, 2-*O*-alkyl, or 2-deoxy-2-alkyl residues by 2-deoxy-2-hexadecanoylamino or 2-deoxy-2-phenyl groups. The acyl chain length (oleic versus octanoic acid) in the *sn*-1 and *sn*-3 position has little effect on stereoselectivity.[13]

TABLE II

STEREOSELECTIVITY OF NATIVE *Rhizopus oryzae* LIPASE IN HYDROLYSIS OF TRIOLEOYLGLYCEROL AND ANALOGS[a]

Substrate	ee (%)	*sn*
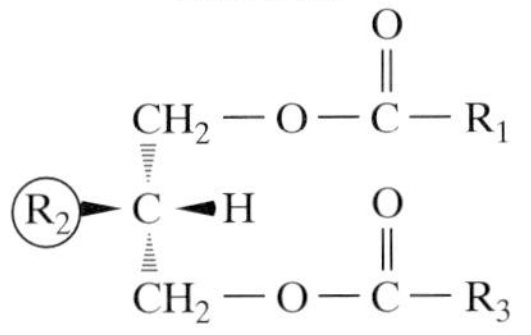 $R_1 = R_2 = C_{17}H_{33}$		
R_2		
$C_{15}H_{31}—C(=O)—N(H)—$	49.0 ± 5.2	3
$C_{17}H_{33}—C(=O)—O—$	33.2 ± 3.0	1
$C_{16}H_{33}—O—$	63.8 ± 4.1	1
$C_{16}H_{33}—$	18.8 ± 2.4	1
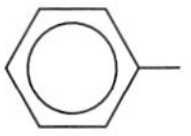	63.8 ± 2.2	3

[a] ee, Enantiomeric excess; *sn*, stereospecific numbering of the preferred *sn* position; values are given as mean ± standard deviation ($n = 3$).

TABLE III
STEREOSELECTIVITY OF NATIVE *Rhizopus oryzae* LIPASE IN ESTERIFICATION OF 2-MONOACYLGLYCEROL AND ANALOGS IN *n*-HEXANE[a]

Substrates

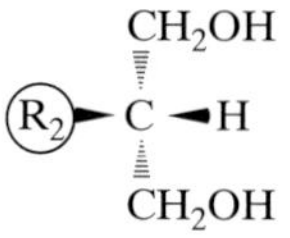

Substrates	ee (%)	*sn*
R_2		
$C_{17}H_{35}-C(=O)-O-$	21.9 ± 8.9	1
$C_{16}H_{33}-O-$	52.6 ± 1.9	1
$C_{16}H_{33}-$	89.7 ± 8.3	1

[a] ee, Enantiomeric excess; *sn*, stereospecific numbering of the preferred *sn* position; values are given as mean ± standard deviation ($n = 3$).

In esterification reactions the same stereopreference (for *sn*-1) was observed as in hydrolysis of 2-acyl-, 2-*O*-alkyl-, or 2-deoxy-2-alkyl substrates (see Tables II and III). In esterification reactions the *sn*-1 stereoselectivity increases concomitantly with a decrease in substrate polarity at the *sn*-2 position. Interestingly, on introducing the most nonpolar (2-deoxy-2-alkyl) residue the enantiomeric excess is much lower in hydrolysis as compared to the esterification reaction. Similar observations were made for the lipases from *Chromobacterium viscosum, Pseudomonas cepacia,* and *Geotrichum candidum.* Different organic solvents (dodecane, toluene, and light petroleum) as well as variation in the chain length of the acylating agent (oleic versus octanoic acid) do not affect the stereoselectivity of ROL (data not shown). As shown in comparative studies the same stereoselectivity is found for the native and the recombinant ROL in hydrolysis and in esterification reactions.

Computer-Aided Modeling of Lipase–Substrate Interaction

Table IV summarizes the modeling strategy that was applied to model the molecular basis of stereopreference of ROL toward triacylglycerols and analogs.[29] Four substrates are compared: trioctanoylglycerol ("ester")

[29] H.-C. Holzwarth, J. Pleiss, and R. D. Schmid, *J. Mol. Catal. B.* in press (1997).

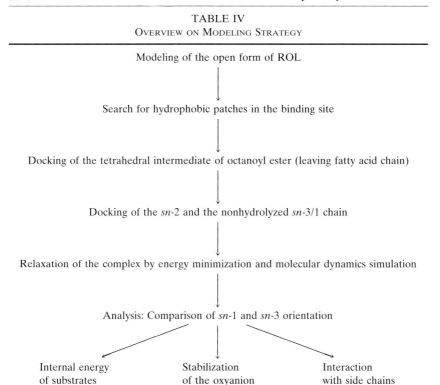

TABLE IV
OVERVIEW ON MODELING STRATEGY

Modeling of the open form of ROL

↓

Search for hydrophobic patches in the binding site

↓

Docking of the tetrahedral intermediate of octanoyl ester (leaving fatty acid chain)

↓

Docking of the *sn*-2 and the nonhydrolyzed *sn*-3/1 chain

↓

Relaxation of the complex by energy minimization and molecular dynamics simulation

↓

Analysis: Comparison of *sn*-1 and *sn*-3 orientation

↙ ↓ ↘

Internal energy of substrates	Stabilization of the oxyanion	Interaction with side chains

and the *sn*-2 analogs 2-*X*-1,3-dioctanoylglycerol, where 2-*X* = 2-*O*-octyl ("ether"), 2-deoxy-2-octanoylamino ("amide"), and 2-deoxy-2-phenyl ("phenyl"). Ester and ether substrates are preferentially hydrolyzed at *sn*-1, whereas amide and phenyl reverse ROL stereopreference to *sn*-3. Stereoselectivity is similar for the 1,3-dioctanoyl and the 1,3-dioleoyl substrates.[13]

Modeling of Open Form of Rhizopus oryzae Lipase

The structure of the open form of ROL was modeled by homology.[7] The structures of two reference proteins are available: lipases from *R. delemar* (RDL)[8,30] and *Rhizomucor miehei* (RML)[6] share 99 and 57% sequence identity with ROL, respectively. The RDL was crystallized in two "semi-open lid" conformations. The structure was determined at a resolution of

[30] L. Swenson, R. Green, R. Joerger, M. Haas, K. Scott, Y. Wei, U. Derewenda, D. Lawson, and Z. S. Derewenda, *Proteins* **18,** 301 (1994).

2.6 Å. The RML was crystallized in the closed form and complexed with two inhibitors: diethylphosphate and hexylphosphonate ethyl ester (resolution, 1.9, 2.6, and 3.0 Å). C_α root-mean-square (RMS) deviation between the two RDL structures (excluding the lid region) is 0.7 Å, between the two RML complexes 0.3 Å, and between RML and the two RDL structures 0.9 and 1.1 Å. The model structure of ROL deviates from RML and the two structures of RDL by 0.8, 1.1, and 1.2 Å. For 19 amino acids in contact with substrate the deviation from RML and RDL is only 0.3 and 0.5 Å, respectively.

Search for Hydrophobic Acyl Chain-Binding Sites

There are two major hydrophobic binding sites in ROL. Searching for hydrophobic regions using the METHYL probe of GRID96[31] revealed a long and deep hydrophobic crevice of about 2 × 16 Å.[29] This crevice was also described in other lipases that are homologous to ROL: lipase from *Humicola lanuginosa*[32] and RML.[6] In the latter, this site binds the hexyl chain of the inhibitor hexylphosphonate ethyl ester; therefore, it may represent the binding site for the leaving fatty acid chain.

Using the DRY[33] probe of GRID, a shallow hydrophobic dent of about 2 × 5 Å was found[29] that is parallel to the hydrophobic crevice at a distance of 4–5 Å. This site exists also in *H. lanuginosa* lipase, and may represent a binding site for a nonhydrolyzed fatty acid chain.[34]

Docking of Tetrahedral Intermediate of 1(3)-Octanoylglycerol

The tetrahedral intermediate of *sn*-1 and *sn*-3 octanoylglycerol is docked to the binding site (Fig. 3). Docking is guided by the position of hexylphosphonate ethyl ester in RML. A covalent bond between Ser-145 and the tetrahedral carbon of the substrate is created. The oxyanion fits into the oxyanion hole formed by the backbone nitrogens of Thr-83 and Leu-146 and the side-chain hydroxyl group of Thr-83. The ester oxygen linked to glycerol and the Ser-145 side-chain oxygen are in hydrogen-bonding distance from the catalytic His-257. The acyl chain is positioned in the hydrophobic crevice.

[31] P. J. Goodford, *J. Med. Chem.* **28,** 849 (1985).

[32] M. Norin, F. Haeffner, A. Achour, T. Norin, and K. Hult, *Protein Sci.* **3,** 1493 (1994).

[33] Molecular Discovery, Ltd., "GRID96 Manual." Molecular Discovery, Ltd., Oxford, England, 1996.

[34] D. M. Lawson, A. M. Brzozowski, S. Rety, C. Vemag, and G. G. Dodson, *Protein Eng.* **7,** 543 (1994).

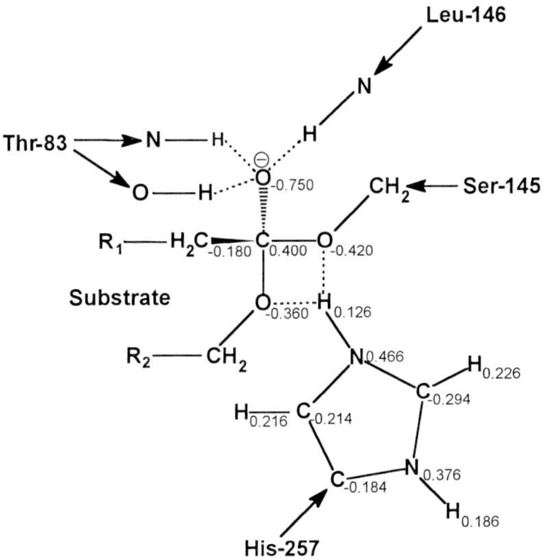

FIG. 3. Partial charges of the catalytic Ser-145 and His-257 and the 1(3)-octanoylglycerol substrate as tetrahedral intermediate (R_1, alkyl chain of leaving fatty acid; R_2, glycerol moiety). The oxyanion is stabilized by three hydrogen bonds from the backbone of Leu-146 and Thr-83 and the side chain of Thr-83.

Docking of sn-2 and Nonhydrolyzed sn-3/1 Chain

It can be assumed that either of the two chains (*sn*-2 and the nonhydrolyzed *sn*-3/1 chain) binds to the hydrophobic dent. So there are four possibilities as to how a triacylglycerol can bind (Fig. 4): depending on whether the *sn*-1 or the *sn*-3 chain binds to the hydrophobic crevice the orientation has been called "*sn*-1" or "*sn*-3 orientation," respectively. If the *sn*-2 chain binds to the hydrophobic dent, the substrate binds in the "*sn*-2 down binding mode," otherwise in the "*sn*-2 up binding mode."[29]

For substrates with an aliphatic chain in *sn*-2 position (ester, ether, amide), the most probable binding mode is *sn*-2 down (Fig. 4a and b). The hydrophobic alkyl chains of the two nonhydrolyzed chains pack favorably to each other as do the polar ester groups of the triacylglycerol substrate. This cooperative effect stabilizes the complex. However, in the *sn*-2 up binding mode the polar ester group of the nonhydrolyzed *sn*-3(1) chain contacts the other two hydrophobic chains (Fig. 4c and d). Molecular dynamics simulations (see below) of ether, ester, and amide substrates in *sn*-2 up binding mode support this intuitive prediction, because the nonhydrolyzed *sn*-3(1) chain tends to move out of the hydrophobic dent, while the complex is stable in the *sn*-2 down binding mode.[29] For the phenyl

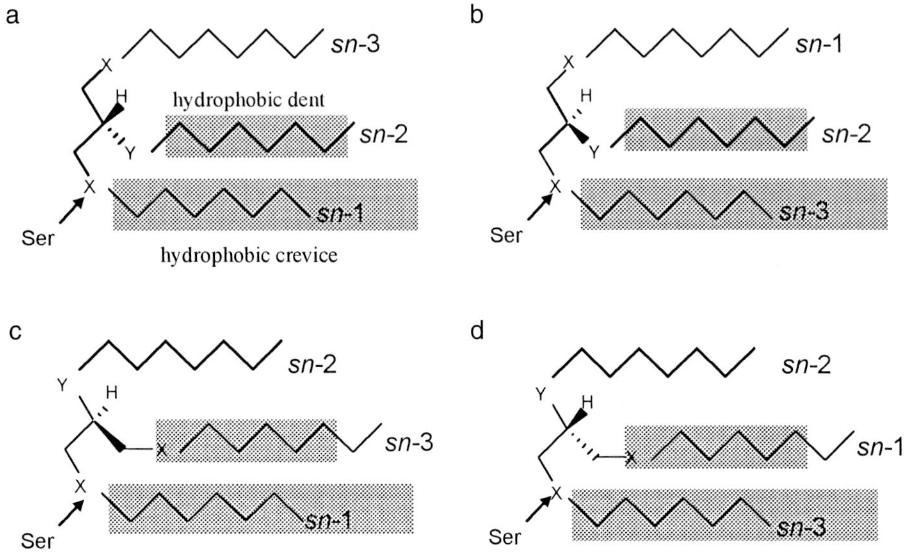

FIG. 4. Four possible ways in which triacylglycerol or analog can bind to ROL. The long and the short gray-shaded regions indicate the hydrophobic crevice and the hydrophobic dent, respectively. The catalytic serine marks the cleavage site. The *sn*-2 chain is marked by a boldface line. X, Ester bond in the *sn*-1 and *sn*-3 chain; Y, *sn*-2 group of ester, ether, or amide. In the phenyl substrate, the Y and the aliphatic chain are replaced by a phenyl ring. The substrate can bind in *sn*-1 (a and c) or *sn*-3 (b and d) orientation. In the "*sn*-2 down" binding mode (a and b) the *sn*-2 chain binds to the hydrophobic dent, whereas in "*sn*-2 up" (c and d) it does not bind to the hydrophobic dent.

substrate, however, the *sn*-2 up binding mode is preferred. The nonhydrolyzed *sn*-3(1) aliphatic chain binds to the hydrophobic dent, and the phenyl group points out of the binding site.

Relaxation of Complex

Energy minimization and molecular dynamics (MD) simulations of the complex of all four substrates each in *sn*-1 and *sn*-3 orientation are carried out using Sybyl 6.1 (Tripos, St. Louis, MO) with the Tripos force field.[35] For the catalytic histidine and the oxyanion the partial charges are modified (Fig. 3) as calculated by the semiempirical method MNDO94/PM3[36] using the Unichem 3.0 interface (Cray Research, Mendota Heights, MN).

Prior to molecular dynamics simulations the initial complexes are energy minimized. They are then gradually heated in four intervals of 2 psec each

[35] M. Clark, R. D. Cramer, III, and N. van Opdenbosch, *J. Comput. Chem.* **10,** 982 (1989).
[36] J. J. P. Stewart, *J. Comput. Chem.* **10,** 209 (1989).

TABLE V
SUMMARY OF MODELING RESULTS[a] AND COMPARISON TO EXPERIMENTAL DATA

	Ether	Ester	Amide	Phenyl
Glycerol conformation	*sn*-1	*sn*-1	*sn*-3	*sn*-3
Thr-83 oxyanion hole	*sn*-1	*sn*-1	*sn*-1	*sn*-3
Leu-258 interaction	*sn*-1	*sn*-3	*sn*-3	*sn*-3
Experimental ee (rounded values)	60% (*sn*-1)	30% (*sn*-1)	50% (*sn*-3)	60% (*sn*-3)

[a] Three factors favor the *sn*-1 or *sn*-3 orientation, depending on the substrate: conformation of the glycerol backbone, participation of Thr-83 side chain in the oxyanion hole, and repulsion of substrate by Leu-258.

to 5, 30, 100, and 200 K and during 6 psec to 300 K. The step size is 1 fsec up to 100 K and 0.5 fsec at 200 and 300 K. The temperature coupling constant is set to 10 fsec,[37] the nonbonded interaction cutoff to 8 Å. After this equilibration phase the system is further simulated for 4 psec. Averaging of structure and torsion angle values is performed only during this production phase.

Comparison of sn-1 and sn-3 Orientation: Internal Energy of Substrates

As a consequence of changing the orientation of a substrate, the configuration at glycerol C-2 inverts (Fig. 4a and b). The local conformation of the glycerol backbone is expected to be different in the two orientations, because the tetrahedral intermediate is fixed at the active site and the position of the *sn*-2 chain in the hydrophobic dent is maintained. To estimate the difference in internal energy of the glycerol backbone in both orientations, a model of the glycerol backbone and the first three atoms of the *sn*-2 moiety is built. Torsion angles are assigned to the respective average values of the substrates during the MD simulation in *sn*-1 and *sn*-3 orientation, and the energy is evaluated by geometry optimization using MNDO94/PM3, constraining the torsion angles. For ether and ester the *sn*-1 orientation is energetically preferred by 5 and 4 kcal/mol, respectively, whereas for amide and phenyl the *sn*-3 orientation is more favorable (10 and 4 kcal/mol, respectively). This result corresponds to the measured stereopreference of the substrates (Table V).

Stabilization of Oxyanion

A second interaction that is different in both orientations can be identified. For ester, ether, and amide substrates in *sn*-1 orientation and phenyl

[37] H. J. C. Berendsen, J. P. M. Postma, and W. F. van Gunsteren, *J. Chem. Phys.* **81,** 3684 (1984).

in *sn*-3 orientation the side-chain hydroxyl of Thr-83 forms a hydrogen bond to the oxyanion (Fig. 3). This residue has previously been suggested to act as additional hydrogen donor to the tetrahedral intermediate of the substrate in RML.[6] The role of this hydroxyl group was further investigated by protein engineering[7,9]: replacing Thr-83 by serine, the activity of the ROL mutant enzyme is still 22% that of wild type. Removing the hydroxyl group by replacing Thr-83 by valine or alanine inactivates the enzyme.

In the unfavored orientation, the Thr-83 side chain is rotated. This may be due to repulsive interaction of the hydroxyl group with glycerol C-3. Instead of stabilizing the oxyanion, the hydroxyl of Thr-83 then points in the direction of Asp-92 (data not shown). Thus, the transition state is less stabilized, which leads to a slower reaction of the substrate in the unfavored orientation.

Interaction with Rhizopus oryzae Lipase Side Chains

Experimentally there is clear evidence that ROL is less *sn*-1 and more *sn*-3 selective as the chemical structure at *sn*-2 becomes more rigid and bulky. So for bulky substrates there should be either an additional stabilizing interaction in the *sn*-3 orientation or a destabilizing interaction in the *sn*-1 orientation. This interaction should be localized near the C-2 glycerol atom and the start of the *sn*-2 chain. In *sn*-1 orientation steric repulsion between the side chain of Leu-258 and the ester (Fig. 5) and amide group at *sn*-2 has been observed.[29] For the ether substrate, which is smaller and more flexible, no difference in local interactions can be found in the two orientations. For the phenyl substrate, the situation is similar: the *sn*-3 orientation is preferred, whereas in *sn*-1 orientation the phenyl ring interferes with Leu-258.

The factors that favor the *sn*-1 or *sn*-3 orientation are summarized in Table V and compared to the experimental data on stereoselectivity. The model is in accordance with the experiments: the more factors favor an orientation, the higher the respective stereoselectivity.

Conclusion

It is evident from the experimental data obtained in hydrolysis and esterification that the substituent in position *sn*-2 of the glycerol- or 1,3-propanediol backbone plays a crucial role in the stereoselectivity observed for ROL. It is noteworthy to mention that modulating polarity and flexibility of the *sn*-2 group not only affects the stereoselectivity expressed as enantiomeric excess, but also results in a reversal of stereopreference (*sn*-1 versus *sn*-3) in hydrolysis. In esterification reactions the stereoselectivity increases

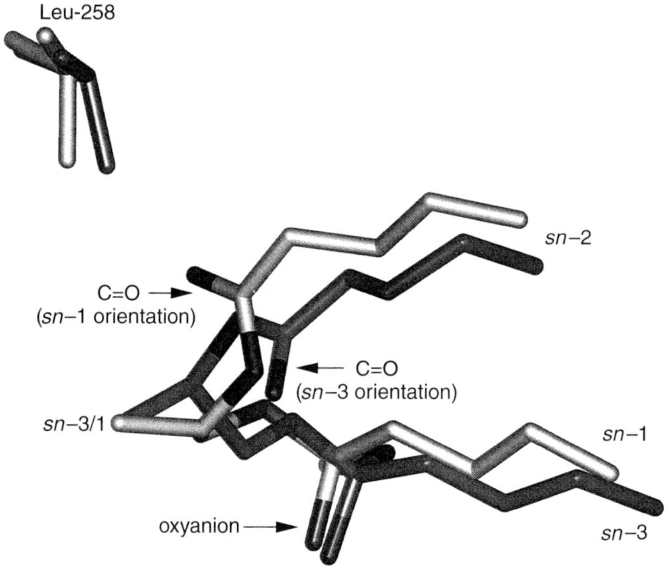

FIG. 5. The *sn*-2 carbonyl group of the ester substrate in *sn*-1 (light gray) and *sn*-3 (dark gray) orientations. The side chain of Leu-258 is pushed away by the substrate in the *sn*-1 orientation.

concomitantly with a decrease of substrate polarity at the *sn*-2 position. However, no explanation can be provided for the different stereoselectivity in hydrolysis and esterification of substrates with 2-deoxy-2-alkyl group in *sn*-2 position.

By now, crystallographic data on binding of acyl chain analogs are available for lipases from *Rhizomucor miehei*,[5] *Candida antarctica*,[38] *Candida rugosa*,[39] and human pancreatic lipase.[40] Although the mechanism of hydrolysis is similar in all lipases, the structure of their binding site and the stereopreference toward triacylglycerols and analogs differ. In contrast to esters of secondary alcohols, for which a common rule has been proposed to predict stereoselectivity,[41] there is no such common rule to predict stereoselectivity toward triacylglycerol substrates. In *R. oryzae* lipase, at least three factors govern stereoselectivity toward these substrates: the internal

[38] J. Uppenberg, N. Oehrner, M. Norin, K. Hult, G. J. Kleywegt, S. Patkar, V. Waagen, T. Anthonsen, and T. A. Jones, *Biochemistry* **34,** 16838 (1995).

[39] P. Grochulski, F. Bouthillier, R. J. Kazlauskas, A. N. Serregi, J. D. Schrag, E. Ziomek, and M. Cygler, *Biochemistry* **33,** 3494 (1994).

[40] M. P. Egloff, F. Marguet, G. Buono, R. Verger, C. Cambillau, and H. van Tilbeurgh, *Biochemistry* **34,** 2751 (1995).

[41] R. J. Kazlauskas, *Trends Biotechnol.* **12,** 464 (1994).

energy of the glycerol backbone of the substrate, the participation of the Thr-83 in the oxyanion hole, and repulsive interaction between Leu-258 and the substrate.

Combining experimental structure information, computer-aided modeling of substrate–lipase interaction, and experimental analysis, this model explains the observed reversal in stereopreference and qualitatively ranks substrates by stereoselectivity.

Acknowledgments

Financial support by the German Research Foundation, grants SP 135/11-1 and SCHM 1240/1-1, and by the Fund for the Promotion of Scientific Research in Austria, grant F102, are gratefully acknowledged.

Author Index

Numbers in parentheses are footnote reference numbers and indicate that an author's work is referred to although the name is not cited in the text.

A

B

C

D

E

H

I

L

M

S

T

U

V

W

X

Y

Z

Subject Index

A

B

C

D

E

R

S

T

V

W

X

ISBN 0-12-182185-4
9 780121 821852 90038